Lecture Notes in Computer Science   **11331**

Commenced Publication in 1973
Founding and Former Series Editors:
Gerhard Goos, Juris Hartmanis, and Jan van Leeuwen

More information about this series at http://www.springer.com/series/7409

Giuseppe Nicosia · Panos Pardalos
Giovanni Giuffrida · Renato Umeton
Vincenzo Sciacca (Eds.)

Machine Learning, Optimization, and Data Science

4th International Conference, LOD 2018
Volterra, Italy, September 13–16, 2018
Revised Selected Papers

 Springer

Editors
Giuseppe Nicosia
University of Catania
Catania, Italy

and

University of Reading
Reading, UK

Panos Pardalos
University of Florida
Gainesville, FL, USA

Giovanni Giuffrida
University of Catania
Catania, Italy

Renato Umeton
Harvard University
Cambridge, MA, USA

Vincenzo Sciacca
IBM, Tivoli Research Lab
Rome, Italy

ISSN 0302-9743   ISSN 1611-3349   (electronic)
Lecture Notes in Computer Science
ISBN 978-3-030-13708-3   ISBN 978-3-030-13709-0   (eBook)
https://doi.org/10.1007/978-3-030-13709-0

Library of Congress Control Number: 2019931948

LNCS Sublibrary: SL3 – Information Systems and Applications, incl. Internet/Web, and HCI

This Springer imprint is published by the registered company Springer Nature Switzerland AG
The registered company address is: Gewerbestrasse 11, 6330 Cham, Switzerland

Preface

LOD is the international conference embracing the fields of machine learning, optimization, and data science. The fourth edition, LOD 2018, was organized during September 13–16, 2018, in Volterra (Pisa) Italy, a stunning medieval town dominating the picturesque countryside of Tuscany.

The International Conference on Machine Learning, Optimization, and Data Science (LOD) has established itself as a premier interdisciplinary conference in machine learning, computational optimization, and big data. It provides an international forum for the presentation of original multidisciplinary research results, as well as the exchange and dissemination of innovative and practical development experiences.

The LOD Conference Manifesto is the following:

The problem of understanding intelligence is said to be the greatest problem in science today and "the" problem for this century – as deciphering the genetic code was for the second half of the last one. Arguably, the problem of learning represents a gateway to understanding intelligence in brains and machines, to discovering how the human brain works, and to making intelligent machines that learn from experience and improve their competences as children do. In engineering, learning techniques would make it possible to develop software that can be quickly customized to deal with the increasing amount of information and the flood of data around us.
The Mathematics of Learning: Dealing with Data
Tomaso Poggio and Steve Smale

LOD 2018 attracted leading experts from industry and the academic world with the aim of strengthening the connection between these institutions. The 2018 edition of LOD represented a great opportunity for professors, scientists, industry experts, and postgraduate students to learn about recent developments in their own research areas and to learn about research in contiguous research areas, with the aim of creating an environment to share ideas and trigger new collaborations.

As chairs, it was an honor to organize a premiere conference in these areas and to have received a large variety of innovative and original scientific contributions.

During LOD 2018, five plenary talks were presented:

"Advances in Inference and Generation with Hierarchical Latent Variable Models"
Jorg Bornschein, DeepMind, London, UK

"The Value of Evaluation: What Does My Machine Learning Metric Tell Me?"
Peter Flach, University of Bristol, UK

"Recent Advances in Recommender Systems: Sets, Local Models, Coverage, and Errors"
George Karypis, University of Minnesota, USA

"Dynamics of Financial Markets"
Panos Pardalos, University of Florida, USA

"Several Problems in Optimization and Learning: From Pure Mathematics to Industrial Applications"
Andrey Raygorodsky, Moscow Institute of Physics and Technology, Russia

LOD 2018 received 126 submissions from 47 countries in five continents, and each manuscript was independently reviewed by a committee formed by at least five members through a blind review process. These proceedings contain 46 research articles written by leading scientists in the fields of machine learning, artificial intelligence, reinforcement learning, computational optimization, and data science presenting a substantial array of ideas, technologies, algorithms, methods, and applications.

At LOD 2018, Springer LNCS generously sponsored the LOD Best Paper Award. This year, the paper by Andrea Patané and Marta Kwiatkowska titled "Calibrating the Classifier: Siamese Neural Network Architecture for End-to-End Arousal Recognition from ECG" received the LOD 2018 Best Paper Award.

This conference could not have been organized without the contributions of exceptional researchers and visionary industry experts, so we thank them all for participating. A sincere thank you goes also to the Program Committee, formed by more than 400 scientists from academia and industry, for their valuable and essential work of selecting the scientific contributions.

Finally, we would like to express our appreciation to the keynote speakers who accepted our invitation, and to all the authors who submitted their research papers to LOD 2018.

September 2018   Giuseppe Nicosia
                                                       Panos Pardalos
                                                       Giovanni Giuffrida
                                                       Renato Umeton
                                                       Vincenzo Sciacca

Organization

General Chairs

Renato Umeton   MIT and Dana-Farber Cancer Institute - Harvard Medical School, USA

Vincenzo Sciacca   IBM, Italy

Conference and Technical Program Committee Co-chairs

Giuseppe Nicosia   University of Reading, UK and University of Catania, Italy

Panos Pardalos   University of Florida, USA

Giovanni Giuffrida   University of Catania, Italy

Special Sessions Chair

Aris Anagnostopoulos   University of Rome La Sapienza, Italy

Tutorial Chair

Giuseppe Narzisi   New York University Tandon School of Engineering, USA

Publicity Chair

Stefano Mauceri   NCRA, University College Dublin, Ireland

Industrial Session Chairs

Ilaria Bordino   UniCredit R&D, Italy

Marco Firrincieli   UniCredit R&D, Italy

Fabio Fumarola   UniCredit R&D, Italy

Francesco Gullo   UniCredit R&D, Italy

Organizing Committee

Alberto Castellini   University of Verona, Italy

Piero Conca   CNR, Italy

Jole Costanza   Italian Institute of Technology, Milan, Italy

Giuditta Franco   University of Verona, Italy

Giorgio Jansen   University of Catania, Italy

Giuseppe Narzisi   New York University Tandon School of Engineering, USA

Steering Committee

Giuseppe Nicosia	University of Reading, UK and University of Catania, Italy
Panos Pardalos	University of Florida, USA

Technical Program Committee

Hector-Gabriel Acosta-Mesa	University of Veracruz, Mexico
Jason Adair	University of Stirling, UK
Agostinho Agra	Universidade de Aveiro, Portugal
Kerem Akartunali	University of Strathclyde, UK
Richard Allmendinger	The University of Manchester, UK
Paula Amaral	University Nova de Lisboa, Portugal
Aris Anagnostopoulos	University of Rome La Sapienza, Italy
Davide Anguita	University of Genoa, Italy
Roberto Aringhieri	University of Turin, Italy
Takaya Arita	Nagoya University, Japan
Jason Atkin	The University of Nottingham, UK
Martha L. Avendano-Garrido	Universidad Veracruzana, Mexico
Chloe-Agathe Azencott	Institut Curie Research Centre, Paris, France
Kamyar Azizzadenesheli	University of California, Irvine, USA
Ozalp Babaoglu	University of Bologna, Italy
Jaume Bacardit	Newcastle University, UK
James Bailey	University of Melbourne, Australia
Marco Baioletti	University of Perugia, Italy
Baski Balasundaram	Oklahoma State University, USA
Elena Baralis	Politecnico di Torino, Italy
Xabier E. Barandiaran	University of the Basque Country, Spain
Cristobal Barba-Gonzalez	University of Malaga, Spain
Helio J. C. Barbosa	Laboratório Nacional de Computação Científica, Brazil
Mikhail Batsyn	Higher School of Economics, Russia
Roberto Battiti	University of Trento, Italy
Lucia Beccai	Istituto Italiano di Tecnologia, Italy
Aurélien Bellet	Inria Lille, France
Gerardo Beni	University of California at Riverside, USA
Khaled Benkrid	The University of Edinburgh, UK
Katie Bentley	Harvard Medical School, USA
Peter Bentley	University College London, UK
Heder Bernardino	Universidade Federal de Juiz de Fora, Brazil
Daniel Berrar	Tokyo Institute of Technology, Japan
Adam Berry	CSIRO, Australia
Luc Berthouze	University of Sussex, UK
Martin Berzins	SCI Institute, University of Utah, USA

Manuel A. Betancourt Odio	Universidad Pontificia Comillas, Spain
Mauro Birattari	IRIDIA, Université Libre de Bruxelles, Belgium
Arnim Bleier	GESIS - Leibniz-Institute for Social Sciences, Germany
Leonidas Bleris	University of Texas at Dallas, USA
Christian Blum	Spanish National Research Council, Spain
Ilaria Bordino	UniCredit R&D, Italy
Anton Borg	Blekinge Institute of Technology, Sweden
Paul Bourgine	École Polytechnique Paris, France
Anthony Brabazon	University College Dublin, Ireland
Paulo Branco	Instituto Superior Técnico, Portugal
Juergen Branke	University of Warwick, UK
Alexander Brownlee	University of Stirling, UK
Marcos Bueno	Radboud University, The Netherlands
Larry Bull	University of the West of England, UK
Tadeusz Burczynski	Polish Academy of Sciences, Poland
Robert Busa-Fekete	Yahoo! Research, NY, USA
Adam A. Butchy	University of Pittsburgh, USA
Sergiy I. Butenko	Texas A&M University, USA
Luca Cagliero	Politecnico di Torino, Italy
Stefano Cagnoni	University of Parma, Italy
Yizhi Cai	University of Edinburgh, UK
Guido Caldarelli	IMT Lucca, Italy
Alexandre Campo	Université Libre de Bruxelles, Belgium
Angelo Cangelosi	University of Plymouth, UK
Salvador Eugenio Caoili	University of the Philippines Manila, Philippines
Timoteo Carletti	University of Namur, Belgium
Jonathan Carlson	Microsoft Research, USA
Celso Carneiro Ribeiro	Universidade Federal Fluminense, Brazil
Alexandra M. Carvalho	Universidade de Lisboa, Portugal
Alberto Castellini	University of Verona, Italy
Michelangelo Ceci	University of Bari, Italy
Adelaide Cerveira	Universidade de Trás-os-Montes e Alto Douro, Portugal
Uday Chakraborty	University of Missouri – St. Louis, USA
Xu Chang	University of Sydney, Australia
W. Art Chaovalitwongse	University of Washington, USA
Antonio Chella	University of Palermo, Italy
Rachid Chelouah	Université Paris-Seine/EISTI, France
Haifeng Chen	NEC Labs, USA
Keke Chen	Wright State University, USA
Steven Chen	University of Pennsylvania, USA
Ying-Ping Chen	National Chiao Tung University, Taiwan
Gregory Chirikjian	Johns Hopkins University, USA
Silvia Chiusano	Politecnico di Torino, Italy
Miroslav Chlebik	University of Sussex, UK
Sung-Bae Cho	Yonsei University, South Korea

Stephane Chretien	National Physical Laboratory, UK
Anders Christensen	Lisbon University Institute, Portugal
Carlos Coello Coello	CINVESTAV-IPN, Mexico
George Coghill	University of Aberdeen, UK
Sergio Consoli	Philips Research, The Netherlands
David Cornforth	University of Newcastle, UK
Luís Correia	University of Lisbon, Portugal
Chiara Damiani	University of Milano-Bicocca, Italy
Thomas Dandekar	University of Würzburg, Germany
Ivan Luciano Danesi	Unicredit Bank, Italy
Christian Darabos	Dartmouth College, USA
Kalyanmoy Deb	Michigan State University, USA
Nicoletta Del Buono	University of Bari, Italy
Jordi Delgado	Universitat Politècnica de Catalunya, Spain
Mauro Dell'Amico	University of Modena – Reggio Emilia, Italy
Ralf Der	MPG, Germany
Clarisse Dhaenens	Université Lille, France
Barbara Di Camillo	University of Padua, Italy
Gianni Di Caro	IDSIA, Switzerland
Luigi Di Caro	University of Turin, Italy
Luca Di Gaspero	University of Udine, Italy
Matteo Diez	National Research Council - IME, Rome, Italy
Stephan Doerfel	Kassel University, Germany
Devdatt Dubhashi	Chalmers University, Sweden
Juan J. Durillo	Leibniz Supercomputing Centre, Germany
Omer Dushek	University of Oxford, UK
Nelson F. F. Ebecken	University of Rio de Janeiro, Brazil
Marc Ebner	Ernst Moritz Arndt Universität Greifswald, Germany
Tome Eftimov	Jožef Stefan Institute, Slovenia
Pascale Ehrenfreund	The George Washington University, USA
Gusz Eiben	VU Amsterdam, The Netherlands
Aniko Ekart	Aston University, UK
Talbi El-Ghazali	University of Lille, France
Michael Elberfeld	RWTH Aachen University, Germany
Michael	Leiden University, The Netherlands
T. M. Emmerich	
Andries Engelbrecht	University of Pretoria, South Africa
Anton Eremeev	Sobolev Institute of Mathematics, Russia
Harold Fellermann	Newcastle University, UK
Chrisantha Fernando	Queen Mary University, UK
Cèsar Ferri	Universidad Politécnica de Valencia, Spain
Paola Festa	University of Naples Federico II, Italy
José Rui Figueira	Instituto Superior Técnico, Lisbon, Portugal
Grazziela Figueredo	The University of Nottingham, UK
Alessandro Filisetti	Explora Biotech Srl, Italy
Christoph Flamm	University of Vienna, Austria

Arjen Hommersom	Radboud University, The Netherlands
Vasant Honavar	Pennsylvania State University, USA
Fabrice Huet	University of Nice Sophia Antipolis, France
Hiroyuki Iizuka	Hokkaido University, Japan
Takashi Ikegami	University of Tokyo, Japan
Hisao Ishibuchi	Osaka Prefecture University, Japan
Peter Jacko	Lancaster University Management School, UK
Christian Jacob	University of Calgary, Canada
Hasan Jamil	University of Idaho, USA
Yaochu Jin	University of Surrey, UK
Colin Johnson	University of Kent, UK
Gareth Jones	Dublin City University, Ireland
Laetitia Jourdan	University of Lille/CNRS, France
Narendra Jussien	Ecole des Mines de Nantes/LINA, France
Janusz Kacprzyk	Polish Academy of Sciences, Poland
Theodore Kalamboukis	Athens University of Economics and Business, Greece
George Kampis	Eotvos University, Hungary
Dervis Karaboga	Erciyes University, Turkey
George Karakostas	McMaster University, Canada
Istvan Karsai	ETSU, USA
Zekarias T. Kefato	University of Trento, Italy
Jozef Kelemen	Silesian University, Czech Republic
Graham Kendall	Nottingham University, UK
Navneet Kesher	Facebook, Seattle, WA, USA
Didier Keymeulen	NASA – Jet Propulsion Laboratory, USA
Daeeun Kim	Yonsei University, South Korea
Zeynep Kiziltan	University of Bologna, Italy
Elena Kochkina	University of Warwick, UK
Min Kong	Hefei University of Technology, China
Erhun Kundakcioglu	Ozyegin University, Turkey
Jacek Kustra	Philips, The Netherlands
C. K. Kwong	The Hong Kong Polytechnic University, SAR China
Renaud Lambiotte	University of Namur, Belgium
Doron Lancet	Weizmann Institute of Science, Israel
Pier Luca Lanzi	Politecnico di Milano, Italy
Alessandro Lazaric	Facebook Artificial Intelligence Research (FAIR), France
Sanja Lazarova-Molnar	University of Southern Denmark, Denmark
Doheon Lee	KAIST, South Korea
Eva K. Lee	Georgia Tech, USA
Jay Lee	Center for Intelligent Maintenance Systems – UC, USA
Tom Lenaerts	Université Libre de Bruxelles, Belgium
Rafael Leon	Universidad Politécnica de Madrid, Spain
Lei Li	Florida International University, USA
Shuai Li	Cambridge University, UK
Xiaodong Li	RMIT University, Australia
Joseph Lizier	The University of Sydney, Australia

Palaniappan Ramaswamy	University of Kent, UK
Jan Ramon	Inria, France
Vitorino Ramos	Technical University of Lisbon, Portugal
Shoba Ranganathan	Macquarie University, Australia
Cristina Requejo	Universidade de Aveiro, Portugal
Paul Reverdy	University of Arizona, USA
John Rieffel	Union College, USA
Francesco Rinaldi	University of Padua, Italy
Laura Anna Ripamonti	University of Milan, Italy
Humberto Rocha	University of Coimbra, Portugal
Eduardo Rodriguez-Tello	Cinvestav-Tamaulipas, Mexico
Andrea Roli	University of Bologna, Italy
Vittorio Romano	University of Catania, Italy
Pablo Romero	Universidad de la República, Uruguay
Andre Rosendo	University of Cambridge, UK
Samuel Rota Bulò	Mapillary Research, Austria
Arnab Roy	Fujitsu Laboratories of America, USA
Alessandro Rozza	Parthenope University of Naples, Italy
Kepa Ruiz-Mirazo	University of the Basque Country, Spain
Florin Rusu	University of California Merced, USA
Jakub Rydzewski	N. Copernicus University, Poland
Nick Sahinidis	Carnegie Mellon University, USA
Lorenza Saitta	University of Piemonte Orientale, Italy
Andrea Santoro	Queen Mary University London, UK
Francisco C. Santos	Instituto Superior Técnico Lisboa, Portugal
Claudio Sartori	University of Bologna, Italy
Fréderic Saubion	Université d'Angers, France
Andrea Schaerf	University of Udine, Italy
Oliver Schuetze	CINVESTAV-IPN, Mexico
Luís Seabra Lopes	Universidade of Aveiro, Portugal
Natalia Selini Hadjidimitriou	University of Modena and Reggio Emilia, Italy
Alexander Senov	Saint Petersburg State University, Russia
Andrea Serani	CNR INM, Italy
Roberto Serra	University of Modena and Reggio Emilia, Italy
Marc Sevaux	Université de Bretagne-Sud, France
Nasrullah Sheikh	University of Trento, Italy
Leonid Sheremetov	Mexican Petroleum Institute, Mexico
Ruey-Lin Sheu	National Cheng Kung University, Taiwan
Hsu-Shih Shih	Tamkang University, Taiwan
Kilho Shin	University of Hyogo, Japan
Patrick Siarry	Université de Paris 12, France
Sergei Sidorov	Saratov State University, Russia
Alkis Simitsis	HP Labs, USA

Alina Sirbu   University of Pisa, Italy
Johannes Söllner   Emergentec Biodevelopment GmbH, Germany
Ichoua Soumia   Embry-Riddle Aeronautical University, USA
Giandomenico Spezzano   CNR-ICAR, Italy
Antoine Spicher   LACL University of Paris Est Creteil, France
Claudio Stamile   Université de Lyon 1, France
Pasquale Stano   University of Salento, Italy
Thomas Stibor   GSI Helmholtz Centre for Heavy Ion Research, Germany
Catalin Stoean   University of Craiova, Romania
Reiji Suzuki   Nagoya University, Japan
Domenico Talia   University of Calabria, Italy
Kay Chen Tan   National University of Singapore, Singapore
Letizia Tanca   Politecnico di Milano, Italy
Charles Taylor   UCLA, USA
Maguelonne Teisseire   Irstea - UMR TETIS, France
Fabien Teytaud   Université Littoral Côte d'Opale, France
Tzouramanis Theodoros   University of the Aegean, Greece
Jon Timmis   University of York, UK
Gianna Toffolo   University of Padua, UK
Michele Tomaiuolo   University of Parma, Italy
Joo Chuan Tong   Institute of High Performance Computing, Singapore
Jaden Travnik   University of Alberta, Canada
Nickolay Trendafilov   Open University, UK
Sophia Tsoka   King's College London, UK
Shigeyoshi Tsutsui   Hannan University, Japan
Ali Emre Turgut   IRIDIA-ULBf, France
Karl Tuyls   University of Liverpool, UK
Gregor Ulm   Fraunhofer-Chalmers for Industrial Mathematics, Sweden
Jon Umerez   University of the Basque Country, Spain
Renato Umeton   MIT and Dana-Farber Cancer Institute, USA
Ashish Umre   University of Sussex, UK
Olgierd Unold   Politechnika Wroclawska, Poland
Rishabh Upadhyay   Innopolis University, Russia
Giorgio Valentini   Università degli Studi di Milano, Italy
Sergi Valverdev   Pompeu Fabra University, Spain
Werner Van Geit   EPFL, Switzerland
Pascal Van Hentenryck   University of Michigan, USA
Ana Lucia Varbanescu   University of Amsterdam, The Netherlands
Carlos Varela   Rensselaer Polytechnic Institute, USA
Eleni Vasilaki   University of Sheffield, UK
Richard Vaughan   Simon Fraser University, Canada
Kalyan Veeramachaneni   MIT, USA
Vassilios Verykios   Hellenic Open University, Greece
Mario Villalobos-Arias   Univesidad de Costa Rica, Costa Rica
Marco Villani   University of Modena and Reggio Emilia, Italy
Susana Vinga   INESC-ID/IST-UL, Portugal

Best Paper Awards

LOD 2018 Best Paper Award
"Calibrating the Classifier: Siamese Neural Network Architecture for End-to-End Arousal Recognition from ECG"
Andrea Patané[*] and Marta Kwiatkowska[*]
[*]University of Oxford, UK

Springer sponsored the LOD 2018 Best Paper Award with a cash prize of EUR 1,000.

MOD 2017 Best Paper Award
"Recipes for Translating Big Data Machine Reading to Executable Cellular Signaling Models"
Khaled Sayed[*], Cheryl Telmer[**], Adam Butchy[*], and Natasa Miskov-Zivanov[*]
[*]University of Pittsburgh, USA
[**]Carnegie Mellon University, USA

Springer sponsored the MOD 2017 Best Paper Award with a cash prize of EUR 1,000.

MOD 2016 Best Paper Award
"Machine Learning: Multi-site Evidence-based Best Practice Discovery"
Eva Lee, Yuanbo Wang and Matthew Hagen
Eva K. Lee, Professor Director, Center for Operations Research in Medicine and HealthCare H. Milton Stewart School of Industrial and Systems Engineering, Georgia Institute of Technology, Atlanta, GA, USA

MOD 2015 Best Paper Award
"Learning with discrete least squares on multivariate polynomial spaces using evaluations at random or low-discrepancy point sets"
Giovanni Migliorati
Ecole Polytechnique Federale de Lausanne – EPFL, Lausanne, Switzerland

Contents

Calibrating the Classifier: Siamese Neural Network Architecture for End-to-End Arousal Recognition from ECG

Andrea Patanè[✉] and Marta Kwiatkowska

Department of Computer Science, University of Oxford, Oxford, UK
{andrea.patane,marta.kwiatkowska}@cs.ox.ac.uk

Abstract. Affective analysis of physiological signals enables emotion recognition in mobile wearable devices. In this paper, we present a deep learning framework for arousal recognition from ECG (electrocardiogram) signals. Specifically, we design an end-to-end convolutional and recurrent neural network architecture to (i) extract features from ECG; (ii) analyse time-domain variation patterns; and (iii) non-linearly relate those to the user's arousal level. The key novelty is our use of a shared-parameter siamese architecture to implement user-specific feature calibration. At each forward and backward pass, we concatenate to the input a user-dependent template that is processed by an identical copy of the network. The siamese architecture makes feature calibration an integral part of the training process, allowing modelling of general dependencies between the user's ECG at rest and those during emotion elicitation. On leave-one-user-out cross validation, the proposed architecture obtains +21.5% score increase compared to state-of-the-art techniques. Comparison with alternative network architectures demonstrates the effectiveness of the siamese network in achieving user-specific feature calibration.

Keywords: Emotion recognition · Electrocardiogram ·
Siamese neural network · Convolutional and recurrent neural network

1 Introduction

Driven by applications in mobile mental health and human-computer interaction [1], affective analysis of physiological signals has recently grown in popularity. Since the pioneering use of electrodermal activity for arousal detection, the research has evolved to cater for a range of physiological signals, such as electrocardiogram (ECG), electroencephalogram, electromyogram, breath rhythm and skin temperature [1]. However, while much effort has focused on multi-modal sensor fusion, model performance on single signal sources is still sub-optimal. At the same time, achieving performance improvement for single sensors can push accuracy boundaries for the overall model architecture even further, potentially leading to increased wearability of emotion recognition systems.

© Springer Nature Switzerland AG 2019
G. Nicosia et al. (Eds.): LOD 2018, LNCS 11331, pp. 1–13, 2019.
https://doi.org/10.1007/978-3-030-13709-0_1

The ECG signal, in particular, has become a focus of investigations because of its unobtrusiveness, low cost and widespread availability of ECG sensors, as well as sensitivity to both arousal and valence component of emotions [2]. Existing state-of-the-art machine learning pipelines for emotion recognition from ECG signals usually proceed by extracting the HR (Heart Rate) signal and applying sophisticated HRV (Heart Rate Variability) analysis techniques in a multi-step process. This is mainly composed of: (i) HRV feature extraction; (ii) automatic feature selection; (iii) user-specific feature calibration; (iv) hyper-parameter optimisation; and (v) model fitting. While steps (iv) and (v) are those actually involved in model estimation, the overall performance of the resulting model mainly depends upon the effectiveness of steps (i) to (iii), as testified by the extensive literature on feature extraction, selection and calibration for HRV analysis [2–6]. The feature extraction and selection steps focus on extracting the most informative features from the HR signal. On the other hand, user-specific feature calibration crucially strives to enforce *relative* variation of feature values in the model, rather than *absolute* variation, as the former are related to changes in the user's affective state. Furthermore, the features based on HRV are the *only* type of features extracted from the ECG signal, and thus affective information carried by most of the ECG signal is completely neglected [8–10].

In this work we pose the arousal recognition problem as a supervised classification problem and investigate the use of deep learning for arousal recognition from ECG. For this purpose, we design a deep Convolutional and Recurrent Neural Network (CRNN) architecture that (through end-to-end training) automatically extracts general non-linear and time-domain features from the time-series ECG signal and non-linearly relates those to specific arousal classes based on common variation patterns found. Inspired by state-of-the-art HRV-based machine learning pipelines, we propose the use of *shared-parameter siamese* neural network architecture [15], called the Siamese CRNN (S-CRNN), as a systematic way to extend and generalise feature calibration techniques into the deep learning framework. By making feature calibration an integral part of the end-to-end learning process, we allow the neural network to model general nonlinear dependencies between the user's ECG signal at rest and that during emotion elicitation experiments. Namely, at each forward and backward pass through the network one branch of the S-CRNN processes a new data sample, while the other S-CRNN branch analyses a *template* sample specific to the user's neutral affective state. We use truncated back-propagation through time and stochastic gradient descent to train the network in the classification problem associated to the user's arousal level.

We compare the S-CRNN architecture against state-of-the-art HRV analysis pipelines on the classification task associated to a dataset for arousal recognition during a real-world driving task [14]. The results obtained empirically demonstrate the advantages of the end-to-end approach for arousal recognition from the ECG signal. Namely, on leave-one-user-out cross validation settings the S-CRNN architecture obtains average AUCs percentage increase of +21.5% on the best results obtained by HRV analysis (that is, from 0.659 to 0.801). We further analyse the proposed S-CRNN against alternative architectures and approaches for feature calibration and find that the approach based on shared-parameter siamese neural networks leads to a +7.5% performance increase compared to the corresponding CRNN, at the cost of negligible increase in network parameters.

Contributions. The paper makes the following contributions.

- We propose an end-to-end classification framework for arousal recognition from ECG. We design a CRNN that automatically extracts features from ECG and analyses time patterns among them, relating them to arousal classes.
- We investigate the use of siamese neural networks as a systematic way to implement feature calibration techniques into the deep learning framework.
- We empirically compare the S-CRNN architecture against state-of-the-art HRV analysis methods, observing a +21.5% performance improvement.
- We compare S-CRNN, models based on HRV analyses and alternative network architectures in terms of generalisation performance to new users when very few users are included in the training set. We assess the advantages of the siamese architecture in achieving personalised feature calibration.

Organisation. The remainder of the paper is organised as it follows. In Sect. 2 we analyse related work in emotion recognition from the ECG signal and the use of deep learning in affective computing. In Sect. 3 we present the S-CRNN architecture designed for arousal recognition from ECG. Empirical results evaluating the effectiveness of the S-CRNN architecture are discussed in Sect. 4. Finally, Sect. 5 completes the paper with a discussion on the method presented, and outlines future work directions.

2 Related Work

In this section, we give a brief overview of machine learning methods developed for HRV analysis and applications of deep learning for affective computing.

2.1 Heart Rate Variability Analysis for Arousal Recognition

Table 1 lists a collection of 31 features generally extracted from HR signal and used for HRV analysis for arousal recognition [2–6]. Machine learning methods based upon HRV analysis are multi-step, and include feature selection and user-dependent feature calibration as crucial steps of the model learning.

In fact, Ollander et al. [5] investigate extensive feature selection for emotion recognition from biosignals. They extract a number of HRV features, which are then calibrated using mean and standard deviation computed from a set of user-specific neutral affective state measurements. Few of the selected HRV

Table 1. Types of HRV feature analysis employed for emotion recognition. Full details, including feature extraction algorithms, can be found in [2–5]

Domain	Name
Time	Mean, Median, SDNN, pNN50
	RMSSD, SDNNi, meanRate, sdRate
Geometrical	TINN, RRTI, HRVTi
Frequency	Welch PSD: LF/HF, (LF+MF)/HF, peakLF, peakHF
	Burg PSD: LF/HF, (LF+MF)/HF, peakLF, peakHF
	L-S PSD: LF/HF, (LF+MF)/HF, peakLF, peakHF
Poincaré	SD_1, SD_2, SD_2/SD_1
Nonlinear	$SampEn_1$, $SampEn_2$, DFA_{all}, DFA_1, DFA_2

features actually survive the feature selection step. Zhao *et al.* [2] extract several features from participants' HR signals and perform feature calibration on them. An SVM model is then trained on the data by using l_1 regularisation for automatic feature selection. Reportedly, only 10 out of 26 extracted features were actually used by the SVM model. Melillo *et al.* [4] extracted 13 HRV features, and applied exhaustive feature selection procedure and linear discriminant analysis. Surprisingly, the resulting classifier relied only upon three of the extracted features. In order to partially overcome the feature selection problem, Gjoreski *et al.* [7] train a multi-layer perceptron to predict arousal level from a PSD of the HR signal. They report improvements over models trained on top of HRV features, albeit the neural network proposed is constrained to use only frequency domain features, and no feature calibration procedure is implemented.

Finally, though most of the above works extract HR from ECG, HRV analysis is the only systematic method used to compute features. Thus, potentially relevant information from most of the ECG signal is ignored [8–10].

2.2 Deep Learning for Affective Computing

Many works have investigated the use of deep learning for face expression classification from images, as well as sentiment analysis of text, with deep learning approaches systematically outperforming other techniques [12].

Martinez *et al.* [11] were among the first to apply end-to-end deep learning for physiological signals' affective processing. They developed a CNN for preference learning from galvanic skin response and blood volume pulse data, and empirically demonstrated the advantages of deep features over manually designed ones. Tripathi *et al.* [16] applied CNNs for arousal recognition from EEG. Empirical results show up to ≈14% improvement against methods based on manual feature extraction. Cho *et al.* [22] present a CNN architecture for stress recognition from breathing patterns. Emphasising data augmentation as a crucial step for model training, they obtain substantial improvements over competitive methods.

Our work is a continuation of the latter works that bring deep learning to the field of emotion recognition from physiological signals. The key novelty is the use of siamese networks as a systematic way to implement feature calibration, which is usually overlooked in deep learning frameworks for emotion recognition.

3   Methods

This section discusses our design of the neural network architecture for arousal recognition from ECG signal. First, we describe data pre-processing and augmentation used. We then present the CRNN architecture we designed for feature extraction, and describe the shared-parameter siamese version of the latter.

3.1   Preprocessing

As pre-processing steps we apply a baseline remover filter and standardisation to each ECG signal. We thus segment the signals into fixed size time windows with 50% overlap. Based on empirical results from ultra-short term HRV analysis [17], we use time windows of 15 s, as these provide just enough information to extract significant features from the ECG signal. Though windows of greater size would increase model sensitivity to small feature variations, they would conflict with the practical limitations of the back-propagation through time training algorithm (i.e., increased training time and the vanishing gradient problem).

3.2   Data Augmentation

Datasets for emotion recognition from physiological signals are typically of small size, and thus deep models applied to them tend to overfit [7]. Furthermore, real-world datasets related to health applications are notoriously unbalanced, with the class associated to the absence of the disorder usually greatly over-represented in the training data. This makes stochastic gradient descent somewhat challenging, as it will likely get stuck in a local optimum corresponding to a trivial majority classifier. We thus heavily rely upon data augmentation techniques in order to train our CRNN model.

First, we re-balance class labels of the training set by making multiple copies of random representatives from the minorities class until the dataset is perfectly balanced (that is, until each class is equally represented in the dataset). Then, from each signal slice, we generate n training samples. Namely, we randomly sub-sample n times the signal to a fixed size time window of m time points. In doing so, we keep the time-stamps associated to the sub-sampled signal. Hence, loss of information due to sub-sampling is mitigated, as the neural network is potentially able to partially interpolate the missing pieces of the signal. Unless otherwise specified, in the experiments of Sect. 4 we use $n = 20$ and $m = 1024$.

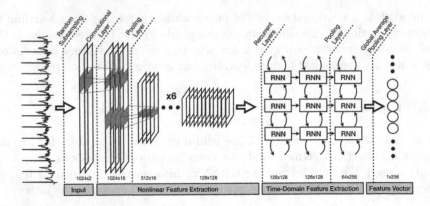

Fig. 1. Proposed CRNN architecture, consisting of 6 convolutional blocks and three stacked bidirectional recurrent neural networks.

3.3 Convolutional Recurrent Neural Networks

The proposed model architecture is sketched in Fig. 1 and summarised in Table 2. Inspired by state-of-the-art HRV features, the CRNN employs a 3-layer bidirectional RNN to summarise temporal patterns on top of a one-dimensional 6-layer CNN. The CRNN is designed to first extract non-linear features from the ECG signal, and then to analyse temporal information of feature variations.

Each convolutional block consists of a convolutional layer and a non-linear activation function layer. After every other block, we use a one-dimensional max-pooling layer to extract salient points from feature maps and compress temporal information. Crucially, we employ Parametric ReLU [18] activation functions in between convolutional layers to avoid dead ReLU problems. Parametric ReLU allows automatic learning of the activation slope for negative input, effectively avoiding the issue of fast death of units slowing down the learning procedure. Notice that, because of data augmentation applied to the training set, data distributions for the training set and the test set are systematically different, and hence we cannot rely on batch-normalisation layers (usually used to circumvent dead ReLU problems).

We use vanilla RNN units, as we experimentally observe that gated recurrent layers quickly lead to overfitting problems. We speculate that this is due to the small size of the dataset used here compared to datasets usually employed to train deep LSTM and GRU recurrent networks [13, 19]. We use a one-dimensional global average pooling layer to summarise temporal patterns extracted by the recurrent layers. Finally, we interleave dropout layers in between each pair of layers, and only for the non-recurrent connections.

The final output of the CRNN is a vector of nonlinear and time domain features extracted from each time window of the ECG signal. Next, we will discuss how this is used to predict an arousal class from each signal window.

Table 2. Details of the architectures and hyper-parameters of the CRNNs designed for arousal recognition from ECG. Ticks (respectively crosses) indicates that the layer is included (not included) in the layer block.

	Layer 1	Layer 2	Layer 3	Layer 4	Layer 5	Layer 6
1-D Conv. filters	16	32	32	64	128	128
1-D Conv. kernel	11	9	9	7	7	7
1-D Max-pooling	✓	✗	✓	✗	✓	✗
Bi-RNN units	128	128	256	✗	✗	✗
1-D Max-pooling	✗	✓	✗	✗	✗	✗

3.4   Siamese Neural Networks

We implement the CRNN inside a shared-parameter siamese architecture [15]. The outline of the siamese network is sketched in Fig. 2. At each forward and backward pass through the network, a user-specific template is fed into the network along with the signal window currently analysed. The latter, and the user-specific template, are independently processed by the CRNN, which extracts two separate feature vectors from them. The resulting feature vectors are concatenated into a unique feature map and altogether processed by a fully connected layer. By relying on the fully connected layer, the siamese architecture has the capability to use features extracted from the user's template to systematically calibrate those extracted from the current signal sample. Finally, a soft-max layer estimates the probability of the user being in an arousal state.

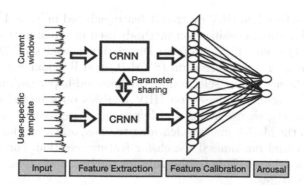

Fig. 2. Shared-parameter siamese architecture for arousal recognition. The current ECG window and the user-specific template are passed through the CRNN. The two feature maps are then concatenated and used to estimate arousal level.

Analogously to methods based on HRV analysis, for the user-specific template we employ a sample recorded from the user before the beginning of the experiment, which is assumed to be representative of the user's neutral affective state. Notice that, in order to mitigate overfitting, we apply data augmentation techniques outlined in Subsect. 3.2 also to the users' templates.

4   Results

In this section we describe experiments related to the following key points:

- Comparison of HRV and S-CRNN on arousal recognition.
- Evaluation of the siamese architecture capabilities to implement feature calibration, comparing the S-CRNN with alternative network architectures.
- Analysis of the number of users included in the training set (population size) to asses the effect on the feature calibration layer.
- Sensitivity analysis of hyper-parameters included in our methodology, focusing on data-augmentation and number of convolutional/recurrent layers.

4.1   Dataset

We perform comparisons on the classification task associated to a dataset for arousal recognition made publicly available by Schneegass et al. [14]. Briefly, a set of physiological signals were recorded from 10 users during a real-world driving task. Data samples were then subjectively labelled by each user for arousal/driving workload. Among the signals included in the dataset, we focus on ECG and use the arousal labels to define a binary classification problem (low vs. high arousal).

4.2   HRV-Based Analyses

We train models based on HRV on the 31 features listed in Table 1 and provide results for a selection of classification methods used in the literature [2–5], that is, k-Nearest Neighbours (K-NN), Linear Discriminant Analysis (LDA), Support Vector Machine with l_1 regularisation (SVM-l1) and Random Forest (RF). We apply state-of-the-art feature selection algorithms and hyper-parameter optimisation to all the techniques based on HRV analysis on a nested cross validation setting. Namely, we use fitting and hyper-parameter optimisation routines implemented in the Matlab machine learning toolbox, and apply forward search, backward search and randomised search for feature selection. For space limitation, for each model we include results only for the best performing combination of parameters/features.

4.3   Experimental Setup

Because of strong class imbalance (only $\approx 6\%$ of samples are representative of the arousal class) we compare the results based on AUC score. Results are presented for leave-one-user-out cross validation. We use Keras [20] with TensorFlow [21] backend for implementation and training of neural networks. We train the networks using Adam optimiser [23] up to a maximum of 100 epochs, and use early stopping on a validation set. We do not investigate exhaustive hyper-parameter optimisation for the S-CRNN, as it is nested in a cross validation and would thus lead to prohibitive computational times. Instead, we perform a local hyper-parameter analysis on the most sensitive hyper-parameters (Sect. 4.6).

We train HRV analysis models on a 2 GHz Intel Core i5 processor with a RAM of 8 GB @1867 MHz. Computational time for a full round of hyper-parameter optimisation and cross validations for each HRV model varied between about 1 and 12 h. We train deep learning models on NVIDIA Tesla K80 GPU. Computational time for a full round of cross validation took about 60 h.

4.4   Comparison of HRV and S-CRNN

In Fig. 3a we compare average AUCs obtained by different classification models learnt on top of HRV features with the results obtained by the S-CRNN, for an increasing number of users included in the training set. Results for population sizes between 1 and 7 are averaged over 10 randomly chosen combinations of users included in the training set (consistently among models).

As expected, we observe an overall trend for all the methods to perform better as the number of users included in the training set increases. However, the performance boost obtained for all the models when increasing the number of training users from 1 to 5 seems to saturate for HRV-based methods, which fail to take advantage of such increases. On the other hand, the S-CRNN obtains additional AUC boosts when more users are included in the training set. For the largest size of the training set allowed by the dataset used here (i.e. 9 users), the S-CRNN obtains average AUCs percentage increase of +21.5% compared to the best results obtained by HRV analysis (i.e. from 0.659 to 0.801 AUC). Finally, notice that all the methods based on HRV analysis perform similarly to each other. This suggests that the low AUC reached is not related to the actual classification model used, but to the weak correlation between the HRV features extracted and the user arousal level.

4.5   Variations on the Architecture

In Fig. 3b we compare the S-CRNN with variants of its architecture, namely, with the CRNN model that does not benefit from the feature calibration layer, and with a M-CRNN (Merged-CRNN). Similarly to the S-CRNN, the latter is based on two separate CRNN branches, but they do not share parameters.

Again, there is an overall trend of AUC increase as the number of training users increases. Contrary to what happens for HRV -based methods, here all the models systematically get performance boost every time new users are included in the training set. This is likely to be related to the greater capacity of neural networks to use information from more data compared to manual feature extraction pipelines. Interestingly, the CRNN slightly outperforms the S-CRNN for population sizes of 1, 3 and 5. We speculate that this is because, with small population sizes, the feature calibration layer overfits to the specific training users characteristics. However, as the number of users increases, the S-CRNN is

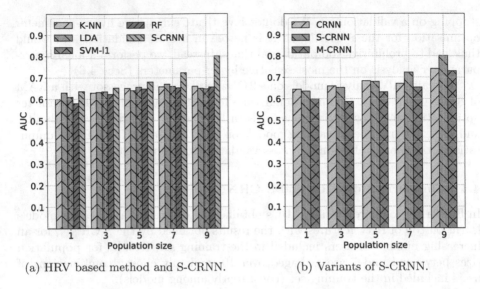

(a) HRV based method and S-CRNN.   (b) Variants of S-CRNN.

Fig. 3. AUCs for increasing the number of users included in the training set.

able to take full advantage of the information carried by new users' data. In fact with population size of 9, by proper calibration of the features extracted by the CRNN, the S-CRNN obtains a +7.5% percentage increase on the corresponding CRNN. Notice that, even though the M-CRNN model is more general than the S-CRNN, it fails to improve even on the score obtained by the CRNN. This could be due to the almost double number of parameters of the M-CRNN.

4.6   Hyper-parameters' Analysis

In Fig. 4a we plot AUCs obtained for different numbers of recurrent and convolutional layers included in the S-CRNN. We analyse the effect of changing the number of layers of one type (either convolutional or recurrent), while keeping the other type of layers fixed to its nominal value. Notice that the x and y axis are normalised with respect to the S-CRNN architecture. The strongest effect is given by the convolutional layers, with the fully recurrent network obtaining only about 60% of the S-CRNN AUC. After an initial rapid increase, the AUC score saturates around the nominal S-CRNN architecture.

Figure 4b shows the analysis results for the two hyper-parameters involved in the data augmentation phase. As expected, there is an overall trend of AUC increase as the number of copies made from each training sample is increased. However, the benefit from having more copies saturates around 15. Analogously, the more samples given as input to the S-CRNN, the higher is the AUC obtained.

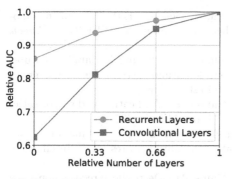
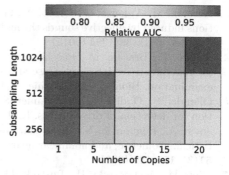

(a) Convolutional and recurrent layers.   (b) Data augmentation hyper-parameters.

Fig. 4. Hyper-parameter analysis for S-CRNN.

5   Conclusions

We proposed a siamese CRNN architecture for arousal detection from ECG. The CRNN is explicitly designed to extract non-linear features from the ECG signal and analyse relevant time patterns using a 3-layer RNN stacked on top of a 6-layer CNN. Relying on a shared-parameter siamese architecture, we implemented feature calibration in the deep learning framework itself, which allows the neural network to model non-linear relationship between users' ECG at rest and that during emotion elicitation. We demonstrated the advantages of our approach compared to state-of-the-art HRV based methods, obtaining up to +21.5% percentage improvement on the AUC score. Further, we showed that the siamese architecture obtains +7.5% score increase compared to the CRNN.

As future work we plan to extend the S-CRNN to long-term analysis settings, and perform comparison with medium and long-term HRV techniques. We emphasise that, though the siamese architecture was introduced for ECG, it can be generalised to most of the physiological signals used for affective state recognition. As feature calibration has proven to be a crucial step for manual feature extraction pipelines, future work will investigate whether affective computing based on deep learning can benefit from the siamese network paradigm.

Acknowledgements. This project was funded by the EU's Horizon 2020 research and innovation programme under the Marie Skłodowska-Curie grant agreement No 722022.

References

1. Picard, R.W.: Affective computing. Massachusetts Institute of Technology (1995)
2. Zhao, M., Adib, F., Katabi, D.: Emotion recognition using wireless signals. In: 22nd International Conference on Mobile Computing and Networking, pp. 95–108 (2016)

3. Nardelli, M., Valenza, G., Greco, A., Lanata, A., Scilingo, E.P.: Recognizing emotions induced by affective sounds through heart rate variability. IEEE Trans. Affect. Comput. **6**(4), 385–394 (2015)
4. Melillo, P., Bracale, M., Pecchia, L.: Nonlinear Heart Rate Variability features for real-life stress detection. Case study: students under stress due to university examination. Biomed. Eng. Online **10**(1), 96 (2011)
5. Ollander, S., Godin, C., Charbonnier, S., Campagne, A.: Feature and sensor selection for detection of driver stress. In: PhyCS, pp. 115–122 (2016)
6. Jovic, A., Bogunovic, N.: Electrocardiogram analysis using a combination of statistical, geometric, and nonlinear heart rate variability features. Artif. Intell. Med. **51**(3), 175–186 (2011)
7. Gjoreski, M., Gjoreski, H., Luštrek, M., Gams, M.: Deep affect recognition from RR intervals. In: 2017 ACM International Joint Conference on Pervasive and Ubiquitous Computing and Symposium on Wearable Computers, pp. 754–762 (2017)
8. Andrassy, G., Szabo, A., Ferencz, G., Trummer, Z., Simon, E., Tahy, A.: Mental stress may induce QT-interval prolongation and T-wave notching. Ann. Noninvasive Electrocardiol. **12**(3), 251–259 (2007)
9. Paoletti, N., Patanè, A., Kwiatkowska, M.: Closed-loop quantitative verification of rate-adaptive pacemakers. ACM Trans. Cyber-Phys. Syst. **2**, 33 (2018)
10. Heslegrave, R.J., Furedy, J.J.: Sensitivities of HR and T-wave amplitude for detecting cognitive and anticipatory stress. Physiol. Behav. **22**(1), 17–23 (1979)
11. Martinez, H.P., Bengio, Y., Yannakakis, G.N.: Learning deep physiological models of affect. IEEE Comput. Intell. Mag. **8**(2), 20–33 (2013)
12. Kahou, S.E., et al.: Emonets: multimodal deep learning approaches for emotion recognition in video. J. Multimodal User Interfaces **10**(2), 99–111 (2014)
13. Rosa, S., Patané, A., Lu, X., Trigoni, N.: CommonSense: collaborative learning of scene semantics by robots and humans. In: Proceedings of the 1st International Workshop on Internet of People, Assistive Robots and Things, pp. 1–6 (2018)
14. Schneegass, S., Pfleging, B., Broy, N., Heinrich, F., Schmidt, A.: A data set of real world driving to assess driver workload. In: 5th International Conference on Automotive User Interfaces and Interactive Vehicular Applications, pp. 150–157 (2013)
15. Bromley, J., Guyon, I., LeCun, Y., Sackinger, E., Shah, R.: Signature verification using a "siamese" time delay neural network. In: Advances in Neural Information Processing Systems, pp. 737–744 (1994)
16. Tripathi, S., Acharya, S., Sharma, R.D., Mittal, S., Bhattacharya, S.: Using deep and convolutional neural networks for accurate emotion classification on DEAP Dataset. In: AAAI, pp. 4746–4752 (2017)
17. Salahuddin, L., Cho, J., Jeong, M.G., Kim, D.: Ultra short term analysis of heart rate variability for monitoring mental stress in mobile settings. In: 29th International Conference of the Engineering in Medicine and Biology Society, pp. 4656–4659 (2007)
18. He, K., Zhang, X., Ren, S., Sun, J.: Delving deep into rectifiers: surpassing human-level performance on imagenet classification. In: Proceedings of the IEEE International Conference on Computer Vision, pp. 1026–1034 (2015)
19. Panzner, M., Cimiano, P.: Comparing hidden Markov models and long short term memory neural networks for learning action representations. In: Pardalos, P.M., Conca, P., Giuffrida, G., Nicosia, G. (eds.) MOD 2016. LNCS, vol. 10122, pp. 94–105. Springer, Cham (2016). https://doi.org/10.1007/978-3-319-51469-7_8
20. Chollet, F., et al.: Keras. GitHub (2015). https://github.com/keras-team/keras

21. Abadi, M., Barham, P., Brevdo, E., Chen, Z. et al.: TensorFlow: large-scale machine learning on heterogeneous systems (2015). https://www.tensorflow.org/
22. Cho, Y., Bianchi-Berthouze, N., Julier, S.J.: DeepBreath: deep learning of breathing patterns for automatic stress recognition using low-cost thermal imaging in unconstrained settings. arXiv:1708.06026 (2017)
23. Kingma, D.P., Jimmy, B: Adam: a method for stochastic optimization. arXiv preprint arXiv:1412.6980 (2014)

Simple Learning with a Teacher via Biased Regularized Least Squares

Sergio Decherchi$^{(\boxtimes)}$ and Andrea Cavalli

Fondazione Istituto Italiano di Tecnologia, Computational Sciences,
16163 Genoa, Italy
sergio.decherchi@iit.it

Abstract. In the paradigm of learning with a teacher, introduced by
Vapnik, a supervised learner is trained on an augmented features space,
and a student is requested to match the teacher accuracy as much as
possible in a reduced feature space. In particular, in the transfer learning mode proposed by Vapnik, a method was formalized to move the
knowledge from the teacher to the student. In this paper, we use biased
regularized least squares as a simple yet effective method to transfer the
knowledge from one learner to another, and to assess its accuracy. We
achieve this by further generalizing a semi-supervised learning method,
which we previously introduced. We will show that, with this approach,
the teacher can be any classifier. In particular, we will employ the Relevance Vector Machine (RVM) as teacher to assess the method's capability in transferring the knowledge in terms of classification accuracy,
and in reproducing the probabilities coming from RVM. We validate the
method against standard UCI datasets and systematically compare it
with Vapnik's original method in terms of accuracy and execution time.
We thus demonstrate the feasibility and speed of this new approach.

1 Introduction

Supervised learning is an extremely well-studied topic in terms of theory
and algorithms. Several variations of the supervised learning paradigm have
appeared over the years, including semi-supervised learning [3–5], transfer learning, domain adaptation [12,13], multiple output learning, deep learning, and,
more recently, the learning with a teacher paradigm [1].

In classical supervised learning for classification, the learning algorithm is
fed with an input matrix X (where rows represent the samples, and columns
represent the features) and the desired labels vector y represents the class membership. The aim is the learning of a function f that can predict the labels y.
In contrast, in the learning with a teacher paradigm, it is hypothesized that a
teacher, namely a classifier, can access a privileged space D^*, whereas a student can only access a reduced unprivileged space D from which to replicate the
teacher classification accuracy. By *privileged*, we mean that the D^* space has an
extended set of features that are not accessible in the space D. For instance, D^*
has ten features, whereas D has access to only five of them. Implicitly, we are

© Springer Nature Switzerland AG 2019
G. Nicosia et al. (Eds.): LOD 2018, LNCS 11331, pp. 14–25, 2019.
https://doi.org/10.1007/978-3-030-13709-0_2

assuming that the additional features in D^* are not noisy but significant. From now on, x^* and x will indicate a sample in the full teacher space D^* and student space D respectively, while the data matrices will be indicated by X^* and X.

In this setting, Vapnik proposed the SVM+ and SVM$_\Delta$+ [1]: SVM$_\Delta$+ functional is a nonlinear optimization problem, which Vapnik showed could be approximated by a convex one with linear constraints. To cope directly with knowledge transfer, which is the object of this paper, Vapnik additionally proposed a learning scheme in which the student learns the kernel rows of the kernel matrix as functions computed on the space D^*, and tries to reproduce this *hidden layer* by learning proper basis functions in the space D. This approach requires the learning of a number of approximate basis functions equal to the number of kernel bases of the original space. Even if the number of landmarks N_l (samples used in the kernel expansion) can be reduced with respect to the original number of samples N, N_l regression problems still have to be solved. If the Regularized Least Squares (RLS) method is used to solve the regression problems, then the overall scaling of the learning procedure is $\mathcal{O}(N^3 N_l)$ because of the (RLS) cubic scaling. In particular, if we choose $N_l = N$ (which is the most agnostic choice), the overall cost is $\mathcal{O}(N^4)$. This makes the procedure computationally intensive. This situation requires simpler and faster methods for learning with privileged information. The recent literature contains works at the theoretical and algorithmic level. At the theoretical level, Hypothesis Transfer Learning (HTS) is another way to indicate learning with hidden/privileged information. In [6], the authors use stability bounds to prove that HTS is convenient and can be successfully used. Interestingly, this group used regularized least squares for their analysis. However, the biasing is placed in the loss function and not in the regularizer. This means that, during the prediction phase, the original function must be evaluated in the augmented space too. This makes the method unsuitable for out-of-sample and disjoint spaces D^* and D, as in our case. In [7,8], some methods are presented that can obtain knowledge transfer. However, they are restricted to mimicking linear classifiers or mimicking nonlinear classifiers in the same function space of the learnt final function. These limits considerably restrict the applicability of these methods.

Here, we propose a biased form of regularized least squares as a simple and effective solution. Furthermore, we show that the proposed functional generalizes a previous one, which we proposed in [3,4] for semi-supervised learning. Indeed, the functional we define has several advantages, including (a) independence of the original classifier space with respect to the mapped one; (b) a simple learning system to be solved to obtain the solution (so SVD can be used to accelerate the search for an optimal regularization parameter); and (c) it offers a simple Bayesian interpretation.

In the following, we introduce the regularized least squares algorithm (RLS), the knowledge transfer scheme proposed by Vapnik, and our mimicking functional and its properties. Then, we apply the method to the Relevance Vector Machine (RVM) as teacher and the proposed functional as student. We thereby assess the method's accuracy and execution speed compared to the knowledge transfer method proposed by Vapnik. In the same section, we discuss the method's feasibility in reproducing RVM probabilities. Finally, we draw some conclusions.

2   Regularized Least Squares

The regularized least squares algorithm is a widely used method for classification and regression [9]. Althought it uses a regularization term identical to SVM, it uses a squares loss instead of the hinge loss. The mathematical problem is:

$$\min_f ||f - y||^2 + \lambda ||f||_{\mathcal{H}}^2 \tag{1}$$

where $f \in \mathcal{H}$ and $\mathcal{H}$ is a Reproducing Kernel Hilbert space. Due to the Representer Theorem, it can be shown that this functional admits as minimizer the following linear expansion in terms of kernel functions:

$$f(x) = \sum_{i=1}^{N} \alpha_i K(x_i, x) \tag{2}$$

Substituting this expansion into the functional, the following matrix form can be obtained:

$$\min_\alpha ||K\alpha - y||^2 + \lambda \alpha^t K\alpha \tag{3}$$

In turn, computing the gradient and nullifying it, the optimal solution can be obtained by solving a linear system of equations or inverting a matrix:

$$\alpha = (K + \lambda I)^{-1} y \tag{4}$$

To compute the regularization path in an efficient way, namely the set of solutions changing the regularization parameter, it is convenient to compute the SVD of the kernel matrix K as per USU^t. Indeed, it is easy to show that, by using the SVD, computing the full regularization path boils down to computing a trivial inversion of the S matrix and two matrix multiplications ($U^t y$ can be precomputed) as per:

$$\alpha = U(S + \lambda I)^{-1} U^t y \tag{5}$$

RLS will be our starting point and we will show that this property is preserved upon the proposed modification.

3   Knowledge Transfer

In [1], the knowledge transfer process is formalized as the process of starting from the teacher representation in kernel space and moving it to the space of the student. In particular, suppose the teacher has learnt a rule as per:

$$f(x^*) = \sum_i^N \delta_i K(x_i^*, x^*) \tag{6}$$

where x_i^* are samples belonging to the full teacher space D^*. We want the student to mimic this function by just employing x instead of x^*, where x now belongs to the reduced student space D. To this end, Vapnik proposes a strategy whereby we directly mimic $f(x^*)$ by *copying* the kernel behaviour on x^*. That procedure can be formalized in three points.

- First, we find the fundamental elements of knowledge in the space D^* called u_i^*. In other words, we define some landmarks a priori, or some support vectors a posteriori in order to reduce the computational burden. This step is computationally relevant only. Conceptually, it is not different than using all the samples. In the following, we will use all the available samples to maximize accuracy.
- Second, we find frames (m functions) $K^*(u_1^*, x^*), ..., K^*(u_m^*, x^*)$ in space D^*. We identify the *hidden layer* elements that we want to transfer.
- Third, we find the basis functions $\phi_1(x), ..., \phi_m(x)$ such that $\phi_k(x_i) \approx K^*(u_k^*, x_i^*)$. This means finding the basis functions that, for the space D, the student space, mimic the behaviour of the full kernel function in space D^*.

Once obtained, the m functions, then the student function becomes:

$$f(x) = \sum_k^m \delta_k \phi_k(x) \tag{7}$$

From a neural network perspective, this strategy is equivalent to mimicking the output of the hidden layer of a single hidden layer (pretrained) neural network (as SVM is, where the number of neurons is equal to the number of samples, i.e., $N = m$).

With this method, the functions $\phi(x)$ are the main source of practical difficulty because, for each of the m landmarks, we have to learn a regression function. As a result, the method is not particularly efficient. In turn, as usual, learning a function means conducting a model selection on it, which makes the algorithm even slower.

To avoid the m regression and the m model selection problems, we propose a much simpler method where just one regression is needed to transfer the knowledge, thus dramatically reducing the computing time.

4 Mimicking the Teacher

After looking at the strategy proposed in [1], the key question is if we really need to mimic the hidden layer to transfer the knowledge. Is it sufficient to mimic the output of the network? To transfer the knowledge, most methods propose the mimicking of the hypothesis space. For instance, in the linear case this means defining the following biased Tikhonov functional:

$$\min_w ||Xw - y||^2 + \lambda ||w - w_t||^2 \tag{8}$$

where we denoted the teacher solution with w_t. The problem with this intuitive approach is that we are implicitly assuming that w_t (the teacher model) and w (the student model) live in the same space. This can be a limiting factor. Here, we propose an alternate approach which directly mimics the teacher and indirectly mimics the hypothesis space. Indeed, the final aim of the learning with a teacher approach is to have a good generalization performance in the student space. This can be achieved by decoupling the two hypothesis spaces.

This can be obtained in RLS by slightly changing its biased form and switching from a hypothesis space bias to a bias in the output of the decision functions. In the linear case, this corresponds to substituting $||w - w_t||^2$ with $\sum_{i=1}^{N}(wx_i - w_t x_i^*)^2$ or in general nonlinear terms:

$$\min_f \sum_{i=1}^{N}(f(x_i) - y_i)^2 + \lambda_1||f||_{\mathcal{H}}^2 + \lambda_2 \sum_{i=1}^{N}(f(x_i) - f_t(x_i^*))^2 \tag{9}$$

where now the mimicking is not done in the hypothesis space but simply by a Euclidean norm in the output of the functions. Note that we here introduced the *mimicking* regularizer λ_2, which tells us to what extent the teacher should be mimicked. This means that we are no longer mimicking the hypothesis space directly. Rather, we are only mimicking the output of the teacher. Philosophically speaking, our student is not understanding the lesson (the hypothesis, the model w), but just mimicking the teacher's behaviour (f values). One might expect this approach to be less effective. However, we will show empirically that this is not the case. This simplification is somehow reminiscent of the transductive approach [10] when compared to full inference where the query points (the questions to the student) are known a priori. Here, we also squeeze the problem to its essence, namely just imitating the teacher. In addition to the philosophical difference, even if the output is just *copied*, the learnt functions generalize well when a proper λ_1 regularizer is chosen. Indeed, imitating is sufficient to shape a sufficiently good hypothesis space for the student.

Such a functional has some very nice properties:

- It decouples completely the f_t space from f, the student function.
- It does not make any assumption about the source of f_t. It could be any classifier that gives some scalar output; even a multiclass classifier could be mimicked by this approach. Later, we will show how this property can be used to mimic class probabilities.
- It can be minimized by solving a linear system.
- It has an obvious Bayesian intepretation when looking at the second regularizer, in that this one could be interpreted as a Gaussian prior centered on the teacher function f_t.
- It is easily generalizable to a multi-teacher or even a multi-aim approach (e.g. teacher+semi-supervised learning).

Those properties are all self-evident except two, namely the linear system and the generalizability.

First, we show that the system minimization can be achieved as a linear system solution, and that SVD can be conveniently used again to quickly compute the regularization path. Writing down the proposed function in matrix terms, we obtain the following functional to be minimized in the α space

$$\min_{\alpha} ||K\alpha - y||^2 + \lambda_1 \alpha^t K\alpha + \lambda_2 ||K\alpha - f_t||^2 \tag{10}$$

Computing the gradient and nullifying it yields the solution:

$$\alpha = (K + \lambda_1 I + \lambda_2 K)^{-1}(y + \lambda_2 f_t) \tag{11}$$

where I is the identity matrix. By employing the SVD on K, we obtain $K = USU^t$. Due to its positive definiteness and using the property that $U^t = U^{-1}$, we can conveniently write:

$$\alpha = U((1 + \lambda_2)S + I)^{-1}U^t(y + \lambda_2 f_t) \tag{12}$$

where it is clear that changing the regularization parameters just means inverting a diagonal matrix that can be trivially obtained by inverting all its diagonal elements. As such, the computation of the regularization path involves just one SVD and matrix multiplications, hence its speed. In terms of computational complexity without considering the cost of model selection, Vapnik's approach scales as per $\mathcal{O}(N^3 N_l)$ (where N_l is a possibly a priori or a posteriori subset of samples), whereas our protocol only needs one RLS problem to be solved, thus obtaining $\mathcal{O}(N^3)$. For large sample sizes, the speed-up can be dramatic.

As already anticipated, it is interesting to note that the proposed functional is very similar to that used in [3,4] to support semi-supervised learning. In contrast to that work, the *mapping* step is not needed here because the f_t is already a classification function. Moreover, semi-supervised learning could be achieved here by only letting f_t represent a learnt function on the full set (supervised + unsupervised) after the Mapping step (see [3]) and applying the square loss only on labeled samples.

In the most general terms, one could envision the following functional:

$$\mathcal{L}(f(x), y) + \lambda_1 \Omega(f) + \lambda_2 \sum_{i=1}^{nt} ||f - \hat{f}_i||^2 \tag{13}$$

where a multi-teacher approach could be used by properly assigning the various $\hat{f}_i$. Interestingly, the amount of computations needed for transfer learning is independent of the number of teachers or reference functions in general. Additionally, if the loss $\mathcal{L}$ and the regularizer Ω are all quadratic, then a simple linear system must be solved. Further, if $\hat{f}_i$ comes from a mapping step as in [3], then the method is flexible enough to allow simultaneous semi-supervised and teacher learning, simultaneously leveraging the space D^* and the unlabelled samples possibly coming from D^*.

Another interesting potential interpretation of (13) is a way of performing boosting on different features. Indeed, one could assign to each $\hat{f}_i$ a specific subset of features, using (13) to boost all of them. Thus $\hat{f}_i$ would play the role of the weak learners. However, these intriguing possibilities are beyond the scope of the current paper, where we focus on learning with a single teacher only.

4.1 The Complete Learning Scheme

As anticipated, we will use the Relevance Vector Machine [11] as teacher in that it can directly deliver the class probabilities. In particular, for computing the probabilities from RVM and from biased RLS, we compute the sigmoid:

$$p(x) = 1/(1 + \exp(-f(x))) \tag{14}$$

where x is a generic sample, f is the learnt function by the learner, and p is the corresponding class probability. Overall, the complete learning scheme can be summarized in the following pseudocode:

Input: X^*,X,y
Output: α
1: Teacher learns by solving RVM(X^*,y)
2: Compute teacher probabilities via eq.14, name them p^*
3: Student learns by solving eq. 9 in space D, do model selection on λ_1 and get α
4: Compute student probabilities via eq. 14, name them p
5: **return** α, p, p^*

Here, we assume that λ_2 is kept fixed during the model selection. Indeed, we found empirically that a sufficiently large (e.g. 1e2) value is sufficient to enforce the correspondence between p^* and p as much as possible. In contrast, for the model selection and as usual for RLS, we found that the first regularization parameter λ_1 value is critical.

5 Experiments

In the following, we discuss experiments conducted on some widely used UCI datasets to demonstrate and study the method's feasibility. We chose the following datasets from UCI: ionosphere, A vs B letter recognition, mammographic masses, musk, sonar and diabetic retinopathy. We split the datasets into two halves, one for training and one for testing. The kernel was always the Gaussian kernel with width automatically fixed to: $\sigma = 8 \max_{ij}(d_{ij}/\sqrt{2N})$, where d_{ij} is the pairwise distance between a pair of samples. Data were always normalized in the $[-1, +1]$ domain, and the data order was randomly permuted. All experiments were conducted with Matlab R2016a on an Intel Xeon 2630 workstation and Linux Ubuntu. To emulate the learning with a teacher setting, we suppressed some dataset features to simulate the student space D. In particular: we suppressed variables 1 to 32 for ionosphere, variables 5 to 160 for musk, variables 1 to 15 for sonar, columns 4, 5 for mammographic data, and columns 1 to 4 for diabetes. This column suppression allows us to fairly emulate a significant difference in spaces between D^* and D. The λ_2 parameter was always kept fixed to 1e2. We performed model selection for λ_1 in the interval $[10^{-10}, 10^3]$ in 25 equispaced values in the exponent space. For learning the m frames in Vapnik's method, we again used RLS. We used the same λ_1 interval to perform model selection, but with fewer values (only 13) to speed up the computation. We did not use the SVD trick for model selection. We limited the total amount of samples to random 1000 samples overall for the training and test sets.

First, we compare our method to Vapnik's method to assess the algorithm's feasibility in terms of accuracy and computing time. Next, we characterize the method's behaviour when changing the training/test sample size. Finally, we comment on the method's ability to reproduce the class probabilites learnt by the RVM teacher.

5.1   Comparison with Vapnik's Algorithm

In this first set of experiments, we sought to understand how the proposed method compares to the knowledge transfer scheme in [1]. In Fig. 1, we report the results of Vapnik's method versus the regularization parameter λ used to learn the regression functions. It is evident from the figures that a proper tuning of the regularization parameter is needed to obtain a good accuracy. This confirms the idea that not only must m regression functions be learnt, but proper model selection must also be conducted on them. The same kind of analysis is performed for our method. Table 1 shows that the accuracy values of the two methods are almost identical. Interestingly, however, our method is not only conceptually much simpler but also much faster. In particular, in the table ν_t indicates the teacher error, ν_r the error in the reduced space D, ν_s the student error after the biasing, ν_{lt} is the student error using the algorithm in [1], t_s and t_{lt} are the respective execution times in seconds for our method and Vapnik's method, and Δp is the probability absolute difference $\Delta p = |p(x) - p(x^*)|$ between the teacher and the student (this last quantity is also reported in the figures as the cyan line with star marks). Clearly, Δp is dataset-dependent and accuracy-dependent. It is likely that a higher classification error induces a higher difference in probability. Nevertheless, we found that, on average, the probability is maintained under a tolerance of about 10%. This means that, on average, even probability estimates are reasonably accurate. By looking at the probability tolerance curves, we found that the minimum is often attained (or very nearly) where the best regularization parameter λ_1 is found. This interesting property allows us to empirically check that, where the best model is located, the estimated probabilities also tend to match the original ones under a certain tolerance (Fig. 2).

5.2   Behaviour by Changing Sample Size

In this second set of experiments, we sought to understand when it is worth transferring the knowledge in terms of the sample regime. We tested our algo-

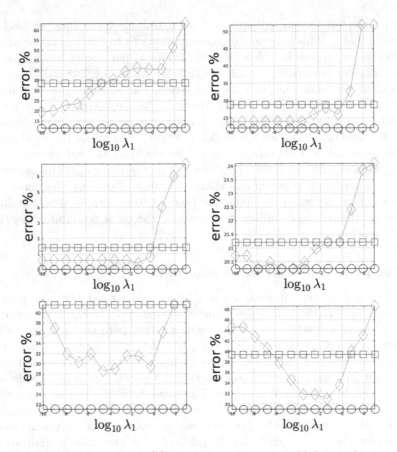

Fig. 1. Results for the method in [1]. From top to bottom and left to right: ionosphere, sonar, letter, mammographic masses, musk, retinopathy. The line with squares (red) indicates error in the D reduced space, the line with circles (black) indicates the teacher error (in D^*), and the line with rhomboids (green) indicates the student errors (Color figure online)

rithm on a subset of datasets and collected the test errors at varying sample sizes. We used only musk and ionosphere because they were the only two where we could guarantee that the teacher error was lower then the error in the reduced space. For certain sample sizes, removing some features can often be beneficial instead of disadvantageous. For this reason, we selected only the datasets that consistently allowed us to obtain a teacher error lower than the student error at all sample sizes. From Fig. 3, it is evident that the student error is consistently significantly better than the error in the reduced set. This confirms that the method is advantageous regardless of sample size, provided that the teacher error is better than the reduced error.

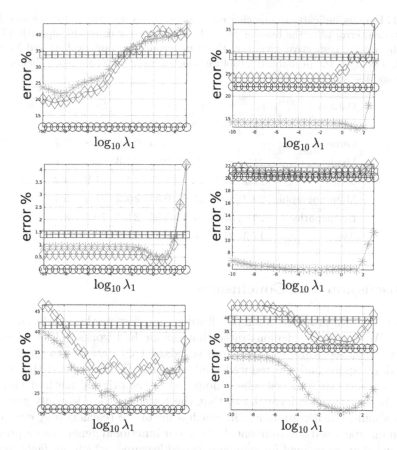

Fig. 2. Results for the proposed method. From top to bottom and left to right: iono-sphere, sonar, letter, mammographic masses, musk, retinopathy. The line with squares (red) indicates error in the D reduced space, the line with circles (black) indicates the teacher error (in D^*), the line with rhomboids (green) indicates the student errors, and the line with stars is Δp (Color figure online)

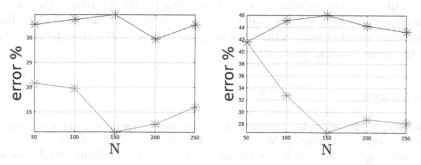

Fig. 3. Analyzing the accuracy at varying sample sizes. The cyan line is the student error. The red line is the error on the reduced dataset. The first figure is the ionosphere dataset. The second figure is the musk dataset. Learning with a teacher is convenient regardless of the sample size. (Color figure online)

Table 1. Here, ν_t indicates the teacher error, ν_r is the error in the reduced space D, ν_s is the student error after the biasing, ν_{lt} is the student error using the algorithm in [1], t_s and t_{lt} are the respective execution times in seconds for our method and Vapnik's method, and Δp is the probability absolute difference $\Delta p = |p(x) - p(x^*)|$ between the teacher and the student

Dataset	ν_t	ν_r	ν_s	t_s	ν_{lt}	t_{lt}	Δp
Iono	11.4	33.7	**18.9**	0.25	19.4	5.2	22.1
Letter	0	1.4	**0.4**	0.36	**0.4**	77.8	0.7
Musk	21	41.6	**28.6**	0.26	**28.6**	9.6	23.1
Sonar	22.1	28.8	**24**	0.22	**24**	1.8	14.2
Mammographic	20.2	20.2	21.2	0.31	**20.2**	31.6	5.3
Retinopathy	29	39.4	31.4	0.35	**31**	79.5	8.4
Avg	17.3	27.5	20.8	0.3	20.6	34.3	12.3

6   Discussion and Conclusions

In this paper, we proposed a simple, flexible, and fast method to learn from a teacher. We empirically demonstrated that the method works well in a scenario where we removed some features from standard UCI datasets to emulate the enriched D^* space and the reduced student space D. We showed the method to be both accurate and fast when compared to the original method proposed in [1]. Furthermore, we have shown that the tested method is easily generalizable to a multi-teacher approach, in which some of the teachers may even come from an unsupervised environment. Indeed, our functional generalizes a previous functional that we defined for semi-supervised learning, which, *de facto*, was an instance of transfer learning. In spirit, our functional resembles that defined in [2]. Interestingly, the three-step procedure proposed in that paper is conceptually analogous to the four steps we proposed earlier in [3]. However, the procedure proposed in [3] is more flexible in that the mapping step allows an easy move from an unsupervised solution to a function f in a RKHS. Moreover, our method is always convex. Future works will systematically analyze the generalized functional, and will mix teachers from the augmented space D^* and unlabelled data D in the same learning process.

References

1. Vapnik, V., Izmailov, R.: Learning using privileged information: similarity control and knowledge transfer. J. Mach. Learn. Res. **16**, 2023–2049 (2015)
2. Lopez-Paz, D., Bottou, L., Scholkopf, B., Vapnik, V.: Unifying distillation and privileged information. In: ICLR 2016 (2016)
3. Bisio, F., Gastaldo, P., Zunino, R., Decherchi, S.: Semi-supervised machine learning approach for unknown malicious software detection. In: IEEE INISTA, 2014, pp. 52–59 (2014)

4. Bisio, F., Decherchi, S., Gastaldo, P.: Inductive bias for semi-supervised extreme learning machine. In: ELM 2014, vol. 1, pp. 61–70 (2014)
5. Decherchi, S., Ridella, S., Zunino, R., Gastaldo, P., Anguita, D.: Using unsupervised analysis to constrain generalization bounds for support vector classifiers. IEEE Trans. Neural Netw. **21**, 424–438 (2010)
6. Kuzborskij, I., Orabona, F.: Stability and hypothesis transfer learning. In: ICML 2013 (2013)
7. Wang, Z., Wang, X., Ji, Q.: Learning with hidden information. In: ICPR 2014 (2014)
8. Niu, L., Shi, Y., Wu, J.: Learning using privileged information with L-1 support vector machine. In: IEEE International Conferences on Web Intelligence and Intelligent Agent Technology (2012)
9. Rifkin, R.: Everything old is new again: a fresh look at historical approaches in machine learning, Ph.D. thesis, Massachusetts Institute of Technology (2002)
10. Gammerman, A., Vovk, V., Vapnik, V.: Learning by transduction. In: UAI 1998, Morgan Kaufmann Publishers Inc., pp. 148–155 (1998)
11. Tippimg, M.E.: Sparse Bayesian learning and the relevance vector machine. JMLR **5**, 211–244 (2001)
12. Pan, S.J., Tsang, I.W., Kwok, J.T., Yang, Q.: Domain adaptation via transfer component analysis. IEEE Trans. Neural Netw. **22**(2), 199–210 (2011)
13. Daume III, H.: Frustratingly easy domain adaptation. In: Proceedings of the 45th Annual Meeting of the Association of Computational Linguistics, pp. 256–263. Association for Computational Linguistics (2007)

Feature Based Multivariate Data Imputation

Alessio Petrozziello$^{(\boxtimes)}$ and Ivan Jordanov

School of Computing, University of Portsmouth, Portsmouth, UK
{alessio.petrozziello,ivan.jordanov}@port.ac.uk

Abstract. We investigate a new multivariate data imputation approach for dealing with variety of types of missingness. The proposed approach relies on the aggregation of the most suitable methods from a multitude of imputation techniques, adjusted to each feature of the dataset. We report results from comparison with two single imputation techniques (*Random Guessing* and *Median Imputation*) and four state-of-the-art multivariate methods (*K-Nearest Neighbour Imputation, Bagged Tree Imputation, Missing Imputation Chained Equations*, and *Bayesian Principal Component Analysis Imputation*) on several datasets from the public domain, demonstrating favorable performance for our model. The proposed method, namely *Feature Guided Data Imputation* is compared with the other tested methods in three different experimental settings: *Missing Completely at Random, Missing at Random* and *Missing Not at Random* with 25% missing data in the test set over five-fold cross validation. Furthermore, the proposed model has straightforward implementation and can easily incorporate other imputation techniques.

Keywords: Missing data · Multivariate data imputation ·
Multitude of imputation models · Data mining

1 Introduction

Dealing with missing data is an important step in dataset pre-processing since most statistical analysis techniques, data reduction tools, and machine learning methods require complete datasets. There are many techniques that can be used to deal with the missingness, but the common approach during imputation is to make the most of the available data through minimizing the loss of statistical power and the bias inevitably brought by the missing data inferred values. The mechanisms of missingness are usually categorized into three groups [1]: MCAR (Missing Completely at Random); MAR (Missing At Random); and MNAR (Missing Not At Random). In the first case, the missingness is generally due to external factors, not correlated to the other variables in the dataset, while in the last two, the cause is related to the other variables; therefore, the risk of bringing bias due to the imputation should be carefully considered.

The approaches of dealing with missingness can be also divided into three categories [1]: deletion; univariate imputation; and multivariate imputation. In the first category fall the list-wise deletion (the patterns with missing values are simply removed), attribute deletion (the features with missing values are excluded) and pair-wise deletion (where, in presence of missing values, the pattern is not dropped, and its other values are still used during the analysis). The methods from the second category

© Springer Nature Switzerland AG 2019
G. Nicosia et al. (Eds.): LOD 2018, LNCS 11331, pp. 26–37, 2019.
https://doi.org/10.1007/978-3-030-13709-0_3

do not consider the correlation between the missing value and the other variables in the dataset, and impute the data using only information of the same attribute. Good examples of this group are: the *Random Guessing*, where the values are substituted randomly, sampling from the other values of the same attribute; and the *Mean (Median) Imputation*, where the values are replaced with the mean (median) of the considered attribute. The last category includes methods that consider the correlation of the different attributes. Four different algorithms of this family are usually considered [1]: *Multiple Imputation Chained Equations* (MICE); *Bagged Tree Imputation* (BTI); *K-Nearest Neighbour Imputation* (KNNI) and *Bayesian Principal Component Analysis Imputation* (bPCA).

These methods have been widely investigated and compared in the past years, showing discordant results [2, 3]. Most approaches of dealing with missingness would select a single method that outperforms the others based on a given performance metrics. However, while a given approach might have a good performance across the whole dataset, it does not mean that its performance will be superior at the level of each individual feature. In the proposed approach, instead of selecting a single method which outperforms the others on the whole dataset, a column-wise selection is used to choose the best imputation method for each individual attribute.

The proposed method, namely *Feature Guided Data Imputation* (FGDI) is extensively tested and validated on thirteen publicly available datasets. Its performance is assessed and compared with other techniques using *Wilcoxon Signed-rank* test for statistical significance [4].

The remainder of the paper is organized as follows. Section 2 describes the considered imputation methods, while Sect. 3 proposes the FGDI method. Section 4 discusses the empirical study carried out. The results of this investigation are discussed in Sect. 5 and in Sect. 6 conclusion given.

2 Imputation Techniques

Baselines. The most common techniques used as baselines for comparison and analysis of data imputation are *Random Guessing, Mean Imputation* and *Median Imputation* [5]. The *Random Guessing* is a very simple benchmark to estimate the performance of a prediction method. It takes as input the missing data with random value drawn from the known values of the same feature. The *Mean (Median) Imputation* replaces every missing value with the mean (median) of the attribute. However, these techniques fall into the single imputation category (the correlation between the variables is ignored), which is the reason for being rejected by the scientific community [6], hence, they are only used here to perform initial fast sanity check of the proposed approach.

Bagged Tree Imputation. The BTI with gradient boosting [7] is a machine learning technique for solving regression problems, which produces a robust prediction model using a vote (ensemble) among weak ones. The method follows few basic steps for each feature with missing data: (1) train several tree models using the other features; (2) for each tree, impute the data using a regression function; (3) use a vote among the trees to select the data that will be imputed in the original dataset. Bagging predictors are used for generating multiple versions of a predictor to get an aggregated one. The aggregation uses the average over the predictor versions when predicting a numerical outcome, and employs a plurality vote when predicting a class. Bagging proved to be more efficient in the presence of label noise when compared to boosting and randomization [8]; it is also robust to outliers and can impute the data very accurately using surrogate splits [9]. Another important feature of the tree model is its flexibility: different models can be trained with the random forests and the prediction deferred to a system vote among them. In this work, we employ gradient boosting technique for the regression values, which uses an ensemble of weak decision trees.

K-Nearest Neighbors Imputation. In the KNNI the missing values are usually imputed applying the mean of the K most similar patterns found by minimizing the *Euclidean Distance* between a pattern with missing values and the complete subset [10]. The KNNI approach comprises three steps: (1) take only the rows of the dataset without missing data and use this subset as a prototype dataset to select the nearest neighbours; (2) choose a distance metric and compute the nearest neighbours between each pattern with missing data and the complete subset; (3) impute the data, using the mean or the mode of the chosen neighbours. An important parameter to select is the number of neighbours K. There are discordant opinions in the literature, some suggesting a low value of 1 or 2 for small datasets [11]. [12] advise a value of 10 for large datasets, and in [10] is argued that the method is insensitive to the choice of the number of neighbours. In all simulations carried out in this work, we used a value of $K = 10$. The K-Nearest Neighbours has some advantages: the method can predict both, categorical variables (the most frequent value among the KNN) and continuous variables (the average among the KNN); and when using this imputation, there is no need to build a model (as in the Bagged Tree Imputation).

Missing Imputation Chained Equations. MICE [13] is a method from the multiple imputation family. In the MICE process, a series of regression models are run modeling each variable with missing data as dependent variable relying on all the other variables in the dataset. This guarantees that each variable is modeled independently to its distribution [13]. The MICE method is divided into four stages: (1) a simple imputation (Mean) is performed for every missing value in the dataset to be used as placeholders; (2) the placeholders for one variable are set back to miss; (3) the missing variable is used as the dependent variable in a regression model and regressed using the other variables. The procedure is followed for every variable with missing entries and repeated many times until the convergence is reached. Practical guide on how to select the number of imputations is given in [14], however, sometimes due to the size of the dataset, it is not feasible to run the procedure many times. Therefore, 10 iterations are usually considered enough for the convergence of the algorithm [15], which number is also adopted in this investigation.

Bayesian Principal Component Analysis Imputation. The bPCA imputation [16] is an evolution of the Single Value Decomposition Imputation [10] (since the SVD is a PCA applied to normalized datasets with a 0 row-mean) with the additional Bayesian estimation, using a known prior distribution. An advantage of this approach is that no hyper-parameters tuning is needed, and the number of components is self-determined by the algorithm at the expense of a higher computational time. The bPCA can be summarized as: (1) apply Principal Component Regression on the initial dataset; (2) perform a Bayesian Estimation; (3) use an EM algorithm until convergence to a specified tolerance.

3   The Proposed Method

All methods described in the previous section have been widely applied for solving missing data problems [2]. However, while a given approach may produce low estimation error for the whole dataset at hand, this does not mean that the method outputs the best result (smaller error) for every individual feature (usually, for some of the features other methods may give better estimates).

The investigated here *Feature Guided Data Imputation* (FGDI) is an imputation approach which aggregates models in a feature-wise fashion (choosing the best model for each feature (column) of the dataset, while allowing it at the same time to be inferior for the rest of the features). In other words, when training the model, the best imputation method for each feature of the dataset is selected among the considered techniques. At the imputation phase, each selected method is sequentially used to impute the features for which its performance was the best during the training stage.

During the learning phase, the algorithm is trained on artificially introduced missing data (e.g., 25% of MCAR, MAR or MNAR) for each feature. A combination of the best performed methods (based on a given error metrics, e.g., RMSE, MAE) is used to impute the missing values in the original dataset. To cope with the random nature of the algorithm and to ensure more robust choice, this process is iterated a given number of times, and the technique that produced the lowest median overall error for each feature is then chosen. For example, let's assume a set of m imputation methods ($M_1, ..., M_m \in S$) and dataset (X) composed of v variables (features) and n samples, where k of them ($0 < k < n$) contain at least one missing value. Once the $n-k$ complete samples are separated (X' subset), a percentage of missingness is added to each variable of X' (e.g., 25%). The missing data in X' are separately imputed using all methods of S, and the estimation error (e.g., RMSE) is calculated for each feature (variable). This process is repeated I times (e.g., $I = 5$), and for every variable in X', the imputation algorithm scoring the lowest median error is selected and included in a set E, ($E \subseteq S$).The selected techniques are then used to estimate the missing values of the whole set X. In particular, $\forall M_i \in E$, $i = 1,.., j$, (where $j \leq m$), the dataset X is entirely imputed, and only the imputed values for the features where M_i scored the lowest error are saved, discarding the others. Since X is imputed independently using each technique, the order of imputation is irrelevant, enabling the process to be parallelized.

4   Empirical Study

In previous works [3, 17], extensive review and experimentation was done in an effort to identify correlation between imputation methods performance and the type of datasets with missingness, which concluded with discordant results (confirming the

'No free lunch theorem'). These findings led to the current investigation, based on the aggregation of different models.

The proposed method (FGDI) is compared with known univariate baselines and multivariate state-of-the-art imputation methods (i.e., KNN, BTI, MICE and bPCA) to assess its performance on the missing data imputation task. The experiments are executed for all the three missing data mechanisms: MCAR, MAR and MNAR. Lastly, a run time analysis is carried to observe the computational cost needed during the training and imputation phases. The results are reported in Sect. 5.

Thirteen publicly available datasets from KEEL repositories [18] are used in this work, namely *Contraceptive, Yeast, Red wine, Car, Titanic, Abalone, White Wine, Page Block, Ring, Two Norm, Pen Based, Nursery*, and *Magic04*. The selection of these datasets was driven by the intent to cover different application domains and data characteristics. They differ in the number of instances (from several hundreds to several thousands), the number of features (from 3 to 20), and in the range and type of the features (real, integer and categorical). The used datasets do not have missing values by default, guaranteeing total control over the experiments and the assessment and evaluation of the results.

From the variety of metrics employed for comparing and evaluating data imputation and prediction models found in the literature, *Mean Squared Error* (MSE) and *Mean Absolute Error* (MAE) are the most widely used [16, 19]. *MSE* measures the difference between predicted and actual values while MAE their absolute difference. The *Mean Absolute Error* (MAE) is argued to be more accurate and informative than the RMSE [20], successively refuted by [21], where it is stated that the two measures picture different aspects of the error and therefore they should both be used to assess the results. As suggested in [20] and [21], RMSE and MAE are implemented to compare the estimated missing values and the original ones, reflecting the average performance of the imputation method. Furthermore, the RMSE is employed as error function for the training phase of the FGDI. The *Standard Accuracy* (SA) and *Variance Relative Error* (RE*) are assumed to be good baseline estimation measures [22]. SA and RE* are used to compare the proposed model with the univariate baseline imputation techniques (discussed earlier). In particular, SA which compares the prediction against the mean of a random sampling of the training response values $SA = 1 - RMSE(predicted, actual)/RMSE(randGuess, actual)$ and the $RE* = \sigma^2(predicted - actual)/\sigma^2(actual)$ which gives score of 1 for a model predicting values with 0 variance. It is considered an appropriate baseline error measure since any model producing RE* greater than 1 would be assumed weak, independently of the dataset [22].

To validate the proposed method, a k-fold cross validation is applied, splitting the dataset into independent training and test sets. The test set is generated using a uniform sampling without repetitions, and the rest of the data is left as a training set. Since the *Shapiro Test* showed that many of our patterns came from non-normally distributed populations, the statistical *Wilcoxon Signed Rank Test* was used to prove which method is giving better performance [4]. Furthermore, the used test does not make any assumptions about the underlying distribution of the data. In order to check the statistical significance of the difference in model performance, we test the following *NULL* hypothesis: "Given a pair of models (M_i, M_j) with $i, j \in \{1, .., n\}, i \neq j$, the RMSEs (MAEs) obtained by model M_i are significantly smaller than the errors produced by model M_j", using confidence level $\alpha = 0.05$.

When simulating *Missing Completely at Random* (MCAR) mechanism, for each feature value in the dataset, a number is drawn from a uniform distribution in the (0, 1) interval. If this number is smaller than assumed missing data threshold (e.g., 0.25), the feature value is set as missing in the original dataset. For the *Missing at Random* (MAR) mechanism, a variance-covariance matrix is built for the considered dataset. For each variable, the probability of missingness is governed by the most correlated feature in the matrix (i.e., the bigger the value of the correlated feature, the higher the probability of introducing missingness). To generate the *Missing Not at Random* (MNAR) mechanism, we draw values (used as thresholds) from a uniform distribution in (0, 1) interval, and sort them in decreasing order. We do the same for the variable values and pair them with the sorted random numbers. For each threshold, we draw a new random number in the (0, 1) interval and if it is smaller than the threshold, we erase the feature value (this way the pairs with higher random numbers are more likely to be set as missing).

5   Results and Discussion

Three different experiments are carried out: MCAR, MAR, and MNAR mechanisms with 25% of missing data and 5-fold cross validation (80% training and 20% testing). To calibrate the model during the training phase, 25% of missing data is added to each attribute of the training set, subsequently imputed using the five imputation techniques and the accuracy is evaluated using both MAE and RMSE. This process is run 5 times and for each attribute, the imputation model achieving the lowest median error (preferred to the mean due to robustness to outliers) is selected. Lastly, the selected techniques are used to impute the data on the independent test set and the results are compared to all the other methods.

The first set of experiments is performed imputing the missing data under the MCAR mechanism. As the MCAR occurs when the missingness is unrelated to anything in the study, the missingness is simulated using a Bernoulli random variable removing values with 25% chance of success. The *SA* values given in Table 1 show superior results for the imputation carried out with our model. It outperformed the baseline methods *Random Guessing* (SA_{Random} is always 0) and the *Median Imputation* ($SA_{FGDI} > SA_{Median}$). The *Mean Imputation* was omitted in favor of the *Median Imputation*, since the latter is considered less biased to outliers. Furthermore, Table 2 presents the *RE** results over five different imputation methods and again, as it can be seen from the values, our FGDI method outperformed the *Median Imputation*, with $RE_{FGDI} < 1$ in almost all case studies. It can be also seen from the table that the $RE_{MICE} > 1$, which means high variance in the imputed values, problem already discussed in [23]. The RE_{KNNI}, instead, shows high variance (from 0.19 to 1.24) depending on the considered dataset and feature. In the *Yeast* dataset, two variables (*Erl* and *Pox*) are removed during the RE* calculation since the variance in the denominator is 0. To finally assure that the proposed method is outperforming the baselines, a *Wilcoxon* test for statistical significance is run, testing the *NULL hypothesis* "The RMSEs provided by FGDI are significantly smaller than the errors produced by the models *Random Guessing* and *Median Imputation*". The results proved FGDI being better than both with *p-value* < *0.05* over all 13 datasets. The *Standard Accuracy* analysis (Table 1) shows that the FGDI method not only outperforms the baselines, but it is also comparable, and even better than the state-of-the-art algorithms. As it can be seen from the table, the SA_{FGDI} is higher than the SA of the other methods in 41 out of the 52 cases, comparable in 9 out of the 52 cases, and worse in only 2 cases. To validate the significance of the difference, the *Wilcoxon* test is run justifying the *NULL* hypothesis "The RMSEs provided by FGDI are significantly smaller than the errors achieved by the state-of-the-art methods".

Table 1. Standard Accuracy (SA) values achieved by FGDI, the baseline (Median Imputation) and state-of-the-art (KNNI, BTI, MICE, and bPCA) techniques over the 13 datasets for 5-fold cross validation with 25% MCAR. Higher values represent better estimation over the random guess

Dataset	FGDI			KNNI			BTI			MICE			bPCA			Median		
	MCAR	MAR	MNAR	MCAR	MAR	MNAR	MCAR	MAR	MNAR	MCAR	MAR	MNAR	MCAR	MAR	MNAR	MCAR	MAR	MNAR
Contraceptive	**0.39**	**0.27**	**0.31**	0.24	0.11	0.17	0.36	0.26	0.30	0.18	-0.02	0.03	0.38	0.23	**0.31**	0.26	0.23	0.27
Yeast	**0.33**	**0.37**	**0.27**	0.24	0.29	0.09	0.32	0.36	0.24	0.06	0.04	-0.02	**0.33**	0.36	0.22	0.28	0.36	0.22
Red Wine	**0.37**	**0.28**	**0.28**	0.33	0.13	0.14	0.33	0.18	0.25	0.23	0.01	-0.08	0.32	0.15	0.25	0.30	**0.28**	0.26
Car	**0.29**	**0.32**	**0.29**	0.12	0.21	0.16	**0.29**	0.31	0.15	-0.01	0.01	-0.07	**0.29**	0.31	0.14	0.25	0.29	**0.29**
Titanic	**0.35**	**0.27**	**0.28**	0.26	0.18	0.00	0.34	**0.27**	0.26	0.05	-0.06	-0.05	0.34	**0.27**	0.23	0.28	0.25	0.26
Abalone	0.68	**0.28**	**0.27**	0.62	-0.32	-0.05	0.57	0.27	0.18	0.66	0.08	-0.10	**0.72**	0.08	-0.10	0.28	**0.27**	**0.27**
White Wine	**0.36**	**0.29**	**0.30**	0.34	0.11	0.12	0.34	0.18	0.18	0.16	-0.01	0.00	0.34	0.18	0.19	0.28	**0.29**	0.29
Page Block	**0.49**	**0.26**	0.22	0.41	0.16	0.17	0.43	**0.26**	0.20	0.39	0.12	0.03	0.46	0.22	0.16	0.25	**0.26**	**0.23**
Ring	**0.31**	**0.30**	**0.29**	0.24	0.24	0.25	0.29	0.29	**0.29**	-0.02	-0.02	0.00	0.29	0.29	**0.29**	0.28	0.29	**0.29**
Two Norm	**0.34**	**0.30**	**0.30**	0.24	0.18	0.21	0.32	0.29	0.29	0.07	0.01	0.01	**0.34**	0.25	0.27	0.30	0.29	0.29
Pen Based	0.54	**0.27**	**0.28**	**0.59**	0.02	0.00	0.49	0.22	0.22	0.47	0.00	-0.01	0.45	0.17	0.20	0.27	**0.27**	**0.28**
Nursery	**0.30**	**0.30**	**0.28**	0.09	0.18	0.13	0.25	0.24	0.25	0.00	0.01	0.00	0.29	0.29	**0.28**	0.23	0.23	0.22
Magic04	**0.47**	**0.26**	**0.22**	0.42	0.13	0.06	0.41	0.22	0.18	0.32	0.07	-0.06	0.45	0.20	0.13	0.28	0.25	**0.22**

Table 2. RE* metric of FGDI and four state-of-the-art imputation methods for the 13 datasets. Each entry represents the number of times that given algorithm scored RE* < 1 (good estimator) on a total of 138 used features. The median imputation is not reported since it always scores RE* = 1

Dataset (# features)	FGDI	KNNI	BTI	MICE	bPCA
Contraceptive (9)	9	3	9	1	8
Yeast (6)	5	1	6	0	4
Red Wine (11)	10	6	11	4	9
Car (6)	6	0	5	0	0
Titanic (3)	3	1	3	0	3
Abalone (8)	8	7	8	7	8
White Wine (11)	10	7	10	1	8
Page Block (10)	10	10	10	4	9
Ring (20)	16	0	20	0	14
Two Norm (20)	20	0	20	0	20
Pen Based (16)	15	16	16	14	16
Nursery (8)	8	0	2	0	0
Magic04 (10)	9	8	9	5	9
Total (138)	129	59	129	36	108

As evidenced in Table 3 (first three columns): the imputation improvement achieved by FGDI is statistically significant (p-value < 0.05) in 40 out of the 52 cases (77%); comparable in 9 cases; and worse in 3 cases only. As suggested in [20], the same *NULL* hypothesis was tested using the MAE metric. The FGDI resulted significantly better in 37 cases (71%), comparable in 12 and worse in only 3 cases. The second-best imputation method (bPCA) for RMSE was significantly better in 31 out of the 52 cases (60%); comparable in 9; and worse in 12 cases, which shows an improvement for FGDI of 17% over the best single method. For the MAE hypothesis, bPCA results were significantly better in 24 out of the 52 cases (46%); comparable in 14; and worse in 14 cases, showing inferior imputation accuracy in 25% of the cases, compared with the FGDI. Furthermore, Table 3 shows the robustness of FGDI when estimating the missing values - lower variance than KNNI, MICE, bPCA, and comparable RE* values with BTI (Table 2).

Table 3. RMSE (MAE) significance test for 5-fold cross validation with 25% MCAR, MAR, and MNAR. Each row shows how many times model M_i is better (win), comparable (tie), or worse (loss) than the other models with the Wilcoxon Signed Rank Test

	MCAR			MAR			MNAR		
	Win	Tie	Loss	Win	Tie	Loss	Win	Tie	Loss
FGDI	**40**	9	3	**41**	10	1	**47**	5	0
	(37)	(12)	(3)	**(47)**	(5)	(0)	**(48)**	(4)	(0)
bPCA	31	9	12	36	8	8	34	7	11
	(24)	(14)	(14)	(31)	(6)	(15)	(31)	(7)	(14)
BTI	26	12	14	28	6	18	23	6	23
	(19)	(15)	(18)	(22)	(11)	(19)	(22)	(8)	(22)
KNNI	15	3	34	11	2	39	14	3	35
	(19)	(11)	(22)	(13)	(6)	(33)	(13)	(6)	(33)
MICE	3	5	44	1	2	49	1	1	50
	(4)	(8)	(40)	(1)	(5)	(46)	(1)	(5)	(46)

The following experiments are considered when the missingness is caused by MAR and MNAR mechanisms.

The *Standard Accuracy* values given in Table 1 for the MAR experiment show slightly superior performance of FGDI when compared with the other imputation techniques. The proposed model outperforms the baseline *Random Guessing* ($SA_{FGDI} > 0$) in all reported cases and the *Median Imputation* ($SA_{FGDI} > SA_{Median}$) in 8 out of 13 datasets. Furthermore, it also shows better accuracy in all 13 cases when compared to KNNI and MICE, and superior than BTI and bPCA results in 11 and 10 cases respectively. It is also worth to notice that the imputation under MAR condition is generally harder task (compared to MCAR), since the missingness is not uniformly distributed across the dataset and depends on the other variables as well (as discussed in Sect. 4). As for all previous experiments, the *Wilcoxon* test is adopted to evaluate the significance in difference for RMSE and MAE metrics. Results in Table 3 (4th to 6th column) show significant imputation improvement of the FGDI for 41 out of the 52 cases (79%); comparable in 10; and worse in only 1 case, when using RMSE. On the other hand, for the MAE metric, the FGDI resulted better in 47 cases (90%); comparable in 5; and never worse. The second-best imputation method (BTI) for RMSE and MAE is significantly better in 36 and 31 out of the 52 cases (69% and 60%); comparable in 8 and 6 cases; and worse in 8 and 15 cases, showing inferior to the FGDI performance in 10% and 30% of the cases respectively.

The same analysis performed under the MNAR condition also suggests that the use of a single imputation method for the whole dataset is not the best option. Again, the SA values (Table 1) are generally lower when compared to the MCAR mechanism as the missingness is caused by the considered variable itself (as explained in Sect. 4), increasing the likelihood of introducing bias when imputing the values. In the MNAR case, Table 1 also shows superior results for our method in 10 out of 13 datasets. The reported SA_{FGDI} is better than SA_{KNNI} and SA_{MICE} for all considered datasets, while being never worse than SA_{BTI} and SA_{bPCA}. When compared to the baselines, the FGDI is always superior to the *Random Guess* ($SA_{FGDI} > 0$), better than the *Median Imputation* in 7 out of 13 cases, and worse only in 1 of the cases. The Wilcoxon analysis Table 3 (columns 7 to 9) shows the FGDI being better than the second best method (BTI) in 25% and 33% of the cases for RMSE and MAE respectively. Comparing the proposed method with the other imputation techniques, the FGDI is better than bPCA, KNNI and MICE in 46%, 64% and 89% of the cases for the RMSE and 50%, 67% and 90% for the MAE metrics. Despite being generally not recommended [6], the *Median Imputation* showed comparable and even better results than the bPCA, BTI, KNNI, and MICE in both MAR and MNAR settings. At first sight, this result is contradictory to the MCAR experiment (Table 1). This could be explained by the fact that the multivariate model can benefit from the uniformly distributed missingness across the dataset (like in the MCAR mechanism), while for the MAR and MNAR (where the missingness depends on a single variable), the use of a univariate model (baselines) could be reducing the noise in the prediction (because of not considering uncorrelated features). However, as it can be seen from the carried experiments, the use of combination of baselines and state-of-the-art techniques (as in our approach) can improve the accuracy in almost all proposed scenarios with a very low risk of worsening the imputation.

Last point to note is that while the FGDI is superior in all setups, the bPCA and BTI are competing for the second position in the three scenarios (bPCA for MCAR; and BTI for MAR and MNAR). All the experiments presented in this work have been done on a 16-core machine with 32gb RAM and 64 Gb SSD of storage. Figure 1 shows the training time for the four state-of-the-art techniques (KNNI, BTI, MICE, and bPCA) and the proposed FGDI method over the 13 datasets, given in seconds. Due to the FGDI parallelization (each imputation algorithm can be run independently from the others), its training execution time is never significantly higher than the time needed for any other single technique. FGDI training time (blue bar in Fig. 1) is always comparable with the slowest technique, plus an overhead due to the different scheduled threads. Furthermore, the proposed method shows a consistent time execution overhead with datasets of different volume and features size. This behavior can be observed from the percentage change between the FGDI and the slowest compared model. he percentage change results are smaller for bigger datasets (7.69, 6.15, 14.37, 11.76, 5.84,

10.74 and 4.76 for *White Wine, Page Block, Ring, Two Norm, Pen Based, Nursery* and *Magic04* respectively) and larger for the small ones (22.5, 20, 43.90, 59.09, 56.25, 46.34 for *Contraceptive, Yeast, Red Wine, Car, Titanic* and *Abalone* respectively).

This finding supports the recommendation of using the FGDI regardless the size of the dataset (as long as the imputation is feasible for the single models employed in the FGDI). For the prediction run-time (applied on the test set), FGDI showed to be comparable with the slowest method selected during the training phase.

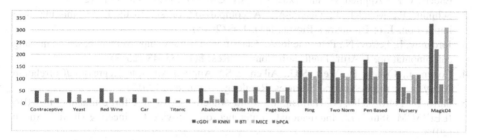

Fig. 1. Training time in seconds (y-axis) of the five considered imputation methods over the 13 datasets (x-axis). The Median Imputation is omitted having always a training time less than 1 s (Color figure online)

6 Conclusion

The investigated FGDI method initially extracts the complete subset (without missing values), and selects through a learning process the most suitable imputation method for each feature. The FGDI imputation performance is evaluated with four widely used metrics for such tasks (SA, RE*, RMSE, and MAE). The results are statistically assessed using the *Shapiro Test* to check the distribution normality, and the non-parametric *Wilcoxon Signed Rank Test*, for statistical significance, using confidence level $\alpha = 0.05$.

Under the MCAR mechanism, the *Standard Accuracy* analysis demonstrates that the proposed model is always more accurate than the baselines and produces better estimation than the state-of-the-art methods in 41 out of 52 cases. The *Wilcoxon* shows improvements of 17% and 25% for the FGDI over the second best performing algorithm (bPCA) over the two metrics. In addition, FGDI and BTI impute values with higher stability (RE* < 1) for 129 out of 138 tested features, followed by bPCA with 108 out of 138.

Although the prediction under MAR and MNAR mechanisms is generally less accurate than the one under MCAR, the FGDI still shows better performance when compared with the baselines and the state-of-the-art techniques. In particular, in the MAR case, the FGDI is more accurate than the second best model (BTI) in 10% and 30% of the cases for RMSE and MAE respectively. Under the MNAR mechanism the proposed model is again better than BTI in 25% and 33% respectively.

Finally, the performed imputation run time analysis proves the approach feasibility regarding the needed training and testing time. The reported results strongly support the efficiency of the proposed method when implementing multivariate imputation as a way of dealing with missingness. Another advantage is that the FGDI can be easily parallelized, having straightforward implementation allowing other imputation methods to be easily incorporated.

References

1. Enders, C.K.: Applied Missing Data Analysis. Guildford Press, Guidford (2010)
2. Schmitt, P., Mandel, J., Guedj, M.: A comparison of six methods for missing data imputation. J. of Biometrics Biostat. **6**(1), 1–6 (2015)
3. Jordanov, I., Petrov, N., Petrozziello, A.: Classifiers accuracy improvement based on missing data imputation. J. Artif. Intell. Soft Comput. Res. **8**(1), 33–48 (2018)
4. Cohen, J., Cohen, P., West, S.G., Aiken, L.S.: Applied Multiple Regression/Correlation Analysis for the Behavioral Sciences. Routledge, Abingdon (2013)
5. Sarro, F., Petrozziello, A., Harman, M.: Multi-objective software effort estimation. In: 2016 IEEE/ACM 38th IEEE International Conference on Software Engineering (ICSE), Austin (2016)
6. Osborne, J., Overbay, A.: Best practices in data cleaning. Best Pract. Quant. Methods **1**(1), 205–213 (2008)
7. Rahman, G., Islam, Z.: A decision tree-based missing value imputation technique for data pre-processing. In: Proceedings of the 9th Australasian Data Mining Conference (2011)
8. Frènay, B., Verleysen, M.: Classification in the presence of label noise: a survey. IEEE Trans. Neural Netw. Learn. Syst. **5**(5), 845–869 (2014)
9. Valdiviezo, C., Van Aelst, S.: Tree-based prediction on incomplete data using imputation or surrogate decisions. Inf. Sci. **311**, 163–181 (2015)
10. Troyanskaya, O., et al.: Missing value estimation methods for DNA microarrays. Bioinformatics **17**(6), 520–525 (2001)
11. Cartwright, M., Shepperd, M.J., Song, Q.: Dealing with missing software project data. In: Proceedings of the 9th International Software Metrics Symposium (2003)
12. Batista, G., Monard, M.: A study of K-nearest neighbour as a model-based method to treat missing data. In: Argentine Symposium on Artificial Intelligence (2001)
13. Lee, M.C., Mitra, R.: Multiply imputing missing values in data sets with mixed measurement scales using a sequence of generalised linear models. Comput. Stat. Data Anal. **95**(1), 24–38 (2016)
14. Graham, J.W.: Missing data analysis: making it work in the real world. Annu. Rev. Psychol. **60**, 549–576 (2009)
15. Bartlett, J., Seaman, S., White, I., Carpenter, J.: Multiple imputation of covariates by fully conditional specification: accommodating the substantive model. Stat. Methods Med. Res. **24**(4), 462–487 (2015)
16. Oba, S., Sato, M.-A., Takemasa, I., Monden, M., Matsubara, K.-I., Ishii, S.: A Bayesian missing value estimation method for gene expression profile data. Bioinformatics **19**(16), 2088–2096 (2003)
17. Petrozziello, A., Jordanov, I.: Column-wise guided data imputation. Proc. Comput. Sci. **108** (1), 2282–2286 (2017)
18. Alcalá-Fdez, J., et al.: KEEL data-mining software tool: data set repository, integration of algorithms and experimental analysis framework. J. Multiple-Valued Log. Soft Comput. **17** (2–3), 255–287 (2011)

19. Pan, X.-Y., Tian, Y., Huang, Y., Shen, H.-B.: Towards better accuracy for missing value estimation of epistatic miniarray profiling data by a novel ensemble approach. Genomics **97** (5), 257–264 (2011)
20. Willmott, C.J., Matsuura, K.: Advantages of the mean absolute error (MAE) over the root mean square error (RMSE) in assessing average model performance. Clim. Res. **30**(1), 79–82 (2005)
21. Chai, T., Draxler, R.: Root mean square error (RMSE) or mean absolute error (MAE)?–Arguments against avoiding RMSE in the literature. Geosci. Model Dev. **7**(3), 1247–1250 (2014)
22. Whigham, P.A., Owen, C.A., Macdonell, S.G.: A baseline model for software effort estimation. ACM Trans. Softw. Eng. Methodol. (TOSEM) **24**(3), 20 (2015)
23. Gòmez-Carracedo, M., Andrade, J., Lòpez-Mahìa, P., Muniategui, S., Prada, D.: A practical comparison of single and multiple imputation methods to handle complex missing data in air quality datasets. Chemometr. Intell. Lab. Syst. **134**(1), 23–33 (2014)

Optimization of Neural Network Training with ELM Based on the Iterative Hybridization of Differential Evolution with Local Search and Restarts

David Sotelo[1], Daniela Velásquez[1], Carlos Cobos[1(✉)],
Martha Mendoza[1], and Luis Gómez[2]

[1] Information Technology Research Group (GTI),
Universidad del Cauca, Popayán, Colombia
{davidfsotelom, angiedanielav, ccobos,
mmendoza}@unicauca.edu.co
[2] Systems and Information Technologies (STI),
Universidad Industrial de Santander, Bucaramanga, Colombia
lcgomezf@uis.edu.co

Abstract. An Extreme Learning Machine (ELM) performs the training of a single-layer feedforward neural network (SLFN) in less time than the back-propagation algorithm. An ELM defines the input weights and biases of the hidden layer with random values, and then analytically calculates the output weights. The use of random values causes SLFN performance to decrease significantly. The present work carries out the adaptation of three continuous optimization algorithms of high dimensionality (IHDELS, DECC-G and MOS) and compares their performance to each other and with the state-of-the-art method, a memetic algorithm based on differential evolution called M-ELM. The results of the comparison show that IHDELS using a validation model based on retention (Training/Testing) obtains the best results, followed by DECC-G and MOS. All three algorithms obtain better results than M-ELM. The experimentation was carried out on 38 classification problems recognized by the scientific community, while Friedman and Wilcoxon nonparametric statistical tests support the results.

Keywords: Extreme Learning Machine · IHDELS ·
Self-adaptive differential evolution ·
Multiple trajectory search with local search · Memetic algorithm

1 Introduction

An Extreme Learning Machine (ELM) is a method of training a single-layer feedforward neural network (SLFN) [1]. ELM, as opposed to the back-propagation algorithm (BP), avoids the iterative process (epochs) in the learning process [2] and has been successfully used in various areas of knowledge such as biomedical engineering, computer vision, and identification systems, among others. A SLFN is a feedforward neural network (FNN) with a single hidden layer. FNNs are a type of neural network

© Springer Nature Switzerland AG 2019
G. Nicosia et al. (Eds.): LOD 2018, LNCS 11331, pp. 38–50, 2019.
https://doi.org/10.1007/978-3-030-13709-0_4

that carry out information processing unidirectionally. The neurons of a layer connect only with the neurons of the layer immediately following and no cycles are allowed in it. In other words, the connections are not made to previous layers.

The theory behind ELM indicates that the input weights (interconnection between the input layer and the hidden layer) and the biases of the hidden layer of an SLFN can be defined in a uniform random manner, while the output weights (interconnection between the hidden layer and the output layer) must be assigned analytically by means of the Moore-Penrose pseudoinverse [1], provided that: (1) the activation function of the output layer neurons is linear [2]; and (2) the activation function of the neurons of the hidden layer is continuous and infinitely differentiable [4].

Experimental results show that defining input weights and biases with random values affects the performance of a SLFN trained with ELM and sometimes causes more neurons to be required in the hidden layer [5]. Bearing in mind that evolutionary algorithms and meta-heuristics have frequently been used as methods for solving complex optimization problems [3], Zhang et al. in 2016 proposed an evolutionary algorithm called Memetic ELM (M-ELM) [1] that uses Differential Evolution (DE) as a global search algorithm, and Simulated Annealing (SA) as a local search algorithm, to better define input weights and biases of the hidden layer of a SLFN trained with ELM. This work was found to obtain better results for accuracy than when the weights are defined with random values or with other state-of-the-art methods.

The current research was carried out taking into account that: (1) M-ELM presented good results compared to the state-of-the-art methods; (2) of the "no free lunch" theorems [6] it follows that given a type of optimization problem, the only way to know which optimization meta-heuristic is the best one to approach it, is through evaluation and experimental comparison; (3) optimization algorithms specialized in continuous multimodal problems of high dimensionality have not been used to solve the definition problem of weights and biases of an SLFN trained with ELM; and (4) IHDELS [7] was one of the algorithms that presented the best results in the continuous large-scale optimization competition in the 2015 IEEE CEC (Congress on Evolutionary Computation) where continuous problems of 1000 and more dimensions are taken into account, along with unimodal, multimodal, separable and non-separable optimization problems.

The work involved adaptation of IHDELS, DECC-G and MOS, high dimensionality, continuous optimization algorithms to the SLFN training problem using ELM, and their comparison with each other in addition to comparison with the state-of-the-art M-ELM method, and a baseline using a Random Walk (RW). The experimentation shows that IHDELS obtains the best results. This conclusion is supported by the statistical analysis based on Friedman and Wilcoxon nonparametric tests.

The rest of the document is organized as follows: Sect. 2 presents the theoretical context that supports SLFN training with ELM. Section 3 then presents the adaptation of the IHDELS algorithm for joint use with ELM to define the weights of the input layer and the biases of the hidden layer of an SLFN. Section 4 describes the experiments carried out, results obtained and their analysis. Finally, conclusions are presented, along with future work that the research team hopes to perform in the short term.

2 Theoretical Context

ELM is a training algorithm for a SLFN, whose learning speed can reach up to three hundred times faster than traditional algorithms such as BP [8]. In the following, current theory relating to ELM is explained, taken from [8]. To train the SLFN, input weights ω (Eq. 1) and biases b (Eq. 2) of the hidden layer were first randomly defined.

$$\omega = \begin{bmatrix} \omega_{11} & \omega_{12} & \cdots & \omega_{1m} \\ \omega_{21} & \omega_{22} & \cdots & \omega_{2m} \\ \vdots & \vdots & \ddots & \vdots \\ \omega_{n1} & \omega_{n2} & \cdots & \omega_{nm} \end{bmatrix} = \begin{bmatrix} \omega_1 \\ \omega_2 \\ \vdots \\ \omega_n \end{bmatrix}; \; \omega_i = \begin{bmatrix} \omega_{i1} & \omega_{i2} & \cdots & \omega_{im} \end{bmatrix} \quad (1)$$

$$b = \begin{bmatrix} b_1 \\ b_2 \\ \vdots \\ b_n \end{bmatrix} \quad (2)$$

where ω_{ij} is the weight between neuron i of the hidden layer and neuron j of the input layer, i being the sub index of the row and j the sub index of the column, b_i is the bias of the i-th neuron of the hidden layer, and n corresponds to the number of neurons in the hidden layer.

With X (Eq. 3) being the training data where each column is a training record and each row is the entry to the i-th neuron of the input layer. This matrix has m rows that correspond to the input variables of the dataset and k columns that correspond to the total number of records in the dataset. And with Y (Eq. 4) representing the output matrix of the training data. In this matrix, each column represents the output of the j-th record of X where only one row per column must have the value of 1, and l corresponds to the total number of classes in the dataset. If y_{ij} takes the value of 1 then it means that the i-th classification is correct and if it takes the value of -1 it means that the record does not belong to the i-th classification.

$$X = \begin{bmatrix} x_{11} & x_{12} & \cdots & x_{1k} \\ x_{21} & x_{22} & \cdots & x_{2k} \\ \vdots & \vdots & \ddots & \vdots \\ x_{m1} & x_{m2} & \cdots & x_{mk} \end{bmatrix} = \begin{bmatrix} x_1 & x_2 & \cdots & x_k \end{bmatrix}, \; x_j = \begin{bmatrix} x_{1j} \\ x_{2j} \\ \vdots \\ x_{mj} \end{bmatrix} \quad (3)$$

$$Y = \begin{bmatrix} y_{11} & y_{12} & \cdots & y_{1k} \\ y_{21} & y_{22} & \cdots & y_{2k} \\ \vdots & \vdots & \ddots & \vdots \\ y_{l1} & y_{l2} & \cdots & y_{lk} \end{bmatrix}, \; \forall y_{ij} \in \{1, -1\} \quad (4)$$

The output matrix of the hidden layer for the training data is H (Eq. 5), where $g(x)$ is the activation function of the hidden layer, which corresponds to the sigmoidal, sine, multi-quadratic, Gaussian, hardlim, and triangular functions, among others, all continuous and infinitely differentiable.

To obtain the weights between the neurons of the hidden layer and the output layer, Eq. 6 is applied.

where H^+ is the Moore-Penrose pseudoinverse of H, Y^T is the transpose of Y and β_{jk} represents the weight between neuron j of the hidden layer and neuron k of the output layer. Finally, the output of the neural network, denoted by T is obtained by applying Eq. 7.

$$H = \begin{bmatrix} g(\omega_1 x_1 + b_1) & g(\omega_2 x_1 + b_2) & \cdots & g(\omega_n x_1 + b_n) \\ g(\omega_1 x_2 + b_1) & g(\omega_2 x_2 + b_2) & \cdots & g(\omega_n x_2 + b_n) \\ \vdots & \vdots & \ddots & \vdots \\ g(\omega_1 x_k + b_1) & g(\omega_2 x_k + b_2) & \cdots & g(\omega_n x_k + b_n) \end{bmatrix} \tag{5}$$

$$\beta = H^+ Y^T \tag{6}$$

$$T = (H\beta)^T \tag{7}$$

To find out the accuracy of the SLFN in classification problems, T and Y are checked, verifying for each column that the highest value of T corresponds to the same row in Y, that is to say that the classification of each training record (which is described in Y) is the same one that predicts the neural network. Each hit is counted, and the final figure divided by the total training records to obtain the accuracy (Eq. 8). In regression problems, T and Y are also compared and mean square error between the expected output and the output of the neural network is calculated.

$$accuracy = \frac{hits}{k} \tag{8}$$

3 IHDELS for Training a SLFN Using ELM

3.1 Representation of the Solution

A solution (individual) integrates the input weights and biases of the hidden layer of a SLFN into a vector where parts of each of them are interspersed so that the data of a hidden neuron remain together (see Fig. 1). In this figure, the first block $\omega_{11}, \omega_{12}, \ldots, \omega_{1m}, b_1$ corresponds to all the weights between the input neurons and the

ω_{11}	ω_{12}	$\cdots$	ω_{1m}	b_1	$\cdots$	ω_{n1}	ω_{n2}	$\cdots$	ω_{nm}	b_n

Fig. 1. Representation of the solution

first neuron of the hidden layer and includes their bias. The same is then done for the second neuron of the hidden layer, and so on.

The theory of the ELM indicates that the input weights and biases are generated randomly in the range $[-1, 1]$. Therefore, each of the values of this vector must remain in this range.

The quality of a solution is evaluated according to the accuracy obtained during the training process. If two solutions obtain the same accuracy, which of the two is better is decided based on the norm two of the output weights β, which was used in [9]. Therefore, if there is a tie in accuracy for two solutions, the one with the lowest norm ($\|\beta\|$) is selected as the best solution.

3.2 Adaptations Made to IHDELS

The iterative hybridization of differential evolution with local search and restarts (IHDELS) is an algorithm specialized to solve high dimensional problems. It uses as an exploratory method an algorithm based on differential evolution (DE) and for exploitation it implements two local search (LS) methods that are used according to their performance [7]. In each iteration, IHDELS executes the Self-adaptive Differential Evolution (SaDE) algorithm together with a LS method, with the objective of complementing them. The LS is selected according to the quality of the individuals produced in its last execution. When executing SaDE or a LS, their adaptive parameters start with the values of their last execution. When no significant improvement is detected, a restart mechanism is applied.

IHDELS (see Algorithm 1) starts from a random population and an initial individual that is constructed with the upper and lower constraints of each variable of the problem. Selection of the LS is made as follows: initially all the LS are executed as indicated in line 5 of Algorithm 1 and the improvement rate is stored for each one (Eq. 9).

$$improvementRate_{BL} = \frac{previousEvaluation - newEvaluation}{previousEvaluation} \qquad (9)$$

In each iteration, the LS with the highest improvement rate is applied. IHDELS keeps a record of the best solution found, named *best*. Meanwhile, a solution called *currentBest* is used to calculate the improvement rate. This corresponds to the best individual found on the current iteration. IHDELS defines a threshold that makes it possible to identify when the improvement rate has been significant. For this to happen, the improvement rate must be greater than or equal to the threshold value [7]. The IHDELS restart mechanism affects the adaptive parameters of LS or its population in the following cases [7]: (i) On applying SaDE and the LS during a number of *restarts* in the immediately preceding iterations and improvement rate has not been sufficiently significant, then: (1) the *currentBest* individual is altered as indicated by line 25 of Algorithm 1 where $rand(-0.05, 0.05)$ returns a random number between $[-0.05, 0.05]$, (2) the population is reset randomly, and (3) the adaptive parameters of the LS return to their original value. (ii) If after applying SaDE there is no improvement, the population is restarted, and (iii) If in the execution of the LS nothing improves, its adaptive parameters are restarted.

The local searches used were MTS-LS1 and Hill Climbing (HC); MTS-LS1 as a local search takes a *currentBest* as its best individual found and starts the optimization from the population indicated. As HC is not a population-based method, it divides the evaluations that have been assigned between each of the individuals of the population and optimizes them separately. Finally, it selects the best individual between the optimized population and the given *currentBest* and returns it.

The L-BFGS-B algorithm, original LS method of IHDELS, was replaced by HC because it is a Quasi-Newton algorithm based on the gradient method. Calculation of this consumes high quantities of evaluations of the objective function in a short time, making it difficult to take full advantage of the algorithm in the context of ELM.

HC is a local search algorithm related to the gradient rise without directly using the gradient, instead, it evaluate individuals around a current individual, the individual with the best fitness replaces the current one [10]. Calculation of the neighbors consists of

---

Algorithm 1 IHDELS meta-heuristic

1: *evaluations* $\leftarrow$ 0
2: *population* $\leftarrow$ *createRandomPopulation(populationSize)*
3: *evaluatePopulation(population)*
4: *initialSolution* $\leftarrow$ *(upperLimit + lowerLimit)/2*
5: *currentBest* $\leftarrow$ *applyLocalSearches (initialSolution)*
6: *best* $\leftarrow$ *currentBest*
7: *LScounter* $\leftarrow$ 0
8: **while** *evaluations* < *maxEvaluations*
9:     *previous* $\leftarrow$ *currentBest.evaluation*
10:     *currentBest* = *SaDE(currentBest, population, FE_DE)*
11:     *best* = *(previous− currentBest.evaluation)/previous*
12:     **if** *improves* = 0 **then**
13:         *population* = *createRandomPopulation(populationSize)*
14:         *evaluatePopulation(population)*
15:     **end if**
16:     *LS* = *selectLS()*
17:     *currentBest* = *LS(population, FE_LS, currentBest)*
18:     *best* = *selectBest(best, currentBest)*
19:     **if** *LS.improvementRate* = 0 **then**
20:         *LS.restartParameters()*
21:     **end if**
22:     **if** *LS.improvementRate* < *threshold* **then**
23:         *LScounter* = *LScounter* + 1
24:         **if** *LScounter* = *restarts* **then**
25:             *currentBest[i]* = *best[i]* + *rand(−0.05, 0.05)* ∗ 0.1 ∗ *(b−a)*
26:             *population* = *createRandomPopulation(populationSize)*
27:             *evaluatePopulation(population)*
28:             *localSearches.restartParameters()*
29:             *LScounter* = 0
30:         **end if**
31:     **Else**
32:         *LScounter* = 0
33:     **end if**
34: **end while**
35: **return** *best*

making small changes to the variables of the current individual. In this case, the limited uniform convolution method was used, which makes changes on a variable if a probability is met [10].

MTS-LS1 is also a local search algorithm. This begin from a search range for each variable of the individuals of the population, which is initially calculated according to the restrictions of the upper and lower limits of the variables, performs a process of exploitation by altering the variables of each individual in different percentages, seeking to obtain a better solution.

The global search algorithm implemented by IHDELS is SaDE. This is an algorithm that generally behaves as a differential evolution, but that uses two mutation strategies and a probability p to select one of these; adaptation of this algorithm lies in the value of p that is updated according to the number of individuals that have been generated with a mutation strategy and have successfully entered the next generation and the number of individuals that have not entered the next generation.

SaDE applies a new scale factor to each individual generated and this value is obtained by making $F = N(0.5, 0.3)$, where $N(0.5, 0.3)$ is a normal distribution with mean 0.5 and standard deviation 0.3 [11]. The crossover rate is calculated for each solution, obtaining this value from a normal distribution and SaDE self-adapts the mean value at the end of each learning period.

4 Experiments

4.1 Datasets and Preprocessing

The datasets used were extracted from the official repository of the University of California at Irvine (UCI). Selection of the classification datasets was made considering those presented in [1]. Each dataset was divided into two files, one for training and another for testing. The division was made following the relationship presented in [1], which is approximately 70% of the instances for training and the remaining 30% for testing. The characteristics of the 38 datasets used are expressed in Table 1. For each dataset, its name, number of features or variables, and number of classes are shown. Variety can be seen in number of features and classes.

The datasets were processed in the following way: (1) features not necessary for the learning of the neural network were eliminated such as instance number or instance id, (2) categorical values were defined as an integer number, and (3) continuous features (variables) were normalized between the values [−1, 1].

Table 1. Description of the datasets

Name	#Features	#Classes	Name	#Features	#Classes
Banknote	4	2	Knowledge	5	4
Blood	4	2	Leaf	14	36
Car	6	4	Letter	16	26
Cardiotocography	21	10	Libras	90	15
Chart	60	6	Optdigits	64	10
ClimateSimulation	18	2	Pen	16	10
Connectionist	60	2	Planning	12	2
Contraceptive	9	3	QSARBiodegradation	41	2
Dermatology	34	6	Seeds	7	3
Diabetes	8	2	Shuttle	9	7
Ecoli	7	8	SPECTF	44	2
Fertility	9	2	Vertebral(2C)	6	2
Glass	9	6	Vertebral(3C)	6	3
Haberman	3	2	Wdbc	30	2
Hayes	5	3	Wilt	5	2
Hill	100	2	Wine	13	3
Indian	10	2	WineRed	11	6
Ionosphere	34	2	Yeast	8	10
Iris	4	3	Zoo	16	7

4.2 Configuration of the Experiment

The IHDELS and M-ELM algorithms were implemented in JAVA in the JMetal framework [12], which is specialized in mono and multi-objective meta-heuristic algorithms. The source code and datasets are available at https://goo.gl/GX8pXF. Execution of the meta-heuristic algorithms was performed on computers with the following characteristics: AMD A-10 at 3.2 GHz, 8 GB RAM and Win 10.

Three thousand evaluations of the objective function (Evaluations of Fitness Objectives, EFOs) were carried out, bearing in mind that an ELM was conceived to perform the training of a SLFN in less time than the back-propagation algorithm. The results were obtained for the cross validation (CV) and retention (training/testing or TT) validation models. Thirty runs were performed for each dataset using an algorithm and a validation model. Thirty represents the minimum number for calculating an average with high convergence at the central point. Each run was performed with a different seed.

The IHDELS parameters were configured as indicated in Table 2, where TT refers to the parameter values used with the retention method (Training/Testing), while CV represents cross-validation. These values were obtained from a tuning process of parameters carried out using covering arrays.

The parameters used in M-ELM were populationSize = 50, evaluations = 3000 and numberLocalOptimizations = 70. This configuration was used for both types of validation and these values were taken from [1]. It was further defined that: (i) Problems evaluated with cross-validation would have ten folders (10-folders cross-validation) and for that reason the evaluation of a model implies the use of 10 EFOs, (ii) The ELM uses 50 neurons in the hidden layer for training the neural network this number was chosen taking the original value from [1], without considering the dataset, and (iii) The activation function of the neurons in the hidden layer is the sigmoidal one.

Table 2. Parameters used in IHDELS according to the validation model

Parameter	Classification	
	TT	CV
(1) Number of evaluations for DE in IHDELS (FE_DE)	90	60
(2) Number of evaluations for each BL (FE_LS)	60	45
(3) Lower limit of search domain (a)	−0.60	−0.20
(4) Upper limit of search domain (b)	0.60	0.20
(5) Number of restarts	8	12
(6) Threshold value	0.40	0.01
(7) SaDE crossover rate 1	0.50	0.50
(8) SaDE scale factor 1	0.95	0.50
(9) SaDE crossover rate 2	0.70	0.20
(10) SaDE scale factor 2	0.95	0.70
(11) Probability of executing Hill Climbing	0.20	0.50
(12) Hill Climbing noise radius	0.6	0.1
(13) Population size	10	10

4.3 Results

The results present the average of the thirty executions carried out and their respective standard deviation. Comparison was also made with the DECC-G [13] and MOS [14] algorithms, meta-heuristic algorithms that had performed well in the 2015 IEEE CEC competition in optimization of high dimensionality problems. Comparison was further made with a random walk (RW) that executes the same number of EFOs, as a baseline. Table 3 presents the results obtained with the validation model of the retention (Training/Testing) method and Table 4 with the cross-validation model. The algorithm with the best result for each dataset is presented in bold.

Table 3. Results with the validation model of the retention method (TT)

Dataset	DECC-G	MOS	IHDELS	M-ELM	RW
Banknote	0.9999 ± 0.0004	0.9998 ± 0.0007	**1.0000 ± 0.0000**	0.9999 ± 0.0006	0.9998 ± 0.0007
Blood	**0.7540 ± 0.0094**	0.7528 ± 0.0097	0.7508 ± 0.0092	0.7491 ± 0.0068	0.7527 ± 0.0106
Car	0.8398 ± 0.0188	**0.8466 ± 0.0250**	0.8465 ± 0.0218	0.8411 ± 0.0224	0.8399 ± 0.0227
Cardiotocography	0.7678 ± 0.0100	0.7623 ± 0.0112	**0.7742 ± 0.0094**	0.7675 ± 0.0092	0.7671 ± 0.0110
Chart	0.9063 ± 0.0193	**0.9170 ± 0.0190**	0.9170 ± 0.0233	0.9078 ± 0.0218	0.9092 ± 0.0228
Climatesimulation	0.8491 ± 0.0033	0.8472 ± 0.0012	**0.8534 ± 0.0090**	0.8494 ± 0.0039	0.8515 ± 0.0053
Connectionist	**0.7629 ± 0.0465**	0.7600 ± 0.0461	0.7567 ± 0.0395	0.7529 ± 0.0501	0.7505 ± 0.0431
Contraceptive	0.5276 ± 0.0156	0.5307 ± 0.0126	**0.5345 ± 0.0123**	0.5298 ± 0.0134	0.527 ± 0.01350
Dermatology	**0.9710 ± 0.0121**	0.9672 ± 0.0144	0.9648 ± 0.0110	0.9678 ± 0.0106	0.9678 ± 0.0108
Diabetes	**0.7578 ± 0.0128**	0.7535 ± 0.0149	0.7531 ± 0.0128	0.7569 ± 0.0165	0.7509 ± 0.0119
Ecoli	**0.8568 ± 0.0152**	0.8494 ± 0.0167	0.8568 ± 0.0169	0.8470 ± 0.0138	0.8488 ± 0.0126
Fertility	**0.7111 ± 0.0395**	0.6722 ± 0.0626	0.7033 ± 0.0697	0.6556 ± 0.0717	0.6644 ± 0.0655
Glass	0.5224 ± 0.0422	**0.531 ± 0.04090**	0.5248 ± 0.0503	0.5229 ± 0.0542	0.5252 ± 0.0422
Haberman	0.7036 ± 0.0172	**0.7101 ± 0.0138**	0.7026 ± 0.0200	0.7098 ± 0.0140	0.7029 ± 0.0146

(continued)

Table 3. (*continued*)

Dataset	DECC-G	MOS	IHDELS	M-ELM	RW
Hayes	0.6826 ± 0.0545	0.6712 ± 0.0516	**0.6864 ± 0.0555**	0.6515 ± 0.0646	0.6553 ± 0.0557
Hill	0.6655 ± 0.0146	0.6518 ± 0.0198	0.6587 ± 0.0223	0.6556 ± 0.0200	**0.6696 ± 0.0263**
Indian	**0.6444 ± 0.0157**	0.6439 ± 0.0160	0.6440 ± 0.0166	0.6416 ± 0.0155	0.6432 ± 0.0150
Ionosphere	**0.8635 ± 0.0264**	0.8561 ± 0.0274	0.8493 ± 0.0282	0.8456 ± 0.0286	0.8476 ± 0.0341
Iris	0.8953 ± 0.0406	0.8853 ± 0.0390	**0.9127 ± 0.0424**	0.8716 ± 0.0406	0.8760 ± 0.0312
Knowledge	0.8256 ± 0.0269	**0.8298 ± 0.0211**	0.8236 ± 0.0257	0.8271 ± 0.0234	0.8279 ± 0.0246
Leaf	0.7767 ± 0.0300	0.7758 ± 0.0245	**0.7830 ± 0.0192**	0.7715 ± 0.0215	0.7709 ± 0.0200
Letter	**0.7056 ± 0.0040**	0.6898 ± 0.0057	0.7011 ± 0.0062	0.6879 ± 0.0052	0.6947 ± 0.0057
Libras	**0.7777 ± 0.0334**	0.7713 ± 0.0434	0.7690 ± 0.0340	0.7597 ± 0.0308	0.7690 ± 0.0372
Optdigits	0.8911 ± 0.0117	0.8823 ± 0.0099	**0.8965 ± 0.0168**	0.8868 ± 0.0131	0.8858 ± 0.0132
Pen	**0.9305 ± 0.0076**	0.9163 ± 0.0078	0.9229 ± 0.0084	0.9152 ± 0.0090	0.9191 ± 0.0086
Planning	0.5196 ± 0.0437	0.5174 ± 0.0518	0.5256 ± 0.0440	**0.5365 ± 0.0445**	0.5146 ± 0.0529
QSARBiodegradation	0.8608 ± 0.0125	0.8612 ± 0.0120	**0.8622 ± 0.0094**	0.8564 ± 0.0107	0.8614 ± 0.0111
Seeds	**0.9810 ± 0.0100**	0.9776 ± 0.0126	0.9805 ± 0.0101	0.9757 ± 0.0153	0.9790 ± 0.0164
Shuttle	0.5059 ± 0.2992	**0.5626 ± 0.2804**	0.5040 ± 0.2991	0.5303 ± 0.2399	0.4587 ± 0.2781
SPECTF	0.3213 ± 0.1378	**0.3408 ± 0.1290**	0.3012 ± 0.1209	0.3202 ± 0.1455	0.3250 ± 0.1114
Vertebral2C	0.8597 ± 0.0152	0.8537 ± 0.0204	**0.8670 ± 0.0197**	0.8573 ± 0.0197	0.8537 ± 0.0196
Vertebral3C	0.8133 ± 0.0176	0.8160 ± 0.0227	**0.8170 ± 0.0159**	0.8133 ± 0.0162	0.8147 ± 0.0182
Wdbc	0.9625 ± 0.0078	0.9607 ± 0.0089	**0.9632 ± 0.0083**	0.9621 ± 0.0091	0.9600 ± 0.0108
Wilt	0.9694 ± 0.0016	0.9689 ± 0.0018	0.9700 ± 0.0014	0.9690 ± 0.0017	**0.9704 ± 0.0016**
Wine	**0.9761 ± 0.0170**	0.9550 ± 0.0276	0.9722 ± 0.0203	0.9611 ± 0.0199	0.9617 ± 0.0232
WineRed	**0.5891 ± 0.0134**	0.5846 ± 0.0136	0.5889 ± 0.0110	0.5875 ± 0.0105	0.5866 ± 0.0140
Yeast	**0.5792 ± 0.0067**	0.5764 ± 0.0084	0.5752 ± 0.0081	0.5764 ± 0.0056	0.5782 ± 0.0080
Zoo	**0.8796 ± 0.0280**	0.8527 ± 0.0376	0.8602 ± 0.0372	0.8108 ± 0.0563	0.8065 ± 0.0507

4.4 Statistical Analysis of the Results

Statistical analysis of the results was made using Friedman and Wilcoxon nonparametric tests with the aim of determining which algorithm generally gives the best performance. These tests were carried out using the KEEL software available at www.keel.es.

Table 4. Results with the validation model of 10-folds cross-validation (CV)

Dataset	DECC-G	MOS	IHDELS	M-ELM	RW
Banknote	**0.9999 ± 0.0004**	0.9998 ± 0.0007	**0.9999 ± 0.0004**	0.9998 ± 0.0007	0.9998 ± 0.0007
Blood	**0.7504 ± 0.0072**	0.7468 ± 0.0088	0.7485 ± 0.0091	0.7470 ± 0.0100	0.7484 ± 0.0085
Car	0.8360 ± 0.0215	0.8425 ± 0.0203	0.8366 ± 0.0226	**0.8456 ± 0.0192**	0.8436 ± 0.0204
Cardiotocography	0.7650 ± 0.0086	0.7617 ± 0.0097	0.7678 ± 0.0122	**0.7682 ± 0.0099**	0.7651 ± 0.0117
Chart	0.9117 ± 0.0230	0.9057 ± 0.0252	**0.9188 ± 0.0188**	0.9095 ± 0.0205	0.9160 ± 0.0188
Climatesimulation	0.8492 ± 0.0046	0.8478 ± 0.0016	0.8495 ± 0.0054	0.8489 ± 0.0030	**0.8503 ± 0.0050**
Connectionist	**0.7662 ± 0.0466**	0.7371 ± 0.0480	0.7367 ± 0.0550	0.7429 ± 0.0394	0.7576 ± 0.0486
Contraceptive	0.5275 ± 0.0114	0.5287 ± 0.0148	**0.534 ± 0.0125**	0.5297 ± 0.0135	0.5289 ± 0.0141
Dermatology	0.9615 ± 0.0151	0.9661 ± 0.0123	0.9612 ± 0.0125	**0.9693 ± 0.0102**	0.9642 ± 0.0093
Diabetes	0.7549 ± 0.0116	0.7542 ± 0.0187	0.7543 ± 0.0125	**0.7575 ± 0.0151**	0.7569 ± 0.0154
Ecoli	0.8491 ± 0.0173	**0.8521 ± 0.0130**	0.8488 ± 0.0163	0.8506 ± 0.0103	0.8488 ± 0.0151
Fertility	0.6567 ± 0.0717	0.6433 ± 0.0588	**0.6689 ± 0.0711**	0.6448 ± 0.0720	0.6600 ± 0.0674

(*continued*)

Table 4. (*continued*)

Dataset	DECC-G	MOS	IHDELS	M-ELM	RW
Glass	0.5490 ± 0.0357	**0.5552 ± 0.0438**	0.5305 ± 0.0439	0.5350 ± 0.0314	0.5419 ± 0.0492
Haberman	**0.7265 ± 0.0144**	0.7261 ± 0.0174	0.7258 ± 0.0170	0.7230 ± 0.0186	0.7222 ± 0.0163
Hayes	**0.6780 ± 0.0563**	0.6561 ± 0.0450	0.6750 ± 0.0646	0.6712 ± 0.0465	0.6583 ± 0.0517
Hill	0.6384 ± 0.0278	0.6383 ± 0.0252	**0.6564 ± 0.0249**	0.6467 ± 0.0218	0.6459 ± 0.0232
Indian	0.6405 ± 0.0178	0.6381 ± 0.0167	0.6386 ± 0.0167	**0.6423 ± 0.0160**	0.6358 ± 0.0183
Ionosphere	0.8493 ± 0.0298	**0.8510 ± 0.0257**	0.8410 ± 0.0265	0.8459 ± 0.0252	0.8413 ± 0.0292
Iris	0.8827 ± 0.0409	**0.8920 ± 0.0409**	0.8920 ± 0.0431	0.8677 ± 0.0510	0.8780 ± 0.0394
Knowledge	**0.8461 ± 0.0234**	0.8314 ± 0.0253	0.8434 ± 0.0236	0.8408 ± 0.0213	0.8380 ± 0.0225
Leaf	0.7745 ± 0.0194	0.7721 ± 0.0257	0.7755 ± 0.0195	**0.7800 ± 0.0251**	0.7779 ± 0.0217
Letter	0.6887 ± 0.0055	0.6859 ± 0.0058	**0.6917 ± 0.0065**	0.6903 ± 0.0042	0.6887 ± 0.0054
Libras	0.7783 ± 0.0323	0.7697 ± 0.0330	**0.7843 ± 0.0346**	0.7667 ± 0.0397	0.7767 ± 0.0355
Optdigits	0.8835 ± 0.0120	0.8767 ± 0.0121	**0.8886 ± 0.0142**	0.8883 ± 0.0142	0.8825 ± 0.0114
Pen	0.9139 ± 0.0110	0.9057 ± 0.0101	0.9153 ± 0.0100	**0.9155 ± 0.0075**	0.9152 ± 0.0102
Planning	0.5306 ± 0.0424	**0.5402 ± 0.0425**	0.5224 ± 0.0443	0.5292 ± 0.0502	0.5242 ± 0.0316
QSARBiodegradation	0.8613 ± 0.0115	0.8608 ± 0.0115	**0.8629 ± 0.0118**	0.8582 ± 0.0095	0.8551 ± 0.0077
Seeds	**0.9733 ± 0.0109**	0.9695 ± 0.0183	0.9705 ± 0.0169	0.9705 ± 0.0195	0.9714 ± 0.0143
Shuttle	0.4123 ± 0.2680	**0.5787 ± 0.2897**	0.5082 ± 0.3006	0.5261 ± 0.2757	0.4390 ± 0.2886
SPECTF	**0.3517 ± 0.1419**	0.3513 ± 0.1584	0.3471 ± 0.1179	0.3267 ± 0.1076	0.3387 ± 0.1193
Vertebral2C	0.8580 ± 0.0202	0.8597 ± 0.0156	0.8603 ± 0.0192	0.8643 ± 0.0193	**0.8690 ± 0.0209**
Vertebral3C	0.8057 ± 0.0180	0.8043 ± 0.0191	0.8037 ± 0.0178	**0.8137 ± 0.0168**	0.8067 ± 0.0194
Wdbc	0.9582 ± 0.0081	0.9618 ± 0.0079	0.9611 ± 0.0110	**0.9619 ± 0.0082**	0.9591 ± 0.0106
Wilt	0.9687 ± 0.0016	0.9687 ± 0.0015	0.9695 ± 0.0013	0.9694 ± 0.0017	**0.9699 ± 0.0014**
Wine	0.9606 ± 0.0230	0.9628 ± 0.0218	0.9611 ± 0.0273	0.9517 ± 0.0217	**0.9639 ± 0.0244**
WineRed	0.5849 ± 0.0115	0.5816 ± 0.0101	0.5851 ± 0.0088	**0.5867 ± 0.0078**	0.5861 ± 0.0091
Yeast	0.5756 ± 0.0068	**0.5780 ± 0.0073**	0.5724 ± 0.0065	0.5734 ± 0.0097	0.5733 ± 0.0080
Zoo	0.7462 ± 0.1568	0.7387 ± 0.1330	0.6763 ± 0.1927	**0.7634 ± 0.1422**	0.7441 ± 0.1213

The results of the Friedman test are presented in Table 5. This test establishes a ranking among the algorithms. Here, IHDELS TT is seen to be the winner, followed by DECC-G TT and thirdly MOS TT. The p-value of the test was equal to 5.80E−6 (less than 0.05) and a chi-square value with 9 degrees of freedom equal to 40.646, which makes this ranking statistically significant. Then, as a Friedman post hoc test, the Wilcoxon test was executed. This test establishes the level of dominance of the results of one algorithm over another (see Table 6). The black dot indicates that the algorithm of the row dominates the algorithm of the column and the white dot indicates that the algorithm of the column dominates that of the row; the empty box indicates that it is not possible to establish a dominance of one algorithm over another. The results below the diagonal have a significance of 0.95 and above the diagonal of 0.90. Table 6 shows that the algorithm with the highest dominance is IHDELS TT, with a significance of 0.95 against all the algorithms except DECC-G TT, which occupies second place in dominance, followed by MOS TT.

Table 5. Results of the Friedman test

Algorithm	Ranking	Algorithm	Ranking
IHDELS TT	3.5526 (1)	Random TT	5.9605 (6)
DECC-G TT	3.8026 (2)	DECC-G CV	6.0000 (7)
MOS TT	5.3421 (3)	Random CV	6.1974 (8)
IHDELS CV	5.5395 (4)	M-ELM TT	6.3289 (9)
M-ELM CV	5.5658 (5)	MOS CV	6.7105 (10)

Table 6. Results of the Wilcoxon test

	(1)	(2)	(3)	(4)	(5)	(6)	(7)	(8)	(9)	(10)
DECC-G TT (1)	-		•	•	•	•	•	•	•	•
IHDELS TT (2)		-	•	•	•	•	•	•	•	•
MOS TT (3)	O	O	-	•	•					
M-ELM TT (4)	O	O	O	-						
RW TT (5)	O	O	O		-					
DECC-G CV (6)	O	O				-				
IHDELS CV (7)	O	O					-			
MOS CV (8)		O						-		
M-ELM (9)	O	O							-	
RW CV (10)	O	O								-

4.5   Analysis of Results

From the results, the IHDELS meta-heuristic algorithm with the retention validation model (Training/Testing) presents a better performance, in optimization of input weights and biases of the hidden layer of an SLFN using ELM, than the M-ELM algorithm from the state-of-the-art. In addition, IHDELS TT obtains better results than other specialized algorithms for the solution of continuous problems of high dimensionality, namely DECC-G and MOS. The best results came from using the retention method, these being more accurate and reliable. The Friedman and Wilcoxon non-parametric statistical test supports these conclusions with 95% confidence.

5   Conclusions and Future Work

The present research work adapted the IHDELS algorithm to the problem of training an SLFN using ELM, considering that this problem is continuous and of high dimensionality. The IHDELS algorithm originally had two local searches, MTS-LS1 and L-BFGS-B, but the second was changed for Hill Climbing since it used gradient information and consumed too many EFOs.

Experimentation was conducted on classification problems recognized by the academic and scientific community, using a specific number of evaluations of the objective function that maintains the main concept of an ELM, which is to carry out the training of the neural network in a shorter time than doing so using back-propagation algorithm. It was determined that IHDELS TT (Training/Testing) presents better results than the other algorithms with which it was compared: M-ELM, DECC-G, MOS and RW except against DECC-G TT. These results are supported by the Friedman and Wilcoxon nonparametric statistical tests. In addition, the results show that it is better to use the retention model (training/testing) than the cross-validation model, showing that the algorithms obtain better results when they can carry out more EFOs.

The working group anticipates adapting the algorithms that presented the best results in the 2017 IEEE CEC competition to the problem of training a SLFN using ELM. Additionally, it is hoped to carry out the experiments by optimizing the number of neurons in the hidden layer together with the values of the input weights and biases for each dataset. Finally, it is expected to evaluate other local search methods in IHDELS.

References

1. Zhang, Y., Wu, J., Cai, Z., Zhang, P., Chen, L.: Memetic extreme learning machine. Pattern Recognit. **58**, 135–148 (2016)
2. Matias, T., Souza, F., Araújo, R., Antunes, C.H.: Learning of a single-hidden layer feedforward neural network using an optimized extreme learning machine. Neurocomputing **129**, 428–436 (2014)
3. Zhu, Q.-Y., Qin, A.K., Suganthan, P.N., Huang, G.-B.: Evolutionary extreme learning machine. Pattern Recognit. **38**(10), 1759–1763 (2005)
4. Huang, G., Huang, G.B., Song, S., You, K.: Trends in extreme learning machines: a review. Neural Netw. **61**, 32–48 (2015)
5. Cao, J., Lin, Z., Huang, G.B.: Self-adaptive evolutionary extreme learning machine. Neural Process. Lett. **36**(3), 285–305 (2012)
6. Wolpert, D.H., Macready, W.G.: No free lunch theorems for optimization. IEEE Trans. Evol. Comput. **1**(1), 67–82 (1997)
7. Molina, D., Herrera, F.: Hibridación iterativa de DE con búsqueda local con reinicio para problemas de alta dimensionalidad. In: XVI Conferencia CAEPIA, pp. 251–260 (2015)
8. Huang, G.B., Zhu, Q.Y., Siew, C.K.: Extreme learning machine: theory and applications. Neurocomputing **70**(1–3), 489–501 (2006)
9. Kong, H.: Evolving extreme learning machine paradigm with adaptive operator selection and parameter control. Int. J. Uncertainty, Fuzziness Knowl.-Base Syst. **21**(December), 143–154 (2013)
10. Luke, S.: Essentials of Metaheuristics (2013)
11. Qin, A.K., Suganthan, P.N.: Self-adaptive differential evolution algorithm for numerical optimization. In: 2005 IEEE Congress on Evolutionary Computation, pp. 1785–1791 (2005)
12. Nebro, A.J., Durillo, J.J.: jMetal: a Java framework for multi-objective optimization. Adv. Eng. Softw. **42**, 760–771 (2011)
13. Yang, Z., Tang, K., Yao, X.: Large scale evolutionary optimization using cooperative coevolution. Inf. Sci. (Ny) **178**(15), 2985–2999 (2008)
14. LaTorre, A., Muelas, S., Peña, J.M.: Multiple offspring sampling in large scale global optimization. In: IEEE World Congress on Computational Intelligence, WCCI 2012 (2012)

Information-Theoretic Feature Selection
Using High-Order Interactions

Mateusz Pawluk[1][(✉)], Paweł Teisseyre[2][(✉)], and Jan Mielniczuk[1,2]

[1] Faculty of Mathematics and Information Science,
Warsaw University of Technology, Warsaw, Poland
m.pawluk@mini.pw.edu.pl
[2] Institute of Computer Science, Polish Academy of Sciences, Warsaw, Poland
{teisseyrep,miel}@ipipan.waw.pl

Abstract. Feature selection is one of the major challenges in machine learning. In this paper, we focus on mutual information based methods, which attracted a significant attention in recent years. A clear limitation of the most existing methods is that they usually take into account only low-order interactions between features (up to 3rd order). We propose a novel criterion which takes into account both 3-way and 4-way interactions and can be naturally extended to the case of higher order terms. The basic component of our criterion is interaction information which is a measure of interaction strength derived from information theory. We show that our method is able to find interactions which remain undetected when using standard methods. We prove some theoretical properties of the introduced criterion and interaction information.

1 Introduction

Feature selection is one of the major problems in machine learning [1–3]. It is a crucial challenge for several reasons. First it improves the understandability of the considered model and allows to discover the relationship between features and the class (target) variable. Secondly, it helps to devise approaches with better generalization and larger predictive power [4]. Finally, it allows to reduce the computational cost of fitting the model.

In this paper, we focus on mutual information (MI) based feature selection. This approach has several important advantages. First MI, unlike some classical measures (e.g. Pearson correlation), is able to capture both linear and non-linear dependencies among random variables. Secondly MI based criteria do not depend on any particular model which allows to find all features associated with the class variable, not only those which are captured by an employed model. This is particularly important in the domains where feature selection itself is the main goal of the analysis, e.g. in human genetics where finding mutations of genes influencing the disease is a crucial problem. Moreover, some advanced MI based criteria are able to discover interactions between features as well as to take redundancy between features into account. Finally information-theoretic

© Springer Nature Switzerland AG 2019
G. Nicosia et al. (Eds.): LOD 2018, LNCS 11331, pp. 51–63, 2019.
https://doi.org/10.1007/978-3-030-13709-0_5

approach can be used for both classification and regression tasks, i.e. nominal and quantitative class variable as well as for any type of the features. In this work we focus on classification problem, but the method can be easily extended to regression.

In recent years many algorithms based on mutual information have been proposed. A clear limitation of the existing methods is that they usually take into account only low-order interactions (up to 3rd order). This can be a serious drawback when some complex dependencies exist in our data. For example recent studies in genetics indicate that high-order interactions between genes may contribute to many complex traits [5] and it is crucial to identify them in order to efficiently predict the trait. Taylor et al. [5] give two examples of high-order interactions: one example of three-locus interactions that influence body weight in a cross of two chicken lines and another that showed a pair of genetic interactions involving five or more loci that determine colony morphology in a cross of two yeast strains. We propose a novel criterion called Interaction Information Feature Selection (IIFS) that takes into account both 3-way and 4-way interactions and can be possibly extended to the case of higher order terms. The basic component of our contribution is interaction information, which is a nonparametric measure of interaction strength derived from information theory. Our method is a generalization of Conditional Infomax Feature Extraction (CIFE) criterion [6] whose limitation is that it only considers 3-way interaction terms. We show that our method is able to find interactions which remain undetected when using standard approaches. We also prove some theoretical properties of 4-way interaction information and of the novel criterion. Moreover we experiment with two different methods of multivariate entropy estimation: plug-in estimator based on data discretization and knn-based Kozachenko-Leonenko estimator [7].

The paper is structured as follows. In Sect. 2 we recall the definition of interaction information and prove some new theoretical properties of 4-way interaction information. In Sect. 3 we define the problem and review the existing methods. In Sect. 4 we present our method and discuss its theoretical properties, Sect. 5 contains the results of numerical experiments.

2   Interaction Information

First we define basic quantities used in Information Theory. We consider the discrete class variable Y and features $X_1, \ldots, X_p$, which can be either continuous or discrete. For sake of simplicity we write definitions only for discrete variables. We first recall the definition of the entropy for discrete class variable:

$$H(Y) = -\sum_y P(Y = y) \log P(Y = y). \tag{1}$$

Entropy quantifies the uncertainty of observing random values of Y. If large mass of the distribution is concentrated on one particular value of Y then the entropy is low. If all values are equally likely then $H(Y)$ is maximal. Let $S = (X_1, \ldots, X_m)$ be a subset of the original feature set of size $m = 1, \ldots, p$. The entropy of S

is defined analogously to (1), with a difference that multivariate probability is used instead of univariate probability. The conditional entropy of S given class variable Y can be written as

$$H(S|Y) = \sum_y P(Y = y)H(S|Y = y). \tag{2}$$

The joint mutual information between S and class variable Y is

$$I(S,Y) = H(S) - H(S|Y). \tag{3}$$

This can be interpreted as the amount of uncertainty in S which is removed when Y is known which is consistent with the intuitive meaning of mutual information as the amount of information that one variable provides about another. Moreover the conditional mutual information between S and Y given variable Z is defined as

$$I(S,Y|Z) = H(S|Z) - H(S|Y,Z). \tag{4}$$

We recall a definition of m-way interaction information (II) [8,9]

$$II(S) = II(X_1,\dots,X_m) = - \sum_{T \subseteq S} (-1)^{|S|-|T|} H(T), \tag{5}$$

which generalizes the 3-way interaction information proposed in [10]. For $m = 2$, interaction information reduces to mutual information. The definition of interaction information is identical to that of multivariate mutual information $I(S)$ [10] except for a change in sign in the case of an odd number of variables, i.e. $II(S) = (-1)^{|S|}I(S)$. II can be understood as the amount of information common to all variables (or set of variables), but that is not present in any subset of these variables. Interestingly, m-way interaction information can be also defined using recursive formula

$$II(X_1,\dots,X_m) = II(X_1,\dots,X_{m-1}|X_m) - II(X_1,\dots,X_{m-1}), \tag{6}$$

where $II(X_1,\dots,X_{m-1}|X_m) = \sum_x P(X_m = x)II(X_1,\dots,X_{m-1}|X_m = x)$. The next formula (also known as Möbius representation) [11–14] shows the relationship between II and joint mutual information $I(S,Y)$ which will be useful in the context of the proposed feature selection method

$$I(S,Y) = I((X_1,\dots,X_m),Y) = \sum_{k=1}^{m} \sum_{T \subseteq S:|T|=k} II(T \cup Y). \tag{7}$$

To better grasp the concept of II, let us discuss in more detail 3-way and 4-way interactions. It follows from Möbius representation (7) that

$$II(X_1,X_2,Y) = I((X_1,X_2),Y) - I(X_1,Y) - I(X_2,Y), \tag{8}$$

which indicates that interaction information can be interpreted as a part of the mutual information of (X_1,X_2) and Y which is due solely to interaction between

X_1 and X_2 in predicting Y i.e. the part of $I((X_1, X_2), Y)$ which remains after subtraction of individual informations between Y and X_1 and Y and X_2. In other words, II is obtained by removing the main effects from the term describing the overall dependence between Y and the pair (X_1, X_2). Here let us mention that 3-way interaction information is a commonly used measure for detecting interactions between genes in genome-wide case- control studies [15,16]. For 4-way interaction we have from (7) and (8) that

$$
\begin{aligned}
II(X_1, X_2, X_3, Y) = {} & I((X_1, X_2, X_3), Y) \\
& - I((X_1, X_2), Y) - I((X_1, X_3), Y) - I((X_2, X_3), Y) \\
& + I(X_1, Y) + I(X_2, Y) + I(X_3, Y).
\end{aligned}
\tag{9}
$$

Observe that both terms $I((X_1, X_2), Y)$ and $I((X_1, X_3), Y)$ in (9) contain $I(X_1, Y)$ as summands (cf. (8)) and as a result $I(X_1, Y)$ is subtracted twice. To account for it we add $I(X_1, Y)$ in the last line of (9). The remaining pairs are treated analogously. The simplest examples of 3-way and 4-way interactions are XOR problems. In XOR $Y = 1$ when the number of input variables taking value 1 is odd. It is easy to check that input binary variables are mutually independent and marginally independent from a class variable. For 3-dimensional case we have $I(X_1, Y) = I(X_2, Y) = 0$ and $II(X_1, X_2, Y) = I((X_1, X_2), Y) = H(Y) - H(Y|X_1, X_2) = H(Y) = \log(2)$. For 4-dimensional case all terms, except the first one, are zero. i.e. $II(X_1, X_2, X_3, Y) = I((X_1, X_2, X_3), Y) = H(Y) - H(Y|X_1, X_2, X_3) = H(Y) = \log(2)$.

Some properties of 4-way Interaction Information which has not been discussed in the literature are discussed below. For the sake of clarity we assume that all variables are discrete and let $p_{ijkl} = P(X_1 = x_i, X_2 = x_j, X_3 = x_k, Y = y_l)$, where P denotes the distribution of (X_1, X_2, X_3, Y). Moreover, $KL(P||Q)$ stands for Kullback-Leibler divergence between P and Q, defined as $KL(P||Q) = \sum_{i,j,k} p_{ijk} \log(p_{ijk}/q_{ijk})$.

Theorem 1. *We have (i) $II(X_1, X_2, X_3, Y) = KL(P||P_K)$, where P_K corresponds to mass function p^K defined as*

$$
p_{ijkl}^K = \frac{\prod_{S:|S|=3} p_S \prod_{S:|S|=1} p_S}{\prod_{S:|S|=2} p_S} = \frac{p_{ijk} p_{ijl} p_{jkl} p_{ikl} p_i p_j p_k p_l}{p_{ij} p_{ik} p_{il} p_{jk} p_{jl} p_{kl}}.
\tag{10}
$$

(ii) If $X_1 \perp X_2|W$, where W is any subset (including $\emptyset$) of $\{X_3, Y\}$ then $II(X_1, X_2, X_3, Y) = 0$.
(iii) Let $\eta = \sum_{i,j,k,l} p_{ijkl}^K$. If $\eta \leq 1$ and $II(X_1, X_2, X_3, Y) = 0$ then $P = P_K$.

Proof. (i) follows from (5) and definition of Kullback-Leibler divergence. (ii) is a consequence of (10) and assumptions. In order to prove (iii) note that $KL(P||Q) = 0$ implies $P = Q$ not only in the case when Q is probability distribution but also in the case when total mass of Q does not exceed 1. This yields the result when applied to $Q = P_K$.

Observe that P_K is not necessarily probability distribution. Condition $\eta \leq 1$ is sufficient condition which ensures that $P = P_K$ when $II = 0$. P_K is generalization of Kirkwood approximation [17] to four-dimensional case.

3   Problem Formulation and Previous Work

In this work we focus on feature selection based on mutual information (MI). MI-based feature selection is concerned with identifying a fixed-size subset $S \subset \{1, \ldots, p\}$ of the original feature set that maximizes the joint mutual information between S and class variable Y. Finding an optimal feature set is usually infeasible because the search space grows exponentially with the number of features. As a result various greedy algorithms have been developed including forward selection, backward elimination and genetic algorithms. Today sequential forward selection is the most commonly adopted solution. Forward selection algorithms start from an empty set of features and add, in each step, the feature that jointly, i.e. together with already selected features, achieves the maximum joint mutual information with the class. Formally, assume that S is a set of already chosen features, S^c is its complement and $X_k \in S^c$ is a candidate feature. The score for feature X_k is

$$J(X_k) = I(S \cup X_k, Y) - I(S, Y). \tag{11}$$

Obviously the second term in (11) does not depend on X_k and it can be omitted, however it is more convenient to use this form. In each step we add a feature that maximizes $J(X_k)$. Criterion (11) is equivalent to

$$J(X_k) = I(X_k, Y|S), \tag{12}$$

see [18] for the proof. We also refer to [19] who proposed a fast feature selection method based on conditional mutual information and min-max approach. Observe that (12) indicates that we select a feature that achieves the maximum association with the class given the already chosen features. Criterion (11) (or equivalently (12)) is appealing and attracted a significant attention. However in practice the estimation of joint mutual information is problematic even for small set S. This makes a direct application of (11) infeasible. A rich body of work in the MI-based feature selection literature approaches this difficulty by approximating the high-dimensional joint MI with low-dimensional MI terms. These approximations may by accurate provided some additional conditions on data distribution are satisfied. A comprehensive review of the existing methods can be found in [18], here we review some representative methods. One of the most popular methods is Mutual Information Feature Selection (MIFS) proposed in [20]

$$J_{\mathrm{MIFS}}(X_k) = I(X_k, Y) - \sum_{j \in S} I(X_j, X_k). \tag{13}$$

This includes the $I(X_k, Y)$ term to ensure feature relevance, but introduces a penalty to enforce low correlations with features already selected in S. The similar idea is used in Minimum-Redundancy Maximum-Relevance (MRMR) criterion [21]

$$J_{\mathrm{MRMR}}(X_k) = I(X_k, Y) - \frac{1}{|S|} \sum_{j \in S} I(X_j, X_k). \tag{14}$$

with the difference that the second term is averaged over features in S. Both MIFS and MRMR criteria focus on reducing redundancy, however they do not take into account interactions between features. Brown et al. [18] have shown that if the selected features from S are independent and class-conditionally independent given any unselected feature X_k then (11) reduces to so-called CIFE criterion [6]

$$J_{\text{CIFE}}(X_k) = I(X_k, Y) + \sum_{j \in S}[I(X_j, X_k|Y) - I(X_j, X_k)]. \tag{15}$$

In view of (8), the second term in (15) is equal $\sum_{j \in S} II(X_j, X_k, Y)$, so it is seen that CIFE is able to detect 3-way interactions. Yang and Moody [22] have proposed using Joint Mutual Information (JMI)

$$J_{\text{JMI}}(X_k) = \sum_{j \in S} I((X_j, X_k), Y), \tag{16}$$

which is equal up to a constant to

$$J_{\text{JMI}}(X_k) = |S|I(X_k, Y) + \sum_{j \in S}[I(X_j, X_k|Y) - I(X_j, X_k)]. \tag{17}$$

JMI is a similar to CIFE, with the difference that in JMI the marginal relevance term plays more important role than the overall interaction term.

4   Feature Selection Based on Interaction Information

In this Section we describe a proposed approach which can be seen as a generalization of CIFE. Our method considers not only 3-way interactions but also 4-way interactions.

4.1   Proposed Criterion: IIFS

In our method we make use of Möbius representation. Recall that S is a set of already selected features of size m and X_k is a candidate feature. First observe that it follows from Möbius representation (7) that

$$J(X_k) = I(S \cup X_k, Y) - I(S, Y) = \sum_{k=0}^{m} \sum_{T \subset S: |T| = k} II(T \cup X_k \cup Y). \tag{18}$$

In the proposed method IIFS (Interaction Information Feature Selection) we define a score

$$J_{\text{IIFS}}(X_k) = I(X_k, Y) + \sum_{j \in S} II(X_j, X_k, Y) + \sum_{i, j \in S: i < j} II(X_i, X_j, X_k, Y), \tag{19}$$

which is a third order approximation of (18). The first term in (19) takes into account marginal relevance of the candidate feature whereas the second and the third terms describe the 3 and 4-way interactions, respectively. Note that IIFS can be seen as an extended version of CIFE which is a second order approximation of $J(X_k)$, namely

$$J_{\text{IIFS}}(X_k) = J_{\text{CIFE}}(X_k) + \sum_{i,j \in S: i < j} II(X_i, X_j, X_k, Y). \qquad (20)$$

It is possible to consider higher order terms in (18), however it would increase the computational cost and make the estimation even more difficult. Below we state some properties of the introduced criteria.

Theorem 2. *The following properties hold.*

(i) Assume that $X_k \perp Y$. Then

$$J_{CIFE}(X_k) = \sum_{j \in S} I(X_k, Y | X_j). \qquad (21)$$

(ii) Assume that $X_k \perp Y$ and $X_k \perp Y | X_j$ for any $X_j \in S$. Then

$$J_{IIFS}(X_k) = \sum_{i,j \in S: i < j} I(X_k, Y | X_i, X_j). \qquad (22)$$

(iii) Assume that $X_i \perp X_j | X_k$ and $X_i \perp X_j | X_k, Y$, for some $X_i, X_j \in S$. Then $II(X_i, X_j, X_k, Y)$ does not depend on X_k.

(iv) If $|S| = 2$ then $\text{argmax}_{X_k \in S^c} J_{IIFS}(X_k) = \text{argmax}_{X_k \in S^c} J(X_k)$.

Proof. To prove (i) observe that property (6) implies

$$II(X_j, X_k, Y) = I(X_k, Y | X_j) - I(X_k, Y). \qquad (23)$$

Under assumption $X_k \perp Y$ we have $I(X_k, Y) = 0$ which, together with (23) and (15) yields (21). Let us now prove (ii). It follows from (6) that

$$II(X_i, X_j, X_k, Y) = II(X_j, X_k, Y | X_i) - II(X_j, X_k, Y) \qquad (24)$$

and

$$II(X_j, X_k, Y | X_i) = I(X_k, Y | X_j, X_i) - I(X_k, Y | X_i). \qquad (25)$$

Under assumption (ii) we have that $I(X_k, Y) = 0$, $II(X_j, X_k, Y) = 0$ and $I(X_k, Y | X_i) = 0$ and thus $II(X_i, X_j, X_k, Y) = I(X_k, Y | X_j, X_i)$ which yields (22). Let us now prove (iii). Using (6) we can write

$$II(X_i, X_j, X_k, Y) = II(X_i, X_j, Y | X_k) - II(X_i, X_j, Y)$$
$$= I(X_i, X_j | X_k, Y) - I(X_i, X_j | X_k) - II(X_i, X_j, Y). \qquad (26)$$

Assumptions of (iii) implies that $I(X_i, X_j | X_k, Y) = I(X_i, X_j | X_k) = 0$, which yields the assertion in view of (26). Finally note that (iv) follows from the fact that for $|S| = 2$ Eqs. (18) and (19) are equivalent. i.e. Möbius representation gives an exact value of $J(X_k)$.

Let us briefly comment the above statements. Items (i) and (ii) of Theorem 2 indicate that under additional assumptions CIFE and IIFS reduce to simpler and more intuitive forms. Using the forms given in (i) and (ii) one may easily give an example showing the advantage of IIFS over CIFE. Indeed, under assumption (ii) we have $J_{\mathrm{CIFE}}(X_k) = 0$ and we may conclude that $J_{\mathrm{IIFS}}(X_k) > 0$ if there exists a pair $X_i, X_j \in S$ such that $I(X_k, Y | X_i, X_j) > 0$. In this case IIFS recognizes X_k as a relevant whereas CIFE treats X_k as a spurious feature. In addition [18] has showed that if assumptions of (iii) hold for any $k \in S^c$, maximization of $J_{\mathrm{CIFE}}(X_k)$ is equivalent to maximization of $J(X_k)$. In (iii) we confirm that indeed in this case the 4-way interaction term can be omitted.

5   Experiments

The aim of the experiments is to compare the performance of the proposed method IIFS with other popular methods discussed in Sect. 3: MIFS, MRMR, JMI and CIFE.

5.1   Artificial Data

The main advantage of the experiments on artificial data is that we can directly investigate which method is able to detect the particular types of interactions. We consider two simulation models, including 3-way and 4-way interactions, respectively. To make a task more challenging we assume in both cases that features are continuous. To assess the quality of the methods we introduce the following measure. Let t be a set of relevant features influencing Y and $j_1, j_2, \ldots, j_p$ be features sequentially selected by the given method. The selection rate (SR) is defined as

$$SR = \frac{|\{j_1, \ldots, j_{|t|}\} \cap t|}{|t|}, \tag{27}$$

i.e. SR is a fraction of relevant features among first $|t|$ selected. For example if we have two relevant features X_1, X_2 then $t = \{1, 2\}$. When the method produces a list $\{1, 2, 5, \ldots\}$ then $SR = 1$. On the other hand if the method gives $\{1, 5, 2, \ldots\}$ then $SR = 0.5$, as one spurious feature X_5 is ranked higher than the relevant feature X_2. In the following we describe two simulation models.

Simulation Model 1 (3-Way Interaction Model). We consider 50 uniformly distributed features: $X_1 \sim U[0, 3]$, $X_j \sim U[0, 2]$, for $j = 2, \ldots, 50$. Only two first features X_1 and X_2 are relevant, i.e. class variable Y depends only on X_1 and X_2, the remaining features are spurious. Table 1 shows the joint distribution of X_1, X_2, Y. This model is an extension of 2-dimensional XOR; note that $Y = 1$ when $X_1 \in A, X_2 \in B$ or $X_1 \in B, X_2 \in A$. It is easy to verify that for this model we have: $I(X_1, Y) > 0$, $I(X_j, Y) = 0$, for $j = 2, \ldots, 50$ and $II(X_1, X_2, Y) > 0$, thus we have one main effect corresponding to X_1 and one 3-way interaction.

Simulation Model 2 (4-Way Interaction Model). We consider 50 uniformly distributed features: $X_1, X_2 \sim U[0,3]$, $X_j \sim U[0,2]$, for $j = 3, \ldots, 50$. Class variable Y depends on X_1, X_2, X_3 whereas the remaining features are spurious. Table 2 shows the joint distribution of X_1, X_2, Y. This model is an extension of 3-dimensional XOR. It is easy to verify that for this model we have: $I(X_1, Y), I(X_2, Y) > 0$, $I(X_j, Y) = 0$, for $j = 3, \ldots, 50$ and $II(X_1, X_2, X_3, Y) > 0$, thus we have two main effects corresponding to X_1 and X_2 and moreover one 4-way interaction.

Table 1. Simulation model 1 (3-way interaction model). Notation: $A = [0,1]$, $B = (1,2]$, $C = (2,3]$ and constant p equals $1/6$.

	1	2	3	4	5	6
X_1	A	A	B	B	C	C
X_2	A	B	A	B	A	B
Y	0	1	1	0	0	0
$P(X_1, X_2, Y)$	p	p	p	p	p	p

Table 2. Simulation model 2 (4-way interaction model). Notation: $A = [0,1]$, $B = (1,2]$, $C = (2,3]$ and constant p equals $1/16$.

	1	2	3	4	5	6	7	8	9	10	11	12	13	14	15	16
X_1	A	B	A	A	B	B	A	B	C	C	C	C	A	A	B	B
X_2	A	A	B	A	B	A	B	B	A	B	A	B	C	C	C	C
X_3	A	A	A	B	A	B	B	B	A	A	B	B	A	B	A	B
Y	0	1	1	1	0	0	0	1	1	1	1	1	1	1	1	1
$P(X_1, X_2, X_3, Y)$	p	p	p	p	p	p	p	p	p	p	p	p	p	p	p	p

Table 3. Computational times.

Feature selection	CIFE	JMI	MIFS	MRMR	IIFS
MADELON	16.312 s	16.228 s	16.089 s	16.147 s	1.109 min
GISETTE	1.156 h	1.153 h	1.091 h	1.124 h	2.701 h
MUSK	11.719 s	11.048 s	13.587 s	14.217 s	15.746 s
BREAST	0.887 s	0.425 s	0.515 s	0.499 s	0.988 s

Figure 1 shows how selection rate (SR) depends on sample size n. In the case of model 1 the methods which take into account 3-way interactions (JMI,

CIFE, IIFS) produce the same rankings. They detect successfully both relevant features: X_1 and X_2. MIFS and MRMR are able to detect only one relevant feature. In the case of model 2, MIFS, MRMR, JMI and CIFE are able to detect only 2 relevant features X_1, X_2 but they fail to select feature X_3. Selection rate (SR) for MIFS, MRMR, JMI and CIFE converges to 2/3. As expected only IIFS chooses all 3 relevant features, which results in $SR = 1$ for sufficiently large sample size. The above experiment shows that there is no significant difference between IIFS, JMI, CIFE when only 3-way interactions occur. In the case of 4-way interaction model, IIFS is significantly superior to other methods. Moreover we analyse how the method of entropy estimation influences the results. We used two methods: standard plug-in method based on data discretization with b bins (solid line) and knn-based Kozachenko-Leonenko estimator [7], with $k = 10$ (dashed line). For small $b = 2$ it is seen that knn-based method is superior to plug-in method. For $b = 5$, plug-in method works better than knn-based method in the case of model 1, whereas knn-based method is a winner for model 2.

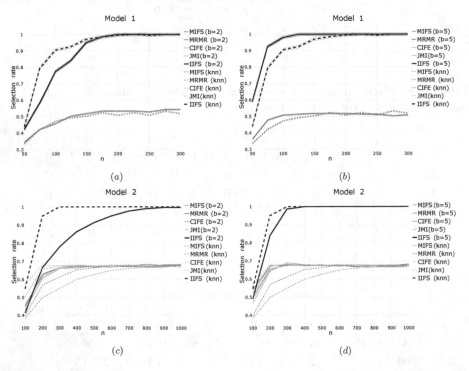

Fig. 1. Selection rate w.r.t. sample size n for simulation models 1 (a)–(b) and 2 (c)–(d). Parameter b corresponds to the number of bins in discretization, 'knn' in brackets corresponds to knn-based entropy estimation.

5.2   Benchmark Data

For more thorough assessment of developed criterion we used datasets from the
NIPS Feature Selection Challenge [23] (MADELON and GISETTE) and UCI
repository [24] (BREAST and MUSK). NIPS datasets consist of training sets
(2000 observations for MADELON and 6000 for GISETTE) and validation sets
(600 observations for MADELON and 1000 for GISETTE), whereas for UCI
datasets we used 10-fold cross-validation in order to calculate error rates. We
carried out the same experiment as that described in [18, Sect. 6.1]. In addition
to methods considered in [18] we investigate the performance of the proposed
method IIFS. Each criterion was used to generate a ranking for the top features.
Then the original datasets were used to classify the validation data. As in [18]
we used kNN method with $k = 3$ neighbours as a classifier. As an evaluation
measure we considered Balanced Error Rate defined as

$$BER = 1 - 0.5 \cdot (\frac{TP}{TP + FN} + \frac{TN}{TN + FP}), \tag{28}$$

where TP, TN, FP, FN denote true positives, true negatives, false positives and
false negatives, respectively. Results of our experiments are presented in Fig. 2.
We only present curves corresponding to plug-in estimator as knn-based entropy

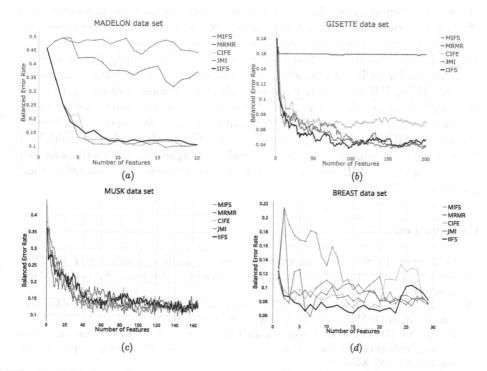

Fig. 2. Validation error curves for MADELON (a), GISETTE (b), MUSK (c) and
BREAST (d) datasets.

estimator worked much worse in this case possibly due to a prior discretization of the original data. For MADELON and MUSK datasets there is no significant improvement of IIFS compared to CIFE and JMI. So we may conclude that considering interactions of order higher than 3 does not improve the performance in this case. Note that for MADELON interactions play an important role; the methods which do not take into account interactions at all (MIFS and MRMR) fail. For GISETTE dataset the proposed criterion IIFS has the lowest error rate when the number of features varies between 20 and 100. For BREAST IIFS is also a winner. This suggests that taking into account high-order interactions helps in these cases. Interestingly, for GISETTE and BREAST, IIFS is significantly better than CIFE, which additionally indicates that including 4-way interaction term improves the performance. The computational times for IIFS are longer than for competitors (see Table 3) which is a price for taking into account high-order interactions. Note however that the times for IIFS, although longer than for CIFE, are of the same order.

6   Conclusions

In this paper we presented a novel feature selection method, named IIFS. Feature selection score in IIFS, based on interaction information, is derived from so-called Möbius representation of joint mutual information. Our method in an extension of CIFE criterion consisting in taking into account 4-way interaction terms. We discussed theoretical properties of 4-way interaction information (Theorem 1) as well as feature selection methods: CIFE and IIFS (Theorem 2). The numerical experiments for artificial datasets show that there is no significant difference between IIFS, JMI and CIFE when only the interactions of order up to 3 are present. This means that estimation of absent 4-way interactions does not cause significant deterioration of IIFS performance. In the case when 4-way interactions occur IIFS is significantly superior to other methods. Future work will include the development of methods considering high-order interactions as well as the comparison of IIFS with such methods, for example with a novel method proposed in [25].

References

1. Li, J., et al.: Feature selection: a data perspective. J. Mach. Learn. Res. 1–73 (2016)
2. Guyon, I., Elisseeff, A.: An introduction to variable and feature selection. J. Mach. Learn. Res. **3**, 1157–1182 (2003)
3. Forman, G.: An extensive empirical study of feature selection metrics for text classification. J. Mach. Learn. Res. **3**, 1289–1305 (2003)
4. Hastie, T., Tibshirani, R., Friedman, J.: The Elements of Statistical Learning: Data Mining, Inference and Prediction. Springer, New York (2009). https://doi.org/10.1007/978-0-387-84858-7
5. Taylor, M.B., Ehrenreich, I.M.: Higher-order genetic interactions and their contribution to complex traits. Trends Genet. **31**(1), 34–40 (2015)

6. Lin, D., Tang, X.: Conditional infomax learning: an integrated framework for feature extraction and fusion. In: Leonardis, A., Bischof, H., Pinz, A. (eds.) ECCV 2006. LNCS, vol. 3951, pp. 68–82. Springer, Heidelberg (2006). https://doi.org/10.1007/11744023_6
7. Kozachenko, L., Leonenko, N.: Sample estimate of the entropy of a random vector. Problemy Peredachi Informatsii **23**(2), 9–16 (1987)
8. Jakulin, A., Bratko, I.: Quantifying and visualizing attribute interactions: an approach based on entropy. Manuscript (2004)
9. Han, T.S.: Multiple mutual informations and multiple interactions in frequency data. Inf. Control **46**(1), 26–45 (1980)
10. McGill, W.J.: Multivariate information transmission. Psychometrika **19**(2), 97–116 (1954)
11. Kojadinovic, I.: Relevance measures for subset variable selection in regression problems based on k-additive mutual information. Comput. Stat. Data Anal. **49**(4), 1205–1227 (2005)
12. Meyer, P., Schretter, C., Bontempi, G.: Information-theoretic feature selection in microarray data using variable complementarity. IEEE J. Sel. Top. Sig. Process. **2**(3), 261–274 (2008)
13. Vergara, J.R., Estévez, P.A.: A review of feature selection methods based on mutual information. Neural Comput. Appl. **24**(1), 175–186 (2014)
14. Brown, G.: A new perspective for information theoretic feature selection. In: Twelfth International Conference on Artificial Intelligence and Statistics, AISTATS-2009, pp. 49–56 (2009)
15. Moore, J., et al.: A flexible computational framework for detecting, characterizing, and interpreting statistical patterns of epistasis in genetic studies of human disease susceptibility. J. Theor. Biol. **241**(2), 256–261 (2006)
16. Mielniczuk, J., Teisseyre, P.: A deeper look at two concepts of measuring gene-gene interactions: logistic regression and interaction information revisited. Genet. Epidemiol. **42**(2), 187–200 (2018)
17. Matsuda, H.: Physical nature of higher-order mutual information: intrinsic correlations and frustration. Phys. Rev. E **62**(3 A), 3096–3102 (2000)
18. Brown, G., Pocock, A., Zhao, M.J., Luján, M.: Conditional likelihood maximisation: a unifying framework for information theoretic feature selection. J. Mach. Learn. Res. **13**(1), 27–66 (2012)
19. Fleuret, F.: Fast binary feature selection with conditional mutual information. J. Mach. Learn. Res. **5**, 1531–1555 (2004)
20. Battiti, R.: Using mutual information for selecting features in supervised neural-net learning. IEEE Trans. Neural Netw. **5**(4), 537–550 (1994)
21. Peng, H., Long, F., Ding, C.: Feature selection based on mutual information: Criteria of max-dependency, max-relevance, and min-redundancy. IEEE Trans. Pattern Anal. Mach. Intell. **27**(8), 1226–1238 (2005)
22. Yang, H.H., Moody, J.: Data visualization and feature selection: new algorithms for nongaussian data. Adv. Neural Inf. Process. Syst. **12**, 687–693 (1999)
23. Guyon, I.: Design of experiments for the NIPS 2003 variable selection benchmark (2003)
24. Dheeru, D., Karra Taniskidou, E.: UCI machine learning repository (2017)
25. Shishkin, A., Bezzubtseva, A., Drutsa, A.: Efficient high-order interaction-aware feature selection based on conditional mutual information. In: Advances in Neural Information Processing Systems, NIPS, pp. 1–9 (2016)

Covering Arrays to Support the Process of Feature Selection in the Random Forest Classifier

Sebastián Vivas, Carlos Cobos[✉], and Martha Mendoza

Information Technology Research Group (GTI), Universidad del Cauca,
Popayán, Colombia
{jusvivas, ccobos, mmendoza}@unicauca.edu.co

Abstract. The Random Forest (RF) algorithm consists of an assembly of base decision trees, constructed from Bootstrap subsets of the original dataset. Each subset is a sample of instances (rows) by a random subset of features (variables or columns) of the original dataset to be classified. In RF, pruning is not applied in the generation of base trees and in the classification process of a new record, each tree issues a vote enabling the selected class to be defined, as that with the most votes. Bearing in mind that in the state of the art it is defined that random feature selection for constructing the Bootstrap subsets decreases the quality of the results achieved with RF, in this work the integration of covering arrays (CA) in RF is proposed to solve this situation, in an algorithm called RFCA. In RFCA, the number N of rows of the CA defines the lowest number of base trees that require to be generated in RF and each row of the CA defines the features that each Bootstrap subset will use in the creation of each tree. To evaluate the new proposal, 32 datasets available in the UCI repository are used and compared with the RF available in Weka. The experiments show that the use of a CA of strength 2 to 7 obtains promising results in terms of accuracy.

Keywords: Classification · Random Forest · Covering arrays ·
Feature selection

1 Introduction

The Random Forest (RF) algorithm, developed by Leo Breiman in 2001 [1], is composed of a collection of M independent trees to which an input is passed. Each emits a unitary vote and RF then selects the most popular class of all the votes received (majority vote). Building each tree starts with a bootstrap sample of the original dataset. Each node of a decision tree is built from a small group of randomly selected features [1]. In RF, pruning is not applied and each of the generated trees functions as a base classifier. RF stands out for its robustness, low sensitivity to noise and low risk of overfitting [2]. As regards limitations, in [3] the authors point to the amount of time it takes to manually fix the hyperparameters (number of trees, M, number of features to be taken into account in each tree, K, and depth of each tree, depth) and the lack of a more suitable feature selection process, since in its original proposal it resorted to a simple random selection. According to [4], the value of K (number of randomly selected

© Springer Nature Switzerland AG 2019
G. Nicosia et al. (Eds.): LOD 2018, LNCS 11331, pp. 64–76, 2019.
https://doi.org/10.1007/978-3-030-13709-0_6

features for each tree) is established arbitrarily or empirically, and often does not have a theoretical or experimental justification.

This paper proposes the integration of covering arrays as a mechanism for selecting the features that are used to construct the Bootstrap samples with which the base trees are built.

A covering array (CA) is a mathematical object that has been used in various areas to evaluate and compare different situations in which many parameters interact with each other and carrying out an exhaustive evaluation is not feasible for reasons of cost, time, or effort. CAs have been used in experimental design, software testing, hardware testing and more recently in clustering [5].

In the selection mechanism proposed in this paper, binary CAs are used, formally defined as CA $(N, P, v = 2, t)$ that can be represented as a matrix of $N \times P$ elements, where N is the number of rows of the CA. In this context, N refers to the number of base trees to be generated. The value of P refers to the number of columns of the CA, which is the number of factors/parameters involved in the problem (dataset, without considering the class column). The $v = 2$ refers to the fact that the CA is binary, and the data found in each cell of the matrix can only take the value of zero (0) or one (1). Finally, parameter t is called strength and defines the degree of interaction between the P factors covered by the CA. Where there is a strength of 2 in a binary CA, it is expected that the values $\{0, 0\}, \{0, 1\}, \{1, 0\}, \{1, 1\}$ will be found in any pair of columns in the CA. In general, in a CA, each $N \times t$ sub-matrix contains all the combinations of the $v = 2$ symbols at least once. Figure 1 shows the CA (6, 6, 2, 2) with 6 rows (N), 6 columns (P), binary alphabet ($v = 2$), and strength 2 (t).

$$CA(N = 6, P = 6, v = 2, t = 2) = \begin{bmatrix} 1 & 1 & 1 & 0 & 0 & 0 \\ 1 & 0 & 1 & 0 & 1 & 1 \\ 0 & 0 & 0 & 1 & 1 & 0 \\ 1 & 1 & 0 & 1 & 1 & 1 \\ 0 & 1 & 1 & 1 & 0 & 1 \\ 0 & 0 & 0 & 0 & 0 & 0 \end{bmatrix}$$

Fig. 1. Example of a covering array: CA (6, 6, 2, 2)

In this paper, each row of the CA defines the features that will be used for building the base trees that RF uses to make decision. A one means that the feature is included and a zero that it is not. The value of t, the strength of the CA, is defined as a hyperparameter of the new algorithm (Random Forest based on Covering Arrays, RFCA) and is established empirically. The proposed RFCA was compared with the original RF algorithm proposed by Breiman [1] implemented in Weka, and one called RF_SQRT, the same original RF but in which the K value (number of features selected for building the nodes of base trees) is defined using a different formula. The evaluation was carried out with 32 datasets using 10-fold cross-validation. The results show that RFCA improves the accuracy in half of the evaluation scenarios with respect to state of the art algorithms. In addition, Friedman and Wilcoxon nonparametric statistical test results are promising.

The rest of the paper is organized as follows: Sect. 2 presents previous research work related to improvements to the RF algorithm; Sect. 3 describes the RFCA classifier, along with an example of its use; Sect. 4 describes the experiments and an analysis of the results; and finally, Sect. 5 presents the conclusions of the work carried out, together with some future work that is expected to be undertaken.

2 Related Work

The original RF algorithm (2001) consists of an assembly of decision trees. Initially in the process of building a RF, bootstrapping is applied on the training dataset to produce many different data subsets [1]. Each subset is then used to construct a decision tree. In the growth process of the tree, the partition of each node depends on the randomly selected features with respect to all the features present in the dataset [4]. No pruning is applied in RF and each of the trees generated works as a base classifier. To define the class of an instance, the vote of each of the base classifiers is received and a weighting is performed that determines the respective class [2, 6].

In 2008 [7] saw the arrival of an algorithm called Forest-RK. Based on Forest-RI [1], Forest-RK, introduced a new method of induction in which an alternative to the arbitrary adjustments of the hyperparameter K is offered. For random feature selection, the K value is chosen randomly for each division of a node, with the aim of generating greater diversity in the trees that make up the forest, in contrast with Forest-RI, in which the value of K is identical for all decision trees. The results show that this new method is statistically more accurate than the Breiman RF [1].

In 2011 [8], the use of oblique tree models as base learners in the algorithm was proposed. The "oblique" RFs focus on the optimal recursive partition of the nodes, so that in each recursive binary division, a new set of features is sampled without replacement, and the optimal division in the sub-space covered by these features is sought. For the search of the optimal division, linear discriminative models are used instead of random coefficients used in the Breiman RF [1]. The results show that RF with orthogonal divisions obtains good results in factor datasets, in numerical and spectral data. This proposal outperformed a wide range of classifiers.

In 2012 [9], a tree regularization framework was proposed that allows many tree models to perform feature selection efficiently. The key idea of the framework is to penalize with λ (a coefficient $\in$ [0, 1] based on information gain), the selection of a feature used for the division of a node, in cases where its quality index is like the features used in previous divisions. It is therefore expected that a regularized tree model contains a set of features that are informative, but not redundant. The results show that the proposed method increases the quality of the RF classifier.

In 2014 [10] a technique was presented for finding the appropriate number for the attribute subspace (K) used in the division of a node into a decision tree. The number of attributes for the subspace is determined by a random number selected in a range that is calculated with the number of samples resulting from the CART partition of a node (size of the bootstrap sample for the root node), and the size of the bootstrap sample of the decision tree. This calculation is performed dynamically for each of the nodes and ensures a diverse range of trees. The results show that the proposed technique can significantly improve the accuracy of the classifier.

In 2016 [11] an RF version was proposed with a cost-sensitive feature selection method called feature-cost-sensitive Random Forest (FCS-RF). In FCS-RF, the cost of the features is incorporated into the building process of the decision tree to produce subsets of low-cost features. The algorithm selects a feature with a probability inversely proportional to its associated cost, instead of being selected randomly. The results show that FCS-RF is mainly useful in cases where there are redundant or higher cost features.

In 2017 [12] an integration of an algorithm called CURE-SMOTE with a hybrid algorithm based on RF was proposed. CURE (Clustering Using Representatives) groups the least representative class samples and SMOTE (synthetic minority oversampling technique) eliminates noise and outliers. The dataset used to solve the classification problem is then generated using random samples between representative points and data from the less representative classes. Elimination of redundant features, feature selection, optimization of parameters, and definition of the number of sub-features is then carried out by means of three hybrid algorithms that use RF: one based on genetic algorithms (GA-RF), another based on particle swarm optimization (PSO-RF), and finally, another based on a swarm of fish (AFSA-RF). The results show that the CURE-SMOTE algorithm minimizes the noise of the original data distribution and that the hybrid algorithms surpass original RF [1] in F-measure, G-mean, AUC, and Out-Of-Bag error.

3 Proposed Random Forest Based on Covering Arrays (RFCA)

In the original Random Forest proposal of Breiman [1], the process of feature selection, denoted as Forest-RI, employs in the division of each node, small groups of randomly selected features. The size, K, of the groups is fixed and is generally equal to the first integer smaller than $\log_2 P + 1$, where P is the total number of attributes of the dataset. The hyperparameter number of trees in the forest, M, is established arbitrarily or empirically, and although an increase in the number of trees can linearly increase the quality of the model, there is a certain point at which increasing the number of trees does not improve and even decreases the accuracy of the model [3]. In this context, covering arrays in RFCA eliminates the need to set and fine-tune the number of trees (hyperparameter M) and improve the feature selection process.

RFCA uses CAs of binary alphabet ($v = 2$). This means that each component of the CA has only values $\{0, 1\}$. The subsets of candidate variables used in building each of the RF trees are determined with the rows of the CA. Parameter P of the CA corresponds to the total number of attributes of the dataset to be classified. Below, an example of RFCA execution is detailed and then the algorithm is presented.

3.1 Example of Model Creation in RFCA

To illustrate the behavior of RFCA, a reduced version of the Churn dataset is used, from the UCI Repository (University of California in Irvine). The term Churn is used to indicate that a client leaves the service of a company to take that of the competition. In this reduced version, only 6 (P columns of the dataset and the CA) of the most important variables are used in 20 of the customer records and the target variable, Churn. In Table 1, the name of the attributes of the dataset is shown and in the upper part a short name appears for each of them, A1, A2, and so on. In this example, 80% of the dataset is used for training and the remaining 20% as the test dataset. The Random Forest algorithm is based on the bootstrapping process, which implies that it generates a set of trees, M in which each tree is generated with its own training dataset that is the result of a random selection of data from the original training dataset.

Table 1. Description of the adapted dataset

Id	A1	A2	A3	A4	A5	A6	Class
	International plan	Voice mail plan	Total minutes morning	Total minutes afternoon	Total minutes night	Total minutes international	Churn
	Training dataset						
1	No	No	178.7	233.7	131.9	9.1	FALSE
2	No	Yes	148.5	114.5	178.3	6.5	FALSE
3	No	Yes	164.1	219.1	220.3	12.3	FALSE
4	Yes	No	197.2	188.5	211.1	7.8	FALSE
5	No	No	124.9	300.5	192.5	11.6	FALSE
6	No	No	115.4	209.9	280.9	15.9	FALSE
7	Yes	No	140	196.4	120.1	9.7	TRUE
...	...	...	...	...	....	...	...
15	No	Yes	156.2	215.5	279.1	9.9	FALSE
16	No	No	231.1	153.4	191.3	9.6	FALSE
	Test dataset						
17	No	No	180.8	288.8	191.9	14.1	FALSE
18	Yes	No	213.8	159.6	139.2	5	FALSE
19	No	Yes	234.4	265.9	241.4	13.7	FALSE
20	No	Yes	265.1	197.4	244.7	10	FALSE

The number of trees to be built in RFCA is determined by the number of rows of the binary CA, chosen according to the number of attributes of the original training dataset. According to the dataset of the example, a binary covering array is required to cover the six input variables. In the present example, the CA of Fig. 1 is used. According to this CA, the number of trees is six (number of rows of the CA). However, because row 6 of the CA contains only zeros, that is, it does not select any attributes, the total number of trees to be built is only five (5).

Each row of the binary CA defines for its corresponding tree which features (attributes of the original training dataset) will be used in building it, where 0 indicates the absence and 1 the presence of a variable in the subset of candidate features. For example, based on row three [0|0|0|1|1|0] it can be said that the tree is built considering features four and five (A4 and A5), which correspond to Total minutes afternoon and Total minutes night.

The original RF has a hyperparameter K that defines the number of attributes that must be considered when each tree is created. This attribute is defined, according to the Breiman proposal [1], as the first integer less than $\log_2(P) + 1$, where P is the number of input variables of the original training dataset, which in this case is equal to three $(K = \lfloor \log_2(6) + 1 \rfloor = 3)$.

As can be seen for the example of row three of the CA, it only selects two attributes. For this reason, the RFCA algorithm adds one randomly to complete the three required by the K parameter. It can also happen that the row of the CA has more selected features than those defined by the K parameter. In this case K features are selected randomly from the subset defined by the line of the CA.

To build $Tree_1$ based on row 1 [1|1|1|0|0|0] of the CA, the attributes selected in the subset are: A1, A2 and A3. Since the number of attributes selected in the CA row matches the value of K (3), all are selected. After sampling the rows, the data sample of Fig. 2a is obtained. In addition, the base tree that is obtained with that input sample is presented on the right side of this figure.

(a)

Data for Tree 1			
A1	A2	A3	Class
no	no	178.7	FALSE
no	yes	148.5	FALSE
yes	no	197.2	FALSE
no	no	124.9	FALSE
no	no	115.4	FALSE
yes	no	140	TRUE
no	no	321.1	TRUE
no	no	193.4	FALSE
no	no	106.6	FALSE
no	yes	156.2	FALSE
no	no	231.1	FALSE

A1 = no
    A3 < 276.1: FALSE (12/0)
    A3 >= 276.1: TRUE (1/0)
A1 = yes
    A3 < 168.6: TRUE (2/0)
    A3 >= 168.6: FALSE (1/0)

(b)

Data for Tree 2			
A1	A3	A5	Class
no	178.7	131.9	FALSE
no	148.5	178.3	FALSE
yes	197.2	211.1	FALSE
no	115.4	280.9	FALSE
yes	140	120.1	TRUE
no	193.9	210.1	FALSE
no	193.4	243.3	FALSE
no	134.7	221.4	FALSE
no	156.2	279.1	FALSE
no	231.1	191.3	FALSE

A5 < 126: TRUE (1/0)
A5 >= 126: FALSE (15/0)

(c)

Data for Tree 3			
A3	A4	A5	Class
178.7	233.7	131.9	FALSE
148.5	114.5	178.3	FALSE
164.1	219.1	220.3	FALSE
197.2	188.5	211.1	FALSE
321.1	265.5	180.5	TRUE
169.8	197.7	193.7	FALSE
193.4	116.9	243.3	FALSE
106.6	284.8	178.9	FALSE
156.2	215.5	279.1	FALSE
231.1	153.4	191.3	FALSE

A4 < 249.6: FALSE (12/0)
A4 >= 249.6
    A3 < 213.85: FALSE (3/0)
    A3 >= 213.85: TRUE (1/0)

Fig. 2. A sample of the trees generated using RFCA.

For building $Tree_2$ based on row 2 [1|0|1|0|1|1] of the CA. The attributes defined by this row are A1, A3, A5 and A6. The number of attributes (4 in total) of the subset selected by this row of the CA is greater than the number K of attributes to be selected ($K = 3$) Therefore, the algorithm selects 3 attributes of the subset randomly and without repetition.

For the example, attributes A1, A3 and A5 were selected. Figure 2b presents the data sample and the base tree obtained. For building $Tree_4$ and $Tree_5$, the same situation occurs (the number of attributes of the subset selected by the row of the CA is greater than the number K of attributes to be selected). The algorithm therefore operates in the same way.

Building $Tree_3$ is based on row 3 [0|0|0|1|1|0] of the CA, which allows the selection of attributes A4 and A5. In this case, the number of attributes of the subset selected by this row of the CA is less than the number K of attributes to be selected ($K = 3$). Consequently, for those cases the algorithm selects the missing attributes randomly and without repetition (in this case one additional attribute). For the example, A3 is selected. Figure 2c presents the data sample and the base tree obtained.

After creating the trees, the classifier test is performed. In this case the test instances are passed to each of the trees, the class label is assigned based on a majority vote. In this case, the classifier obtains 100% of instances correctly classified.

3.2   RFCA Algorithm

Next, Algorithm 1 summarizes the technique of building an ensemble of decision trees using bagging and covering arrays for feature selection (RFCA). The function *trainDT (T', K)* performs training of a decision tree on a bootstrap sample *T' and K features* selected based on each row of the CA. The process of training a decision tree is presented in Algorithm 2.

4   Experiments and Results

4.1   Configuration of the Experiments

Validation was performed using 32 available datasets in the repository at UCI, namely: Banknote, Blood, Car, Chart, Climate, Contraceptive, Dermatology, Diabetes, Ecoli, Fertility, Glass, Haberman, Hayes, Indian, Ionosphere, Iris, Knowledge, Leaf, Libras, Planning, QSARBiodegradation, Seeds, Segment, Sonar/Connectionist, Soy Bean, Spectf, Vowel, Wdbc, Wine, Wine Red, Yeast, and Zoo. The total number of training instances corresponds to approximately 70% of the total data in each dataset. The algorithm was implemented in Java as a package of Weka 3.8. Binary CAs of strength 2 through 7 were shared by CINESTAV-Tamaulipas of Mexico. All the experiments were performed on an Intel Core i7 4510U, 2.0 GHz, 8 GB RAM, Windows 10. The source code and other resources (such as the CAs) required to replicate the experiments are available online at https://github.com/sebasv22/RFCA.

4.2   Parameters of the Algorithms

To evaluate the RFCA algorithm, a comparison was made with the original Random Forest (RF) algorithm proposed by Breiman [1] and a version of Random Forest called RF_SQRT in which the size K of the randomly selected subsets of features is defined as: $K = \sqrt[2]{P}$, where P corresponds to the number of attributes in the dataset [4, 13]. The values of the parameters common to the three algorithms are those designated by default in Weka (see Table 2). The hyperparameter K (numFeatures in Weka) for RF and RFCA is defined, according to the Breiman proposal [1], as the first integer less than $\log_2(P) + 1$.

Algorithm 1. The proposed Random Forest using Covering Arrays

```
inputs:      T        /* Dataset */
             f        /* Covering array strength */
             lsize    /* Leaf size limit, parameter of Random Forest in weka */
output:      Class label for the input data.
begin
    r = numRows (T)          /* Number of rows of the Dataset */
    p = numAttributes (T)    /* Number of attributes of the dataset */
    ca = loadCA (f, p)       /* Loads the covering array (CA) according to f and p */
    M = numRowsCA (ca)       /* Number of trees is equal to the number of rows of the CA */
    K = ⌊log₂(p) + 1⌋        /* Number of attributes to use in tree construction */
    att = listAttributes (T) /* Set of attributes from the Dataset */
    for i = 1 to M do
        /* Select (2*r)/3 points (dataset instances), with replacement, uniformly in T */
        T' = bootstrap (T)
        subAS = SelAttributesCA (ca[i])     /* from row i determines selected attributes */
        Tree = trainDT (T', K, subAS, att, lsize)
        add Tree to RF
    end for
    Once M Trees are created, Test instance will be passed to each tree and class label will
    be assigned based on majority of votes.
End
```

Algorithm 2. Function trainDT (for Training each Decision Tree to the Random Forest)

```
inputs:      T'       /* Bootstrap sample */
             K        /* Number of random features */
             subAS    /* Attribute indices selected in the CA Row */
             att      /* Set of attributes from the Dataset */
             lsize    /* Leaf size limit */
output:      Tree, a trained decision tree
begin
    if numInstances (T') > lsize then
        subK = ∅
        if size (subAS) >= K then
            subK = RandomSelect (subAS, K)   /*Select uniformly, without replacement, a
            subset subK from K attributes in subAS, subK ⊂subAS */
        else
            subK = RandomSelectAttributes (att, subAS, K) /*subK include all attributes in
            subAS plus additionally attributes from att uniformly selected without replace-
            ment that do not belong to subAS. at the end subK has K attributes */
        end if
        /* Select the best split in T' by optimizing the CART-Split criterion */
        (leftT, rightT) = Split (T', subK)
        Tree.add (trainDT (leftT, K, subAS, att, lsize))    /* left child */
        Tree.add (trainDT (rightT, K, subAS, att, lsize))  /*right child */
    else return Tree
    end if
end function
```

Table 2. Default Weka algorithm parameters

Parameters	Value	Parameters	Value
bagSizePercent	100	maxDepth	0
batchSize	100	numExecutionSlots	1
breakTiesRandomly	False		

In RFCA the parameter M (number of trees) is not defined but the strength parameter is defined, with which parameter T of the binary CA used in building the forest is determined. This parameter takes the values of strength 2 to strength 7. To make the comparison between RFCA, RF and RF_SQRT under the same conditions, the parameter M (number of trees) is defined in the last two algorithms equal to the one defined in RFCA with the number of rows of the CA.

4.3   Results and Analysis

A sample of the results of the experiments are shown in Table 3 and were obtained from the average of 30 executions of each algorithm in each dataset using 10-fold cross-validation. The CAs were evaluated using strengths from 2 to 7, all with the binary alphabet. The evaluation was carried out with different strength values to identify which value achieves the best results in terms of accuracy. Each dataset that was evaluated required a CA defined according to the number of variables in the set (parameter P in the CA). The number of trees increases as strength (T) increases.

A total of 192 scenarios were obtained (32 datasets by 6 CA strength values) for evaluating the proposed algorithm with respect to the state of art. As can be seen in Table 4, the RFCA algorithm performs better in approximately half of the evaluation scenarios. The remaining percentage is divided between the two algorithms of the state of the art.

Taking the results of Friedman's non-parametric statistical test, we obtain the ranking in Table 5, which confirms RFCA performing better than RF and RF_SQRT. The p-value of the test was not less than 0.05, which means that the results are not statistically significant. Nevertheless, the Wilcoxon test shows with 90% confidence that the RFCA algorithm outperforms RF and RF_SQRT, and at 95% confidence, that RFCA outperforms RF_SQRT.

In the results, those corresponding to the Car and Leaf datasets predominate: in Car, RF outperforms RFCA by a wide margin, while in Leaf, RFCA outperforms RF by a considerable advantage. On reviewing the structure of the attributes of the Car dataset, it was found to have 6 ordinal attributes, 4 classes and 1,728 instances, data that are like those of other datasets and as such was not able to be identified since in this case RF greatly outperforms RFCA (see the left-hand side of Fig. 3).

In the case of Leaf, it was found to consist of 14 continuous attributes, 1 nominal, 36 classes and 340 instances. Although not conclusive, the high number of classes in this dataset can benefit from the analysis of the interactions of the columns that are made with the CA. In this dataset, from the lowest strength value (2) to the highest (7), there is a considerable difference of RFCA with respect to RF and RF_SQRT (see the right side of Fig. 3).

Table 3. A sample of the results using 10-fold cross-validation (best results in bold)

Dataset	T	M	RFCA	RF	RF_SQRT	Dataset	T	M	RFCA	RF	RF_SQRT
Banknote	2	4	98.5593	**98.9359**	98.8654	Blood	2	4	**74.8396**	74.1355	73.9617
	3	7	99.1764	99.1618	**99.1861**		3	7	**73.6453**	73.066	72.8654
	4	15	99.2347	**99.2541**	99.2444		4	15	**74.3449**	73.3378	73.1907
	5	30	**99.2833**	99.2687	99.2736		5	30	**74.1266**	73.4804	73.369
	6	60	**99.3051**	**99.3051**	99.2687		6	60	**73.9171**	73.057	72.9724
	7	120	**99.3659**	99.3367	99.3076		7	120	**73.9394**	73.0348	73.0704
Car	2	5	86.8538	**91.4892**	89.2091	Chart	2	9	**96.7556**	96.3667	96.6556
	3	11	81.603	**93.4143**	92.338		3	31	98.2222	98.2111	**98.2778**
	4	20	88.2234	**94.1223**	93.5899		4	67	**98.6167**	98.5556	98.5667
	5	31	90.4649	**94.3769**	94.2149		5	135	**98.7722**	98.65	98.7111
	6	63	88.2195	94.6509	**94.8669**		6	675	**98.9556**	98.9222	98.8889
	7	126	88.2851	94.7647	**95.0675**		7	2453	**99**	98.9389	98.9444
Climate	2	7	92.0247	**92.8025**	92.4074	Contraceptive	2	5	**49.6108**	48.959	48.6309
	3	17	91.7223	**93.0741**	92.5864		3	11	**51.2333**	49.9208	49.7873
	4	39	91.5062	**93.1975**	92.5864		4	23	**52.2788**	50.74	50.5295
	5	99	91.4938	**93.2222**	92.5988		5	53	**52.9894**	51.222	51.0138
	6	300	91.4815	**93.0864**	92.6049		6	107	**53.0572**	51.5886	51.437
	7	630	91.4815	**93.0555**	92.6543		7	169	**52.9758**	51.9167	51.6452
Dermatology	2	7	**94.5902**	94.1712	94.2896	Diabetes	2	5	71.9488	**73.138**	72.7083
	3	23	**96.3024**	95.7377	95.8561		3	11	73.2031	**74.5747**	74.0148
	4	62	**96.5118**	96.2113	96.1566		4	23	74.349	75.0304	**75.0564**
	5	133	**96.6484**	96.2751	96.3206		5	52	75.3212	**75.4601**	75.3255
	6	482	**96.7395**	96.3935	96.439		6	84	**75.6727**	75.4644	75.5599
	7	1178	**96.7122**	96.4299	96.4845		7	128	75.6858	75.5642	**75.7726**
Ecoli	2	5	80.9821	**82.8572**	**82.8572**	Fertility	2	5	**86.6333**	86.3667	85.7333
	3	11	83.4127	**84.9206**	84.6329		3	11	**87.6**	87.5	87.4333
	4	23	84.5536	**85.7441**	85.6052		4	23	**87.8333**	87.4	87.5
	5	42	85.5952	85.9921	**86.4286**		5	53	**88.0333**	87.8667	87.9
	6	64	85.248	86.1508	**86.6171**		6	107	**88.2333**	87.4667	87.9333
	7	127	86.3194	86.4385	**86.8353**		7	169	**88.5**	87.1667	87.9
Glass	2	5	71.9003	**73.5358**	73.3022	Haberman	2	3	**70.9477**	66.9499	68.2135
	3	11	75.4829	**76.3396**	75.7788		3	7	**70.3268**	67.048	69.0196
	4	23	**78.1153**	77.4766	77.6947		4	14	**71.1764**	68.3878	69.9782
	5	53	**79.6262**	78.972	79.5639		5	28	**71.7538**	68.5839	70.3703
	6	107	**79.8287**	79.6885	79.6885		6	56	**71.8191**	69.2265	71.3072
	7	169	79.8754	80.1246	**80.2025**		7	112	**72.2004**	69.0087	71.2636
Hayes	2	4	**77.7083**	76.1458	76.1875	Indian	2	5	**69.9257**	69.0566	69.4225
	3	7	77.6667	77.6458	**77.8958**		3	11	**70.9605**	69.5598	70.1658
	4	15	79.1042	79.0833	**79.3125**		4	23	**71.6295**	70.1487	70.6632
	5	30	79.6875	79.0208	**79.8958**		5	55	**71.9897**	70.8005	71.3093
	6	60	79.5	78.6042	**79.875**		6	115	**71.9554**	70.7604	71.2236
	7	120	79.3125	78.8125	**79.9167**		7	219	**72.1555**	70.8005	71.3665
Knowledge	2	5	87.3643	**92.8295**	90.8527	Leaf	2	7	**54.598**	18.7451	28.2745
	3	10	93.5659	**93.7855**	92.429		3	16	**55.4902**	19.8529	32.6275
	4	15	93.708	**94.4315**	93.2688		4	34	**55.9216**	20.5196	34.9412
	5	31	94.031	**94.664**	93.9664		5	79	**57.6176**	21.3529	36.7941
	6	62	94.4702	**94.7674**	94.186		6	128	**61.3137**	21.7843	37.6373
	7	124	94.5736	**94.8966**	94.3669		7	255	**61.3824**	21.6961	37.9902

Table 4. Results summary

	Exceeds			Equals		Total
	RFCA	RF	RF_SQRT	RF y RF_SQRT	RFCA y RF	
# Scenarios	96 (50%)	53 (27.60%)	29 (15.1%)	13 (6.77%)	1 (0.52%)	192

In Fig. 4, the average graph across all strengths is shown. In the graph, the disturbance caused by the atypical results of the Car and Leaf datasets can be seen. Likewise, it is evident that the gradual increase in strength of the CA generates, on average, a better result in terms of accuracy.

Table 5. Ranking of algorithms based on the Friedman test

Algorithm	Ranking
RFCA	**1.8828 (1)**
RF_SQRT	2.0078 (2)
RF	2.1094 (3)
p-value	0.08436 (chi-square with 2 degrees of freedom: 4.9453)

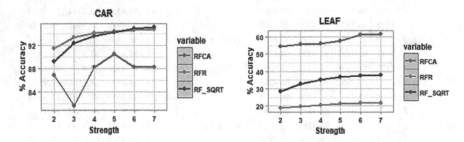

Fig. 3. Accuracy on car and leaf datasets

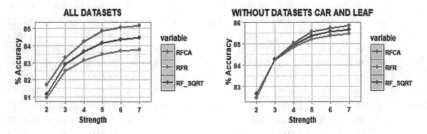

Fig. 4. Average accuracy over all the strengths

5 Conclusions and Future Work

In this work, a new method was proposed and evaluated for the selection of features in the Random Forest algorithm, based on binary covering arrays of strength 2 to strength 7. In the CAs, each row represents a subset of selected variables to build the bootstrap samples and the number of rows of the CA allows adjusting the hyperparameter number of trees, suppressing the random feature selection of the Random Forest originally proposed by Breiman and with it, the need to adjust this hyperparameter. The experiments were performed on 32 datasets evaluated by 10-fold cross-validation and the results obtained are promising, in which an average improvement in accuracy between 0.5% and 2% is achieved. The new RFCA classifier improves accuracy in half of the evaluation scenarios with respect to the state-of-the-art algorithms used. On average the greatest accuracy in the RFCA algorithm is obtained at strength 7. However, considering that a greater strength represents a greater number of trees, strength 5 is considered a suitable value for parameter t of RFCA.

As future work it is expected to study the performance of RFCA on variations of Random Forest in which the bootstrap sample is different in size from two thirds of the training dataset [1], or in which the sampling of the instances is carried out without replacement [14]. Tests will meanwhile be run on other test datasets of the UCI repository or similar, using CAs with different values of strength (t = 2, ..., 10), comparing them against other more recent algorithms. These tests will include additional statistical tests, not only those of Friedman and Wilcoxon. Further tests will be done using values for the hyperparameter K that are different from those used in the present work.

References

1. Breiman, L.: Random forests. Mach. Learn. **45**, 5–32 (2001)
2. Ziegler, A., König, I.R.: Mining data with random forests: current options for real-world applications. Wiley Interdiscip. Rev. Data Min. Knowl. Discov. **4**, 55–63 (2014)
3. Wawre, S.V., Deshmukh, S.N.: Sentimental analysis of movie review using machine learning algorithm with tuned hypeparameter. Int. J. Innov. Res. Comput. Commun. Eng. (ISO) **4**, 12395–12402 (2016)
4. Bernard, S., Heutte, L., Adam, S.: Influence of hyperparameters on random forest accuracy. In: Benediktsson, J.A., Kittler, J., Roli, F. (eds.) MCS 2009. LNCS, vol. 5519, pp. 171–180. Springer, Heidelberg (2009). https://doi.org/10.1007/978-3-642-02326-2_18
5. Timaná-Peña, J.A., Cobos-Lozada, C.A., Torres-Jimenez, J.: Metaheuristic algorithms for building covering arrays: a review. Rev. Fac. Ing. **25**, 31–45 (2016)
6. Verikas, A., Gelzinis, A., Bacauskiene, M.: Mining data with random forests: a survey and results of new tests. Pattern Recogn. **44**, 330–349 (2011)
7. Bernard, S., Heutte, L., Adam, S.: Forest-RK: a new random forest induction method. In: Huang, D.-S., Wunsch, D.C., Levine, D.S., Jo, K.-H. (eds.) ICIC 2008. LNCS (LNAI), vol. 5227, pp. 430–437. Springer, Heidelberg (2008). https://doi.org/10.1007/978-3-540-85984-0_52

8. Menze, B.H., Kelm, B.M., Splitthoff, D.N., Koethe, U., Hamprecht, F.A.: On oblique random forests. In: Gunopulos, D., Hofmann, T., Malerba, D., Vazirgiannis, M. (eds.) ECML PKDD 2011. LNCS (LNAI), vol. 6912, pp. 453–469. Springer, Heidelberg (2011). https://doi.org/10.1007/978-3-642-23783-6_29
9. Deng, H., Runger, G.: Feature selection via regularized trees. In: Proceedings of the International Joint Conference on Neural Networks, pp. 1–8. IEEE (2012)
10. Adnan, M.N.: On dynamic selection of subspace for random forest. In: Luo, X., Yu, J.X., Li, Z. (eds.) ADMA 2014. LNCS (LNAI), vol. 8933, pp. 370–379. Springer, Cham (2014). https://doi.org/10.1007/978-3-319-14717-8_29
11. Zhou, Q., Zhou, H., Li, T.: Cost-sensitive feature selection using random forest: selecting low-cost subsets of informative features. Knowl.-Based Syst. **95**, 1–11 (2016)
12. Ma, L., Fan, S., Haywood, A., Ming-tian, Z., Rigol-Sanchez, J.: CURE-SMOTE algorithm and hybrid algorithm for feature selection and parameter optimization based on random forests. BMC Bioinform. **18**, 169 (2017)
13. Geurts, P., Ernst, D., Wehenkel, L.: Extremely randomized trees. Mach. Learn. **63**, 3–42 (2006)
14. Scornet, E., Biau, G., Vert, J.P.: Consistency of random forests. Ann. Stat. **43**, 1716–1741 (2015)

A New Distributed and Decentralized Stochastic Optimization Algorithm with Applications in Big Data Analytics

Reza Shahbazian[1(✉)], Lucio Grandinetti[2], and Francesca Guerriero[3]

[1] Department of Mathematics and Computer Science,
University of Calabria (UniCal), 87036 Rende, CS, Italy
Reza.Shahbazian@unical.it
[2] Department of Electronics, Informatics, and Systems,
University of Calabria (UniCal), 87036 Rende, CS, Italy
[3] Department of Mechanical, Energy and Management Engineering,
University of Calabria (UniCal), 87036 Rende, CS, Italy

Abstract. The world is witnessing an unprecedented growth of needs in data analytics. Big Data is distinguished by its three main characteristics: velocity, variety and volume. An open issue and challenge faced by the data community is how to scale up analytic algorithms. To address this issue, optimization of large scale data sets has attracted many researchers in recent years. In this paper, we first present the most recent advances in optimization of Big Data analytics. Further, we introduce a fully distributed stochastic optimization algorithm for decision making over large scale data sets. We also propose the optimal weight design for the proposed algorithm and study its performance by considering a practical application in cognitive networks. Experimental results confirm that the proposed method performs well, proven to be distributed, scalable and robust to missing data and communication failures.

Keywords: Distributed · Stochastic · Optimization · Big Data · Decision making

1 Introduction

Enormous amount of data have been continually generated at unprecedented and ever increasing scales. Large-scale data sets are collected and studied in numerous domains, from engineering sciences to social networks, commerce, biomolecular research, and security. Nowadays, the term "Big Data" referring to its modern definition, i.e. information explosion and large sets of data, has truly influenced our lives, at-least by introducing new insights. In 1999, for the first time, the term Big Data appeared in an article, published by the Association for Computing Machinery [1]. The authors of this paper quoted that 'the purpose of computing is insight, not numbers'. Since then, Big Data is becoming more and more popular. In 2001, Doug Laney, analyst at Gartner, defined three terms of volume, velocity

© Springer Nature Switzerland AG 2019
G. Nicosia et al. (Eds.): LOD 2018, LNCS 11331, pp. 77–91, 2019.
https://doi.org/10.1007/978-3-030-13709-0_7

and variety that are commonly accepted as three main characteristics of Big Data [2].

One important subject in this era is Big Data optimization. To understand the value of Big Data optimization, we introduce the distinctive characteristics of Big Data, namely volume, variety and velocity. Interconnection between these characteristics known as the 3 Vs are shown in Fig. 1. The term volume represents the growing amount of data in Exabytes and Zettabytes. The Variety of data is produced by sources such as physical sensors, smart devices and social media in semi-structured, structured or unstructured formats. The velocity describes how quickly the data is retrieved, stored and processed.

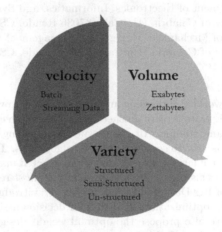

Fig. 1. The interconnection of three main characteristics of Big Data

Today, the Big Data is a common topic and many researchers, in various fields of study, including convex optimization and machine learning, have contributed to the literature. The basic ingredient for every smart and intelligent system, is data. Smarter systems acquire more data to make efficient decision that leads to large scale data sets. This data could be generated from many sensors in smart phones [3], physical sensors attached to cyber-physical systems [4], many objects in Internet of Things (IoT) [5] platforms and smart cities [6]. This data may further be transferred to a center using new technologies such as 5G [7]. Therefore, data gathering is the first challenge of Big Data Era. Other challenges may include data storage and data processing [8]. Many researches are focused on adaptation of existing technologies or inventing new ones to store Big Data. Some examples may be found in [9] for cloud storage of Big Data. However, many researchers believe that the main challenge is still finding efficient and optimal solutions to process the data in appropriate time by considering the Big Data challenges.

In general, data processing can be performed in a centralized or distributed manner. Analysis on very large data sets and Big Data seems infeasible by using central processing and storage units. Considering the streaming data sources, learning must often be performed in real-time or near real-time [10]. Although centralized processing methods usually provide the optimal decision, considering the challenges faced by data storage in the cloud or any distributed file system [12], decentralized methods are still preferred [11]. Therefore, there is an urgent need to scalable methods, capable of efficient data processing, considering the storage, query, and communication challenges. In some cases, privacy and security concerns are critical and prevent accessing the full data. In these cases only partial data or processed output (decision) might be transferred through communication interfaces.

As depicted in Fig. 2, the characteristics of Big Data require an optimization algorithm that is scalable, compatible with missing values of data (robust), performs near real-time and is applicable in distributed platforms such as cloud. These challenges are not properly answered by traditional optimization methods and the final purpose of any modified or new optimization algorithm in Big Data era is to reduce the computational, storage, and communications bottlenecks.

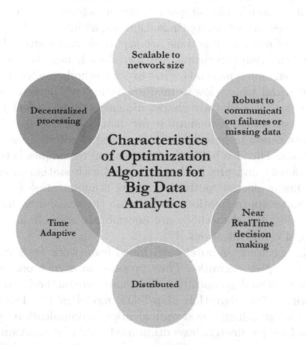

Fig. 2. Different characteristics of optimization algorithms for Big Data analytics

In this paper we investigate optimization techniques for Big Data analytics. We introduce a fully distributed stochastic optimization algorithm for decision making over large scale data sets. We describe the proposed model mathematically. Our method is scalable to any network or data size, works based on cooperation of neighbor processing/storage units and it is adaptive to any dynamic behavior of processing/storage units. We further propose an optimal weighting of cooperation coefficients.

The rest of this paper is organized as follows: In Sect. 2, we introduce the related work and literature review of Big Data optimization. The system model is introduced in Sect. 3 and the proposed method is presented in Sect. 4. The evaluation scenarios and simulation results are presented in Sect. 5 and finally the Sect. 6 concludes the paper.

2   Related Work

Optimization plays a centric role in Big Data analytics. Optimization for Big Data has recently attracted significant attention not only from its own community, but also from the other scientific and engineering communities such machine learning, statistics, and signal processing. In this section, we present the most recent advances in optimization techniques for Big Data analytics.

The convex optimization techniques have attracted many researchers in the last decade due to the rise of new theory for rank minimization, and successful statistical learning models like support vector machines [13]. In [14] authors review the advances of convex optimization algorithms for Big Data. They assume that the defined cost functions are separable and convex. In [14], three methods for optimization algorithms are introduced, namely first order, randomization and distributed methods. First order methods use techniques such as gradient estimates and achieve low or medium accuracy. The first order methods provide convergence rates that are almost dimension independent and theoretically robust. These methods are suitable for distributed and parallel computation [15]. Randomization techniques are introduced to enhance the scalability of first order methods. The idea behind randomization techniques is to replace the deterministic gradient and proximal calculations with statistical estimators to speed up basic linear algebra routines by using randomization [16]. First order methods, with some approximations to increase the scalability, form the third category. These distributed methods are enormously scalable algorithms often with decentralized communications [17].

In [18], authors consider a novel partitioned framework for distributed optimization in peer to peer networks. They propose an asynchronous distributed algorithm, based on dual decomposition and coordinate methods. In [19], authors propose an optimization algorithm of p-DOT model in Big Data computing framework by analyzing high speed optical fiber communication system. They consider the machine parameters, execution mode and cost function as reference variables and aim to improve the efficiency of high speed optical fiber communication system.

In [20,21] authors investigate non-convex and hybrid (convex and non-convex) optimization problems for Big Data. They propose a decomposition framework for the parallel optimization of the sum of a differentiable function and a block separable convex one. In their proposed framework, the (block) variables are updated and chosen according to a mixed random and deterministic procedure. They also present the almost sure convergence of the proposed scheme. Authors in [22] present an algorithmic framework for Big Data optimization, called the block Successive Upper Bound Minimization (BSUM). Their proposed BSUM includes methods such as the Block Coordinate Descent (BCD), the Convex Concave Procedure (CCCP), the Block Coordinate Proximal gradient (BCPG), the Nonnegative Matrix Factorization (NMF) and the Expectation Maximization (EM).

In [23], authors consider the wireless big sensory data networks and propose an accelerated distributed rate control method to minimize the recovery error of big sensory data. Their proposed method is claimed to guarantee the error minimization of reconstructed data and converge to the optimal value with a lower latency. Distributed optimization algorithms based on Alternating Direction Method of Multipliers (ADMM), to solve Big Data optimization problem in smart grid communication networks are presented in [24]. They introduce the canonical formulation of optimization problem and the general form of ADMM. Authors in [25] study Evolutionary Algorithms (EAs) to solve Big Data optimization problems that involve a very large number of variables and need to be analyzed in a short period of time. They consider the issues of EA algorithms such as scalability and propose a heterogeneous framework that integrates a cooperative co-evolution method. Their proposed framework splits the big problem into subproblems in order to increase the efficiency of the solving process. A review of the recent advances in the secure outsourcing of large scale computations for a Big Data analysis is given in [26]. Authors in [26] focuses on linear algebra and its application in Big Data optimization problems. The authors also investigate both iterative and convex solutions for Big Data optimization problems.

In a very simple explanations, Big Data optimization methods try to partition the data so that it is feasible to process, mostly in a centralized manner. Although many research articles are published in this era, there is still a big gap between practice and theory, specially considering the needs for scalability, robustness and characteristics of Big Data. It seems that distributed optimization algorithms are the promising solution so fill this gap, although there still is a long road to go. In next section we introduce the system model of proposed method.

3   System Model

From data point of view, there are two different approaches namely, centralized and distributed. In centralized techniques, the data is transferred to a center for further processing/storage, whereas in the distributed manner, the data is exchanged and processed within the network locally. Transmitting the data to a center may cause network congestion and waste of communication and power resources. It is obvious that any malfunction in the center causes network breakdown. In addition, center requires high computation power to process the large volume of collected data. In comparison, in a distributed approach, the network computational load is divided between processing/storage units using cooperation and no centralized infrastructure is required. In this paper we consider fully decentralized and distributed techniques.

Notation. The following notations are used throughout this paper. Matrices are represented by upper case and vectors by lower case letters. Boldface fonts are reserved for random variables and normal fonts are used for deterministic quantities. Superscript $(.)^T$ denotes transposition for real-valued vectors and matrices while $(.)^*$ denotes conjugate transposition for complex valued vectors and matrices. The symbol $\mathbf{E}[.]$ is the expectation operator, $Tr(.)$ represents the trace of its matrix argument. I_M represents the identity matrix of order M.

We consider a network consisting of N processing/storage units, also called 'node' from now-on. The nodes are assumed to be distributed, each capable of processing and storing limited size of data (at-least during the processing) and may or may not be involved in initial data generation. We assume that neighbor units are able to communicated to each other by using direct connection interfaces. Node l is said to be a neighbor of node k if they can communicate and cooperate with each other. We denote the set of all neighbors of node k by $\mathcal{N}_k$.

In this model we assume that the nodes are generating or receiving continuous data with Big Data characteristics. It is impossible to transfer and process the data in a centralized manner because of the challenges faced by communication, security, time and storage. The objective of the nodes in the network is to make a decision in a fully distributed manner. In other words, the solution is an estimate of an unknown parameter vector ω^o in a distributed manner through stochastic optimization. At every time instant (iteration), i, each node k observes a scalar random process $\mathbf{d}_k(i)$ and a vector random process $\mathbf{u}_{k,i}$ which are related to ω^o via the linear regression model presented as follows [27]:

$$\mathbf{d}_k(i) = \mathbf{u}_{k,i}\omega^0 + \mathbf{v}_k(i) \tag{1}$$

In writing Eq. (1), it is assumed that:

- The regression data $\{\mathbf{u}_{k,i}\}$ is zero mean, independent and identically distributed (i.i.d.) in time and independent over space with covariance matrices $R_{u,k} = \mathbf{E}\left[\mathbf{u}_{k,i}^*\mathbf{u}_{k,i}\right] > 0$.
- The noise $\mathbf{v}_k(i)$ is zero mean, i.i.d. in time and independent over space with variances $\sigma_{v,k}^2$.
- The $\mathbf{u}_{k,i}$ and the noise $\mathbf{v}_k(i)$ are mutually independent.

The network will try to estimate ω^o by searching for the minimized global cost function as presented in Eq. (2).

$$J^{glob}(\omega) = \sum_{k=1}^{N} \mathbf{E}\left|\mathbf{d}_k(i) - \mathbf{u}_{k,i}\omega\right|^2 \tag{2}$$

The most important issue to solve an optimization problem in a distributed manner is to be able to separate the cost function among processing units. Each processing/storage unit should be able to act on its own, while cooperating with neighbor nodes. Moreover, we assume that the cost function is separable among all processing units as presented in Eq. (3).

$$J^{glob}(\omega) = \sum_{k=1}^{N} J_k(\omega) \tag{3}$$

$J_k(\omega)$ is the cost function of k processing/storage units defined as Eq. (4).

$$J_k(\omega) = \mathbf{E}|d_k(i) - \mathbf{u}_{k,i}\omega|^2 \tag{4}$$

The cost function $J_k(\omega)$ can further be written in another form as presented in Eq. (5).

$$J_k(\omega) = \|\omega - \omega^o\|^2_{R_{u,k}} + mmse_k \tag{5}$$

where $\|x\|^2_\Sigma$ denotes the weighted square quantity as $x^*\Sigma x$ for any semi-definite matrix $\Sigma \geq 0$, $R_{u,k} = \mathbf{E}\left[\mathbf{u}^*_{k,i}\mathbf{u}_{k,i}\right] > 0$ and $mmse_k$ is an additional MMSE term that is independent of ω. Therefore, we may conclude Eq. (6).

$$J^{global}(\omega) = J_k(\omega) + \sum_{l \neq k} \left(\|\omega - \omega^o\|^2_{R_{u,l}} + mmse_l \right) \tag{6}$$

It is obvious that the optimum value, ω^o that appears in the quadratic parts is not known. It should also mentioned that the weighting matrices $R_{u,l}$ are not available in general and only those from the neighbors can be assumed to be available. Therefore, we may conclude Eq. (7).

$$J^{dist}_k(\omega) = J_k(\omega) + \sum_{l \in \mathcal{N}_k \setminus \{k\}} \left(\|\omega - \omega^o\|^2_{R_{u,l}} \right) \tag{7}$$

Please note that the term $mmse_l$ is ignored since is independent of ω and have no effects in finding the optimal value, ω^o. The covariance matrices $R_{u,l}$ is not available in practice. Usually, processing/storage units can only observe realizations $\mathbf{u}_{l,i}$ of data arising from distributions whose covariance matrix is unknown $R_{u,l}$. One way to address this issue is to replace each of the weighted norms by a scaled multiple of the form as presented in Eq. (8).

$$\|\omega - \omega^o\|^2_{R_{u,l}} \approx b_{l,k} \|\omega - \omega^o\|^2 \tag{8}$$

where $b_{l,k}$ is a non-negative coefficient. Considering Eq. (8), each node k approximates the moment $R_{u,l}$ from its neighbors by multiples of the identity matrix. This Approximation is reasonable because using the Rayleigh-Ritz characterization of eigenvalues, it holds that:

$$\lambda_{\min}(R_{u,l}) \|\omega - \omega^o\|^2 \leq \|\omega - \omega^o\|^2_{R_{u,l}} \leq \lambda_{\max}(R_{u,l}) \|\omega - \omega^o\|^2 \tag{9}$$

Therefore, we may conclude that:

$$J^{dist}_k(\omega) \approx J_k(\omega) + \sum_{l \in \mathcal{N}_k \setminus \{k\}} b_{l,k} \|\omega - \omega^o\|^2 \tag{10}$$

This recent cost function at node k relies only on available information from neighbor nodes. Now, each node k can apply a steepest-descent iteration to minimize the cost function as presented in Eq. (11).

$$\omega_{k,i} = \omega_{k,i-1} - \mu_k \left[\nabla_\omega J_k^{dist}(\omega) \right]^*$$
$$\omega_{k,i} = \omega_{k,i-1} + \mu_k \left(r_{du,k} - R_{u,k}\omega_{k,i-1} \right) - \mu_k \sum_{l \in N_k \backslash \{k\}} b_{l,k} \left(\omega_{k,i-1} - \omega^o \right) \quad (11)$$

where ∇_ω denotes the gradient vector. The step size parameters μ_k can be constant or variant. Constant step size allows the algorithms to work continuously, while variant step sizes that decay to zero, causes the algorithms to stop after a while. An adaptive implementation of can be obtained by replacing covariance matrices by instantaneous approximations as presented in Eq. (12).

$$r_{du,k} \approx \mathbf{d}_k(i) \mathbf{u}_{k,i}^*$$
$$R_{u,k} \approx \mathbf{u}_{k,i}^* \mathbf{u}_{k,i} \quad (12)$$

Finally, by some substitution of equations, we may conclude that:

$$\omega_{k,i} = \omega_{k,i-1} + \mu_k \mathbf{u}_{k,i}^* \left(\mathbf{d}_k(i) - \mathbf{u}_{k,i}\omega_{k,i-1} \right) - \mu_k \sum_{l \in N_k \backslash \{k\}} b_{l,k} \left(\omega_{k,i-1} - \omega^o \right) \quad (13)$$

The last correction term still depends on the unknown ω^o. Choosing different approximations for ω^o leads to different strategies such as consensus.

4   Proposed Method

In this section we first present the proposed method by its mathematical model. We further propose an optimal weighting and finally discuss on computational complexity of the presented method.

4.1   Mathematical Model

In the proposed method, we apply diffusion adaptation [28] and by defining an intermediate variable, ψ, we have:

$$\psi_{k,i} = \omega_{k,i-1} + \mu_k \mathbf{u}_{k,i}^* \left(\mathbf{d}_k(i) - \mathbf{u}_{k,i}\omega_{k,i-1} \right)$$
$$\omega_{k,i} = \psi_{k,i} - \mu_k \sum_{l \in \mathcal{N}_k \backslash \{k\}} b_{l,k} \left(\omega_{k,i-1} - \omega^o \right) \quad (14)$$

The unknown term ω^o is still shown in the equation. Considering $\psi_{l,i}$ as a substitute for ω^o we have:

$$\psi_{k,i} = \omega_{k,i-1} + \mu_k \mathbf{u}_{k,i}^* \left(\mathbf{d}_k(i) - \mathbf{u}_{k,i}\omega_{k,i-1} \right)$$
$$\omega_{k,i} = \psi_{k,i} - \mu_k \sum_{l \in \mathcal{N}_k \backslash \{k\}} b_{l,k} \left(\psi_{k,i} - \psi_{l,i} \right) \quad (15)$$

It should be noted that in previous methods, $\omega_{l,i-1}$ is usually substituted as ω^o. Defining $a_{l,k}$ as a weighting coefficient as presented in Eq. (16) and μ_k as Eq. (17), we may conclude Eq. (18).

$$a_{l,k} = \begin{cases} 1 - \sum\limits_{j \in \mathcal{N}_k \setminus \{k\}} \mu_k b_{j,k}, l = k \\ \mu_k b_{l,k}, l \in \mathcal{N}_k \setminus \{k\} \\ 0, otherwise \end{cases} \tag{16}$$

$$\sum_{i=1}^{\infty} \mu_k(i) = \infty, \sum_{i=1}^{\infty} \mu_k^2(i) < \infty \tag{17}$$

$$\begin{aligned} \psi_{k,i} &= \omega_{k,i-1} + \mu_k \mathbf{u}_{k,i}^*(\mathbf{d}_k(i) - \mathbf{u}_{k,i}\omega_{k,i-1}) \\ \omega_{k,i} &= \sum_{l \in \mathcal{N}_k} a_{l,k} \psi_{l,i-1} \end{aligned} \tag{18}$$

The Eq. (18) could also be written in another form as presented in Eq. (20). Where $c_{l,k}$ are the entries of the right-stochastic matrix $\mathbf{C}$, satisfying Eq. (19).

$$c_{l,k} \geq 0, \quad \mathbf{C}\mathbf{1}_N = \mathbf{1}_N, \quad c_{l,k} = 0 \quad if \quad l \in \mathcal{N}_k \tag{19}$$

$$\begin{aligned} \psi_{k,i} &= \omega_{k,i-1} + \mu_k \sum_{l \in \mathcal{N}_k} c_{l,k} \mathbf{u}_{l,i}^*(\mathbf{d}_l(i) - \mathbf{u}_{l,i}\omega_{l,i-1}) \\ \omega_{k,i} &= \sum_{l \in \mathcal{N}_k} a_{l,k} \psi_{l,i-1} \end{aligned} \tag{20}$$

4.2 Optimal Weighting

In this section, we assume that the network topology may vary in each iteration, i. It means that the neighborhood of each node changes over time and the static combination rules are not applicable for these dynamic networks. In these networks, the entries of the variant left stochastic matrix $\mathbf{A}_i$ can be expressed as follow:

$$a_{l,k}(i) = \gamma_{l,k} \mathcal{I}_{l,k}(i) \tag{21}$$

where $\gamma_{l,k}$ is positive fixed combination weights that node k assigns to neighbors $l \in \mathcal{N}_{k,i}$, and $\mathcal{I}_{l,k}(i)$ is defined in Eq. (22).

$$\mathcal{I}_{l,k}(i) = \begin{cases} 1, & if \quad l \in \mathcal{N}_{k,i} \\ 0, & otherwise \end{cases} \tag{22}$$

As can be seen in Eq. (22), $\mathcal{I}_{l,k}(i)$ is a random variable with Bernoulli distribution. Considering Eq. (21) the weighting function could be written as:

$$a_{l,k}(i) = \begin{cases} \gamma_{l,k} \mathcal{I}_{l,k}(i), & if \quad l \in \mathcal{N}_{k,i} \setminus \{k\} \\ 1 - \sum\limits_{l \in \mathcal{N}_{k,i} \setminus \{k\}} a_{l,k}(i), & if \quad l = k \\ 0, & otherwise \end{cases} \tag{23}$$

The stability of proposed method in the mean does not depend on the particular choice of combination matrix. Theoretically, other choices are possible as long as $\mathbf{E}\left[A_i\right]$ yields a left stochastic matrix. Intuitively, $\mathbf{E}\left[A_i\right]$ will be a left stochastic matrix if the instantaneous combination matrix A_i satisfies $\sum_{l \in \mathcal{N}_{k,i}} a_{l,k}\left(i\right) = 1$ for all i. Mathematically the claim can be stated as follows. If

$$\forall i \in \{0,1,\ldots\}, \forall k \in \{1,\ldots,N\}, \sum_{l \in \mathcal{N}_{k,i}} a_{l,k}\left(i\right) = 1 \text{ then } \mathbf{E}\left[\sum_{l \in \mathcal{N}_{k,i}} a_{l,k}\left(i\right)\right] = 1.$$

This can be proven as presented in Eq. (24).

$$\mathbf{E}\left[A_i\right] = \mathbf{E}\left[\sum_{l \in \mathcal{N}_{k,i}} a_{l,k}\left(i\right)\mathcal{I}_{l,k}\left(i\right)\right]$$

$$\mathbf{E}\left[A_i\right] = \lim_{i \to \infty} \frac{1}{i} \sum_{j=1}^{i} \sum_{l \in \mathcal{N}_{k,i}} a_{l,k}\left(j\right)\mathcal{I}_{l,k}\left(j\right) \qquad (24)$$

$$\mathbf{E}\left[A_i\right] = \lim_{i \to \infty} \frac{1}{i}\left[\sum_{l \in \mathcal{N}_{k,i}} a_{l,k}\left(1\right)\mathcal{I}_{l,k}\left(1\right) + \cdots + \sum_{l \in \mathcal{N}_{k,i}} a_{l,k}\left(i\right)\mathcal{I}_{l,k}\left(i\right)\right] = 1$$

Therefore, the optimum weight based on Metropolis combination rule in large scale and dynamic networks with variant topology can be presented as Eq. (25).

$$a_{l,k}\left(i\right) = \begin{cases} \frac{1}{\max\{|\mathcal{N}_{k,i}|,|\mathcal{N}_{l,i}|\}} \; if\, l \in \mathcal{N}_{k,i} \setminus \{k\} \\ 1 - \sum_{l \in \mathcal{N}_{k,i} \setminus \{k\}} a_{l,k}\left(i\right), \; if\, l = k \\ 0, \, otherwise \end{cases} \qquad (25)$$

4.3   Computational Complexity

Considering n as the average number of neighbor nodes in each iteration, and assuming that total I iterations is needed for the convergence and having N processing/storage units in the network, the computational complexity is presented in Table 1. Comparing with convex optimization and considering semidefinite programming (SDP) or Second Order Cone Programming (SOCP) the complexity is non-linearly related to number of data generator/processor/storage units, at-least by $o(N^2)$.

Table 1. Computational complexity of proposed distributed method

Number of additions	Number of multiplications
(n+2)NI	2NI

5   Evaluation Results

In this section we evaluate the proposed algorithm using a practical example. In this application, we define an optimization problem and evaluate the performance in terms of accuracy and robustness.

5.1   Cognitive Networks

One practical example of Big Data may be found in wireless sensor networks where the sensors generate huge amount of data in a non-stop manner [29]. In wireless networks, it is shown that only a partial spectrum is used by the users. So, the cognitive systems are proposed as a solution and to improve the spectrum usage efficiency. Such systems include two types of users namely, primary and secondary. The primary users are the owner of spectrum and secondary users should continuously scene the spectrum (called spectrum sensing). When they find the an unused band of spectrum for a period of time, use it based on the network predefined policy. Now, assume that the sensors should sense the data and send their observations to a center. Besides the security, power consumption and data processing issues, the latency introduced in transferring data is not acceptable. This might be a simple application of proposed method to make a decision continuously with high accuracy and reliable against any communication failure [29]. A simple diagram of distributed spectrum sensing procedure is presented in Fig. 3.

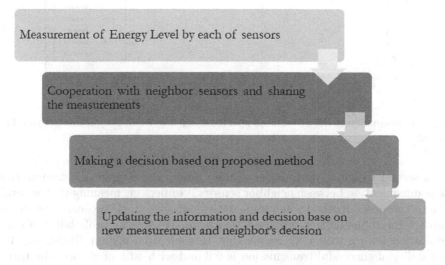

Fig. 3. Instruction of performed tasks in distributed spectrum sensing

As presented in Fig. 3, first each sensor needs to measure the energy level and cooperate with other neighbors to make a distributed decision with local information. We consider a sample network consisting of 15 sensors (secondary user). We consider two scenarios; in the first scenario, we assume that the communication link between sensors (neighbors) is ideal. In this case, each sensor sends the data to its neighbors and makes a decision accordingly. The simulation results are presented in Fig. 4. As illustrated in Fig. 4, the first decision of each sensor is different. It is because each sensor have only access to its local information. Obviously, the first decision of most sensors is wrong. The challenges forced by communication, security, processing power and more important time of decision, makes it impossible to gather the information in a center and process them simultaneously. We assume that information are transferred only to neighbors and nodes perform the proposed optimization algorithm. After a few iterations, the whole network (each node) can reach a correct decision while each has only processed local information. This simulation shows how the continuously generated large amount of data that is impossible to transfer and process in any of nodes is processed in a fully distributed manner.

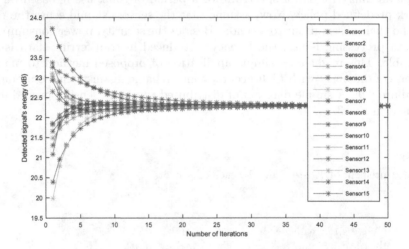

Fig. 4. Performance of the distributed spectrum sensing when the communication link is ideal

In second scenario, we consider a more practical example. We assume that the communication between neighbor sensors is imperfect, meaning that we evaluate the proposed method when some data is missing. In this scenario, we try to evaluate the robustness of proposed algorithm. We set the probability of communication failure to 0.4. It means that in each time instant (iteration), the probability of successful transmission is 0.6 and with 40% of chance, the transmitted data is missing. The result is presented in Fig. 5.

As simulation results indicate, the proposed stochastic optimization method finds the global optimum while only local information is exchanged through the network. The convergence of this network means that, although sensors only process their own cost function, the information diffuses to the network. Considering

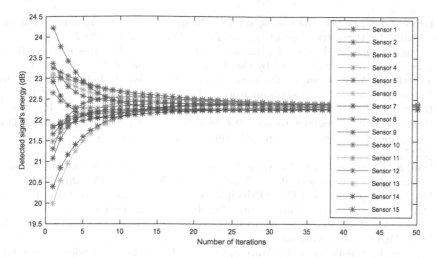

Fig. 5. Performance of the distributed spectrum sensing when the communication link fails with probability of 0.4

the communication link imperfection, the method is robust against missing data. It should be mentioned that the algorithm is capable to be used in all network with arbitrary size.

6   Conclusion

In this paper we investigated the optimization techniques for Big Data analytics. We presented a fully distributed method to make a decision over large scale networks and data sets. We further proposed optimal weighting function for proposed stochastic optimization algorithm. The proposed method is scalable to any network configuration, is near real-time (in each iteration, a solution is provided although it might not be the optimum one) and more important, robust to any missing data or communication failures. We evaluated the proposed method by a practical example and simulations on cognitive networks. Simulation results confirmed that the proposed method is efficient in terms of accuracy and robustness. In this paper we evaluated the proposed algorithm with a simple application of cognitive sensor networks. In future works, the convergence of proposed algorithm and more applications, specially in IoT and Intelligence Transportation Systems could be evaluated.

References

1. Bryson, S., Kenwright, D., Cox, M., Ellsworth, D., Haimes, R.: Visually exploring gigabyte data sets in real time. Commun. ACM **42**(8), 82–90 (1999)
2. Laney, D.: 3D data management: controlling data volume, velocity and variety. META Group Res. Note **6**(70), 1 (2001)

3. Laurila, J.K., et al.: The mobile data challenge: big data for mobile computing research. In: Pervasive Computing, No. EPFL-CONF-192489 (2012)
4. Zhang, Y., Qiu, M., Tsai, C.W., Hassan, M.M., Alamri, A.: Health-CPS: healthcare cyber-physical system assisted by cloud and big data. IEEE Syst. J. (2015)
5. Taherkordi, A., Eliassen, F., Horn, G.: From IoT big data to IoT big services. In: Proceedings of the Symposium on Applied Computing, pp. 485–491 (2017)
6. Braem, B., Latré, S., Leroux, P., Demeester, P., Coenen, T., Ballon, P.: City of things: a multidisciplinary smart cities testbed for IoT, big data and living labs innovation. In: Network Design and Optimization for Smart Cities, pp. 279–292 (2017)
7. Han, S., Chih-Lin, I., Li, G., Wang, S., Sun, Q.: Big data enabled mobile network design for 5G and beyond. IEEE Commun. Mag. **55**(9), 150–157 (2017)
8. Marz, N., Warren, J.: Big Data: Principles and Best Practices of Scalable Realtime Data Systems. Manning Publications Co., Shelter Island (2015)
9. Chang, V., Wills, G.: A model to compare cloud and non-cloud storage of Big Data. Future Gener. Comput. Syst. **57**, 56–76 (2016)
10. Slavakis, K., Giannakis, G.B., Mateos, G.: Modeling and optimization for big data analytics: (statistical) learning tools for our era of data deluge. IEEE Signal Process. Mag. **31**(5), 18–31 (2014)
11. Li, C., Yu, X., Huang, T., He, X.: Distributed optimal consensus over resource allocation network and its application to dynamical economic dispatch. IEEE Trans. Neural Netw. Learn. Syst. **29**(6), 2407–2418 (2018)
12. Patel, A.B., Birla, M., Nair, U.: Addressing big data problem using Hadoop and Map Reduce. In: 2012 Nirma University International Conference on IEEE Engineering, NUiCONE, pp. 1–5 (2012)
13. Witten, I.H., Frank, E., Hall, M.A., Pal, C.J.: Data Mining: Practical Machine Learning Tools and Techniques. Morgan Kaufmann, Burlington (2016)
14. Cevher, V., Becker, S., Schmidt, M.: Convex optimization for big data: scalable, randomized, and parallel algorithms for big data analytics. IEEE Signal Process. Mag. **31**(5), 32–43 (2014)
15. Fountoulakis, K., Gondzio, J.: Performance of first-and second-order methods for big data optimization. Technical report ERGO-15-005. University of Edinburgh (2015)
16. Kambatla, K., Kollias, G., Kumar, V., Grama, A.: Trends in big data analytics. J. Parallel Distrib. Comput. **74**(7), 2561–2573 (2014)
17. Richtárik, P., Takáč, M.: Parallel coordinate descent methods for big data optimization. Math. Program. **156**(1–2), 433–484 (2016)
18. Notarnicola, I., Carli, R., Notarstefano, G.: Distributed partitioned big-data optimization via asynchronous dual decomposition. IEEE Trans. Control Netw. Syst. (2017)
19. Liu, X.: Research on optimization algorithm based on big data calculation framework in high speed optical fiber communication system. Revista de la Facultad de Ingeniería **32**(11) (2017)
20. Facchinei, F., Scutari, G., Sagratella, S.: Parallel selective algorithms for nonconvex big data optimization. IEEE Trans. Signal Process. **63**(7), 1874–1889 (2015)
21. Daneshmand, A., Facchinei, F., Kungurtsev, V., Scutari, G.: Hybrid random/deterministic parallel algorithms for convex and nonconvex big data optimization. IEEE Trans. Signal Process. **63**(15), 3914–3929 (2015)
22. Hong, M., Razaviyayn, M., Luo, Z.Q., Pang, J.S.: A unified algorithmic framework for block-structured optimization involving big data: with applications in machine learning and signal processing. IEEE Signal Process. Mag. **33**(1), 57–77 (2016)

23. Chen, S., Wang, K., Zhao, C., Zhang, H., Sun, Y.: Accelerated distributed optimization design for reconstruction of big sensory data. IEEE Internet Things J. **4**, 1716–1725 (2017)
24. Liu, L., Han, Z.: Multi-block ADMM for big data optimization in smart grid. In: 2015 IEEE International Conference on Computing, Networking and Communications, ICNC, pp. 556–561, February 2015
25. Sabar, N.R., Abawajy, J., Yearwood, J.: Heterogeneous cooperative co-evolution memetic differential evolution algorithm for big data optimization problems. IEEE Trans. Evol. Comput. **21**(2), 315–327 (2017)
26. Salinas, S., Chen, X., Ji, J., Li, P.: A tutorial on secure outsourcing of large-scale computations for big data. IEEE Access **4**, 1406–1416 (2016)
27. Sayed, A.H.: Fundamentals of Adaptive Filtering. Wiley, Hoboken (2003)
28. Sayed, A.H.: Diffusion adaptation over networks. In: Academic Press Library in Signal Processing, vol. 3, pp. 323–454 (2013)
29. Qiu, R., Wicks, M.: Cognitive Networked Sensing and Big Data. Springer, New York (2014). https://doi.org/10.1007/978-1-4614-4544-9

Generating Term Weighting Schemes Through Genetic Programming

Ahmad Mazyad, Fabien Teytaud$^{(\boxtimes)}$, and Cyril Fonlupt

LISIC, Université du Littoral Côte d'Opale,
50 Rue Ferdinand Buisson, 62100 Calais, France
teytaud@univ-littoral.fr

Abstract. Term-Weighting Scheme (TWS) is an important step in text classification. It determines how documents are represented in Vector Space Model (VSM). Even though state-of-the-art TWSs exhibit good behaviors, a large number of new works propose new approaches and new TWSs that improve performances. Furthermore, it is still difficult to tell which TWS is well suited for a specific problem. In this paper, we are interested in automatically generating new TWSs with the help of evolutionary algorithms and especially genetic programming (GP). GP evolves and combines different statistical information and generates a new TWS based on the performance of the learning method. We experience the generated TWSs on three well-known benchmarks. Our study shows that even early generated formulas are quite competitive with the state-of-the-art TWSs and even in some cases outperform them.

1 Introduction

Text Classification (TC) aims to automatically assign a set of predefined categories to a text document based on their content. TC is an important machine learning problem that has been applied to numerous applications such as spam filtering [28], language identification [32], and so on. Generally, the TC approach is to learn an inductive classifier from a set of predefined categories. This approach requires that documents are represented in a suitable format such as the Vector Space Model (VSM) representation [26].

In a VSM, a document d_j is represented by a term vector $d_j = (w_{1,j}, w_{2,j}, ..., w_{t,j})$ where each term is associated with a weight $w_{k,j}$.

The weight represents how much a term contributes to the semantics of a document. The method which assigns a weight to a term is called Term Weighting Scheme (TWS).

Numerous TWS exist and we introduce the most famous in Sect. 2. They are generated according to human a priori and mathematical rules. TWSs are usually simple mathematical expressions. Unfortunately, depending on the application, it is not easy to know a priori which TWS will be effective.

As expression discovery may naturally be addressed by genetic programming [1], we are interested in this paper to study the effectiveness of Genetic Programming (GP) generated formulas for term-weighting and their aspects. We

© Springer Nature Switzerland AG 2019
G. Nicosia et al. (Eds.): LOD 2018, LNCS 11331, pp. 92–103, 2019.
https://doi.org/10.1007/978-3-030-13709-0_8

are also interested to know if a stochastic evolutionary process with no information about the complexity, the shape and the size of the expression can find at
least competitive discriminative TWS.

The paper is organized as follows: Sect. 2 presents the TWSs and related
works. In Sect. 3 we present Genetic Programming and how it is applied to
TWS. Section 4 presents the experiments and the results, and then we conclude
in Sect. 5.

2 Term Weighting Schemes

TC is a supervised learning task. Hence, the training data consists of a set of
labeled documents $D = ((d_1, l_1), ..., (d_N, l_N))$, such that d_j is the term vector
of j-th document, l_j is its label and N is the total number of training documents. As in VSM representation, a document d_j is represented by a term
vector $d_j = (w_{1,j}, w_{2,j}, ..., w_{m,j})$ where $w_{i,j}$ is a weight assigned to the i-th term
of the vocabulary t_i of the document d_j and determined by the TWS.

2.1 Statistical Information

Generally, a multi-labeled classification task is turned into several distinct single-
label binary task, one for each label, using the binary relevance (BR) transformation strategy. That is, given the list of labels $L = \{l_1, l_2, ..., l_m\}$, the original
data set is transformed into m different data sets $D = \{D_1, D_2, ..., D_m\}$. For
each data set D_k, documents having the label l_k will be tagged as the positive
category c_k, and the rest as the negative category $\overline{c_k}$. Weights are then computed
independently for each binary data set.

Based on the BR transformation, given a term t_i and a category c_k, TWS
could be expressed using statistical information a, b, c and d obtained from the
training data:

- a is the number of documents that contain the term t_i and belong to the
  positive category c_k.
- b is the number of documents that don't contain t_i and belong to c_k.
- c is the number of documents that contain t_i and don't belong to c_k.
- d is the number of documents that don't contain t_i and don't belong to c_k.

Besides the statistics described above, Table 1 shows different statistical information that could be extracted from the training data.

2.2 Term Weighting Schemes

Generally, TWSs combines two of three factors pointed out by Salton et al. in
[26] that are believed to improve both recall and precision:

- *Term Frequency (TF) factor*: The TF factor is used to capture the relative
  importance of terms in a document.

Table 1. Statistical information (Terminals) used to evolve a TWS.

Label	Description
N	# documents
C	# categories
C_t	# categories that contain the term t
N_t	# doc that contain t
$\overline{N_t}$	# doc that do not contain t
N_{cat}	# doc in the positive category cat
$\overline{N_{cat}}$	# doc that do not belong to cat

Table 2. Six traditional CF factors.

CF	Defined by
χ^2	$\dfrac{N*(a*d-b*c)*(a*d-b*c)}{(a+c)*(b+d)*(a+b)*(c+d)}$
or	$\log(2 + \frac{a*d}{b*c})$
rf	$\log(2 + \frac{a}{\max(1,c)})$
icf	$\log(\frac{C}{C_t})$
ig	$(\frac{a}{N} \times \log \frac{a \times N}{(a+b)(a+c)})$ $+(\frac{c}{N} \times \log \frac{c \times N}{(c+d)(a+c)})$ $+(\frac{b}{N} \times \log \frac{b \times N}{(a+b)(b+d)})$ $+(\frac{d}{N} \times \log \frac{d \times N}{(c+d)(b+d)})$

- *Collection Frequency (CF) factor*: Also called term discrimination. The importance of words in a document (TF factor) does not provide enough discrimination ability. A common word like 'The' is frequent in almost all documents, and then it could not separate a group of documents from the remainder of the collection. Hence a discrimination factor is needed to favor those terms that are concentrated in a few documents of the collection. Main known CF factors are presented in Table 2.

TWSs could be divided into two sets depending on whether they make use of available information on document membership (Supervised TWSs) or not (Unsupervised TWSs).

Unsupervised TWSs are generally borrowed from Information Retrieval domain [26] and adopted for TC [7,22,23].

Term Frequency-Inverse Document Frequency (TF-IDF) is the most famous term weighting method. This method combines the TF factor and the CF factor and can be formally defined as $w_{i,j} = tf_{i,j} \times \log \frac{N}{N_t}$ where $w_{i,j}$ is the weight of the term t_i in the document d_j, $tf_{i,j} = f_{i,j}$ is the term frequency represented by the raw count of t_i in d_j, and $\log \frac{N}{N_t}$ is the inverse document frequency (idf).

Besides the raw count $(f_{t,d})$ representation of tf, there exist numerous other variants such as binary representation ($w_{i,j} = 1$ if the term t_i occurs in the document d_j and 0 otherwise), $log(f_{i,j}) + 1$, $f_{i,j}/\sum_{t' \in d} f_{t',d}$. All these variants are also used as TWS on their own [7,8,22,26]. The inverse document frequency has also a number of variants such as $\log(N/N_t) + 1$, $\log((N - N_t)/N_t)$ [26].

Supervised TWSs makes use of available information on the membership of training documents by replacing the unsupervised idf component in TF-IDF by another supervised component. Debole et al. and Deng et al. in [7,8] are the first to take advantage of such information by combining the unsupervised TF component with different supervised term discrimination component: χ^2 (TF-CHI), which makes a test of independence between a term and a category. χ^2 alongside with other supervised feature selection metrics, has been tested in several papers, as a term weighting methods for text categorization. For example,

Deng et al. in [8], replaces the idf factor with χ^2 factor, claiming that TF-CHI is more efficient than TF-IDF. In contrast, in a similar test, Debole et al. in [7], compare TF-IDF with three supervised term weightings, namely, TF-CHI, Odds Ratio (TF-OR) and Information Gain (TF-IG). The authors have found no consistent superiority of these new term weighting methods over TF-IDF; Information Gain TF-IG [2] which measures the amount of information obtained for category prediction by knowing the presence or absence of a term in a document [7,8,31]; Gain Ratio (TF-GR) first used in a feature selection method defined as the ratio between the information gain of two variables and the entropy of one of them [7]; Odds Ratio (TF-OR) was first used as a feature selection method by Mladeni'c et al. [24]. It is a measure that describes the strength of association between two random variables. A comparative study on term-weighting for TC is made by Deng et al. in [8]. The study shows a good performance of TF-OR but is outperformed by TF-GR; Relevance frequency (TF-RF) proposed in [20], measures the distribution of a term between the positive and the negative category, and favors those terms that are more concentrated in positive category than in negative categories; Inverse Category Frequency (TF-ICF) is a new supervised TWS proposed by Wang et al. in [30]. The measure aims to favor those terms that appear in fewer categories. More similar methods have appeared in [12,13,15]. Several comparative studies on these TWSs for both term-weighting and feature selection has been reported in [8,22,24,31]. A new approach for term-weighting based on (TF-IG) have been proposed for multi-labeled classification task in [23]. The method computes a score based on all categories and then subtracts it from the original TF-IG weight. The idea is to take into consideration the weights of terms not only in terms of positive and negative categories but also in terms of every single category. Similar approaches have been proposed to learn TWSs via GPin [4–6,11,25,29], however, these studies have focused on information retrieval problem. For TC, a similar approach proposed by Escalante et al. in [9]. However our study differs in two ways: first, Escalante et al. try to generate new TWSs by combining existing TWS, and secondly, they learn a single TWS for each data set whereas we learn a TWS for each category in a data set. In our work, we generate TWSs by combining statistical information at a microscopic level to evolve new TWSs. We also extend the study on the thematic TC. We hope this leads into more robust non human based TWSs.

3   Genetic Programming

Evolutionary computing is based on Darwin's theory of "survival of the fittest". The main scheme of evolutionary algorithms is to evolve a population of individuals that are randomly generated. Each individual represents a candidate solution that undergoes a set of genetic operators that allow to mix and alter partial solutions. One of the key features of evolutionary algorithms is that they are stochastic schemes.

3.1   Introduction

GP belongs to the family of evolutionary algorithms. It was first proposed by Cramer [3] and then popularized by Koza [19]. Unlike genetic algorithms where the aim is to discover a solution, the goal of GP is to find out a computer program that is able to solve a problem.

In GP, a set of random expressions that usually represent computer programs are generated. As in all evolutionary computation algorithms, this set of programs will evolve and change dynamically during the evolution. What makes GP suitable for a number of different applications is that these computer programs can represent many different structures, such as mathematical expressions for symbolic regression [27], decision trees [17], programs that control a robot [18, 21] to fulfill a certain task or programs that are able to predict defibrillation success in patients and so on.

The quality of a candidate solution (i.e. a program) is usually assessed by confronting it with a set of fitness cases. This step is usually the most time-consuming step as the programs may get huge and several thousands of candidate programs are usually evaluated at each generation. These computer programs will undergo one or several evolutionary operators that will alter in a hope-fully beneficial way. The most classical evolutionary operators are usually the crossover operator that allows the exchange of genetic material (in our case sub-trees) and the mutation operator that allows a small alteration to the program.

In the most conventional GP approach, programs are usually depicted by trees. In GP terminology, the set of nodes are split into two sets, inner nodes of the tree are drawn from a set of functions while the terminal nodes (leaves) are drawn from a so-called terminal set. Depending on the problem, the set of functions can be mathematical functions, boolean functions, control flow functions (if, ...), or any functions that may be suitable to solve the given problem. The terminal set is usually the set of inputs of the problem, e.g., parameters and constants for symbolic regression problems, sensors for robot planning, etc.

When the stopping criterion is reached, the best individual is returned, otherwise, the loop continues and the best individuals are selected (according to their fitness). There exist numerous ways for selecting the population, the mutation and the crossover operators. This is beyond the scope of this paper and the reader can refer to [16, 19] for more information.

3.2   Evolving Term Weighting Scheme Using Genetic Programming

A CF factor is a combination of statistical information. It is intended to measure the discriminative power of a term, i.e. it tells how much a term is related to a certain category. These statistics combined by means of mathematical operators and functions.

We are interested in automatically evolving a CF factor (an individual) using GP. In our approach, the learned CF factor combined to the TF factor forms a term weighting method.

In our context of automatically evolving term weighting methods, an individual is a combination of the function set that is built with simple arithmetical operators ($+$, $-$, $*$, $/$, log, ...) and the terminal set (constant values and inputs to our problem).

Table 1 shows the statistical information used as terminal set for generating formulas which represent CF factors. As it can be seen, the function set is made of very simple arithmetical functions while the terminal set includes to the best of our knowledge all the statistical information used to build a TWS.

As previously mentioned, programs (generated TWS) are depicted as trees. In this problem, the terminal nodes consist of statistical information extracted from training data, while the inner nodes are a set of defined operators that combines the statistical information to form a new TWS (Table 3).

Table 3. Parameters used in our genetic program.

Parameter	Value
Population size	100
Initial individual size	20
Number of generations	100
Function set	$+$, $-$, $/$, $*$, $\sqrt{x}$, $log1(x)$, $log2(x)$
Terminal set	a, b, c, d, N, N_t, $\overline{N_t}$, N_{cat}, $\overline{N_{cat}}$, C, C_t
Mutation	OnePointMutation ($P = 1/individual\ size$)
CrossOver	SubtreeCrossover ($P = 0.85$)

Terminals and Function Set. In this study, we try to generate new TWS by evolving the CF factor and then combines it with the TF factor. The CF factor is a combination of constants, statistical information (N, N_t, ...), and mathematical operators. Hence we define the terminals as the statistical information shown in Table 1. Regarding the mathematical operators, they are defined as one of the following ($+$, $-$, $/$, $*$, $\sqrt{x}$, $log1(x) = \log(1 + x)$ and $log2(x) = \log(2 + x)$).

We should note that the statistical information has different types (single value, vector, and matrix). For instance, the number of documents in the training data N is a constant (single value), the number of documents that contains a term t is a vector containing the number of documents for each term and finally, the number of documents that belongs to a category cat and contains a term t is a matrix. Operations on these different types of statistical information are taken care of by Eigen[1] library using element-wise transformations.

Genetic Operators. In GP, a set of individuals is initialized and then evolved according to a set of genetic operators. At first, we randomly generate a random size individuals with a max size of twenty genes (the max size could be

---

[1] http://eigen.tuxfamily.org/.

overpassed during the cross-over operation). As for genetic operators, we use the elite selection and re-insertion, a subtree crossover with a probability of 0.85 and one point mutation with a probability of 1/size of the individual.

Fitness Function. Generally, the performance of a TWS is assessed on known benchmarks by evaluating a classification model on VSM representation of this TWS. Numerous evaluation metrics exist that evaluate the classification model such as f_1 measure. Evaluating the classification model is a vital step that affects the performance of the GP. However, it could be very time-consuming. Hence, it is important to choose a good and fast machine learning algorithm. LibLinear [10] is an open source library for large-scale linear classification. It supports linear support vector machines.

In our study, once a new individual is generated, we perform a 3-fold cross-validation on the training data which generates three disjoint subsets. We use two subsets as the training set and one subset as the test set. The process is repeated three times using each time different subset for testing. The performance is measured using the f_1 measure. The average classification performance is used as the fitness function. The f_1 measure considers both precision p (true positive over true positive plus false positive) and recall r (true positive over true positive plus false negative) and can be formally defined as $f_1(p, r) = \frac{2rp}{r+p}$.

4 Experiments and Results

This section presents an empirical evaluation of the proposed approach. The goal of this study is to assess the effectiveness of the generated TWSs and compare their performances to standard TWSs. The souce code of the implementation needed for our experiments could be found in a public repository[2].

4.1 Experimental Setup

In our experiments, we have used three widely well-known benchmarks in TC: Reuters-21578 Benchmark Corpus[3], Oshumed Benchmark Corpus (see footnote 2) and the 4 Universities data set also called Webkb[4]. The Reuters-21578 data set is one of the most used test collection for TC research. We use the well-known "ApteMod" split [14]. This version of the data set contains ninety categories, however, in our experiments, we report results only for the largest ten categories. Oshumed dataset is extracted from the Oshumed (see footnote 1) collection compiled by William Hersh. It includes 13,929 medical abstracts from the MeSH categories of the year 1991. Each document in this data set belongs to one or more categories from 23 cardiovascular diseases categories. Webkb data

---

[2] https://bitbucket.org/mazyad/eigennlp.

[3] http://disi.unitn.it/moschitti/corpora.htm.

[4] http://www.cs.cmu.edu/afs/cs.cmu.edu/project/theo-20/www/data/.

set contains WWW-pages collected from computer science departments of various universities in January 1997 by the World Wide Knowledge Base (Webkb) project of the CMU text learning group. In this experiment, we kept only the four largest categories ("student", "faculty", "course" and "project"), and we split it into three random folds where two folds are used for the training set and one fold for the test set.

For all three data sets considered in the experiments, a default list of stop words, punctuation and numbers are removed, lower case transformation and Porter's stemming are performed.

Furthermore, for each experiment, a binary transformation is applied. That leads to multiple distinct single-label binary task, one for each label (see Sect. 2.1). Each task could be treated as an independent experiment with its own data set.

As mentioned above, each data set has been split into training and test subsets. Table 4 shows, for each data set, the number of documents in the training and test subsets, the number of classes, the number of terms, the size of smallest category and the size of the largest category.

TWSs are evolved using the training subset (see Sect. 3.2). Finally, the test subset is used to evaluate the performance of the generated TWS. And finally, for each data set, we report the f_1 measure (see Sect. 3.2).

In order to obtain more reliable results, we have performed 20 runs on each task. After having evaluated the generated TWSs, we report the performance average and standard deviation over the 20 runs. In addition, we report the maximum and minimum f_1 score obtained across the 20 runs (for each run, only the last generated TWS is taken into account).

Tables 5, 6 and 7 show the results obtained by the generated TWSs and the best baseline using linearSVM. Table 8 shows the average classification performance of the generated TWSs on the test subset of the training data (Validation) and the performance on the test data (Test). The goal of this experiment is to assess the learning ability and to warn us of eventual overfitting. Table 9 shows the average classification performance of a random learned TWS for a single-label binary task on the complete data set. This is important in order to know whether our GP-Based TWS has good generalization performances.

Table 4. Statistics on the selected data sets used for our experiments (training/test).

	Reuters	Oshumed	Webkb
Number of documents	7769/3019	6286/7643	2803/1396
Number of classes	90	23	4
Number of terms	26000	30198	7890
Size of the smallest category	1/1	65/70	336/168
Size of the largest category	2877/1087	1799/2153	1097/544

4.2   Results

First, a fast study of the Tables 5, 6 and 7 shows that the best baseline TWS is different for each binary task. Therefore, a multi-labeled task requires different TWSs for each category. Using different TWSs could lead to better results. However, the problem is to recognize the best TWS for a specific task. Finding the TWS by cross-validation does not mandatory return the best TWS.

Regarding Reuters-21578, the generated TWSs and the baseline schemes have similar performances. However, on Oshumed and Webkb data sets, the GP-Based TWSs outperform the best baseline schemes. Reuters-21578 is one of the most studied data-set in TC for TWS making the task of finding better TWS harder. Moreover, it is the most unbalanced data set in the study which makes generalization harder.

Table 5. Classification performance on top 10 categories of Reuters-21578 obtained with the generated TWSs and the best standard TWS. Best results are bolded.

	GP			Best TWS	
Label	f1	Min	Max	f1	TWS
Earn	98.34 ± 0.09	98.24	98.54	**98.38**	tf.idf
Acq	96.93 ± 0.23	96.55	97.54	**97.10**	tf.idf
Money-fx	**79.60 ± 0.50**	78.16	80.45	78.63	tf.idf
Grain	**94.25 ± 0.63**	93.10	95.22	93.43	tf.rf
Crude	**90.01 ± 0.81**	88.27	90.94	88.24	tf.rf
Trade	**79.10 ± 1.21**	77.69	80.18	78.03	tf.rf
Interest	75.16 ± 0.50	74.45	76.19	**76.19**	tf.idf
Ship	**80.52 ± 1.54**	77.84	82.93	78.95	tf.or
Wheat	88.11 ± 1.26	86.12	90.96	**90.20**	tf.chi
Corn	92.80 ± 0.27	90.83	93.94	**93.91**	tf.chi
Average	**87.48 ± 0.70**	86.13	88.69	87.30	

From Table 8, we can see that the performance of generated TWSs on the test subset of the training data during the cross-validation (See Sect. 3.2) is very similar to the performance on the test data. In addition, the standard TWSs have different results. This is interesting as it suggests that there is no overfitting and that further learning can improve the performance.

From Table 9, we can see that the average performance (macro-f_1) of the generated TWSs outperforms the best baseline on the three corpora which means that the three learned TWS have good generalization performance.

Finally, compared to the results obtained in [9] on Reuters-21578 and Webkb, we have similar results. Note that, in [9], they used Reuters-10 data set which contains only documents from the top 10 categories of the Reuters-21578 data set, whereas we use Reuters-21578 "ModApte" split which contains documents from 90 categories.

Table 6. Classification performance on Oshumed data set obtained with the generated TWSs and the best baseline of the standard TWSs.

L	GP f1	Min	Max	Best TWS f1	TWS	L	GP f1	Min	Max	Best TWS f1	TWS
C01	**68.19**±1.00	65.91	70.71	64.36	tf.or	C13	**66.48**±0.47	64.72	67.92	63.70	tf.or
C02	**41.28**±1.20	38.45	43.51	36.38	tf.or	C14	**80.08**±0.39	79.22	80.55	77.11	tf.idf
C03	76.54±3.28	72.03	81.21	**78.23**	tf.or	C15	**65.98**±0.71	64.16	67.20	61.53	tf.chi
C04	80.06±1.48	77.67	81.72	**80.06**	tf.chi	C16	**33.54**±0.89	31.14	35.41	28.00	tf.or
C05	**59.48**±0.20	59.05	60.59	52.85	tf.or	C17	**64.85**±0.90	61.87	66.87	59.24	tf.chi
C06	**73.99**±1.29	71.49	75.76	71.44	tf.or	C18	61.21±1.50	57.50	65.12	**61.22**	tf.or
C07	**41.40**±3.35	34.86	47.45	32.6	tf.or	C19	**41.60**±2.04	38.23	45.01	39.84	tf.or
C08	**63.97**±2.51	59.13	67.69	61.34	tf.or	C20	**71.61**±0.28	70.96	72.07	69.62	tf.or
C09	**53.75**±2.63	50.85	58.43	48.00	tf.or	C21	**65.55**±0.32	64.18	67.56	64.37	tf.chi
C10	**57.00**±2.33	51.05	59.53	50.2	tf.rf	C22	**10.31**±0.12	8.33	14.37	4.21	tf.or
C11	**67.78**±1.06	65.52	69.23	66.67	tf.or	C23	**46.77**±0.08	45.59	47.20	46.15	tf.idf
C12	**76.72**±1.10	73.52	78.25	72.86	tf.or	Avg	**59.48**±1.26	56.76	61.89	56.08	

Table 7. Classification performance on Webkb data set obtained with the generated TWSs and the best baseline of the standard TWSs.

L	GP f1	Min	Max	Best TWS f1	TWS
Student	**90.29** ± 0.50	89.05	90.90	90.11	tf.rf
Faculty	**86.62** ± 0.15	85.69	87.81	86.21	tf.rf
Project	**80.82** ± 0.64	77.48	81.76	80.25	tf.rf
Course	**94.47** ± 0.34	93.86	96.08	93.56	tf.rf
Avg	**88.05** ±0.41	86.52	89.14	87.53	

Table 8. Average classification performance for validation phase and test phase.

	Validation	Test
Reuters	**89.15** ± 0.42	**87.48** ± 0.70
Oshumed	**59.74** ± 0.9	**59.48** ± 1.26
Webkb	**87.74** ± 0.31	**88.05** ± 0.41

Table 9. Average classification performance of random TWS learned for a single-label task on its corresponding data set and the best baseline. The selected TWS is randomly chosen between the best generated TWSs for each category.

Data set	GP-Based Prefixed formula	TWS	f_1	Baseline f_1	Best baseline
Reuters	$* * C * //acN\ log2c\ C$	$C * C * (\frac{a}{c*N} * \log(2 + c))$	**86.88**	85.92	tf.rf
Oshumed	$/d/ + N_t log2\ C_t a$	$\frac{a}{d*(N_t + log(2+C_t))}$	**60.30**	57.10	tf.chi
Webkb	$log1\ log2\ a$	$\log(1 + log(2 + a))$	**88.43**	87.53	tf.rf

5 Conclusion

In this paper, we have studied the benefits of using genetic programming for generating term-weighting schemes for text categorization. Unlike previous studies, we generate formulas by combining statistical information at a microscopic level. This kind of generation is new, and we can conclude that :

- Different data sets require different formulas. This means that having a good generic formula is really hard to find.
- Within a corpus, it is even better to use a different formula for each category. The hard task is to find out the best for each one.
- Genetic programming is able to find very good formulas which outperform standard formulas given by experts in the literature.
- Eventually, even if the generated formula is specific to a given category, results show that the best formula for one category is generic enough to be good (but not best) for other categories.

References

1. Cazenave, T.: Nested Monte-Carlo expression discovery. In: ECAI, pp. 1057–1058 (2010)
2. Cover, T.M., Thomas, J.A.: Elements of Information Theory. Wiley, Hoboken (2012)
3. Cramer, N.L.: A representation for the adaptive generation of simple sequential programs. In: Proceedings of the First International Conference on Genetic Algorithms, pp. 183–187 (1985)
4. Cummins, R., O'Riordan, C.: Evolving general term-weighting schemes for information retrieval: tests on larger collections. Artif. Intell. Rev. **24**(3–4), 277–299 (2005)
5. Cummins, R., O'Riordan, C.: Evolved term-weighting schemes in information retrieval: an analysis of the solution space. Artif. Intell. Rev. **26**(1–2), 35–47 (2006)
6. Cummins, R., O'Riordan, C.: Evolving local and global weighting schemes in information retrieval. Inf. Retr. **9**(3), 311–330 (2006)
7. Debole, F., Sebastiani, F.: Supervised term weighting for automated text categorization. In: Sirmakessis, S. (ed.) Text mining and its applications. STUDFUZZ, pp. 81–97. Springer, Heidelberg (2004). https://doi.org/10.1007/978-3-540-45219-5_7
8. Deng, Z.-H., Tang, S.-W., Yang, D.-Q., Li, M.Z.L.-Y., Xie, K.-Q.: A comparative study on feature weight in text categorization. In: Yu, J.X., Lin, X., Lu, H., Zhang, Y. (eds.) APWeb 2004. LNCS, vol. 3007, pp. 588–597. Springer, Heidelberg (2004). https://doi.org/10.1007/978-3-540-24655-8_64
9. Escalante, H.J., et al.: Term-weighting learning via genetic programming for text classification. Knowl.-Based Syst. **83**, 176–189 (2015)
10. Fan, R.E., Chang, K.W., Hsieh, C.J., Wang, X.R., Lin, C.J.: Liblinear: a library for large linear classification. J. Mach. Learn. Res. **9**(Aug), 1871–1874 (2008)
11. Fan, W., Fox, E.A., Pathak, P., Wu, H.: The effects of fitness functions on genetic programming-based ranking discovery for web search. J. Assoc. Inf. Sci. Technol. **55**(7), 628–636 (2004)
12. Guru, D., Suhil, M.: A novel term class relevance measure for text categorization. Proc. Comput. Sci. **45**, 13–22 (2015)
13. Ibrahim, O.A.S., Landa-Silva, D.: Term frequency with average term occurrences for textual information retrieval. Soft Comput. **20**(8), 3045–3061 (2016)
14. Joachims, T.: Text categorization with support vector machines: learning with many relevant features. In: Nédellec, C., Rouveirol, C. (eds.) ECML 1998. LNCS, vol. 1398, pp. 137–142. Springer, Heidelberg (1998). https://doi.org/10.1007/BFb0026683

15. Kadhim, A.I.: Statistical computation and term weighting for feature extraction on Twitter. In: 2018 International Conference on Advance of Sustainable Engineering and its Application (ICASEA), pp. 109–114, March 2018

16. Karakus, M.: Function identification for the intrinsic strength and elastic properties of granitic rocks via genetic programming (GP). Comput. Geosci. **37**(9), 1318–1323 (2011)

17. Koza, J.R.: Concept formation and decision tree induction using the genetic programming paradigm. In: Schwefel, H.-P., Männer, R. (eds.) PPSN 1990. LNCS, vol. 496, pp. 124–128. Springer, Heidelberg (1991). https://doi.org/10.1007/BFb0029742

18. Koza, J.R.: Genetic Programming II, Automatic Discovery of Reusable Subprograms. MIT Press, Cambridge (1992)

19. Koza, J.R.: Genetic programming: on the Programming of Computers by Means of Natural Selection, vol. 1. MIT Press, Cambridge (1992)

20. Lan, M., Tan, C.L., Su, J., Lu, Y.: Supervised and traditional term weighting methods for automatic text categorization. IEEE Trans. Pattern Anal. Mach. Intell. **31**(4), 721–735 (2009)

21. Lewis, M.A., Fagg, A.H., Solidum, A.: Genetic programming approach to the construction of a neural network for control of a walking robot. In: IEEE International Conference on Robotics and Automation, vol. 3, pp. 2618–2623 (1992)

22. Mazyad, A., Teytaud, F., Fonlupt, C.: A comparative study on term weighting schemes for text classification. In: Nicosia, G., Pardalos, P., Giuffrida, G., Umeton, R. (eds.) MOD 2017. LNCS, vol. 10710, pp. 100–108. Springer, Cham (2018). https://doi.org/10.1007/978-3-319-72926-8_9

23. Mazyad, A., Teytaud, F., Fonlupt, C.: Information gain based term weighting method for multi-label text classification task. In: Arai, K., Kapoor, S., Bhatia, R. (eds.) IntelliSys 2018. AISC, vol. 868, pp. 607–615. Springer, Cham (2019). https://doi.org/10.1007/978-3-030-01054-6_44

24. Mladeni'c, D., Grobelnik, M.: Feature selection for classification based on text hierarchy. In: Text and the Web, Conference on Automated Learning and Discovery CONALD-98. Citeseer (1998)

25. Oren, N.: Reexamining tf.idf based information retrieval with genetic programming. In: Proceedings of the 2002 Annual Research Conference of the South African Institute of Computer Scientists and Information Technologists on Enablement Through Technology, pp. 224–234. South African Institute for Computer Scientists and Information Technologists (2002)

26. Salton, G., Buckley, C.: Term-weighting approaches in automatic text retrieval. Inf. Process. Manag. **24**(5), 513–523 (1988)

27. Searson, D.P., Leahy, D.E., Willis, M.J.: GPTIPS: an open source genetic programming toolbox for multigene symbolic regression. In: Proceedings of the International Multiconference of Engineers and Computer Scientists, vol. 1, pp. 77–80. Citeseer (2010)

28. Tretyakov, K.: Machine learning techniques in spam filtering. In: Data Mining Problem-Oriented Seminar, MTAT, vol. 3, pp. 60–79 (2004)

29. Trotman, A.: Learning to rank. Inf. Retr. **8**(3), 359–381 (2005)

30. Wang, D., Zhang, H.: Inverse category frequency based supervised term weighting scheme for text categorization. preprint arXiv:1012.2609v4 (2013)

31. Yang, Y., Pedersen, J.O.: A comparative study on feature selection in text categorization. In: ICML, vol. 97, pp. 412–420 (1997)

32. Zissman, M.A.: Comparison of four approaches to automatic language identification of telephone speech. IEEE Trans. Speech Audio Process. **4**(1), 31 (1996)

Data-Driven Interactive Multiobjective Optimization Using a Cluster-Based Surrogate in a Discrete Decision Space

Jussi Hakanen[1]($\boxtimes$), Jose Malmberg[1], Vesa Ojalehto[1], and Kyle Eyvindson[2]

[1] University of Jyvaskyla, Faculty of Information Technology,
P.O. Box 35 (Agora), FI-40014 University of Jyvaskyla, Finland
{jussi.hakanen,jose.malmberg,vesa.ojalehto}@jyu.fi
[2] University of Jyvaskyla, Dept. of Biological and Environmental Science,
P.O. Box 35, FI-40014 University of Jyvaskyla, Finland
kyle.eyvindson@jyu.fi

Abstract. In this paper, a clustering based surrogate is proposed to be used in offline data-driven multiobjective optimization to reduce the size of the optimization problem in the decision space. The surrogate is combined with an interactive multiobjective optimization approach and it is applied to forest management planning with promising results.

Keywords: Data-driven optimization · Surrogates · Clustering ·
Preference information · Decision maker · Boreal forest management

1 Introduction

Recently, emphasis on optimization has been shifting from model-based to data-driven optimization where the optimization problem is formulated based on available data. The size of the data can sometimes be large which means that the optimization problem(s) to be solved become large as well increasing their solution times. This is especially challenging in multiobjective optimization having a large number of objective functions. In more details, this is because interaction with a human decision maker (DM) is required to find satisfactory solutions to such problems and long solution times can make the interaction less efficient.

Surrogate-assisted optimization approaches are often used to solve computationally expensive optimization problems both for single and multiobjective problems (see, e.g., [2,4]). Typically, computational expensiveness is considered as the time taken to evaluate objective and/or constraint functions since that can take a long time for e.g. simulation or experiment-based models. In data-driven optimization, the expensiveness is typically not in evaluating the objective function values, but in the size of the problems solved (in the decision and/or objective space). The main idea in surrogate-assisted optimization is to use a relatively small sample of expensive function evaluations to train surrogate functions that approximate the expensive functions but are faster to evaluate [2,4].

© Springer Nature Switzerland AG 2019
G. Nicosia et al. (Eds.): LOD 2018, LNCS 11331, pp. 104–115, 2019.
https://doi.org/10.1007/978-3-030-13709-0_9

In this paper, we introduce a surrogate-assisted approach for data-driven multiobjective optimization problems that are based on large data sets motivated by a case study in forest management described later. We assume that all data is available at the beginning of optimization and no new data can be obtained (often referred to as offline data-driven optimization [15]). Further, we consider linear problems with discrete decision space. The method uses clustering in the decision space as a surrogate to decrease the size of the optimization problem by reducing the number of similar variables. The resulting optimization problems are not as accurate as the original problem but are faster to solve. The proposed surrogate is combined with an interactive multiobjective optimization approach that iteratively utilizes preferences of a DM in finding a most preferred solution for the multiobjective problem considered.

In the literature, one approach has been presented that is somewhat similar to what we present, in [15] where the design of a trauma system was optimized. Due to the large amount of data available, the data was first clustered and the cluster centers were then used as data in evolutionary optimization of finding non-dominated solutions for a bi-objective problem. In our approach, we use mathematical programming together with interaction with a DM to find the most preferred PO solution. Furthermore, hierarchical clustering was used in [15] to represent the real hierarchy of the data which is not necessary in our case study. Further, functional analysis of variance decomposition was used in [12] to decompose a multiobjective optimization problem both in objective and decision spaces. Then, solution of the original problem was constructed by solutions of the decomposed problems. A different approach from ours was presented in [1] where clustering was used to find versatile solutions after finding a set of non-dominated solutions by multiobjective optimization. To summarize, there does not exist similar approach in the literature as far as we know.

As a case study to demonstrate the developed approach, we consider a boreal forest management problem where both the economical and biodiversity related objectives are considered. The underlying data gathered from around 30 000 forest stands simulated 50 years into future (with seven management options) was used to formulate a four objective combinatorial optimization problem which was then solved by interacting with an expert DM. Previous considerations of similar problems have included directly using the combinatorial optimization problem together with the epsilon constraint method which optimizes only one of the objectives while considering others as constraints [9,13,14]. When using our proposed approach, it is possible to (1) consider larger problems (i.e., more stands and/or management options) with comparable results in fewer time, and (2) more conveniently handle the conflicting objectives and inherent trade-offs while interacting with an expert DM.

The rest of the paper is organized as follows. First some background information is given in Sect. 2, while the proposed clustering based optimization approach is described in Sect. 3. Our case study and the obtained results are described in Sects. 4 and 5, respectively. Finally, conclusions and future research ideas are given in Sect. 6.

2 Background

2.1 Multiobjective Optimization

When multiple conflicting objectives are concerned, the optimal solutions are often called Pareto optimal (PO) which means none of the objective values can be improved without impairing some other ones [6]. In this paper, we consider multiobjective integer linear programming problems of the form

$$
\text{maximize } \left\{ f_1(\mathbf{x}) = \sum_{i=1}^{n} \sum_{j=1}^{m} c_{ij}^1 x_{ij}, \ldots, f_k(\mathbf{x}) = \sum_{i=1}^{n} \sum_{j=1}^{m} c_{ij}^k x_{ij} \right\}
$$
$$
\text{s.t. } \sum_{j=1}^{m} x_{ij} = 1, \ \forall i = 1, \ldots, n, \tag{1}
$$
$$
x_{ij} \in \{0, 1\}.
$$

The problem includes k objective functions to be maximized. Further, $i \in \{1, 2, \ldots, n\}$ denotes index for the ith decision variable while $j \in \{1, 2, \ldots, m\}$ denotes index for different values for the decision variables. Note that categorical variables having several possible values in the original problem have been converted into binary variables, i.e., $x_{ij} \in \{0, 1\}$. Coefficients c_{ij}^l denote the objective values for the decision variable values x_{ij} for the lth objective function and they are attained from data.

A feasible solution $\mathbf{x}^*$ for problem (1) is called PO if there does not exist another feasible solution $\mathbf{x}$ such that $f_i(x) \geq f_i(x^*)$ for all $i = 1, \ldots, k$ and $f_j(x) > f_j(x^*)$ for at least one j. Note that there can exist infinitely many PO solutions that are mathematically equally good, i.e., none of them is better than others without any additional preference information.

Many different approaches have been developed over the years for solving multiobjective optimization problems (see, e.g. [3,6]). In this paper, we will concentrate on interactive approaches [7], where a DM provides preference information in order to find the most preferred solution for the problem considered. The general idea of interactive approaches is that first some PO solution is computed and shown to the DM for evaluation. The DM indicates how that solution should be improved if she is not satisfied with it by providing preference information. The type of preference information depends on the interactive method used. Then, the preference information is taken into account and new PO solution(s) is computed and again shown to the DM for evaluation. This iterative process continues until the DM is satisfied.

To solve problem (1) with the help of a DM, we will use a surrogate approach based on clustering (described in more details in Sect. 3) combined with the synchronous NIMBUS method. Synchronous NIMBUS [8] is an interactive method based on classification where preference information is indicated by classifying objective functions into different classes at the current PO solution. More precisely, an objective function can be classified either (1) to be improved as much as possible, (2) to be improved until a given aspiration level z^{asp}, (3) to retain its current value, (4) to be allowed to impair until a given bound z^{bnd}, or (5) to change freely (i.e., not interesting at this iteration). A feasible classification is such that there should be at least one objective function in the first two classes and in the last two classes since if any improvements are required, some impairments have to be allowed. Then, the original multiobjective problem together with the preference information are used to formulate up to four different single objective scalarized subproblems that are then solved by using a suitable single objective optimizer. The resulting solutions are proven to be PO [8].

2.2 Forest Management

In Fennoscandia, much of the countries are dominated by Boreal forests, which provide a wide range of ecological, economic, and social values. Most of these forests can be considered to be semi-natural, where limited silvicultural and management actions are done infrequently throughout the development of each forest stand (a relatively homogeneous parcel of forest). A forested stand in Fennoscandia follows rather similar development following a clear felling (the removal of the trees in a specific area). Depending on the site, trees are either planted, seeded, or allowed to grow through natural regeneration (where seeds provided from the forests surrounding the stand, and specific trees left within the stand for this specific purpose). Following this, within 5 to 10 years, tending of the stand may be required to remove grasses and shrubs. Once the forest stand is established it is left to grow. Throughout the forest stands development the forest stand can be thinned (the selected removal of specific trees) several times prior to clear felling, where the process is repeated.

From a forest management perspective, the specific actions conducted in a forest stand can vary according to intensity and timing. For instance, thinnings may or may not be performed, and final felling can be delayed, done years prior to the expected maturity or delayed indefinitely. Each management decision will impact the quantity of timber provided, and ecosystem services provided from the forest stand. At a landscape (500–5000 ha) or regional scale (500–20000 km^2) managing forests becomes a combinatorial optimization problem where the decision variables describe the number of stands and the number of options allowed to for managing each forest stand. Managing the use of forests involves significant conflicts between different objectives. Economic objectives conflict with ecological objectives, and conflicts can arise between different ecological objectives. The quantification of the economic and ecological objectives is done through forecasting future forest growth through forest simulators. In Fennoscandia, there are multiple varieties of forest simulators available, and each software package utilize over four hundred empirically based models to predict forest development and growth.

3 Clustering Based Interactive Multiobjective Optimization Approach

The main idea of the developed surrogate is to cluster the decision variables in such a way that similar variables are represented in the optimization problem through a representative one within the cluster, thus, reducing the size of the optimization problem. In this paper, we consider only discrete decision variables, but our approach can also be extended to mixed variables. To solve the resulting multiobjective optimization problem, we utilize here the synchronous NIMBUS method as already mentioned, which leads us to solve a series of single objective subproblems. By reducing the number of decision variables, the resulting subproblems are easier/faster to solve which reduces the time that the DM needs to wait between interactions.

3.1 Clustering as a Surrogate

The core of forming the surrogate is clustering the discrete decision variables using some hard clustering method: original n variables are assigned to $K \leq n$ clusters according to their similarity in values. To guide how the clustering is performed, it is important to define a similarity measure, i.e., how the similarity of variables is defined. Even though clustering using expert knowledge is possible, the numerical similarities of the variables in each cluster are more important. As the method is used to reduce the computational burden, manual clustering would require extreme human effort due to large number of decision variables.

The clustering based surrogate is built on a large number of round clusters used to approximate the decision space. In the traditional clustering, the number of clusters K is supposed to match the real number of different classes in the data, and it is one of the most important elements of clustering. However, in the clustering based surrogate this aspect is not as important but the focus of designing clusters should be the ability to compress and represent the data accurately and to be sufficient for its purpose. On occasions, it could be profitable to use more clusters than compared to what would be otherwise optimal to improve accuracy.

In traditional clustering, the shapes of the clusters are supposed to capture and separate different classes from the data. In the clustering based surrogate, this does not need to be the case as the focus could be on appropriately approximating and compressing the data. Especially when the number of clusters is "too large", the most suitable shape for clusters is rounded. This enables that all the clusters can be handled similarly as local approximations.

When the n variables have been assigned into K clusters, the most "representative" variable x_i is selected from each cluster $i \in \{1, 2, ..., K\}$ as a proxy variable. As the clusters are rounded, the most representative should be the center of each cluster. If the chosen clustering method is not using existing variables as centers, then the variable closest to the center can be used as proxy. The proxy variables that are representing all the variables of individual clusters are then already existing variables. Note that if variables in the same cluster have different numbers of discrete value alternatives, the proxy variable's ability to represent all the variables in the cluster is greatly impaired.

The chosen proxy variable x_i is denoted by y_i and it is assigned a weight w_i according to the proportion of the variables in the given cluster i. For example, if there are 356 variables in a single cluster i, its corresponding weight is $w_i = \frac{356}{n}$.

In addition, the coefficients c_{ij}^l are renamed to d_{ij}^l, and the previously presented multiobjective integer linear programming problem (1) is transformed to

$$\text{maximize} \quad \left\{ n \sum_{i=1}^{K} \sum_{j=1}^{m} w_i d_{ij}^1 y_{ij}, \ldots, n \sum_{i=1}^{K} \sum_{j=1}^{m} w_i d_{ij}^k y_{ij} \right\}$$

$$\text{s.t.} \quad \sum_{j=1}^{m} y_{ij} = 1, \ \forall i = 1, \ldots, n, \tag{2}$$

$$y_{ij} \in \{0, 1\},$$

where $i \in \{1, 2, ..., K\}$, denotes index for proxy variable, $j \in \{1, 2, ..., m\}$ index for discrete value alternatives for each proxy variable i, and w_i the weighting coefficient for the proxy variable. Value d_{ij}^l denotes the lth objective value of the proxy i when the jth discrete value alternative is chosen. For ith proxy variable, y_{ij} has value 1 if jth value is chosen for proxy variable i, and otherwise 0. The parameter n is the number of original variables. As can be seen, if $K = n$, then $w_i = \frac{1}{n}$ for all $i \in \{1, 2, ..., K\}$ and this formulation is identical with problem (1). Thus, this guarantees the validity of this approach of combining the described surrogate and optimization.

Building the cluster-based surrogate is summarized as follows:

1. Cluster n decision variables into K clusters by using some clustering method.
2. For each K clusters, choose the center of the cluster as the proxy variable if the center is an existing variable. Otherwise, choose the variable closest to the center as the proxy variable.
3. Solve multiobjective optimization problem (2) by using the values of the ith proxy variable for all the variables in the ith cluster.

The proposed surrogate is based only on local approximations of the decision space, so the results of the clustering based multiobjective optimization problem naturally include some approximation error. Due to the structure of the surrogate, the larger the number of the clusters used the more accurate is the surrogate and, thus, the result of optimization. On the other hand, since the idea of clustering is to reduce the number of decision variables, the amount of reduction is dependent on the number of clusters, so that the less clusters there are, the lighter the computational burden. It is thus evident, that the accuracy and the ability to compress the decision space are contradicting features.

In multiobjective optimization, the different objectives are typically contradicting with each other and this is likely to show in the clustering also. In practice, this means that depending on the chosen clustering paradigm, approximation errors for different objectives may be different. When using the clustering based surrogate in multiobjective optimization, this problem becomes more evident as different objectives may reach their real optima to different degrees.

As the scalarized subproblems of problem (1) used here are linear [8] with integer variables, the resulting values in the objective space may be discontinuous in its original state. When using the clustering based surrogate and combining several decision variables, this trait will be emphasized and there will be "bigger holes" in the PO front (i.e. the set of all PO solutions in the objective space).

Finding the most preferred PO solution from this kind of PO front can be quite challenging depending on the multiobjective method used. Therefore, we have decided to use the synchronous NIMBUS method, which uses up to four scalarizing functions [8] that can be used for any kind of PO fronts, even discontinuous, to find different PO solutions using the same preference information. To summarize, we are much more likely to find an acceptable solution even from such a challenging PO front.

For the scalarizing functions used in the synchronous NIMBUS method it is important to attain ranges for all the objectives within the PO front, i.e., to calculate ideal and nadir vectors. This is usually done by computing the optimal solutions for all the single objective optimization problems (forming the ideal objective vector) and then estimating the nadir values by using a so-called pay-off table [6]. When using the clustering based surrogate, these values can be calculated with optimization using the surrogate, but if possible, the optima based on the original variables and problem should be used instead. Even though the scalarizing functions in synchronous NIMBUS were used with the clustering based surrogate, it would still be better to use the original ideal and nadir values in their formulations. The reason is that the surrogate based ideal and nadir values are more averaged because of the approximations used in the surrogate.

The interactive solution process itself remains the same even when using the clustering based surrogate in optimizations. The DM gives her/his preferences, explores different PO solutions, and finally chooses the most preferred PO solution as usual with interactive approaches. The main effect of using the surrogate is that it reduces the computational burden significantly and so enables more seamless and less delayed interaction during the iterative solution process.

When the preferred PO solution is found using the clustering based surrogate, it would be good to know how far it is from the real PO front, i.e., what is the approximation error introduced by using the proposed surrogate. This is required as the usage of any surrogate always introduces some error, which may misguide optimization and, thus, also the selection of the most preferred solution. To overcome this problem, the values of the chosen surrogate based optimal solution can be used as a reference point for the achievement scalarizing function (see, e.g., [8]) and optimize it with the original objective functions. As this would require using the original uncompressed decision space and be potentially computationally very expensive, it may not always be possible to solve the optimization problem in a reasonable time.

3.2   Implementation

The clustering based surrogate approach is not dependent on a specific clustering algorithm, a similarity metric, or a way of choosing the most representative variable, as these are always case specific. As an example, in the following case study the clustering based surrogate is constructed using commonly known K-means algorithm with cosine distance and the variable closest to the Euclidean center of each cluster is chosen as the representative one.

The actual clustering was implemented and verified using Python libraries and Jupyter Notebooks[1]. To solve the resulting multiobjective problem, IND-NIMBUS [10], an implementation of the synchronous NIMBUS method, was used. The single objective subproblems produced were solved with the CPLEX optimizer. Note that all solutions produced by synchronous NIMBUS are PO if the single optimal subproblems are solved to optimality [8].

A screenshot of the graphical user interface of IND-NIMBUS is shown in Fig. 1. On the left hand side, the current PO solution is shown in the *Classification* panel as a bar chart. Each horizontal bar represents an objective function and the end points denote the nadir and ideal values, respectively. For maximized objective functions the colored part starts from right and, thus, the less color the better the value. In this case, all objectives are to be maximized. The DM can indicate preferences by clicking different parts of the bars. If one clicks on the colored part, it means that the objective needs to be improved. On the other hand, if one clicks on the non-colored part, it means that the objective is allowed to impair. All the PO solutions computed during the solution process are shown in the top right panel called *Alternatives* while the most interesting ones found so far can be dragged to the *Best candidates* panel in bottom right.

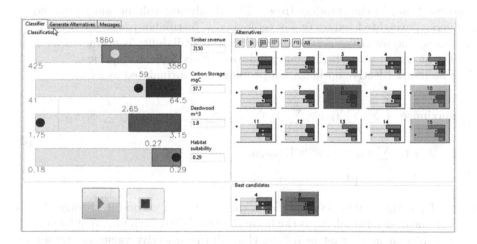

Fig. 1. A screenshot of IND-NIMBUS showing interaction with the DM.

4   Case Study: Multiobjective Forest Management

A forest landscape from Central Finland is used as a demonstrative example of the clustering approach. Information on the current state of the forest was collected by the Finnish Forest Center through field measurements. The forest

---

[1] Code available in https://github.com/josejuhani/gradu-code.

information represents 68700 ha, organized as 29666 stands. To predict the future forest resources, a forest simulator (MOTTI [11]) was used. The forest simulator predicted forest growth for a 50-year period, according to a pre-determined set of management alternatives. Depending on the initial stand characteristics, a range of seven management alternatives were generated. These alternatives ranged from setting the forest aside (doing nothing), conducting the typical management (business as usual), with a variety of extending/shortening the final harvest and including or excluding the option to thin the forest prior to final harvesting. The simulated data is openly available at https://dvn.jyu.fi/dvn/dv/Boreal_forest, and more detailed descriptions of the data and simulations can be found in [9,13,14].

Following the simulation of the set of different management alternatives, indicators representing a range of values were extracted. This set of indicators represented economic and ecological interests, and the set was selected to represent potential interests of specific stakeholders. The set of indicators (i.e., objective functions) was: timber revenue, carbon storage, deadwood volume, and a species habitat availability. The timber revenue was measured as the net present value revenue using a 3% discount rate. Carbon storage was measured as the tonnes of carbon contained within the forest (including the carbon in the soil, in the deadwood and in the standing trees). The deadwood volume was evaluated as a diversity weighted index: this is ecologically justifiable proxy for deadwood-inhabiting biodiversity [5]. The species habitat availability is evaluated as done in [9] which aggregates high quality habitat for six indicator species.

The multiobjective optimization problem was formulated as follows:

$$
\text{maximize} \left\{ \sum_{i=1}^{n} \sum_{j=1}^{7} T_{ij} x_{ij}, \sum_{i=1}^{n} \sum_{j=1}^{7} C_{ij} x_{ij}, \sum_{i=1}^{n} \sum_{j=1}^{7} D_{ij} x_{ij}, \sum_{i=1}^{n} \sum_{j=1}^{7} S_{ij} x_{ij} \right\}
$$
$$
\text{s.t.} \quad \sum_{j=1}^{7} x_{ij} = 1, \forall i = 1, \ldots, n, \tag{3}
$$
$$
x_{ij} \in \{0,1\},
$$

where T_{ij} is the timber revenue, C_{ij} is the amount of carbon in storage, D_{ij} is the volume of deadwood, S_{ij} is the habitat availability, each provided by stand i from management alternative j. Note that all the objective values are presented as per hectare. The decision variable values x_{ij} denote the jth management alternative selected for stand i. The total number of stands $n = 29666$.

This forest management problem has been solved earlier, focusing on various conservation related issues. In [9] the focus was on understanding the impacts conservation has on the profitability of forest management. The range of compromise solutions and the conflicts between various solutions has been explored in [13] and [14]. The common feature between these earlier solutions is the lack of integration with the DM.

For solving this forest management problem, the implementation of the clustering based surrogate presented in Sect. 3.2 was used. When empirically tested, the accuracy of the surrogate increased linearly with the increase of the number of clusters. Based on this, it was decided to choose 600 clusters for the surrogate as that amount kept the time between interactions in about 10 s. For the case study, the ideal and nadir values were obtained by using the original functions as previously suggested. These were verified with the previous research in [14]. For the chosen clustering, the optimal solutions for the four individual objective functions differed from the known real optima by 0.15%, 0.47%, 2.67% and 1.38%. Further, the usefulness of clustering was verified by comparing the approach against random clustering. The accuracy with random clustering in optimizing each objective individually varied between 3.2%–17.8% indicating poor performance of random clustering (results based on 10 independent runs).

5   Results and Discussion

The interactive solution process was performed by using the implementation described in Sect. 3.2. The DM involved has significant experience in both research and implementation of forest management solutions. To start the solution process, a neutral compromise solution with values $(2710, 58.3, 2.76, 0.26)$ (obtained by using the midpoint between ideal and nadir values as a reference point), i.e., a solution where all the objectives were balanced, was shown to the DM. Starting from that solution, the DM wanted in the second iteration to improve carbon storage and habitat suitability while allowing timber revenue and deadwood volume impair. Based on those preferences, four fairly similar new alternative solutions were produced as shown in Table 1. From the new solutions obtained, the DM deduced that he would like to improve timber revenue.

As the current solution for the third iteration, the DM chose the first solution $(2070, 60.4, 3.02, 0.28)$. He wanted to see how solution changes if timber revenue is desired to improve until 2500 and the others left to reach for the values set already in the previous iteration except for small increase for carbon storage. Now, the DM wanted to see two new solutions (i.e. use only two scalarizations) and optimization produced two new alternative solutions shown in Table 1.

The DM was quite happy with both the solutions, slightly preferring the second one which had higher timber revenue (2420) when compared to the first one (2280). He also realized that deadwood volume was not changing much. However, he wanted to see how would a solution in between these too look like and, thus, gave preferences as $(2400, 59.5, 2.81, 0.28)$. After optimization, the solution $(2380, 59.4, 2.87, 0.28)$ was obtained which the DM was happy with. It had a moderate amount of timber revenue and quite high carbon storage and overall it was focusing more on the ecological aspects of forest management.

Finally, the DM wanted to still see what happens to ecological objectives if the timber revenue is maximized while letting the other values change freely. That should produce an alternative solution focusing on the monetary aspect and enable comparison with the preferred solution already found. As expected, the two solutions found maximizing the timber revenue had poor values for all the ecological objectives and, thus, supports the selection of the balanced solution having objective values $(2400, 59.5, 2.81, 0.28)$. The DM was now satisfied and the solution process was finished.

Table 1. Results of iterations in solving the multiobjective problem.

Iter	Issue	Timber revenue [€]	Carbon storage [mgC]	Deadwood volume [m^3]	Habitat suitability
	Ideal	3640.0	64.8	3.18	0.29
	Nadir	450.0	41.2	1.16	0.17
1	Init. Sol.	2710.0	58.3	2.76	0.26
2	Cur. Sol.	2710.0	58.3	2.76	0.26
	Classif	$z_1^{bnd} = 2070.0$	$z_2^{asp} = 59.2$	$z_3^{bnd} = 2.19$	$z_4^{asp} = 0.28$
		2070.0	60.4	3.02	0.28
		2180.0	60.0	2.92	0.28
		2250.0	59.9	2.92	0.28
		2150.0	60.1	2.91	0.28
3	Cur. Sol.	2070.0	60.4	3.02	0.28
	Classif	$z_1^{asp} = 2500.0$	$z_2^{bnd} = 59.9$	$z_3^{bnd} = 2.19$	$z_4^{bnd} = 0.28$
		2280.0	59.9	2.99	0.28
		2420.0	59.3	2.83	0.27
4	Cur. Sol.	2420.0	59.3	2.83	0.27
	Classif	$z_1^{bnd} = 2400.0$	$z_2^{asp} = 59.5$	$z_3^{bnd} = 2.81$	$z_4^{asp} = 0.28$
		2380.0	59.4	2.87	0.28
5	Cur. Sol.	2380.0	59.4	2.87	0.28
	Classif	$z_1^{asp} = 3640.0$	$z_2^{bnd} = 41.2$	$z_3^{bnd} = 1.16$	$z_4^{bnd} = 0.17$
		3630.0	41.2	1.16	0.17
		3630.0	41.8	1.53	0.19
	Final Sol.	**2380.0**	**59.4**	**2.87**	**0.28**

6 Conclusions

Using the developed cluster-based surrogate approach to find nearly optimal solutions, a quick interactive decision process was enabled. Although the DM only went through a small number of iterations, the process was quick enough to maintain interest in the decision making process until a final acceptable solution was found. By using the implemented decision support tool, the DM was able to conveniently steer the solution process towards a final solution emphasizing ecological values while still having moderate amount of timber revenue. In addition, the nature of the conflicts between different objectives considered became more clear to him.

While this forest management problem has been solved in the extensive form earlier, it can be made more realistic. In this case, only a limited number of predefined management alternatives were used, which prevented the problem from being too large. Additionally, we did not explore the temporal sequence of planning outcomes, nor were spatial relationships maintained. As future research is concerned, the proposed cluster-based approach will be extended to mixed variables. In addition, it will be tested with larger and more realistic data sets in forest management as well as applied to different types of applications.

Acknowledgment. This research was supported by the Academy of Finland (projects no. 311877 and 287496) and is related to the thematic research area DEMO of the University of Jyväskylä.

References

1. Aittokoski, T., Äyrämö, S., Miettinen, K.: Clustering aided approach for decision making in computationally expensive multiobjective optimization. Optim. Methods Softw. **24**(2), 157–174 (2009)
2. Chugh, T., Sindhya, K., Hakanen, J., Miettinen, K.: Handling computationally expensive multiobjective optimization problems with evolutionary algorithms: a survey. Soft Comput. (to appear)
3. Ehrgott, M.: Multicriteria Optimization, 2nd edn. Springer, Berlin (2005). https://doi.org/10.1007/3-540-27659-9
4. Jin, Y.: Surrogate-assisted evolutionary computation: recent advances and future challenges. Swarm Evol. Comput. **1**(2), 61–70 (2011)
5. Lassauce, A., Paillet, Y., Jactel, H., Bouget, C.: Deadwood as a surrogate for forest biodiversity: meta-analysis of correlations between deadwood volume and species richness of saproxylic organisms. Ecol. Ind. **11**(5), 1027–1039 (2011)
6. Miettinen, K.: Nonlinear Multiobjective Optimization. Kluwer Academic Publishers, Boston (1999)
7. Miettinen, K., Hakanen, J., Podkopaev, D.: Interactive nonlinear multiobjective optimization methods. In: Greco, S., Ehrgott, M., Figueira, J. (eds.) Multiple Criteria Decision Analysis: State of the Art Surveys, vol. 233, 2nd edn, pp. 931–980. Springer, New York (2016). https://doi.org/10.1007/978-1-4939-3094-4_22
8. Miettinen, K., Mäkelä, M.M.: Synchronous approach in interactive multiobjective optimization. Eur. J. Oper. Res. **170**, 909–922 (2006)
9. Mönkkönen, M., et al.: Spatially dynamic forest management to sustain biodiversity and economic returns. J. Environ. Manag. **134**, 80–89 (2014)
10. Ojalehto, V., Miettinen, K., Laukkanen, T.: Implementation aspects of interactive multiobjective optimization for modeling environments: the case of GAMS-NIMBUS. Comput. Optim. Appl. **58**(3), 757–779 (2014)
11. Salminen, H., Lehtonen, M., Hynynen, J.: Reusing legacy FORTRAN in the MOTTI growth and yield simulator. Comput. Electron. Agric. **49**(1), 103–113 (2005)
12. Tabatabaei, M., Lovison, A., Tan, M., Hartikainen, M., Miettinen, K.: ANOVA-MOP: ANOVA decomposition for multiobjective optimization. SIAM J. Optim. (to appear)
13. Triviño, M., et al.: Managing a boreal forest landscape for providing timber, storing and sequestering carbon. Ecosyst. Serv. **14**, 179–189 (2015)
14. Triviño, M., et al.: Optimizing management to enhance multifunctionality in a boreal forest landscape. J. Appl. Ecol. **54**(1), 61–70 (2017)
15. Wang, H., Jin, Y., Jansen, J.O.: Data-driven surrogate-assisted multiobjective evolutionary optimization of a trauma system. IEEE Trans. Evol. Comput. **20**(6), 939–952 (2016)

Data Science in the Business Environment: Skills Analytics for Curriculum Development

Jing Lu[✉]

University of Winchester, Winchester SO22 5HT, UK
Jing.Lu@winchester.ac.uk

Abstract. Data science is an interdisciplinary field of methods, processes, algorithms and systems to extract knowledge or insights from data. University of Winchester Business School, UK is developing an undergraduate degree programme in Data Science which brings together student-centred and business-driven approaches: positioning the course for the interests of students and requirements of employers. The new programme follows the expectations of relevant subject benchmark statements and is built on activities which focus on different aspects of data science, drawing on some existing modules as a base. It integrates key themes in information management, data mining, machine learning and business intelligence. This paper presents the ongoing development of the Data Science programme through the key aspects in its conception and design. Understanding the employment market while defining specific skills sets associated with potential graduates is always important for courses in higher education. The Skills Framework for the Information Age (SFIA) has been adopted and a novel mapping proposed for the interpretation of employability skills related to data science. These are then linked to an adapted process model as well as the specialist modules across academic levels.

Keywords: Subject benchmarks · Skills frameworks · Business analytics · Data mining · Machine learning · Business intelligence · Analytical tools · SFIA

1 Introduction

Data Science is an emerging field that requires multi-disciplinary principles to guide the extraction of knowledge from data. In the Business context, the ultimate goal of data science is improving decision making and its links to Big Data and other data-driven technologies. Within the University of Winchester (UoW) Business School in the UK, a new BSc (Hons) Data Science programme is under development which recognises the increasing importance to organisations of knowledge as a commodity. The curriculum is adopting a distinctive structure and pedagogy, building on the well-established Digital & Technology Solutions Degree Apprenticeship as well as the newly-validated Computer Science suite of courses. This is articulated particularly through some specialist modules where technology and business-oriented activities are designed to focus on different aspects of data science, namely: information management, data mining, machine learning and business intelligence.

© Springer Nature Switzerland AG 2019
G. Nicosia et al. (Eds.): LOD 2018, LNCS 11331, pp. 116–128, 2019.
https://doi.org/10.1007/978-3-030-13709-0_10

This paper describes the ongoing development of the Data Science programme and starts with the expectations of relevant QAA subject benchmark statements in the UK, including Business and Management; Computing; and Mathematics, Statistics and Operational Research. Understanding the employment market while defining specific skills sets associated with data science is important for corresponding courses in higher education. Within the same Sect. 2, the national Skills Framework for the Information Age (SFIA) has been adopted to provide the necessary underpinning for the programme, which allows a novel interpretation of data science skills.

Section 3 extends the theme of data science in practice by adapting a cross-industry standard process model as a methodology to guide relevant activities and tasks, which are linked to SFIA-related skills from a business-driven perspective. Analytical tools and publicly available data sources have been recommended here in order to facilitate student projects in terms of data pre-processing, visualisation and analytics.

The Data Science curriculum design has been illustrated in Sect. 4 through a graphical representation across the three academic levels, which gives an indication of the specialist modules versus the more diverse. All of the specialist modules are then linked with relevant SFIA skills through a visual mapping. The paper draws to a close with some concluding remarks and a pointer to future work in relation to the EDISON Data Science Framework.

2 Academic and Professional Frameworks

2.1 Subject Benchmark Statements

Considering Part A of the UK Quality Code for Higher Education, which covers *setting and maintaining academic standards*, there is a range of Subject Benchmark Statements which UK universities are required to meet across their undergraduate provision [9]. There is no particular statement for Data Science as yet, but it is relevant to consider three current subject benchmarks in the context of this paper.

The Business and Management benchmark statement from 2015 [10] generally applies to the various honours degree courses in business studies and management studies, including (e.g.) organisational development and strategic management. However, it can also be used to inform a wider provision, including those courses focused on business functions or sectors. A broad, analytical and highly integrated study of business and management is expected within a framework encompassing organisations, business environment and management. Environment here comprises a range of factors, notably the digital and technological, while management includes rational analysis and other processes of decision making within organisations.

Graduates from Business Schools should be able to demonstrate knowledge and understanding in several areas: one of these is information systems and business intelligence. Skills of particular relevance include problem solving and critical analysis; research – ability to analyse and evaluate a range of business data, sources of information and appropriate methodologies … for evidence-based decision making; and numeracy – use of quantitative skills to manipulate data, evaluate, estimate and model business problems [10].

The next subject benchmark considered here is that for Mathematics, Statistics and Operational Research from 2015 [12]. It applies to cognate programmes of study in MSOR including (e.g.) computational mathematics, numerical analysis and statistical modelling. There are so many real-world applications of mathematics, which has its roots in the systematic development of methods to solve practical problems in areas such as construction and commerce. Understanding of the world is facilitated by identifying and codifying patterns, enabling deeper relationships to be found than could otherwise have been possible from observation or unaided reasoning.

Statistics has been characterised in the MSOR statement as the science of drawing conclusions from data. It includes methods for describing and visualising data to reveal patterns within it as well as the underlying processes producing such data, to extract information and predict future outcomes. The subject area of analytics has become increasingly associated with operational research in recent years. While the name OR is generally well understood, some provision has adopted other titles across the sector, notably: management science, business analytics, business decision methods and business systems modelling [12].

It is worth noting that the MSOR subject benchmark advises that its statement is unlikely to apply to teaching outside cognate departments, although it is important that such programmes pay due attention to the place of MSOR within them.

Moving on finally to the Computing benchmark statement from 2016 [11], a range of provision across computer science and information systems is addressed. For the purposes of this paper, courses in data management, information modelling, machine learning and knowledge representation are especially relevant. Computing overlaps with a number of adjacent subjects, including (e.g.) mathematics and business. Information systems in particular is concerned with the modelling, codification and storage of data for subsequent analysis – specific areas of interest again relate to databases and information modelling as well as the interactions between information systems and the more socio-technical systems. Those courses focused on Computing in society fall under this subject benchmark if their content is informed by computer engineering, software engineering, information technology or information systems.

Computing-related cognitive skills include an understanding of scientific method and its application to problem solving; knowledge and understanding of modelling for the purpose of (e.g.) prediction; as well as the deployment of methods and tools for implementation of systems. On the other hand, Computing-related practical skills comprise a range of abilities, notably deploying tools effectively in the solution of real-world applications; and critically evaluating and analysing complex problems, including those with incomplete information. Universities are also required to provide every student the opportunity to acquire more generic skills to enhance employability. They include intellectual skills, self-management, team working and significantly *contextual* awareness here, to understand and meet the needs of (e.g.) business and the community [11].

It can be seen that there are elements of all three benchmarks which are relevant to some extent to Data Science and these are developed further in the following sections. In particular it is clear that, if only one subject benchmark statement was selected, it would be that for Computing.

2.2   Professional Skills Frameworks

The Skills Framework for the Information Age (SFIA) describes the skills expected of professionals in roles involving information and communications technology. It has become the globally accepted common language for skills and competencies required in the digital world [13]. SFIA gives employers a framework which they can use to measure the skills they have against the skills they need, and tells education and training providers what the job market wants. It is supported by key organisations such as: BCS (British Computer Society), Tech Partnership (formerly e-skills UK), IET (Institution of Engineering and Technology), IMIS (Institute for the Management of Information Systems) and the IT Service Management Forum (itSMF).

BCS in conjunction with SFIA offer a skills matrix, called SFIAplus [14], which contains the framework of IT skills plus detailed training and development resources. It provides the most established and widely adopted skills, training and development model reflecting current industry needs. SFIAplus can be viewed as a three-dimensional model which comprises Categories of Work – Strategy and Architecture; Change and Transformation; Development and Implementation; Delivery and Operation; Skills and Quality; Relationships and Engagement – as well as Levels of Responsibility and Task Components.

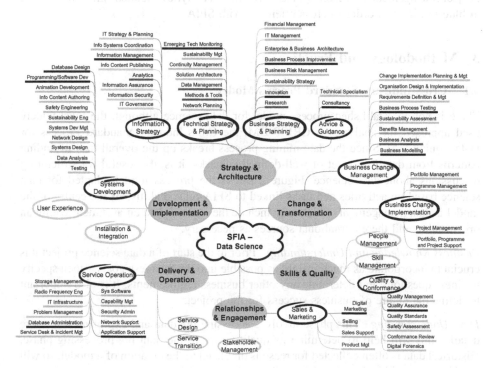

Fig. 1. An interpretation of data science skills using SFIA. (*Colour figure online*)

The Data Science programme proposed at UoW aims to develop how to use data to provide key business insights, helping companies improve their performance and make key decisions. Moreover, the programme is designed to meet employers' needs for innovative expertise as well as students' needs for an engaging and developmental course of study leading ultimately to rewarding employment. While the UK Government has published the essential *capabilities* it needs from Data Scientists [3], describing in some detail the knowledge and experience required, the corresponding essential *competencies* comprise those used in the Civil Service and are more generic.

For the Business environment, SFIA has been chosen as the reference against which employability skills are mapped here. A novel presentation of this framework is given in Fig. 1 – individual skills are displayed across six categories of work and associated subcategories – however, the ones which are considered most relevant to data science are shown in bold. It is interesting to note that a key category in this interpretation is Strategy, although Business Change is relatively significant too.

In terms of wider frameworks for data science, the EU-funded EDISON project [2] has focused on activities to establish the new profession of Data Scientist. This has included development of a Data Science Competence Framework (CF-DS) which provides the basis for other components. CF-DS defines five competence groups as Data Analytics; Data Science Engineering; Data Management; Research Methods and Project Management; and Domain-based Business Analytics. Related skills are labelled in blue in Fig. 1 in order to cross reference with SFIA.

3 Methodology and Practice

3.1 Cross-Industry Standard Process Model

There is not an established process model for data science although the most widely used approach for analytics is CRISP-DM, the Cross-Industry Standard Process for Data Mining [15]. Since the data mining process breaks up the overall task of finding patterns from data into a set of well-defined subtasks, it is also useful for structuring discussions about data science. Figure 2 shows the process model adapted for data science based on activities and tasks linked to SFIA-related skills. At the centre of the model is data management, which may include the internal data environment within an organisation and the external data sources as necessary.

Business Knowledge and Understanding. Prior to the start of a data science project it is crucial to incorporate as much insight as possible into the business goals – then specify business questions and determine any other business requirements. It is also important to define the nature of business success for the project.

Data Understanding. This phase involves accessing the data and exploring it in more detail – this will help to determine its quality prior to the data pre-processing phase. Historical data is often collected for reasons unrelated to the creation of a model, so will need to be considered appropriate to the project.

Data Pre-processing. The data chosen to be included in the analysis may be based on the objectives set at the business understanding stage, the quality of the data determined at the data understanding stage or other practical aspects. Data may be constrained by the analytical technologies used to create the model, e.g. it may be required to be in a different format. Preparation will involve all activities required to construct the final dataset including selecting attributes, cleaning the data to address any data quality issues and transforming data to create derived variables.

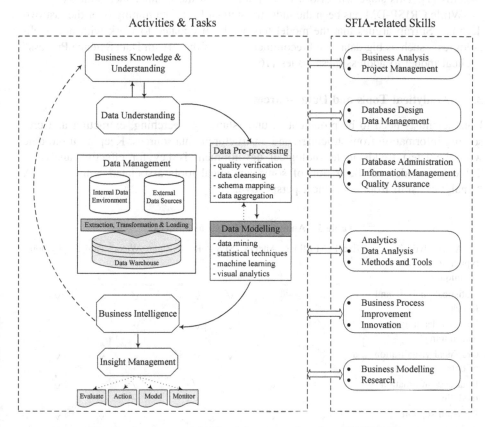

Fig. 2. A process model for data science based on CRISP-DM.

Data Modelling. Various modelling approaches will be deployed based on the business objectives and the dataset which is used [5]. Statistical analysis, data mining and machine learning are fundamentally involved with extracting information from a dataset. Common analytical techniques are classification, clustering, regression and dimension reduction while visual analytics technology combines data analysis with data visualisation and human interaction.

Business Intelligence. The primary goal of data science for business is to support decision making – business intelligence focuses on supporting and improving the decision-making process. The modelling results will need to be evaluated carefully as "various stakeholders have interests in the business decision-making that will be accomplished or supported by the resultant models" [8].

Insight Management. The results from the modelling and subsequent evaluation will determine how the model will be deployed to make improvements in organisations. This could include implementing a predictive model into pre-existing information systems [1]. This stage will also involve planning of the maintenance strategy.

While CRISP-DM has been the industry standard for data mining over the last two decades, Stirrup argues that the model has not been updated to work with new technologies – such as big data – and recommends use of the "Team Data Science Process" cyclical model to address these issues [16].

3.2 Analytical Tools and Data Sources

Data scientists need to be proficient in understanding, searching, extracting and presenting information from structured and unstructured data sources. Keeping up-to-date with the latest trends in technological development is key for effective analytics. Table 1 provides an illustration of some analytical tools associated with the SFIA "Analytics" skill from a technical perspective.

Table 1. Analytical tools and techniques.

SFIA "Analytics" – typical tools and techniques	Excel	XLMiner	Alteryx	SPSS	R	iNZight	Weka	Tableau	Python
Statistical analysis and forecasting	√	√		√	√	√	√	√	√
Machine learning and data mining		√		√	√		√		√
Graphical visualisation of data	√	√		√	√	√	√	√	√
Data and information modelling			√		√		√		√
Decision support systems		√	√					√	√

As a spreadsheet, Excel can be used for data entry, manipulation and presentation, but it also offers a suite of statistical analysis functions and other tools that can be used to run descriptive statistics and perform inferential statistical tests. In addition, XLMiner is the comprehensive data mining plug-in for Excel, now known as Analytic Solver.

Alteryx is a tool especially made to extract, transform and load data into a data warehouse. Its key capabilities for data preparation include: connect to and cleanse data from data warehouses, spreadsheets and other sources; improve quality of data with profiling, advanced data cleansing and data manipulation tools; repeatable workflow design to assist with data integrity during data preparation process.

SPSS is a software package which has been widely used for statistical analysis by social scientists, education researchers, health researchers, market researchers, survey companies, government and other organisations for many years. IBM SPSS Modeler is a data mining and text analytics software application used to build predictive models and conduct other analytical tasks.

R is a language and environment for statistical computing and graphics, with RStudio providing a user-friendly interface to analyse and manipulate data. R is commonly used for big data management and analysis – it is widely accepted in the data science field and has a very active support community. Developed using R, iNZight can also generate insights into real-world data by producing graphs and summaries through statistical analysis.

Weka (Waikato Environment for Knowledge and Analysis) is open source software written in Java [4] which offers a wide range of statistical inference and machine learning algorithms. It contains tools for data pre-processing, classification, regression, clustering, association rules, sequential patterns mining and visualisation. It provides a way to easily test the performance of a comprehensive suite of data mining and machine learning algorithms on real-world problems.

Tableau Software provides a collection of interactive data visualisation products designed for business intelligence. Its advanced analytics functionalities include: cohort analysis through drag-and-drop segmentation; what-if analysis by modifying calculations and testing different scenarios; and predictive analysis using trending and forecasting models. In addition, an R plug-in allows integration with other platforms.

Python has become an even more popular and powerful programming language in the era of data science. Data analysis, machine learning, information visualisation and text analysis techniques can be applied through Python software libraries and toolkits such as pandas, scikit-learn, matplotlib and nltk to gain further insight into data.

Table 2 is a list of some useful resources for data science projects in the areas of data cleansing, visualisation, data mining and machine learning – the data sources column contains hyperlinks to the individual repositories.

Table 2. Data sources for data science projects.

	Data sources	Description
Data cleansing	data.world	A social-based data source that allows users to share/clean/improve data collectively. Can write SQL within the interface to explore data and join multiple datasets
	The world bank	The platform provides several tools like Open Data Catalog, world development indices, education indices etc.
	Reddit	A community discussion site which has a section devoted to sharing interesting datasets

(*continued*)

Table 2. (*continued*)

	Data sources	Description
Data visualisation	FiveThirtyEight	Interactive news and sports site with data-driven articles. Each dataset includes the data, a dictionary and the link to the story
	FlowingData	Catalogue of data sources, described in detail and shown with examples. It explores how statisticians, data scientists and others use analysis and visualisation
	Tableau public	Sample data for visual analytics in the categories of Education, Public, Government, Science, Technology, Health, Business, Sports and Entertainment etc.
Machine learning	UCI machine learning repository	One of the oldest and most famous sources of datasets online. Vast majority are clean and ready for machine learning
	Kaggle	A data science community which hosts machine learning competitions – contains externally-contributed datasets
	Quandl	For financial and economic datasets – useful for building models to predict economic indicators or stock prices

4   Education and Training

4.1   UG Programme Development

Within the UoW Business School, the BSc Digital & Technology Solutions degree apprenticeship and BSc Computer Science suite both inform the BSc Data Science prototype. One way to express the significance of current modules to the new undergraduate programme is to display the relationships graphically. Figure 3 shows the extent to which relevant modules may contribute to Data Science – each of the individual boxes represents a module with the colour-coding across Level 4 to 6. The boxes within the triangle are the specialist modules proposed for data science while the oval shapes outside indicate diverse modules from other programmes, with dotted ovals for optional modules. There is also a Group Project module for Level 5 and a double-credit Data Science Project for Level 6.

A brief description for each specialist taught module is given below – these are linked with Data Science Body of Knowledge areas from the EDISON project [2], which are associated with their CF-DS competence groups.

Database Analysis and Design. Introduces analysis and design concepts (using SQL and UML) that are essential for developing and implementing relational database solutions in given business scenarios [DSDM/DMS: Data management systems].

Quantitative Data Analysis. Introduces quantitative analytics concepts, procedures and software tools (Excel and SPSS) for specific data analysis tasks [DSDA/SMDA: Statistical methods for data analysis].

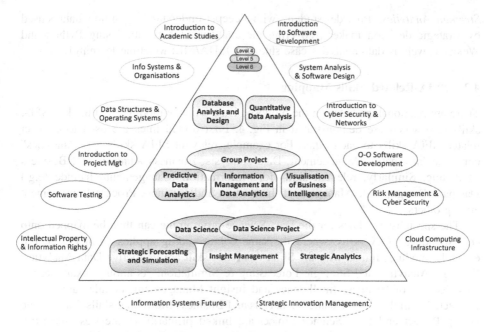

Fig. 3. BSc data science prototype.

Information Management and Data Analytics. Organised around three themes: Database Management and SQL; Data Warehousing and Information Modelling; Data Mining and Knowledge Discovery [DSDA/DM: Data mining].

Predictive Data Analytics. Provides experience of predictive modelling and analytics across a range of domains, acquiring relevant practical skills (using R and Weka) in data science to create data visualisations and carry out analyses [DSDA/PA: Predictive analytics].

Visualisation of Business Intelligence. Focuses on techniques for data extraction and preparation while analysing data in visual ways (using Alteryx and Tableau) to generate insight for business intelligence and decision making [DSENG/IS: Information systems].

Insight Management. Provides knowledge and skills to identify and evaluate a business issue and/or research problem, effectively analyse data and interpret insights (using iNZight and Tableau) so that they can have an impact at managerial levels of organisations [DSBPM/BA: Business analytics].

Strategic Forecasting and Simulation. Covers the data-driven business prediction topics of forecasting and simulation (using R and XLMiner) to develop advanced models and solutions to real-world problems [DSDA/MODSIM: Computational modelling, simulation and optimisation].

Strategic Analytics. Provides students with a deeper understanding of how data is used by strategic decision makers, covering the analysis of big data (using Python and Weka) as well as data analytics case studies [DSDA/ML: Machine learning].

4.2   SFIA-Related Skills Mapping

There are relationships between the proposed Data Science modules and the key SFIA skills too, which are demonstrated in Fig. 4. First dotted lines are used to connect related SFIA skills to each other. For example, relevant SFIA skills for "Analytics" comprise Information Management, Data Analysis, Business Analysis and Business Modelling. Similarly, relevant skills for "Information Management" include (e.g.) Database Design, Data Management, Innovation and Business Process Improvement among others.

The specialist modules for the Data Science programme can then be mapped onto corresponding SFIA skills, where the same colour-coding applies as in Fig. 3. For example, modules linked with the SFIA Analytics skill are Predictive Data Analytics, Strategic Analytics and Strategic Forecasting & Simulation. As another example, the Visualisation of Business Intelligence and Insight Management modules are closely connected with the SFIA Business Analysis and Business Modelling skills. Finally, the Group Project and Data Science Project are linked primarily to the Research skill, although the taught specialist modules will all apply to some degree. Figure 4 represents a novel aspect of skills analytics in the Data Science context.

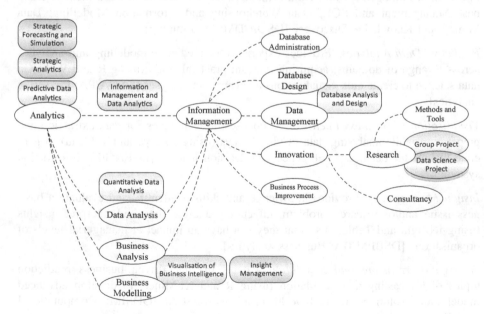

Fig. 4. Linkage between SFIA skills and specialist modules.

5 Conclusion

The digital economy has facilitated an explosion in the data available to the world which has affected businesses, jobs and education. The term "Big Data" refers to datasets so large and complex that it would be impossible to analyse them using traditional methods. Big data was originally defined in terms of the three Vs, namely: "high-*volume*, high-*velocity* and high-*variety*". A 4th V for *veracity* ensued, referring to the trustworthiness of the data. However all this enormous quantity of fast-moving data of different types and confidence levels has to be turned into *value*, which leads to the 5th V for big data [6]. Data Science will help organisations to turn data into valuable insights in order to better understand their customers and optimise their internal processes while identifying cost savings and growth opportunities [7]. Some representative business analytics approaches include for example financial analytics, market analytics, customer analytics, employee analytics and operational analytics alongside the core analytical tools and techniques.

This paper has discussed an overall curriculum design and the skills required for Data Science in the business environment. The new BSc Data Science development is already having a positive impact on other programmes within the University of Winchester Business School, for example: BA Accounting & Finance/Management and their Level 5 Research and Analysis module; MSc Digital Marketing & Analytics and its Analytical Tools for Digital Data module; and the Executive MBA module delivering Insight Management for business professionals.

In terms of the next stage for programme development at Winchester, the EDISON Data Science Framework [2] will be considered further – in particular the detailed Data Science Model Curriculum. An evaluation of the extent to which their recommended learning outcomes and topics would apply in UK higher education will be significant here, especially within a Business School context. The real evidence of what can be achieved by the programme will begin to materialise following its first year of delivery in 2019/20.

References

1. Abbott, D.: Applied Predictive Analytics: Principles and Techniques for the Professional Data Analyst. Wiley, Indianapolis (2014)
2. EDISON: Building the data science profession. http://edison-project.eu/. Accessed 6 Mar 2018
3. GOV.UK: Data scientist – skills they need. https://www.gov.uk/government/publications/data-scientist-skills-they-need/data-scientist-skills-they-need. Accessed 12 Mar 2018
4. Hall, M., Frank, E., Holmes, G., Pfahringer, B., Reutemann, P., Witten, I.H.: The WEKA data mining software: an update. SIGKDD Explor. **11**(1), 10–18 (2009)
5. IBM: CRISP-DM help overview. https://www.ibm.com/support/knowledgecenter/SS3RA7_15.0.0/com.ibm.spss.crispdm.help/crisp_overview.htm. Accessed 28 Feb 2018
6. Marr, B.: Big Data in Practice: How 45 Successful Companies Used Big Data Analytics to Deliver Extraordinary Results. Wiley, Oxford (2016)
7. Marr, B.: Key Business Analytics: The 60 + Business Analysis Tools Every Manager Needs to Know. Pearson, Harlow (2016)

8. Provost, F., Fawcett, T.: Data Science for Business: What You Need to Know About Data Mining and Data-Analytic Thinking. O'Reilly Media, California (2013)
9. QAA Subject Benchmark Statements for subjects studied at honours degree level. http://www.qaa.ac.uk/assuring-standards-and-quality/the-quality-code/subject-benchmark-statements/honours-degree-subjects. Accessed 9 Mar 2018
10. QAA Subject Benchmark Statement: Business and Management. http://www.qaa.ac.uk/publications/information-and-guidance/publication?PubID=2915#.WqKJh2rFLIU. Accessed 9 Mar 2018
11. QAA Subject Benchmark Statement: Computing. http://www.qaa.ac.uk/publications/information-and-guidance/publication?PubID=3043#.WqKIk2rFLIU. Accessed 9 Mar 2018
12. QAA Subject Benchmark Statement: Mathematics, Statistics and Operational Research. http://www.qaa.ac.uk/publications/information-and-guidance/publication?PubID=2952#.WqKJ4mrFLIU. Accessed 9 Mar 2018
13. SFIA: The Skills Framework for the Information Age. http://www.sfia.org.uk. Accessed 28 Feb 2018
14. SFIAplus: BCS. http://www.bcs.org/server.php?show=nav.7849. Accessed 28 Feb 2018
15. Shearer, C.: The CRISP-DM model: the new blueprint for data mining. J. Data Warehouse. **5** (4), 13–22 (2000)
16. Stirrup, J.: What's wrong with CRISP-DM, and is there an alternative? https://jenstirrup.com/2017/07/01/whats-wrong-with-crisp-dm-and-is-there-an-alternative/. Accessed 28 Feb 2018

Adaptive Dimensionality Reduction in Multiobjective Optimization with Multiextremal Criteria

Victor Gergel[ID], Vladimir Grishagin[✉][ID], and Ruslan Israfilov[ID]

Lobachevsky State University, Gagarin Avenue 23, 603950 Nizhni Novgorod, Russia
{gergel,vagris}@unn.ru, ruslan@israfilov.com

Abstract. The paper is devoted to consideration of multicriterial optimization (MCO) problems subject to multiextremality of criteria. Application of convolution techniques for finding partial Pareto-optimal solutions generates under this assumption the multiextremal problems of scalar optimization. For solving these problems it is necessary to use efficient global optimization algorithms. As such the methods the nested schemes of dimensionality reduction in combination with univariate characteristic optimization algorithms are considered. A general description of the scheme is given and its modification accelerating the search is presented. Efficiency of the proposed approach is demonstrated on the base of representative computational experiment on a test class of bi-criterial MCO problems with essentially multiextremal criteria.

Keywords: Multicriterial optimization · Multiextremal criteria · Dimensionality reduction · Global search algorithms

1 Introduction

Mathematical models formulated as multiobjective, or multicriterial optimization (MCO) problems describe complicated decision making processes in which the main factors of complexity are contradictoriness of partial criteria and dimensionality of the problem. The contradictoriness leads to the necessity of consideration a set of compromise solutions (Pareto set) as a general solution of the multicriterial problem investigated. For finding the compromise solutions the initial MCO problem is often reduced to a family of scalar optimization problems in the form of mathematical programming ones, for example, by means of convolution techniques.

Various approaches to investigation of the MCO problems have been described in many fundamental publications (see, for example, the monographs [1–4]). Some theoretical and practical aspects of MCO investigation can be found in [5–12].

The variety of MCO models is the source of different classes of MCO problems determined by the properties of criteria and constraints describing the

© Springer Nature Switzerland AG 2019
G. Nicosia et al. (Eds.): LOD 2018, LNCS 11331, pp. 129–140, 2019.
https://doi.org/10.1007/978-3-030-13709-0_11

model. Among these models the class of multidimensional MCO problems with multiextremal criteria is one of the most difficult ones for the research because reducing to a single-criterion problem that determines a compromise solution generates a global optimization problem. For this class of problems the exponential growth of computational complexity when increasing the dimension (so called "the dimensionality curse" of multiextremal problems) [15] takes place. In order to solve arising problems of multiextremal optimization it is necessary to use efficient global search algorithms. There are many approaches to constructing such the methods oriented at different classes of multiextremal problems (see, for example, the monographs [15–21]). In this paper the methods based on ideas of dimensionality reduction are considered and applied to solving the MCO problems. In the framework of given approach the initial multidimensional problem is reduced to a family of univariate subproblems solved in general theoretical description by the characteristical methods [22] and by the core information global search algorithm [15] in computational experiment. This approach has demonstrated [23] its efficiency in comparison with other global optimization methods, in particular, with the popular method DIRECT [24].

The rest of the paper is organized as follows. Section 2 contains the statement of MCO problems to be investigated, and the general description of dimensionality reduction scheme on the base of recursive nested optimization. Section 3 is devoted to consideration of two modifications of the nested scheme (classical and adaptive) in combination with characteristical algorithms of univariate optimization. Section 4 presents the results of computational experiments and Sect. 5 concludes the paper.

2   Problem Statement and Reduction Schemes

The considered decision making model described as the multicriterial (or multiobjective) optimization (MCO) problem contains functions $w_i(y) : \mathbb{R}^N \to \mathbb{R}^1$, $1 \le i \le p$, $p > 1$, called *partial* criteria of the problem, depending on the vector of arguments $y = (y_1, \ldots, y_N) \in \mathbb{R}^N$ and defined over the domain

$$H = \{y \in \mathbb{R}^N : a_i \le y_i \le b_i, 1 \le i \le N\} \tag{1}$$

being a hyperparallelepiped in N-dimensional Euclidean space $\mathbb{R}^N$.

The statement of the MCO problem is to minimize in the domain (1) the vector function (*vector criterion*)

$$W(y) = (w_1(y), \ldots, w_p(y)). \tag{2}$$

Hereinafter this problem will be written in the form

$$W(y) \to \min, y \in H. \tag{3}$$

Each partial criterion $w_i(y)$, $1 \le i \le p$, is supposed to satisfy in H the Lipschitz condition

$$\left|w_i(y') - w_i(y'')\right| \le L_i \|y' - y''\|, \; y', y'' \in H, \tag{4}$$

with corresponding Lipschitz constant $L_i > 0$ where $\| \cdot \|$ denotes the Euclidean norm. The assumption (4) is required for the purpose of further applying the Lipschitz global optimization methods as tools of analyzing the problem (3).

Moreover, all the partial criteria are considered to possess positive values in the domain H. This requirement is necessary in the framework of the convolution scheme (5)–(7).

If the partial criteria are contradictory, it is impossible to find a point y^* such that it is the global minimum point for all of them. In this situation a compromise solution is considered as a partial solution of the problem and all the compromise solutions are the full solution of the MCO problem. We will deal with efficient (Pareto-optimal) points that form the Pareto set as the full solution of the problem (3).

There are many approaches to finding the partial solutions. For example, it is possible to build the Pareto set by means of reducing the initial MCO problem (3) to solving a parametrized family of mathematical programming (scalar optimization) problems

$$\Phi_\lambda(y) \to \min, y \in H, \tag{5}$$

with the objective function (convolution)

$$\Phi_\lambda(y) = \max_{1 \leq i \leq p} \left(\lambda_i w_i(y) \right) + \alpha \sum_{i=1}^{p} \lambda_i w_i(y), \tag{6}$$

where parameters λ belong to the set

$$\Lambda = \left\{ \lambda \in \mathbb{R}^p : \lambda_i \geq 0, \ 1 \leq i \leq p, \ \sum_{i=0}^{p} \lambda_i = 1 \right\} \tag{7}$$

and α is a small positive number [13, 14].

Under the Lipschitz condition (4) for partial criteria, the function (6) is Lipschitzian as well and, in general case, multiextremal. This circumstance necessitates applying efficient algorithms of global optimization for solving the problems (5). One of the known approaches to creating the qualitative global search methods is based on ideas of dimensionality reduction. This approach has been developing for many years and it is the source of many efficient global optimization algorithms [15, 19, 25–33]. Application of dimensionality reduction methods to the MCO problems and investigation of their efficiency for this goal is a novel research described in the present paper.

The main dimensionality reduction scheme considered hereinafter is the scheme of recursive nested optimization (another reduction approach based on Peano-type space filling curves can be found in the works [15, 19]). The nested optimization scheme reduces the multidimensional problem (5) to a family of univariate subproblems in the following way.

Let us introduce a family of reduced functions as

$$\Phi_\lambda^N(y) \equiv \Phi_\lambda(y), \tag{8}$$

$$\Phi^q_\lambda(y_1,\ldots,y_q) = \min\left\{\Phi^{q+1}_\lambda(y_1,\ldots,y_{q+1}) : a_{q+1} \le y_{q+1} \le b_{q+1}\right\}, 1 \le q < N. \tag{9}$$

Then, following the relation [15, 25, 26]

$$min_{y \in H}(y) = \min_{a_1 \le y_1 \le b_1} \min_{a_2 \le y_2 \le b_2} \cdots \min_{a_N \le y_N \le b_N} \Phi_\lambda(y), \tag{10}$$

instead of the multidimensional problem (5) one can solve the univariate problem

$$\Phi^1_\lambda(y_1) \to \min, \ y_1 \in [a_1, b_1]. \tag{11}$$

However, when solving this problem it is necessary to evaluate the function $\Phi^1_\lambda(y_1)$ at points of the interval $[a_1, b_1]$ but any evaluation at a given point $\tilde{y}_1$ leads to solving the problem

$$\Phi^2_\lambda(\tilde{y}_1, y_2) \to \min, \ y_2 \in [a_2, b_2] \tag{12}$$

being one-dimensional as well, and so on up to solving the univariate problem

$$\Phi^N_\lambda(y) \equiv \Phi_\lambda(y) \to \min, \ y_N \in [a_N, b_N], \tag{13}$$

where the coordinates $y_1, \ldots, y_{N-1}$ are fixed (obtained from preceding levels of one-dimensional optimization).

Thus, the described scheme allows one to substitute solving the multidimensional problem (5) for solving the family of nested univariate subproblems

$$\Phi^q_\lambda(y_1, \ldots, y_{q-1}, y_q) \to \min, \ y_q \in [a_q, b_q], \ 1 \le q \le N. \tag{14}$$

If the objective function $\Phi_\lambda(y)$ from (5) satisfies the Lipschitz condition (in our case this property is provided by the assumptions (4) the one-dimensional objective functions $\Phi^q_\lambda(y)$ in (14) also meet the Lipschitz condition (see [33]) and for solving subproblems (14) the methods of Lipschitz global optimization can be used.

3   Characteristical Algorithms and Adaptive Nested Optimization

Combining the nested scheme (14) with different one-dimensional optimization methods enables to design a wide spectrum of multidimensional algorithms. In particular, for solving the subproblems (14) the methods belonging to the wide class of characteristical algorithms [22] can be taken which many well-known global optimization algorithms [15, 19, 26, 33–35] belong to. The use of these algorithms inside the nested scheme allows one to modify the classical nested scheme for getting improvements of its functioning. Before returning to this modification and its further explanation, the general computational structure of characteristical algorithms should be described.

To simplify the description let us present the subproblems (14) in the following unified form

$$\varphi(x) \to \min, \ x \in [a, b]. \tag{15}$$

Then a numerical method for solving the optimization problem is characteristical one if its computational scheme consists in the following.

First $n \geq 1$ trials (evaluations of the objective function $\varphi(x)$) are executed at arbitrary trial points $x^1, \dots, x^n$ of the interval $[a, b]$ and the function values $z^1, \dots, z^n$ are evaluated at these points, i.e., $z^j = \varphi(x^j), 1 \leq j \leq n$. For obtaining a point x^{s+1} of any subsequent $(s+1)$-th trial for $s \geq n$ it is required to realize the following steps:

1. The points $x^1, \dots, x^s$ of preceding trials and the points a and b (if they were not the trial points earlier) are ordered in increasing order and renumbered by subscripts, i.e.,

$$x_0 = a \leq x_1 < \dots < x_{\nu-1} < x_\nu = b. \tag{16}$$

The values $z_j = \varphi(x_j)$ are juxtaposed to the points x_j from (16) belonging to the sequence $x^1, \dots, x^s$.

2. The ordering (16) splits the search region $[a, b]$ into ν subintervals (x_{j-1}, x_j), $1 \leq j \leq \nu$, for each of those a numerical value $R(j)$ (called *characteristic* of this interval) is assigned.

3. The subinterval (x_{k-1}, x_k), $1 \leq k \leq \nu$, such that

$$R(k) = \max_{1 \leq j \leq \nu} R(j), \tag{17}$$

is chosen among all the subintervals formed by the ordering (16).

4. The new $(s+1)$-th trial is carried out at a point $x^{s+1} \in (x_{k-1}, x_k)$, the value $z^{s+1} = \varphi(x^{s+1})$ is computed and the iteration number s is increased by 1.

General conditions of convergence for characteristical algorithms (including convergence to global minima) are presented in the paper [22]. These general results substantiate the stopping rule in the form

$$x_k - x_{k-1} \leq \varepsilon, \tag{18}$$

where $\varepsilon > 0$ is a given coordinate accuracy, i.e., the search is completed if the length of the subinterval with maximal characteristic from (17) is less than the accuracy ε.

As an example of characteristical method let us describe the core information global search algorithm [15, 33] using for it hereafter the short denotation GSA.

When solving the problem (15) two first trials GSA are executed at the end points of the search region, namely, $x^1 = a$, $x^2 = b$ and, consequently, $s = 2$ and $\nu = s - 1$ for $s \geq n$. Next trials are carried out in accordance with Steps 1–4, where the characteristics of the subintervals (x_{j-1}, x_j), $1 \leq j \leq \nu$, are calculated as

$$R(j) = m\delta_j + \zeta_j^2/(m\delta_j) - 4(z_{j-1} + z_j) \tag{19}$$

and the point of new trial

$$x^{s+1} = (x_{k-1} + x_k)/2 - \zeta_k/(2m). \tag{20}$$

Here $\delta_j = x_j - x_{j-1}; \zeta_i = z_i - z_{i-1}$, $i = j, k$; index k from (17), the factor m is evaluated as

$$m = \begin{cases} rM, & M > 0, \\ 1, & M = 0, \end{cases} \tag{21}$$

where

$$M = \max\{|\zeta_j|/\delta_j : 1 \le j \le \nu\} \tag{22}$$

and $r > 1$ is the parameter of GSA.

Applying characteristical algorithms in the nested optimization scheme makes possible to accelerate the search. A brief explanation of this effect consists in the following (more detailed information can be found in the papers [23,31]). In classical implementation of the nested scheme at any moment only one univariate subproblem of the level can be active; the others either have been solved already or will be solved after completing the current subproblem. Moreover, the information obtained in the course of optimization in the completed subproblems is not used during solving the current one. This loss of information slows up the multidimensional optimization.

In the paper [31] a new version of the nested scheme called *adaptive dimensionality reduction* has been proposed and theoretically substantiated. The core of the adaptive scheme consists in *simultaneous* consideration of all the subproblems (14) arising in the course of multidimensional optimization and in the choice for realization of a certain subproblem with some "best" features. It means that it is necessary to introduce a quality criterion for the subproblems. If a characteristical algorithm is used for solving the problems (14), for each subproblem its current maximal characteristic (17) is taken as the quality criterion of the whole subproblem.

The results of large-scale experimental comparison on complicated test classes of essentially multiextremal functions for several global optimization methods presented in [23] demonstrate significant advantage of the adaptive nested optimization over its classical prototype and the other methods compared.

Taking into account the results mentioned above and the confirmation of efficiency of GSA obtained earlier in other researches [15,22,31], in this paper the classical and adaptive optimization schemes combined with GSA are considered for the study of efficiency of the dimensionality reduction approach for solving the MCO problems.

4   Numerical Experiments

For efficiency assessment of the nested optimization schemes in classical and adaptive variants a class of bi-criterial MCO problems is constructed in the following way. 100 functions of dimension 2 (see [22,31])

$$\varphi^l(y) = \left\{ \left(\sum_{i=1}^{7}\sum_{j=1}^{7} \left[A_{ij}^l a_{ij}(y) + B_{ij}^l b_{ij}(y) \right] \right)^2 + \right.$$

$$\left. + \left(\sum_{i=1}^{7}\sum_{j=1}^{7} \left[C_{ij}^l a_{ij}(y) + D_{ij}^l b_{ij}(y) \right] \right)^2 \right\}^{\frac{1}{2}}, \quad (23)$$

where $a_{ij}(y) = \sin(\pi i y_1)\sin(\pi j y_2)$, $b_{ij}(y) = \cos(\pi i y_1)\cos(\pi j y_2)$, $1 \le l \le 100$, are taken with the coefficients A_{ij}^l, B_{ij}^l, C_{ij}^l, D_{ij}^l, which are independent random numbers, distributed uniformly over the interval $[-1,1]$. The functions of this class are essentially multiextremal and they are widely used for experimental testing in global optimization (see [22,27,29,31,33,35]).

Each function was converted to the criterion

$$w^k(y) = 15 - \varphi^k(y) \qquad (24)$$

being positive in the square

$$H = \left\{ y \in \mathbb{R}^2 : 0 \le y_1, y_2 \le 1 \right\}. \qquad (25)$$

There was considered 100 bi-criterial problems according to the following rule. In the l-th MCO problem ($1 \le l \le 99$) the function $w^l(y)$ was taken as the first criterion and the function $w^{l+1}(y)$ as the second one. For the last problem the function $w^{100}(y)$ was chosen as the first criterion and the function $w^1(y)$ as the second. The square (25) was used as the admissible domain (1) in all the problems.

For numerical building the Pareto set each bi-criterial problem was reduced to a family of scalar subproblems (5) corresponding to 160 values of the parameter λ_1 taken as nodes of the uniform grid in the interval $[0,1]$ (the coefficient $\lambda_2 = 1 - \lambda_1$ because of conditions (7)).

An example of convolution (6) (level curves and surface) is presented in Fig. 1. The image of the domain (25) mapped onto the plane of criteria is shown in Fig. 2, where the Pareto boundary is marked with red colour.

The global optimization subproblems (5) were solved by the algorithms on the base of classical and adaptive nested schemes combined with the method GSA. For comparison of the described algorithms with a method of other nature, the subproblems (5) were solved by the known and very popular global optimization method DIRECT [24] as well. For assessment of efficiency of the nested schemes and the method DIRECT the methodology of operational characteristics [15,23, 31] has been used. Briefly, this methodology consists in the following.

After solving a collection of optimization problems by a method with some fixed parameters we can evaluate the average number K of trials spent by the method and the number Q of problems solved successfully. Repeating this experiment with different method's parameters we get a set of pairs (K,Q) which is called the operational characteristic of the method. Operational characteristics

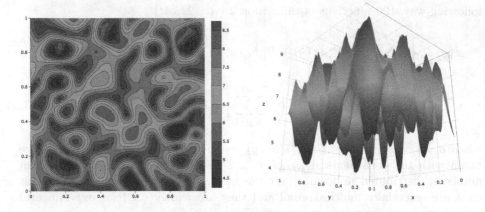

Fig. 1. Level curves and surface of convolution

of several methods presented in graphical form on the plane (K, Q) enable visual comparing the methods' efficiency. Namely, if for given value K the operational characteristic of one method is placed above the characteristic of the other, the first method is better because it has solved more problems.

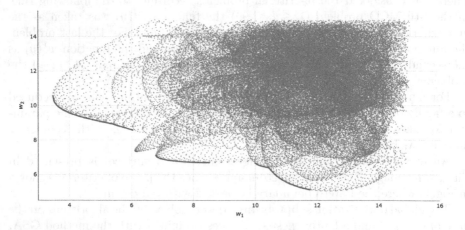

Fig. 2. Criterial plane and Pareto set (Color figure online)

The compared nested schemes optimized the convolutions (6) for different values of accuracies ε from (18) and used the GSA parameter $r = 3.2$ from (21) that provided the sufficient conditions of global convergence. The operational characteristics of the dimensionality reduction schemes are presented in Fig. 3.

In the figure the number K of trials is considered as the average number of evaluations spent for solving one subproblem (5) and this indicator is plotted on the abscissa axis in the logarithmic scale.

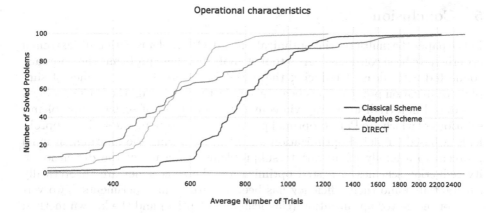

Fig. 3. Operational characteristics of the compared methods

The indicator corresponding to the vertical axis reflects the number of multicriterial problems solved successfully with a given tolerance. A problem (3) is considered to have been solved, if an approximation P of the Pareto set evaluated by the method via solving partial subproblems (5) corresponding to all the chosen parameters λ is sufficiently close to the "ideal" Pareto set P^*. As the measure of closeness the criterion $\omega(P) = 1 - \mathrm{hv}(P)/\mathrm{hv}(P^*)$ was used, where $\mathrm{hv}(P)$ is the hypervolume index [10,11] introduced for evaluating the quality of approximation. In the experiment the multicriterial problem was supposed to have been solved if $\omega(P) < 0.02$.

Another experiment has been carried out for 5-dimensional MCO problem with two multiextremal criteria taken from the test class GKLS [36] (subclass of hard complexity). GKLS is widely used for testing the global optimization methods. For building the Pareto set 100 scalar convolutions (6) corresponding to the different coefficients λ_1 uniformly distributed in the interval $[0,1]$ have been minimized.

In the nested schemes the parameter r from (21) was equal to 4.5 and $\varepsilon = 0.02$ in the stopping rule (18). Both the nested schemes and DIRECT have built the Pareto set with accuracy $\omega(P) < 0.02$, but the classical nested scheme has spent on average 248 745 trials (evaluations of convolution $\Phi_\lambda(y)$) per one scalar problem (5), DIRECT 100 258 trials and the adaptive scheme 45 155 evaluations.

The results of the experiment demonstrate the successful applicability of the global optimization methods based on the dimensionality reduction schemes to solving the MCO problems in the case of multiextremal criteria. As it follows from Fig. 3 both the nested optimization schemes are more efficient for high levels of reliability Q than DIRECT and the use of the adaptive nested scheme is more preferable than the classical one, while in 5-dimensional case the best efficiency demonstrates the adaptive reduction of dimensionality and DIRECT is better than the classical nested scheme.

5   Conclusion

In the paper the multicriterial optimization (MCO) problems with multiextremal criteria have been considered. As a tool of analyzing these problems the approach connected with ideas of reducing the initial MCO problem to families of simpler optimization problems has been taken. At the beginning, the MCO problem is reduced in a traditional way via convolutions to a set of scalar subproblems, solutions of which are Pareto-optimal points. Further, solving the scalar subproblems is based on global optimization algorithms reducing the multidimensional problem to a family of univariate subproblems by means of the dimensionality reduction schemes of nested optimization. These schemes are theoretically substantiated and their efficiency has been confirmed in experiments. Two versions of the nested optimization (classical and adaptive) and the known method DIRECT have been considered for comparison.

The general description of the mentioned approach has been done and the results of numerical testing on a test set of bi-criterial MCO problems with essentially multiextremal criteria have been presented. The results of the experiment have demonstrated that the adaptive dimensionality reduction can be used as an effective tool for solving the multiextremal MCO problems.

It is worth to note that the algorithms considered in the paper can be developed in directions of search acceleration connected with the use of additional information about the studied problems and with designing their parallel versions.

Acknowledgements. The research has been supported by the Russian Science Foundation, project No 16-11-10150 "Novel efficient methods and software tools for time-consuming decision make problems using superior-performance supercomputers."

References

1. Miettinen, K.: Nonlinear Multiobjective Optimization. Springer, Heidelberg (1999). https://doi.org/10.1007/978-1-4615-5563-6
2. Marler, R.T., Arora, J.S.: Multi-Objective Optimization: Concepts and Methods for Engineering. VDM Verlag, Riga (2009)
3. Ehrgott, M.: Multicriteria Optimization, 2nd edn. Springer, Heidelberg (2010)
4. Collette, Y., Siarry, P.: Multiobjective Optimization: Principles and Case Studies (Decision Engineering). Springer, Heidelberg (2011)
5. Mardani, A., Jusoh, A., Nor, K., Khalifah, Z., Zakwan, N., Valipour, A.: Multiple criteria decision-making techniques and their applications—a review of the literature from 2000 to 2014. Econ. Res.-Ekonomska Istraživanja **28**(1), 516–571 (2015). https://doi.org/10.1080/1331677X.2015.107513911
6. Marler, R.T., Arora, J.S.: Survey of multiobjective optimization methods for engineering. Struct. Multi. Optim. **26**, 369–395 (2004)
7. Hillermeier, C., Jahn, J.: Multiobjective optimization: survey of methods and industrial applications. Surv. Math. Ind. **11**, 1–42 (2005)
8. Figueira, J., Greco, S., Ehrgott, M. (eds.): Multiple Criteria Decision Analysis: State of the Art Surveys. Springer, Heidelberg (2005). https://doi.org/10.1007/978-1-4939-3094-4

9. Eichfelder, G.: Scalarizations for adaptively solving multi-objective optimization problems. Comput. Optim. Appl. **44**, 249–273 (2009)
10. Evtushenko, Y.G., Posypkin, M.A.: A deterministic algorithm for global multi-objective optimization. Optim. Methods Softw. **29**(5), 1005–1019 (2014)
11. Zilinskas, A., Zilinskas, J.: Adaptation of a one-step worst-case optimal univariate algorithm of bi-objective Lipschitz optimization to multidimensional problems. Commun. Non-linear Sci. Numer. Simul. **21**, 89–98 (2015)
12. Gergel, V.P., Kozinov, E.A.: Accelerating parallel multicriterial optimization methods based on intensive using of search information. Procedia Comput. Sci. **108**, 1463–1472 (2017)
13. Krasnoshekov, P.S., Morozov, V.V., Fedorov, V.V.: Decomposition in design problems. Eng. Cybern. **2**, 7–17 (1979). (in Russian)
14. Wierzbicki, A.: The use of reference objectives in multiobjective optimization. In: Fandel, G., Gal, T. (eds.) Multiple Objective Decision Making, Theory and Application, vol. 177, pp. 468–486. Springer-Verlag, New York (1980)
15. Strongin, R., Sergeyev, Y.: Global Optimization with Non-convex Constraints: Sequential and Parallel Algorithms. Kluwer Academic Publishers, Dordrecht (2000)
16. Horst, R., Pardalos, P.M.: Handbook of Global Optimization. Kluwer Academic Publishers, Dordrecht (1995)
17. Pintér, J.D.: Global Optimization in Action (Continuous and Lipschitz Optimization: Algorithms, Implementations and Applications). Kluwer Academic Publishers, Dordrecht (1996)
18. Zhigljavsky, A.A., Žilinskas, A.: Stochastic Global Optimization. Springer, New York (2008). https://doi.org/10.1007/978-0-387-74740-8
19. Sergeyev, Y.D., Strongin, R.G., Lera, D.: Introduction to Global Optimization Exploiting Space-Filling Curves. Springer, Heidelberg (2013). https://doi.org/10.1007/978-1-4614-8042-6
20. Paulavičius, R., Žilinskas, J.: Simplicial Global Optimization. Springer, New York (2014). https://doi.org/10.1007/978-1-4614-9093-7
21. Sergeyev, Y.D., Kvasov, D.E.: Deterministic Global Optimization: An Introduction to the Diagonal Approach. Springer, New York (2017). https://doi.org/10.1007/978-1-4939-7199-2
22. Grishagin, V.A., Sergeyev, Y.D., Strongin, R.G.: Parallel characteristic algorithms for solving problems of global optimization. J. Glob. Optim. **10**, 185–206 (1997)
23. Grishagin, V., Israfilov, R., Sergeyev, Y.: Convergence conditions and numerical comparison of global optimization methods based on dimensionality reduction schemes. Appl. Math. Comput. **318**, 270–280 (2018)
24. Jones, D.R.: The DIRECT global optimization algorithm. In: Floudas, C., Pardalos, P.M. (eds.) Encyclopedia of Optimization, pp. 431–440. Kluwer Academic Publishers, Dordrecht (2001)
25. Carr, C., Howe, C.: Quantitative Decision Procedures in Management and Economic: Deterministic Theory and Applications. McGraw-Hill, New York (1964)
26. Piyavskij, S.: An algorithm for finding the absolute extremum of a function. Comput. Math. Math. Phys. **12**, 57–67 (1972)
27. Sergeyev, Y., Grishagin, V.: Parallel asynchronous global search and the nested optimization scheme. J. Comput. Anal. Appl. **3**, 123–145 (2001)
28. Dam, E.R., Husslage, B., Hertog, D.: One-dimensional nested maximin designs. J. Glob. Optim. **46**, 287–306 (2010)
29. Grishagin, V., Israfilov, R.: Multidimensional constrained global optimization in domains with computable boundaries. In: CEUR Workshop Proceedings, vol. 1513, pp. 75–84 (2015)

30. Gergel, V., Sidorov, S.: A two-level parallel global search algorithm for solution of computationally intensive multiextremal optimization problems. In: Malyshkin, V. (ed.) PaCT 2015. LNCS, vol. 9251, pp. 505–515. Springer, Cham (2015). https://doi.org/10.1007/978-3-319-21909-7_49

31. Gergel, V., Grishagin, V., Gergel, A.: Adaptive nested optimization scheme for multidimensional global search. J. Global Optim. **66**, 35–51 (2016)

32. Grishagin, V.A., Israfilov, R.A.: Global search acceleration in the nested optimization scheme. In: AIP Conference Proceedings, vol. 1738, p. 400010 (2016)

33. Strongin, R.: Numerical Methods in Multiextremal Problems (Informational Statistical Algorithms). Nauka, Moscow (1978). (in Russian)

34. Kushner, H.J.: A new method of locating the maximum point of an arbitrary multipeak curve in the presence of noise. Trans. ASME Ser. D. J. Basic Eng. **86**, 97–106 (1964)

35. Gergel, V.P., Grishagin, V.A., Israfilov, R.A.: Local tuning in nested scheme of global optimization. Procedia Comput. Sci. **51**, 865–874 (2015)

36. Gaviano, M., Kvasov, D.E., Lera D., Sergeyev Y.D.: Software for generation of classes of test functions with known local and global minima for global optimization. ACM TOMS **29**, 469–480 (2003)

REFINE: Representation Learning
from Diffusion Events

Zekarias T. Kefato$^{(\boxtimes)}$, Nasrullah Sheikh, and Alberto Montresor

University of Trento, Trento, Italy
{zekarias.kefato,nasrullah.sheikh,alberto.montresor}@unitn.it

Abstract. Network representation learning has recently attracted considerable interest, because of its effectiveness in performing important network analysis tasks such as link prediction and node classification. However, most of the existing studies rely on the knowledge of the complete network structure. Very often this is not the case, unfortunately: the network is either partially or completely hidden. For example, due to privacy and competitive market advantage, the friendship and follower networks of Facebook and Twitter are hardly accessible. User activity logs (also known as cascades), instead, are usually available. In this study we propose REFINE, a representation learning algorithm that does not require information about the network and simply utilizes cascades. Nodes embeddings learned through REFINE are optimized for network reconstruction. Towards this end, it utilizes the global interaction patterns exposed by reaction times and co-occurrences. We present an extensive experimentation using two OSN datasets and show that our approach outperforms existing baselines. In addition, we empirically show that REFINE can be used to predict cascades as well.

Keywords: Network inference · Representation learning · Cascade prediction

1 Introduction

Network representation learning (NRL) has recently attracted considerable research attention. In particular, the ubiquitous success of deep learning has inspired social network scientists to exploit neural networks to automatically learn representation of nodes, that could later be used for several social analysis tasks. A number of existing studies have assumed that the network structure is completely known. Very often, however, this is not the case; instead, information about the network is either partial or completely absent. For instance, companies seeking a marketing campaign through Facebook or Twitter desire access to the structural properties of the social graph; such information, however, is usually not accessible due to privacy and competitive market advantage [1].

Some information is available, though. For example, extensive logs of events occurring on the social graph can be easily obtained, e.g. through public APIs.

© Springer Nature Switzerland AG 2019
G. Nicosia et al. (Eds.): LOD 2018, LNCS 11331, pp. 141–153, 2019.
https://doi.org/10.1007/978-3-030-13709-0_12

These logs represent the propagation of information over the latent network, for example by recording the instant in which a user shares a meme or a piece of fake news. The process of propagation is known as a *cascade*; it is usually triggered by a few sources (*seeds*) and spreads over the graph through its edges [2–4].

In other words, we can observe who shares a meme and when this happens, but not the edge through which the meme has been transported. The goal of this study is to learn a representation of nodes optimized for reconstructing the latent network by simply using the cascades.

Related Work. Several studies [2–7] have been proposed towards the network reconstruction task. In general, we can divide them into two broad categories, which are (i) delay-aware and (ii) delay-agnostic. Some of the existing delay-aware models, such as NETINF [2], NETRATE [6], INFOPATH [5], and KERNEL-CASCADE [3], exploit infection rates based on delay patterns between infection timestamps. The main assumption is that if a pair of nodes tend to get infected right after each other, then there is a diffusion pattern that is a likely indicator of connections. Some of them [5,6] assume a fixed parametric form (e.g. exponential) of influence model or transmission rate on the edges of the network. Nonetheless, a particular study [3] has argued and empirically demonstrated that such an assumption is too strong for capturing the complex diffusion patterns and user infection dynamics in real networks.

On the other hand, some studies [4,7] follow a delay-agnostic approach simply based on the order and/or context of infection events. Furthermore, they have argued that delay-aware models are likely to miss out several diffusion patterns, even in the presence of recurring ones, because of the delay intervals of such models that could potentially be too large or too small. This problem is normally caused by explicitly pre-defined infection rates (delay patterns) and fixed parametric forms of influence models, as argued by [3].

In the area of network representation learning, there are also quite a number of studies [8–15]. The algorithms vary from classical techniques that rely on matrix factorization to recent techniques using deep neural networks. Their goal is usually to embed nodes of the network in a low-dimensional latent space in such a way that the embedding preserves different properties of the network, for example local neighborhoods. Our work is essentially different from the above techniques, because we lack the knowledge of the network structure.

Current Work. In this study, we propose REFINE, an delay-aware algorithm for network reconstruction based on representation learning. Contrary to [4,7], we argue that delay-aware models can also perform as well as delay-agnostic models if they are properly designed. Therefore, REFINE utilizes the delays between infection events; unlike some of the existing methods [2,5,6], however, it avoids any assumption regarding the influence model and infections rates. Instead, it directly embeds users according to the inherent interaction patterns exposed by them.

REFINE is established on the premise that closely connected users, for example members of a community, expose interaction patterns that are expressed by reaction time and frequency. In terms of reaction time, given a post by a certain member of a community, it is very likely for another member to share the post faster than non-members. In terms of frequency, it is more likely for a member of a community to co-occur with another member in cascades more frequently than with other non-members. REFINE learns a low-dimensional embedding of nodes that capture such interaction patterns and use the learned embedding to estimate pairwise edge probabilities towards reconstructing the network.

We have performed extended evaluations of our approach and compared it against strong baselines. Besides utilizing the embedding to reconstruct the latent network, we have also evaluated the capability of our representation learning approach to predict the cascades themselves.

The rest of the paper is organized as follows. Section 2 introduces the notation which is used in the rest of the paper. Section 3 describes the REFINE algorithm. Section 4 presents the results of our experiments and we conclude the paper in Sect. 5.

2   Model and Problem Definition

We assume that cascades occur over a *hidden* graph $H = (U, E)$, where U is a set containing n vertexes, each vertex corresponding to a user, and E is a set containing m edges (connections) between users. We will use the term vertex and user as synonyms, preferring the former when referring to human being, and the latter when referring to graph-theoretic concepts. Interactions between users occur over the network; while the set of users is normally well-defined, the set of connections among them can be partially or completely unknown.

The spread of multiple contagions across the network H generates a collection $\mathcal{C}$ of cascades. A contagion can be considered as any piece of online content, such as, a tweet, meme, video, that spread through online networks as a result of re-sharing activities. A cascade $C \in \mathcal{C}$ is a sequence that captures both the order and the time instant in which users have been infected by a given contagion. More formally, it is defined as: $C = [(u_1, t_1), (u_2, t_2), \ldots, (u_c, t_c)]$ where t_i is the timestamp associated to user u_i. We assume that $i < j \Rightarrow t_i \leq t_j$.

We use $C(i)$ to denote the i-th user of C; and $C_t(i)$ to denote the corresponding timestamp. We also use $\mathcal{C}_u \subseteq \mathcal{C}$ to denote the subset of all cascades that user u is involved in $\mathcal{C}_u = \{C : \exists i \wedge 1 \leq i \leq |C| \wedge C(i) = u\}$ with $\mathcal{C}_u \neq \emptyset$, meaning that all users in U have been involved in at least one cascade.

Given a cascade C, we define a function $r_C : U \times U \to \mathbb{R}^+$ measuring the reaction time between the infection events of u and v, if both have been infected in C, or ∞ otherwise:

$$r_C(u, v) = \begin{cases} |C_t(i) - C_t(j)| & \exists i, j : u = C(i) \wedge v = C(j) \\ +\infty & \text{otherwise} \end{cases}$$

In addition, we define the *co-infection frequency* function $f(u, v) = |\mathcal{C}_u \cap \mathcal{C}_v|$ that computes the number of cascades that involve both u and v.

The problem we want to solve is the following: given a set of observed cascades $\mathcal{C}$ over a hidden network $H = (U, E)$, we want to infer a network $G = (U, E')$ such that E' approximates E as much as possible.

To evaluate the performance of our algorithms, similar to [8] we use the *precision-at-K* (*P@K*) metric. Our approach will produce an edge probability for every pair of vertexes; we can thus rank pairs of vertexes according to such probability. We cut this rank at different thresholds K and we compute the precision on the top-K pairs, i.e. the fraction of those pairs that are true edges on the ground-truth network.

3 The REFINE Algorithm

REFINE considers global interaction patterns expressed through users reaction time and co-occurrences in cascades. For a given user $u \in U$, REFINE computes (i) a reaction time summary between the infection time of u and all other users and (ii) the relative co-occurrence frequency between u and all other users, both measured over the entire collection $\mathcal{C}$. Our assumption is that if two users u and v exhibit a strong interaction pattern, then they are likely to be connected.

A straightforward approach towards reconstruction is to compute similarity between users according to their global interaction representation. However, this leads to poor performances as this representation is very sparse. Rather, we first learn an embedding of users in such a way that their interaction patterns in the input representation space is preserved. Finally, we estimate the pairwise edge probabilities between every pair of nodes to reconstruct the latent network.

3.1 Interaction Pattern Summarization

REFINE is a delay-aware model based on the global interaction delays (reaction-time) and frequency (co-occurrence) in cascades. We start by computing a reaction time distribution for each user. Given a cascade $C \in \mathcal{C}$ and a user u appearing in C (e.g., $\exists i : u = C(i)$), we compute, for the sake of numerical convenience, an inverted reaction time function $r_C^{-1}(u, v)$ defined as follows:

$$
r_C^{-1}(u, v) = \begin{cases} 0 & r_C(u, v) = \infty \\ 1 & r_C(u, v) = 0 \\ e^{-r_C(u,v)} & \text{otherwise} \end{cases} \tag{1}
$$

$r_C^{-1}(u, v)$ is a well-defined function from pairs of nodes to $[0, 1]$, given that $r_C^{-1}(u, v)$ approaches 0 when $r_C(u, v)$ grows to infinity, and $r_C^{-1}(u, v)$ approaches 1 when $r_C(u, v)$ tends to 0.

REFINE utilizes the function r_C^{-1} to compute an (inverted) *reaction time summary vector* $\boldsymbol{R}'(u)$ for each user $u \in U$, aggregated over all cascades $\mathcal{C}$, where each entry $\boldsymbol{R}'(u)[v]$, $v \in U$, is defined as follows:

$$
\boldsymbol{R}'(u)[v] = \frac{\sum_{C \in \mathcal{C}_u \cap \mathcal{C}_v} r_C^{-1}(u, v)}{\sum_{C \in \mathcal{C}_u} \sum_{i=1}^{|C|} r_C^{-1}(u, C(i))} \tag{2}
$$

Equation 2 computes the (inverted) average reaction time between u and v, normalized over all the cascades pertinent to u, $\mathcal{C}_u$.

One can easily notice that the reaction time summary vector $\mathbf{R}'(u)$ captures a reaction time distribution for each user u. Nonetheless, it fails to account for the co-infection frequency between u and every other node v, which we consider to be another strong signal for the existence of an edge between u and v. For example, let v and w be two nodes with equal values in their respective entries in the reaction time summary vector of u, i.e. $\mathbf{R}'(u)[v] = \mathbf{R}'(u)[w]$. If $f(u,v) \gg f(u,w)$, it is obvious that u and v have a stronger interaction tendency than u and w, which is not modeled by $\mathbf{R}'$.

To compensate for that, we first compute the relative co-infection frequency vector $\mathbf{F}(u)$, where each $\mathbf{F}(u)[v]$, $v \in U$, is defined as follows:

$$\mathbf{F}(u)[v] = \frac{f(u,v)}{\sum_{w \in U} f(u,w)} \tag{3}$$

Finally, we combine $\mathbf{R}'$ and $\mathbf{F}$ to obtain the *interaction pattern summary* $\mathbf{I}(u) = \mathbf{F}(u) \times \mathbf{R}'(u)$ for each user u. The vectors $\mathbf{I}(u)$ can be summarized in a matrix $\mathbf{I} = [\mathbf{I}(u_1), \ldots, \mathbf{I}(u_n)] \in [0,1]^{n \times n}$ that contains a row for each user.

Now, even though two users v and w have a tie for u in terms of $\mathbf{R}'(u)$, i.e, $\mathbf{R}'(u)[v] = \mathbf{R}'(u)[w]$, $\mathbf{F}(u)$ breaks such tie by putting more weight on the user with a stronger co-infection frequency with u.

A naïve approach towards reconstructing the hidden network could be to compute the similarity between each pair of users u,v based on $\mathbf{I}(u)$ and $\mathbf{I}(v)$, for example by computing their distance over $[0,1]^n$. This approach, however, leads to a poor performance as $\mathbf{I}$ is very sparse. We apply instead a learning phase to embed $\mathbf{I}$ in a low and dense latent embedding space, in such a way that the patterns encoded in $\mathbf{I}$ are preserved. In other words, we intend to identify a mapping function $\Phi : [0,1]^{n \times n} \to \mathbb{R}^{n \times d}$, with $d \ll n$.

Finally, we utilize Φ to effectively learn the probability for an edge between a pair of nodes to exist, in order to reconstruct the hidden network.

3.2 User Embedding

The hidden network structure that we seek to reconstruct lives in a highly non-linear space [8]. Therefore, one has to identify a mapping $\Phi \in \mathbb{R}^{n \times d}$ that enables her to recover the non-linear network structure. Towards this goal, REFINE uses a deep *autoencoder*, an unsupervised neural network model.

An autoencoder enables us to embed $\mathbf{I}$ in a low-dimensional latent space by composing several non-linear functions (layers), as shown in Fig. 1. The input is given by the matrix $\mathbf{I}$. The *user embedding* module of Fig. 1 has two components, the *encoder*

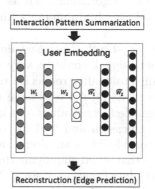

Fig. 1. The REFINE framework (Color figure online)

(blue layers) and the *decoder* (black layers). The former transforms the input into an *embedding* (white layer), while the latter tries to regenerate and output the original input from the embedding.

Formally, the encoder $\mathcal{E} : [0,1]^{n \times n} \rightarrow \mathbb{R}^{n \times d}$ and the decoder $\mathcal{D} : \mathbb{R}^{n \times d} \rightarrow [0,1]^{n \times n}$ are a composition of non-linear functions defined as follows:

$$\mathcal{E}(I) = e_l(\ldots e_\ell(\ldots (e_1(I \cdot W_1) \cdot W_2) \ldots) \ldots) = \Phi \tag{4}$$

$$\mathcal{D}(\Phi) = d_l(\ldots d_\ell(\ldots (d_1(\Phi \cdot \widehat{W}_1) \cdot \widehat{W}_2) \ldots) \ldots) = \tilde{I} \tag{5}$$

where e_ℓ and d_ℓ are the non-linear functions (e.g., *relu, tanh*) of the $\ell-th$ encoder and decoder layers, respectively. Each layer of an autoencoder is fully connected, meaning that it is a linear transformation of the output of the previous layer $\ell - 1$, i.e. $f_{\ell-1}(\cdot) \cdot W_\ell$, and f_ℓ is either e_ℓ or d_ℓ.

Optimization. The weights are the main parameters of the model that needs to be trained. Normally this is achieved by minimizing the cost function of Eq. 6.

$$L = \arg \min_W \| I - \tilde{I} \|_F^2 \tag{6}$$

where I is the input matrix and $\tilde{I}$ is the regenerated output matrix. The mere optimization of Eq. 6 leads to a poor performance due to I's sparsity. To deal with this, we adopt Wang's strategy [8] and reformulate Eq. 6 as

$$L = \arg \min_{W, \widehat{W}} \| (I - \tilde{I}) \oplus S \|_F^2 + \lambda \xi \tag{7}$$

where $\oplus$ is the Hadamard product and $S \in \mathbb{R}_+^{n \times n}$ a term to avoid the sparsity problem, is associated with I, i.e if $I(u,v) = 0$, then $S(u,v) = 1$ otherwise $S(u,v) = \mu > 1$ and μ is an alias for $S(u,v)$. The second term in Eq. 7, $\xi = \sum_{\ell=1}^l \| W_\ell \|_F^2 + \| \widehat{W}_\ell \|_F^2$, is a regularization term to avoid over-fitting and $\lambda \in (0,1)$ is the regularization constant. Finally, Eq. 7 can be optimized using classical algorithms such as gradient descent. Then, once the optimization is solved, we obtain an embedding $\Phi(u)$ of each user $u \in U$.

Speeding-Up the User Embedding. For a very large value of n, training an autoencoder using I could be very expensive. Thus, we propose an intermediate step of dimensionality reduction using truncated (partial) singular value decomposition (T-SVD) for very large matrices [16]. T-SVD utilizes a few of the highest or smallest eigenvalues of a large matrix. As a result, we can efficiently reduce I's dimension and feed the reduced I_r to the autoencoder. Moreover, this can be considered as an alternative solution to tackle the sparsity problem with I. Note that when employing this component there is no need for the sparsity term in the loss function of Eq. 7. We have observed that including this optimization provides similar or better results, with a significant reduction in memory and computational time.

3.3   Reconstruction

Once Φ is computed as in Sect. 3.2, we exploit it to predict the probability that an edge exists between a pair of users. We assume that if a pair of users never co-occur in any cascade, they have a very small chance of being connected. Therefore, we discard such pairs and analyze the remaining ones.

Let $p(u, v) = 1/(1 + e^{-(\Phi(u)^T \cdot \Phi(v))})$ be a function that predicts the probability that an edge exists between u and v. We build a network $G = (U, E'), E' \approx E$ by adding an edge (u, v) to E' with probability $p(u, v)$. E' can be refined by pruning edges (u, v) where $p(u, v) < \tau$ for some threshold τ.

4   Experiments and Results

Dataset Description. Our experiments are performed on the following datasets, whose characteristics are summarized in Table 1.

Twitter [17] contains a set of Twitter users with a reciprocal follower relationships, collected from March 24th to April 25th, 2012. The follower network is considered as a ground truth. Two kinds of cascades are present: (1) Hashtag (HT): Cascades collected from user activity when using/adopting hashtags; (2) Retweet (RT): Cascades collected from user retweeting tweets.

MemeTracker (MT) [5], contains users represented by a collection of news media and blog sites. Cascades are formed based on the spread of memes. A contagion occurs when a particular meme is used by a site for the first time. The sequence of all the infected sites form a cascade. The ground truth network is built based on hyper-links found in each site.

Settings. In order to tune the hyper-parameters of REFINE, we use the random grid search strategy; its weights are initialized according to [18] for uniform distribution. To implement our models, we adopted the TENSORFLOW[1] and SCIPY[2] Python-based libraries. In all the experiments, both the encoder and decoder of REFINE use the *tanh* activation function.

Table 1. Dataset summary. Number of users, number of edges, number of cascades, number of users after removing large cascades.

Dataset	$\|U\|$	$\|E\|$	$\|C\|$	$\|U'\|$
HT	595,460	14,273,311	1,345,913	34,371
RT	595,460	14,273,311	226,488	11,700
MT	3,836,314	15,540,787	71,568	52,088

---

[1] https://www.tensorflow.org/.
[2] https://www.scipy.org/.

Results. In the first set of experiments, we have compared REFINE with two strong baselines INFOPATH [5] (delay-aware) and DEEPINFER [7] (delay-oblivious). To perform a fair comparison, we have selected four topics of the Memetracker dataset that have been evaluated in the INFOPATH original paper. The cascades derived from these topics are associated with 5000 users.

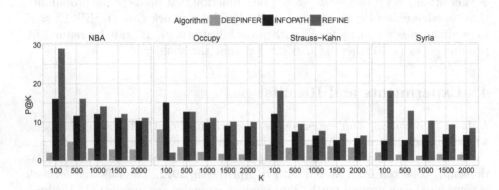

Fig. 2. Comparison of REFINE with the baselines over four topics from the Memetracker dataset for different value of K for the $P@K$ metric. For all datasets, REFINE applies T-SVD and $I_r \in \mathbb{R}^{n \times 1024}$. Cascade length: for *Syria* and *Occupy*, between 3 and 100; for *NBA* and *Strauss Kahn*, between 3 and 1000. (1) *Syria* $n = 1,207$, and $|\mathcal{C}| = 615,176$; REFINE: layer sizes = $[1024, 700, 300, 200]$, learning rate $\alpha = 0.005$, regularization constant $\lambda = 0.0005$. (2) *Occupy*: $n = 1,875$, $|\mathcal{C}| = 655,183$; REFINE: layer sizes = $[1024, 900, 400, 200]$, $\alpha = 0.001$, $\lambda = 0.009$. (3) *NBA*: $n = 2,087$, and $|\mathcal{C}| = 1,543,630$; REFINE: layer sizes = $[1024, 700, 300, 200]$, $\alpha = 0.003$, $\lambda = 0.0005$; (4) *Strauss-Kahn*: $n = 1,263$, and $|\mathcal{C}| = 204,238$; REFINE: layer sizes = $[1024, 800, 500, 200]$, $\alpha = 0.005$, $\lambda = 0.01$. For DEEPINFER: $s = 15$, and $d = 200$. For INFOPATH, we have adopted the exponential influence model, as it performs slightly better than the others.

The results are reported in Fig. 2. REFINE performs better than the baselines in almost all of the cases, by up to an order of magnitude. Apart from this, it is worthwhile to note that a single-threaded version of INFOPATH would require several days to complete. In fact, the original paper reports 4 h of computation to infer 38 different time-varying networks for 38 different topics, in a cluster equipped with 1000 CPU cores and 6 TB total RAM [5]. REFINE has been executed on a 48-core, 128 GB machine and takes at most 10 min to reconstruct the topic-associated networks for each of the four topics.

In the same figure, it is possible to observe the poor performance of DEEPINFER; this is due to the fact that we only consider 5000 users. To detect patterns, DEEPINFER relies on frequent co-occurrence of users in close contexts; however, we do not have any guarantee that the 5000 users will occur in such manner, hence the poor performance. This would not be an issue for REFINE and INFOPATH, as they rely on reaction time and/or mere co-occurrence patterns rather than context proximity.

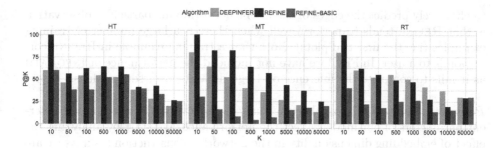

Fig. 3. REFINE vs DEEPINFER. REFINE applies T-SVD, $I_r \in \mathbb{R}^{n \times 1024}$. REFINE: HT – layer sizes $[1024, 700, 500, 300, 100]$, learning rate $\alpha = 0.0001$, regularization constant $\lambda = 0.0002$; RT & MT – layer sizes $[1024, 900, 700, 500, 200]$, $\alpha = 0.001$. RT – $\lambda = 0.004$, and MT – $\lambda = 0.001$. DEEPINFER configuration: window size $s = 10$ and $d = 200$

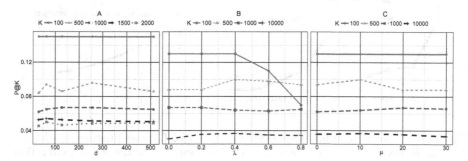

Fig. 4. Parameter sensitivity analysis with respect to (A) Embedding size (#dimensions - d), (B) Regularization constant (λ), and (C) Sparsity penalizer (μ) using Strauss-Kahn.

In all of the above experiments, the T-SVD step of REFINE has been executed. As shown in Fig. 5, handling the sparsity issue through T-SVD gives better result than the formulation in Eq. 7. REFINE with T-SVD is more robust than REFINE when K increases. However, one could ask if simply using the T-SVD method as an embedding technique could be sufficient. In the following experiment we show that a variant of REFINE, referred to as REFINE-BASIC which simply considers the T-SVD output as node embedding, is not sufficient. For this experiment, we have chosen cascades of minimum length 5 and maximum length 200. In fact, it has been argued that users belonging to large cascades are usually not similar, as such cascades tend to be viral and include almost all users [17]. By discarding cascades which are too large in order to reduce noise, the number of users decreases, as shown in column $|U'|$ of Table 1.

Figure 3 shows how poorly REFINE-BASIC performs when it is compared against REFINE and DEEPINFER. Recall that the network structure is highly non-linear and our main goal for designing the complete REFINE solution is to capture such non-linearity. REFINE-BASIC is a linear model, and hence it fails to effectively predict the edges of the latent network. One particular observation is that REFINE tends to perform well when there is a large number of training examples (i.e. the first two plots). Note that a training example in REFINE corresponds to a user. In Fig. 3 we have not included the performance of INFOPATH as it fails to complete the inference on large datasets after several days.

Parameter Analysis. To complete the analysis, we investigate now how the different parameters of our models affect the performance. We start by analyzing the effect of embedding dimensionality in the network reconstruction task. As we are interested in understanding the effect of the parameters, in the following experiments we only set the minimum size of cascades to be 3, i.e. $\{C : |C| \geq 3, C \in \mathcal{C}\}$.

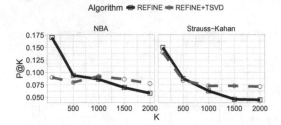

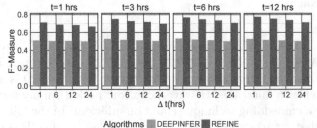

Fig. 5. The effect of using the T-SVD step in REFINE using two topics, NBA and Strauss-Kahn, from the Memetracker dataset

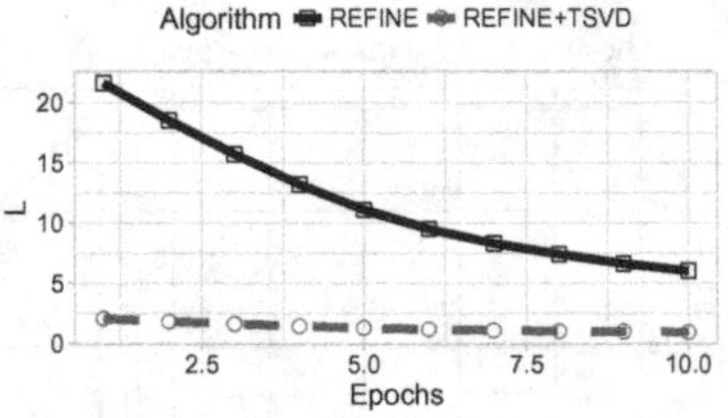

Fig. 6. The progress of the loss function at the end of 10 iterations for REFINE and REFINE with the T-SVD step

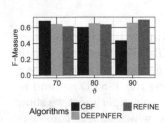

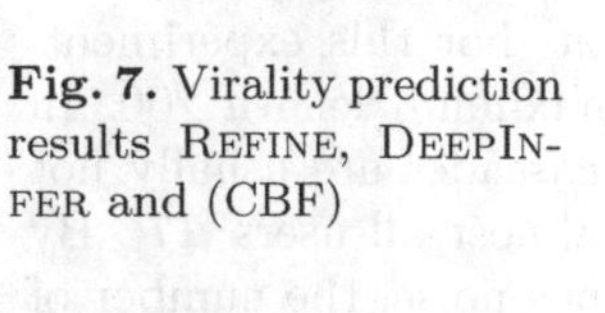

Fig. 7. Virality prediction results REFINE, DEEPINFER and (CBF)

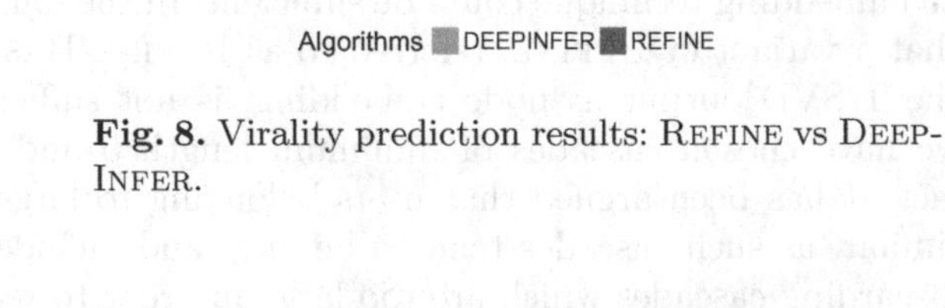

Fig. 8. Virality prediction results: REFINE vs DEEPINFER.

The first plot of Fig. 4 shows the effect of increasing the embedding dimensionality in the network reconstruction task. As one might expect, increasing this parameter up to a given threshold improves the results, because we can encode more information. However, beyond a certain point the performance either reaches a plateau or decreases. Our experiments show that in most of the cases, the best results occur when the embeddings size is in the range 150–200. In the second plot of Fig. 4, the effect of the regularization constant λ (introduced in Eq. 7) is analyzed. In line with previous findings [8], our experiments show that in most of the cases, the best results are obtained when λ is between 0.0 and 0.4; after that point, the performance usually decreases. Finally, in the third plot of Fig. 4 we analyze the effect of the sparsity factor μ, introduced in Eq. 7. Our experiments show that in most of the cases, the best results are obtained when μ is between 0 and 10.

Earlier we have shown the advantage of using T-SVD in terms of the quality of the result; here, we analyze the effect from the convergence of the loss function L, Eq. 7. Figure 6 shows that the loss function converges much faster (after a couple of iterations) for REFINE with T-SVD rather than REFINE without T-SVD.

Cascade Prediction. Besides its effectiveness in network reconstruction, our approach can be extended to perform other tasks, such as *cascade prediction*: given the state of a cascade C up to a certain time t, we want to predict whether the cascade will go viral by time $t + \Delta t$. This is a practically relevant problem and a crucial challenge in social networks analysis [17,19,20].

In this study, we formulate the virality prediction problem similarly to Weng et al. [17]. Let $S_t(C) = \{u : u = C_t(i) \wedge i \leq t\}$ be the number of users who participated in a cascade up to a discrete time t. Let ϑ be a *virality threshold*; we seek to predict whether the cascade will affect a number of users which is larger than $\vartheta\%$ of the recorded cascades. We utilize the embeddings proposed in Sect. 3.2. We compute a feature vector $\mathbf{f} \in \mathbb{R}^d$ that encodes the current state of the cascade based on $S_t(C)$ as follows. Let $p = |S_t(C)|$, and let $\mathcal{E} \in \mathbb{R}^{p \times d}$ be an embedding matrix constructed from the set of p starting users at time t, $u \in S_t(C)$. We then compute $\mathbf{f}$ by aggregating $\mathcal{E}$, *i.e.* the $j - th$ component $\mathbf{f}_j$ for $j = 1, \ldots, d$ is computed as $\mathbf{f}_j = \frac{1}{p} \sum_{i=1}^{p} \mathcal{E}_{ij}$.

Once we automatically build the feature vectors, we assign binary labels for each cascade according to their state at $t + \Delta t$ and ϑ. That is, a cascade C is labeled as *viral* if its size at $t + \Delta t$ is greater than the size of $\vartheta\%$ of the cascades; otherwise, it is labeled *non-viral*. Finally, we follow a standard *machine learning* approach by splitting the data into training (60%) and test (40%). To make a fair comparison with community-based features (CBF) [17], we follow the same techniques and settings. As we have a rare-class classification task, we use F-Measure with $\beta = 3$ [19].

We use the same dataset as [17] (Twitter-HT). We compare REFINE with CBF and DEEPINFER; for CBF only, features are manually extracted from the underlying network.

Figure 7 shows that REFINE is no better than the baselines for $\vartheta = \{70, 80\}$. However it is much better for $\vartheta = 90$ (REFINE = **69.7%**, DEEPINFER = 65.5%, CBF = 43%), and in virality prediction it is crucial to have an effective prediction at higher values of ϑ [19].

A vital task in this problem is to predict virality as early as possible. Therefore, in the following experiments we seek to predict virality of a cascade C at different $t + \Delta t$ based on the observation of C at different values of t with a fixed ϑ. In this experiment, we compare the two strong algorithms REFINE and DEEP-

INFER, and for both algorithms d is equal to 200. As shown in Fig. 8, REFINE is a clear winner for this task. In particular, note that the prediction quality for REFINE improves as we increase t, and this provides a strong case for the delay-aware approach. As it is difficult to predict far in the future, performance decreases as we increase Δt.

5   Conclusions

This study addresses the problem of network reconstruction from diffusion events through node embedding, and proposes a novel algorithm called REFINE.

One of our objectives is to argue against some existing studies [4] and show that, if carefully designed, delay-aware models are as good as or even better than delay-oblivious models in reconstructing the hidden network.

REFINE is based on user embeddings learned from cascade logs, that are leveraged to predict edge probabilities between pairs of users. Unlike some existing techniques that assume a parametric form of influence model, we make no assumption regarding the transmission rates over edges. Instead, we simply embed the interaction patterns between users in a low-dimensional space and utilize that for reconstructing the edges. We show the effectiveness of this technique by comparing it against existing delay-aware and delay-agnostic methods.

Moreover, we have also demonstrated the technique presented in this study can be used for cascade prediction. Compared to existing manual or automatic feature extraction techniques, our algorithm shows a significant performance gain. Our study is limited to inferring the existence of edges between a pair of users, and in a future work we seek to infer the direction of edges as well.

References

1. Barbieri, N., Bonchi, F., Manco, G.: Cascade-based community detection. In: Proceedings of WSDM 2013, pp. 33–42. ACM (2013)
2. Gomez Rodriguez, M., Leskovec, J., Krause, A.: Inferring networks of diffusion and influence. In: Proceedings of KDD 2010. ACM (2010)
3. Du, N., Song, L., Smola, A., Yuan, M.: Learning networks of heterogeneous influence. In: Proceedings of NIPS 2012. Curran Associates Inc., Red Hook (2012)
4. Lamprier, S., Bourigault, S., Gallinari, P.: Extracting diffusion channels from real-world social data: a delay-agnostic learning of transmission probabilities. In: Proceedings of ASONAM 2015. ACM (2015)
5. Gomez-Rodriguez, M., Leskovec, J., Schölkopf, B.: Structure and dynamics of information pathways in online media. CoRR, vol. abs/1212.1464 (2012)
6. Gomez-Rodriguez, M., Balduzzi, D., Schölkopf, B.: Uncovering the temporal dynamics of diffusion networks. In: Proceedings of ICML 2011. Omnipress (2011)
7. Kefato, Z.T., Sheikh, N., Montresor, A.: Deepinfer: diffusion network inference through representation learning. In: Proceedings of MLG 2017. ACM, August 2017
8. Wang, D., Cui, P., Zhu, W.: Structural deep network embedding. In: Proceedings of KDD 2016. ACM (2016)

9. Perozzi, B., Al-Rfou, R., Skiena, S.: Deepwalk: online learning of social representations. In: Proceedings of KDD 2014. ACM (2014)
10. Grover, A., Leskovec, J.: Node2vec: scalable feature learning for networks. In: Proceedings of KDD 2016. ACM (2016)
11. Hamilton, W.L., Ying, R., Leskovec, J.: Inductive representation learning on large
graphs. CoRR, vol. abs/1706.02216 (2017)
12. Kipf, T.N., Welling, M.: Semi-supervised classification with graph convolutional
networks. CoRR, vol. abs/1609.02907 (2016)
13. Pan, S., Wu, J., Zhu, X., Zhang, C., Wang, Y.: Tri-party deep network representation. In: Proceedings of the IJCAI 2016, pp. 1895–1901. AAAI Press (2016)
14. Kefato, Z.T., Sheikh, N., Montresor, A.: Mineral: multi-modal network representation learning. In: Proceedings of MOD 2017. ACM, September 2017
15. Sheikh, N., Kefato, Z., Montresor, A.: gat2vec: representation learning for attributed
graphs. Computing (2018). https://doi.org/10.1007/s00607-018-0622-9
16. Baglama, J., Reichel, L.: Augmented implicitly restarted Lanczos bidiagonalization
methods. SIAM J. Sci. Comput. **27**(1), 19–42 (2005)
17. Weng, L., Menczer, F., Ahn, Y.-Y.: Virality prediction and community structure
in social networks. Sci. Rep. **3** (2013). Article no 2522
18. Glorot, X., Bengio, Y.: Understanding the difficulty of training deep feedforward
neural networks, May 2010
19. Subbian, K., Prakash, B.A., Adamic, L.: Detecting large reshare cascades in social
networks. In: Proceedings of WWW 2017 (2017)
20. Cheng, J., Adamic, L., Dow, P.A., Kleinberg, J.M., Leskovec, J.: Can cascades be
predicted? In: Proceedings of WWW 2014. ACM (2014)

Augmented Design-Space Exploration by Nonlinear Dimensionality Reduction Methods

Danny D'Agostino[1,2], Andrea Serani[1(✉)], Emilio Fortunato Campana[1], and Matteo Diez[1]

[1] CNR-INM, National Research Council-Institute of Marine Engineering, Rome, Italy
andrea.serani@insean.cnr.it
[2] Department of Computer, Control, and Management Engineering "Antonio Ruberti", Sapienza University of Rome, Rome, Italy

Abstract. The paper presents the application of nonlinear dimensionality reduction methods to shape and physical data in the context of hull-form design. These methods provide a reduced-dimensionality representation of the shape modification vector and associated physical parameters, allowing for an efficient and effective augmented design-space exploration. The data set is formed by shape coordinates and hydrodynamic performance (based on potential flow simulations) obtained by Monte Carlo sampling of a 27-dimensional design space. Nonlinear extensions of the principal component analysis (PCA) are applied, namely kernel PCA, local PCA and a deep autoencoder. The application presented is a naval destroyer sailing in calm water. The reduced-dimensionality representation of shape and physical parameters is set to provide a normalized mean square error smaller than 5%. Nonlinear methods outperform the standard PCA, indicating significant nonlinear interactions in the data structure. The present work is an extension of the authors' research [1] where only shape data were considered.

Keywords: Shape optimization · Hull-form design · Nonlinear dimensionality reduction · Kernel methods · Deep autoencoder

1 Introduction

The simulation-based design (SBD) analysis and optimization paradigm has demonstrated the capability of supporting the design decision process, not only providing large sets of design options but also exploring operational spaces by assessing design performance for a large number of operating and environmental conditions. The recent development of high-performance computing systems has driven the SBD towards integration with global optimization (GO) algorithms and uncertainty quantification (UQ) methods, moving the SBD paradigm to automatic deterministic and stochastic SBD optimization (SBDO) possibly

G. Nicosia et al. (Eds.): LOD 2018, LNCS 11331, pp. 154–165, 2019.
https://doi.org/10.1007/978-3-030-13709-0_13

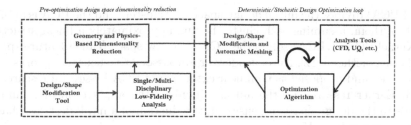

Pre-optimization design space dimensionality reduction   *Deterministic/Stochastic Design Optimization loop*

Fig. 1. SBDO scheme, including pre-optimization strategy.

aiming at global solutions to the design problem. In shape design, SBDO consists of three main elements: (i) a deterministic and/or stochastic simulation tool (integrating physics-based solvers, such as computational fluid dynamics, CFD, with UQ), (ii) an optimization algorithm, and (iii) a shape modification tool (see Fig. 1, right box). In this context, both GO and UQ are affected by the *curse of dimensionality* as the algorithms' complexity and computational cost rapidly increase with the problem dimension. This is generally also true if metamodels are applied. Therefore, the assessment and breakdown of the design-space dimensionality are key elements for the efficiency and affordability of SBDO [2].

On-line linear design-space dimensionality reduction techniques have been developed, requiring the evaluation of the objective function or its gradient. Specifically, principal component analysis (PCA) or proper orthogonal decomposition (POD) methods have been applied for local reduced-dimensionality representations of feasible design regions [3]. A PCA/POD-based approach is used in the active subspace method [4] to discover and exploit low-dimensional monotonic trends in the objective function, based on the evaluation of its gradient.

Off-line linear methods have been developed with focus on design-space variability and dimensionality reduction for efficient optimization procedures. A method based on the Karhunen-Loève expansion (KLE, equivalent to POD) has been formulated in [2] for the assessment of the shape modification variability and the definition of a reduced-dimensionality global model of the shape modification vector and applied to a fast catamaran. The method has been applied to a naval destroyer [5], a small waterplane area twin hull [6], and a hydrofoil [7], showing significant reduction of the design space dimensionality with great benefit to the efficiency of the shape SBDO. Nevertheless, significant physical phenomena induced by small shape modifications may be overlooked, since no physical information is processed by the method. Furthermore, linear methods such as KLE/POD/PCA may not be efficient when complex nonlinear relationship between design variables are involved.

An extension to augmented design-space dimensionality reduction methods by combining shape and physics based data was introduced in earlier research [8–10]. This extension improved the effectiveness of the dimensionality reduction, bringing physics based information (provided by low-fidelity hydrodynamic computations) into the variability breakdown analysis (see Fig. 1 left box).

In order to address data with nonlinear structures, nonlinear dimensionality reduction methods have been developed and investigated. Among others, local

PCA (LPCA) divides the initial design space in k clusters and the PCA is applied to each of them, assuming each cluster having approximately a linear structure [11]. Kernel PCA (KPCA) solves the PCA in a new space (called feature space) using kernel methods [12]. Autoencoders or autoassociative neural networks have been studied and proposed as nonlinear extension of PCA by several researchers [13,14]. Earlier research by the authors includes the application of nonlinear methods to the design-space dimensionality reduction of a naval destroyer based on shape data only [1,15,16].

The objective of the present work is to solve the dimensionality reduction of combined shape and physical data using nonlinear methods and assessing their efficiency and effectiveness.

Nonlinear methods include LPCA, KPCA, and DAE and are demonstrated for the DTMB 5415 model (an early and open-to-public version of an USS Arleigh Burk-class destroyer) in calm water at 18 kn. The data set is formed by the results of 9,000 potential flow simulations obtained by the Monte Carlo sampling of a 27-dimensional design space. Data include three heterogeneous distributed and suitably discretized parameters (geometry modification vector, pressure distribution on the hull, and wave elevation) and one lumped parameter (wave resistance coefficient). The reduced-dimensionality representation of shape and physical parameters is set to provide a mean square error smaller than 5%, normalized with the overall data variance. The efficiency and effectiveness of nonlinear methods are assessed considering their compression capability and associated reconstruction error compared to PCA. Current formulations and methods go beyond design-space dimensionality reduction for shape optimization and can be extended to large sets of heterogeneous physical data from simulations, experiments, and real operation measurements. An extended version of the current paper has been presented in [17].

2   Dimensionality-Reduction Formulation and Methods

Global optimization tries to find the best design exploring the entire design space. This solution is obviously unknown *a priori* and therefore the problem can be considered as affected by an epistemic type of uncertainty. Consequently,

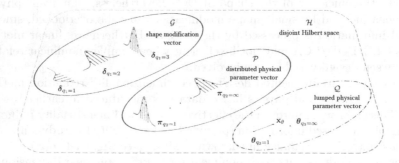

Fig. 2. Domains for shape modification, distributed physical parameter, and lumped (or global) physical parameter vectors in a disjoint Hilbert space.

the design-variable vector can be associated to a probability density function and studied as a random variable [2].

General definitions and assumptions are presented in the following, along with the solution of data reduction by PCA, LPCA, KPCA and DAE.

2.1 Combined Shape- and Physics-Based Formulation

Assume that $\mathbf{u} \in \mathcal{U} \subset \mathbb{R}^M$ is the design-variable vector, defining the shape modification vector $\boldsymbol{\delta} \in \mathbb{R}^{q_1}$, $q_1 = 1, ..., 3$, along with a distributed physical parameter vector $\boldsymbol{\pi} \in \mathbb{R}^{q_2}$, $q_2 = 1, ..., \infty$ (representing, e.g., velocity, pressure distribution, wave elevation, etc.), and a lumped (or global) physical parameter vector $\boldsymbol{\theta} \in \mathbb{R}^{q_3}$, $q_3 = 1, ..., \infty$ (representing, e.g., resistance, motion RMS, etc.). For the sake of simplicity, consider one set of coordinates $\mathbf{x} \in \mathbb{R}^n$, and assume $\mathcal{G}$, $\mathcal{P}$, and $\mathcal{Q}$ as the domain of $\boldsymbol{\delta}$, $\boldsymbol{\pi}$, and $\boldsymbol{\theta}$ respectively, as schematized in Fig. 2. Note that $\mathcal{Q}$ has a null measure and corresponds to an arbitrary point $\mathbf{x}_\theta$ where the lumped physical parameter vector is virtually defined. Also note that, in general, $\mathcal{H} \equiv \mathcal{G} \cup \mathcal{P} \cup \mathcal{Q}$ is not simply connected. Finally, consider $\mathbf{u}$ as a random variable with associated probability density function $p(\mathbf{u})$. Consider a combined geometry and physics based vector $\boldsymbol{\gamma} \in \mathbb{R}^q$ with $q = \max\{q_1, q_2, q_3\}$

$$\gamma(\mathbf{x}, \mathbf{u}) = \begin{cases} \boldsymbol{\delta}'(\mathbf{x}, \mathbf{u}) / \langle \|\boldsymbol{\delta}'\|^2 \rangle & \text{if } \mathbf{x} \in \mathcal{G} \\ \boldsymbol{\pi}'(\mathbf{x}, \mathbf{u}) / \langle \|\boldsymbol{\pi}'\|^2 \rangle & \text{if } \mathbf{x} \in \mathcal{P} \\ \boldsymbol{\theta}'(\mathbf{x}, \mathbf{u}) / \langle \|\boldsymbol{\theta}'\|^2 \rangle & \text{if } \mathbf{x} \in \mathcal{Q} \end{cases} \qquad (1)$$

as belonging to a disjoint Hilbert space $\mathcal{H}$, where each component (generically called $\boldsymbol{\psi}' = \boldsymbol{\psi} - \langle \boldsymbol{\psi} \rangle$) is centered and normalized by the associated variance

$$\sigma^2 = \langle \|\boldsymbol{\psi}'\|^2 \rangle = \int_\mathcal{U} \int_\mathcal{H} \boldsymbol{\psi}'(\mathbf{x}, \mathbf{u}) \cdot \boldsymbol{\psi}'(\mathbf{x}, \mathbf{u}) p(\mathbf{u}) \mathrm{d}\mathbf{x} \mathrm{d}\mathbf{u} \qquad (2)$$

with $\langle \cdot \rangle$ the ensemble average over $\mathbf{u}$.

The aim of the dimensionality reduction is to identify a reduced dimensionality representation $\hat{\boldsymbol{\gamma}}(\mathbf{x}, \boldsymbol{\alpha})$ of the vector $\boldsymbol{\gamma}$, for which its modification depends on a new reduced order design variable $\boldsymbol{\alpha} \in \mathcal{A} \subset \mathbb{R}^N$ with $N < M$. $\hat{\boldsymbol{\gamma}}(\mathbf{x}, \boldsymbol{\alpha})$ is estimated during a process of encoding/decoding by the dimensionality reduction methods. Figure 3 shows an example for shape modification ($\boldsymbol{\delta}$) only, with $n = 1$ and $q = 2$.

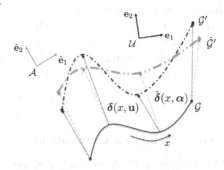

Fig. 3. Scheme and notation for the current formulation, example for shape modification only with $n = 1$ and $q = 2$.

A convenient metric to evaluate the goodness of $\hat{\boldsymbol{\gamma}}(\mathbf{x}, \boldsymbol{\alpha})$ to fit $\boldsymbol{\gamma}(\mathbf{x}, \mathbf{u})$ is the mean square error (MSE) normalized to the design-space original variance (σ^2) as

$$\text{NMSE} = \frac{\text{MSE}}{\sigma^2} = \frac{\iint\limits_{\mathcal{U}\times\mathcal{A},\mathcal{H}} \|\hat{\gamma}(\mathbf{x},\boldsymbol{\alpha}) - \gamma(\mathbf{x},\mathbf{u})\|^2 p(\mathbf{u},\boldsymbol{\alpha})\mathrm{d}\mathbf{x}\mathrm{d}\mathbf{u}\mathrm{d}\boldsymbol{\alpha}}{\iint\limits_{\mathcal{U},\mathcal{H}} \|\gamma(\mathbf{x},\mathbf{u})\|^2 p(\mathbf{u})\mathrm{d}\mathbf{x}\mathrm{d}\mathbf{u}} \tag{3}$$

where $p(\mathbf{u},\boldsymbol{\alpha})$ is an unknown joint probability distribution over the product space $\mathcal{U}\times\mathcal{A}$. Discretizing $\mathcal{H}$ by elements of equal measure $\Delta\mathcal{H} = 1$ and sampling $\mathcal{U}$ by a statistically convergent number of Monte Carlo realizations S, so that $\{\mathbf{u}_k\}_{k=1}^S \sim p(\mathbf{u})$, the discretization $\mathbf{g}(\mathbf{u}_k)$ of $\gamma(\mathbf{x},\mathbf{u}_k)$ are organized in a $[L\times S]$ data matrix as

$$\mathbf{D} = \begin{bmatrix}\mathbf{g}(\mathbf{u}_1) \mid \ \dots \ \mid \mathbf{g}(\mathbf{u}_S)\end{bmatrix} \text{ with } \mathbf{g}(\mathbf{u}_k) = \{\mathbf{d}^\mathsf{T}, \mathbf{p}^\mathsf{T}, \mathbf{t}^\mathsf{T}\}_k^\mathsf{T}, \tag{4}$$

where L is the dimensionality of $\mathbf{g}$ and with $\mathbf{d}$, $\mathbf{p}$, and $\mathbf{t}$ the discrete form of the vectors $\boldsymbol{\delta}(\mathbf{x},\mathbf{u})$, $\boldsymbol{\pi}(\mathbf{x},\mathbf{u})$, and $\boldsymbol{\theta}(\mathbf{x},\mathbf{u})$, respectively. Equation 3 can be approximated as

$$\text{NMSE} = \frac{\text{MSE}}{\sigma^2} = \frac{\sum_{k=1}^S \|\hat{\mathbf{g}}(\boldsymbol{\alpha}_k) - \mathbf{g}(\mathbf{u}_k)\|^2}{\sum_{k=1}^S \|\mathbf{g}(\mathbf{u}_k)\|^2} \tag{5}$$

Details of formulation and numerical discretization can be found in [8].

2.2 Principal Component Analysis

PCA allows to reduce the dimensionality of the data matrix by representation in a linear subspace defined by the eigenvectors of the $[L\times L]$ sample covariance matrix $\mathbf{C} = \mathbf{D}\mathbf{D}^\mathsf{T}/S$. Thus, PCA reduces to the solution of the eigenproblem

$$\mathbf{C}\mathbf{Z} = \mathbf{Z}\boldsymbol{\Lambda} \tag{6}$$

where $\mathbf{Z}$ and $\boldsymbol{\Lambda}$ collect the L eigenvectors and eigenvalues of $\mathbf{C}$, respectively. The eigenvalues represent the variance resolved along the corresponding eigenvectors. The linear subspace formed by the N eigenvectors (collected in $\hat{\mathbf{Z}}$) associated to the largest N eigenvalues resolves the largest variance, compared to any other linear subspace of dimension N [18]. The cumulative sum of the eigenvalues is used to assess the variance resolved by the linear subspace of dimension N. The associated reconstruction of $\mathbf{D}$ is given by

$$\hat{\mathbf{D}} = \hat{\mathbf{Z}}\hat{\mathbf{Z}}^\mathsf{T}\mathbf{D} \tag{7}$$

2.3 Local Principal Component Analysis

LPCA performs a PCA for each disjoint region of the input space $\mathcal{H}$. If local regions are small enough the associated data manifold will not curve much over the extent of the region and the linear model is assumed to be a good fit [11]. The first step in LPCA is clustering the data in k sets, such that $\mathbf{D} = \{\mathbf{D}_1, \dots, \mathbf{D}_k\}_{i=1}^k$. Here, the k-means algorithm [19] is used. After k clusters are defined, k PCA eigenproblems are solved

$$\mathbf{C}_i\mathbf{z}_i = \lambda_i\mathbf{z}_i \qquad \forall i = 1, \dots, k \tag{8}$$

LPCA results are highly dependent on the clustering method and the number of clusters. The number of clusters needs to be defined carefully to avoid increasing the computational cost and data overfitting.

2.4 Kernel Principal Component Analysis

KPCA [12] finds directions of maximum variance in a higher (possibly infinite) dimensional feature space $\mathcal{F}$, mapping data points from the input space $\mathcal{H}$ by a (possibly) nonlinear function $\Phi : \mathcal{H} \to \mathcal{F}$

$$\mathbf{g}(\mathbf{u}_k) \to \Phi(\mathbf{g}_k), \qquad \forall k = 1, \ldots, S \tag{9}$$

The PCA is computed in the feature space $\mathcal{F}$. Assuming $\sum_k \Phi(\mathbf{g}_k) = 0$, the kernel principal component $\{\mathbf{z}_p\}_{p=1}^P$ can be found solving the eigenproblem

$$\Sigma_\Phi \mathbf{z}_p = \lambda_p \mathbf{z}_p \tag{10}$$

where Σ_Φ is the $[P \times P]$ covariance matrix in the feature space $\mathcal{F}$, defined as

$$\Sigma_\Phi = \frac{1}{S} \sum_{k=1}^S \Phi(\mathbf{g}_k)\Phi(\mathbf{g}_k)^\mathsf{T} \tag{11}$$

Defining $K(\mathbf{g}_i, \mathbf{g}_k) = \Phi(\mathbf{g}_i)^\mathsf{T}\Phi(\mathbf{g}_k)$ and $\mathbf{z}_p = \sum_{k=1}^S c_{pk}\Phi(\mathbf{g}_k)$ Eq. 10 can be rewritten as

$$\mathbf{K}\mathbf{c}_p = \lambda_p S \mathbf{c}_p \tag{12}$$

where $\mathbf{K}$ is the symmetric and positive-semidefinite $[S \times S]$ kernel matrix, with $\mathbf{K}_{ik} = K(\mathbf{g}_i, \mathbf{g}_k)$. The length of the S-component vector $\mathbf{c}_p$ is chosen such that $\mathbf{z}_p^\mathsf{T}\mathbf{z}_p = \lambda_p S \mathbf{c}_p^\mathsf{T}\mathbf{c}_p = 1$. Once the eigenproblem of Eq. 12 is solved, the new parametrization can be found projecting $\Phi(\mathbf{g})$ on $\mathbf{z}_p$ as

$$\alpha = \Phi(\mathbf{g})\mathbf{z}_p = \sum_{k=1}^S c_{pk}\Phi(\mathbf{g})^\mathsf{T}\Phi(\mathbf{g}_k) = \sum_{k=1}^S c_{pk}K(\mathbf{g}, \mathbf{g}_k) \tag{13}$$

The reconstruction of the original data from the feature space $\mathcal{F}$ in KPCA is more problematic than PCA. Here, the approximate pre-images technique proposed in [20] is used.

2.5 Deep Autoencoders

An autoencoder is a feedforward ANN that performs two main tasks: (i) an encoder function $\mathcal{E}$ maps the input data $\mathbf{g}(\mathbf{u}_k)$ into compressed data α_k; (ii) a decoder function $\mathcal{D}$ maps from the compressed data α_k back to $\hat{\mathbf{g}}(\alpha_k)$. The overall operation is performed setting the same number of neurons (L) in the input and output layer. The hidden layer is set to have $N < M$ neurons and is responsible for the data compression.

Consider a single hidden layer autoencoder and assume no bias vector. New design variables can be expressed as $\boldsymbol{\alpha}_k = \mathcal{E}(\mathbf{H}_{(1)}\mathbf{g}(\mathbf{u}_k))$ where $\mathbf{H}$ is a weight matrix and subscript "(1)" indicates the encoding operation. The reconstruction vector can be expressed as $\hat{\mathbf{g}}(\boldsymbol{\alpha}_k) = \mathcal{D}(\mathbf{H}_{(2)}\boldsymbol{\alpha}_k)$ where subscript "(2)" indicates the decoding operation. Finally, the network parameters $\mathcal{N} = \{\mathbf{H}_{(1)}, \mathbf{H}_{(2)}\}$, are evaluated by the (non trivial) minimization of the MSE in the form:

$$\mathrm{MSE}(\mathcal{N}) = \frac{1}{S}\sum_{k=1}^{S}\|\hat{\mathbf{g}}(\boldsymbol{\alpha}_k) - \mathbf{g}(\mathbf{u}_k)\|^2 = \frac{1}{S}\sum_{k=1}^{S}\|\mathcal{D}(\mathbf{H}_{(2)}\mathcal{E}(\mathbf{H}_{(1)}\mathbf{g}(\mathbf{u}_k))) - \mathbf{g}(\mathbf{u}_k)\|^2$$

(14)

Using nonlinear activation functions and multiple hidden layers, DAE provides a nonlinear generalization of the PCA. The DAE compression capability is represented by the number of neurons N in the central hidden layer and defined based on parametric minimization of the MSE, varying N.

3   Application

Figure 4 shows a schematic representation of the heterogeneous data set. Hull and performance details of the orginal geometry can be found in [21]. The shape parameter vector used for design-space dimensionality reduction collects the $y-$component (δ_y) of the shape modification vector $(\boldsymbol{\delta})$. The shape modification is defined using a combination of $M = 27$ basis functions over a hyper-rectangle embedding the demi hull. Details of equations and setting parameters may be found in [22]. The distributed (heterogeneous) physical parameter vector collects values of the pressure distribution (p) and wave elevation (η), whereas the lumped physical parameter vector includes the wave resistance coefficient (C_w). Physical parameters are based on calm-water potential flow solution at Fr = 0.25. Hydrodynamic simulations are conducted using the code WARP (Wave Resistance Program), developed at CNR-INSEAN. Wave resistance computations are

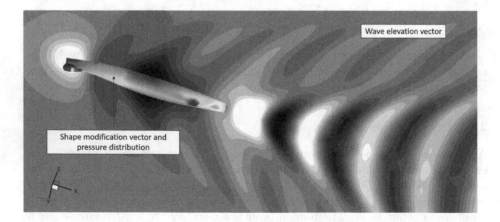

Fig. 4. Distributed shape and physical parameters for current application.

based on linear potential flow theory using Dawson (double-model) linearization [23]. The frictional resistance is estimated using a flat-plate approximation, based on the local Reynolds number [24]. Details of equations, numerical implementations, and validation of the numerical solver are given in [25]. Simulations are performed for the right demi-hull, taking advantage of symmetry about the xz-plane. The computational domain for the free-surface is defined by a 75×20 grid nodes. The associated hull grid is defined by 90×25 nodes. The design-space dimensionality reduction is performed combining together all geometric and physical parameters.

3.1   Numerical Results

The original design space is sampled using a uniform random distribution of $S = 9,000$ hull-form designs. The reduced-dimensionality models are validated using 10-fold cross-validation repeated 6 times to compute the hypothesis test (t-test). The reduced-dimension N is set so as to achieve a maximum NMSE equal to 5%. A number of cluster $k = 45$ is used for LPCA. A quadratic polynomial kernel is used for the KPCA. Three hidden layers are used for DAE (composed by 600-N-600 neurons) with an exponential linear units [26] activation function for each hidden layer. A linear activation function is used for the output layer. The DAE training is performed by the Adam optimization algorithm [27], using a minibatch size of 512 data point for gradient evaluation by the backpropagation algorithm [28]. For the implementation of the DAE the open-source python library [29] is used.

Table 1 shows the dimensionality-reduction results in terms of number of components N required by the methods to reconstruct successfully the data set along with the associated NMSE (averaged on the training and test datasets). The number of components N also indicates the reduced-dimensionality parametrization of the shape modification vector for future SBDO. The non-linear methods outperform linear PCA. Specifically, LPCA and KPCA are found the most effective methods for the current problem in terms of dimensionality reduction capability ($N = 14$). DAE ($N = 17$) also shows a sufficient compression capability, whereas PCA is found the least effective method requiring $N = 19$ principal components. This suggests the presence of significant nonlinear structures into the data set.

Table 1. Methods' compression capability (N), dimensionality reduction (DR), and training and test NMSE (p-value ≤ 0.05).

Method	N	DR%	NMSE% (training)	NMSE% (test)
PCA	19	29.6	4.5	4.6
LPCA	14	48.2	3.6	4.6
KPCA	14	48.2	4.1	4.6
DAE	17	37.0	4.3	4.5

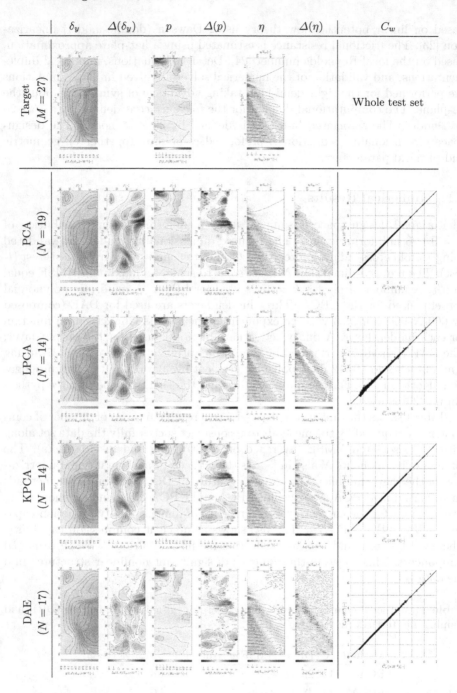

Fig. 5. Reconstruction of hull shape (δ_y), pressure distribution (p), wave elevation (η), and corresponding errors $\Delta(\cdot)$ for a target design (results are shown versus $I-$ and $J-$nodes of the computational grid); reconstruction of the wave resistance coeffiecient (C_w) for the whole test set.

Figure 5 shows the reconstruction of the hull shape (δ_y) and the distributed physical (p and η) parameters vector for an example design in the test set. A good agreement between the target and reconstructed data is achieved by all methods. Furthermore, Fig. 5 shows the reconstruction of the wave resistance coefficient (C_w) for the whole test set, showing a remarkable agreement.

4   Conclusions and Future Work

Nonlinear dimensionality reduction methods have been applied to the design space assessment of the DTMB 5415 hull form in calm water at Fr = 0.25. Nonlinear extensions of principal component analysis (PCA) have been applied, namely local PCA (LPCA), kernel PCA (KPCA), and a deep autoencoder (DAE). The data matrix under investigation was formed by the results of potential flow simulations coming from the MC sampling of a 27-dimensional design space associated to a shape-optimization problem. The dataset includes the geometry as well as two heterogeneous physical distributed parameters (pressure and wave elevation) and one lumped parameter (wave resistance coefficient). The reduced-dimensionality representation of shape and physical variables was sets to achieve an NMSE smaller than 5% of the data variance.

The standard (linear) PCA meets the requirement using 19 principal components/parameters. DAE shows here the least promising compression capability among the nonlinear methods with 17 components required by the reduced-dimensionality parametrization. Finally, LPCA and KPCA provides the most promising compression capability with 14 components. Reconstructed data for shape, pressure, wave elevation, and wave resistance coefficients were presented, showing a remarkable agreement to target values.

The current results are promising, representing a first step towards data compression and reduced-order model prediction of complex physical phenomena. Current formulation goes beyond shape optimization and can be applied to large sets of heterogeneous physical data from simulations, experiments, and real operation measurements.

Future work includes extensions to multi-physics heterogeneous data from multiple design conditions [10]. The possibility of using higher-fidelity analysis solver with metamodels will be addressed. In parallel, a similar approach is being applied to particle image velocimetry data of complex flows to assess the compression capability of nonlinear extensions of the proper orthogonal decomposition (POD) technique [30].

Acknowledgments. The work is supported by the US Office of Naval Research Global, NICOP grant N62909-18-1-2033, under the administration of Dr. Salahuddin Ahmed and Dr. Woei-Min Lin, and by the Italian Flag-ship Project RITMARE.

References

1. D'Agostino, D., Serani, A., Campana, E.F., Diez, M.: Nonlinear Methods for design-space dimensionality reduction in shape optimization. In: Nicosia, G., Pardalos, P., Giuffrida, G., Umeton, R. (eds.) MOD 2017. LNCS, vol. 10710, pp. 121–132. Springer, Cham (2018). https://doi.org/10.1007/978-3-319-72926-8_11
2. Diez, M., Campana, E.F., Stern, F.: Design-space dimensionality reduction in shape optimization by Karhunen-Loève expansion. Comput. Methods Appl. Mech. Eng. **283**, 1525–1544 (2015)
3. Raghavan, B., Breitkopf, P., Tourbier, Y., Villon, P.: Towards a space reduction approach for efficient structural shape optimization. Struct. Multi. Optim. **48**, 987–1000 (2013)
4. Lukaczyk, T., Palacios, F., Alonso, J.J., Constantine, P.: Active subspaces for shape optimization. In: Proceedings of the 10th AIAA Multidisciplinary Design Optimization Specialist Conference, National Harbor, Maryland, USA, 13–17 January 2014
5. Diez, M., Serani, A., Campana, E.F., Volpi, S., Stern, F.: Design space dimensionality reduction for single- and multi-disciplinary shape optimization. In: AIAA/ISSMO Multidisciplinary Analysis and Optimization (MA&O), AVIATION 2016, Washington D.C., USA, 13–17 June (2016)
6. Pellegrini, R., Serani, A., Broglia, R., Diez, M., Harries, S.: Resistance and payload optimization of a sea vehicle by adaptive multi-fidelity metamodeling. In: 56th AIAA Aerospace Sciences Meeting, SciTech 2018, Gaylord Palms, Kissimmee, Florida, USA, 8–12 January (2018)
7. Volpi, S., Diez, M., Stern, F.: Multidisciplinary design optimization of a 3D composite hydrofoil via variable accuracy architecture. In: 19th AIAA/ISSMO Multidisciplinary Analysis and Optimization Conference (MA&O), AVIATION 2018, Atlanta, GA, USA, 25–29 June (2018)
8. Diez, M., Serani, A., Stern, F., Campana, E.F.: Combined geometry and physics based method for design-space dimensionality reduction in hydrodynamic shape optimization. In: Proceedings of the 31st Symposium on Naval Hydrodynamics, Monterey, CA, USA (2016)
9. Serani, A., Campana, E.F., Diez, M., Stern, F.: Towards augmented design-space exploration via combined geometry and physics based Karhunen-Loève expansion. In: 18th AIAA/ISSMO Multidisciplinary Analysis and Optimization Conference (MA&O), AVIATION 2017, Denver, USA, 5–9 June (2017)
10. Serani, A., Diez, M.: Reliability-based robust design optimization by design-space augmented dimensionality reduction. In: 19th AIAA/ISSMO Multidisciplinary Analysis and Optimization Conference (MA&O), AVIATION 2018, Atlanta, GA, USA, 25–29 June (2018)
11. Kambhatla, N., Leen, T.K.: Dimension reduction by local principal component analysis. Neural Comput. **9**(7), 1493–1516 (1997)
12. Schölkopf, B., Smola, A., Müller, K.R.: Nonlinear component analysis as a kernel eigenvalue problem. Neural Comput. **10**(5), 1299–1319 (1998)
13. Bourlard, H., Kamp, Y.: Auto-association by multilayer perceptrons and singular value decomposition. Biol. Cybern. **59**(4), 291–294 (1988)
14. Kramer, M.A.: Nonlinear principal component analysis using autoassociative neural networks. AIChE J. **37**(2), 233–243 (1991)

15. D'Agostino, D., Serani, A., Campana, E.F., Diez, M.: Deep autoencoder for off-line design-space dimensionality reduction in shape optimization. In: 56th AIAA Aerospace Sciences Meeting, SciTech 2018, Gaylord Palms, Kissimmee, Florida, USA, 8–12 January 2018

16. D'Agostino, D., Serani, A., Diez, M.: On the combined effect of design-space dimensionality reduction and optimization methods on shape optimization efficiency. In: 19th AIAA/ISSMO Multidisciplinary Analysis and Optimization Conference (MA&O), AVIATION 2018, Atlanta, GA, USA, 25–29 June 2018

17. Serani, A., D'Agostino, D., Campana, E.F., Diez, M.: Assessing the interplay of shape and physical parameters by nonlinear dimensionality reduction methods. In: Proceedings of the 32nd Symposium on Naval Hydrodynamics, Hamburg, Germany (2018)

18. Hotelling, H.: Analysis of a complex of statistical variables into principal components. J. Educ. Psychol. **24**(6), 417 (1933)

19. Lloyd, S.: Least squares quantization in PCM. IEEE Trans. Inf. Theor. **28**(2), 129–137 (1982)

20. Bakır, G.H., Weston, J., Schölkopf, B.: Learning to find pre-images. Adv. Neural Inf. Process. Syst. **16**, 449–456 (2004)

21. Stern, F., Longo, J., Penna, R., Olivieri, A., Ratcliffe, T., Coleman, H.: International collaboration on benchmark CFD validation data for surface combatant DTMB model 5415. In: Proceedings of the Twenty-Third Symposium on Naval Hydrodynamics, Val de Reuil, France, 17–22 September (2000)

22. Serani, A., et al.: Ship hydrodynamic optimization by local hybridization of deterministic derivative-free global algorithms. Appl. Ocean Res. **59**, 115–128 (2016)

23. Dawson, C.W.: A practical computer method for solving ship-wave problems. In: Proceedings of the 2nd International Conference on Numerical Ship Hydrodynamics, Berkeley, pp. 30–38 (1977)

24. Schlichting, H., Gersten, K.: Boundary-Layer Theory. Springer-Verlag, Berlin (2000)

25. Bassanini, P., Bulgarelli, U., Campana, E.F., Lalli, F.: The wave resistance problem in a boundary integral formulation. Surv. Math.Ind. **4**, 151–194 (1994)

26. Clevert, D.A., Unterthiner, T., Hochreiter, S.: Fast and accurate deep network learning by exponential linear units (ELUs). arXiv preprint arXiv:1511.07289 (2015)

27. Kingma, D., Ba, J.: Adam: a method for stochastic optimization. arXiv preprint arXiv:1412.6980 (2014)

28. Rumelhart, D.E., Hinton, G.E., Williams, R.J., et al.: Learning representations by back-propagating errors. Cogn. Model. **5**(3), 1 (1988)

29. Chollet, F., et al.: Keras. https://github.com/fchollet/keras (2015)

30. Serani, A., et al.: PIV data clustering of a buoyant jet in a stratified environment. In: 57th AIAA Aerospace Sciences Meeting, SciTech 2019, Manchester Grand Hyatt San Diego, San Diego, 7–11 January (2019)

Classification and Survival Prediction in Diffuse Large B-Cell Lymphoma by Gene Expression Profiling

Pierangela Bruno$^{(\boxtimes)}$ ⓘ, Francesco Calimeri ⓘ, and Aldo Marzullo

Department of Mathematics and Computer Science, University of Calabria, Rende, Italy
{bruno,calimeri,marzullo}@mat.unical.it

Abstract. We present a novel framework for developing a risk model for class prediction from high-dimensional gene expression data; we define a new model that relies on several already known classification methods. We make use of the model for a survival analysis of tumor and immune subtype from Diffuse Large B-cell Lymphoma patients. Experimental analyses show good level of accuracy in the detection of Cell-of-Origin of diseases.

Keywords: Classification · Gene expression · Lymphoma

1 Introduction

The univocal identification of cancer and the understanding of its composition are crucial in medicine; however, they represent non-trivial challenges. In order to extrapolate features of single cells from complex tumor admixtures, non-trivial approaches and accurate statistics analyses are required.

Among the different kinds of cancer, the treatment of Lymphoma requires some of the most difficult tasks; indeed, a proper understanding conditions in which it arises is still an open problem, as well as the definition of the specific kind of genetic mutation causing its growth [1]. Furthermore, we know that DNA changes related to Lymphoma are usually acquired after birth, rather than being inherited [2]; nevertheless, even if they may result from several causes, such as exposure to radiation, cancer-causing chemicals or infections, changes occur for no apparent reason, in general.

In order to effectively tackle these challenges, new techniques have been recently developed that enhance already existing immune profiling technology [3]. In this context, statistical analysis of gene expression [4] plays a crucial role, and it can be of help for immune profiling, therapeutic design, treatment strategies and also for studying and understanding the unusual growth and/or

The work is partially funded by Dottorato innovativo a caratterizzazione industriale PON R&I FSE FESR 2014–2020.

© Springer Nature Switzerland AG 2019
G. Nicosia et al. (Eds.): LOD 2018, LNCS 11331, pp. 166–178, 2019.
https://doi.org/10.1007/978-3-030-13709-0_14

the migration of cells into organs or tissues from their sources of origin. For instance, in malignant tumors, levels of cellular infiltration are associated with tumor growth, cancer progression and patient outcome.

Several methods for prognosis prediction of Diffuse Large B-cell Lymphoma and analysis of gene expression profiling have been proposed, based on Fuzzy Neural Networks [5], statistical techniques [6], survival analysis [7] or microarray manipulation [8], among others. In recent years, a new research trend has been arising, mostly based on discovering Cell-of-Origin (COO) into two distinct molecular subtypes, identified by gene expression profiling: the activated B-cell-like (ABC) and the germinal center B-cell-like (GCB)[9]. Indeed, the assignment of Diffuse Large B-cell Lymphoma into COO groups has become increasingly important with the emergence of novel therapies that have selective biological activity in GBC or ABC groups [10]. Many studies take advantage of different feature extraction methods to discover independent components from gene expression profile, such as Principal Component Analysis (PCA) [11], Linear Discriminant Analysis (LDA) [12] and Locally Linear Discriminant Embedding (LLDE) [13], and Prediction Analysis for Microarrays (PAM) [14]. Although such methods have solid biomedical support, there are a great number of gene subsets with the same predictive performance which could lead to the arbitrariness selection of candidate gene subsets. In fact, each method suffers from some drawbacks, and many factors such as normalization, small sample size, noisy data, improper evaluation methods, and too many model parameters can lead to the overfitting of the resulting models, the bias of results and even false discovery [15]. Among these methods, promising results has been attained by the work of Dabney et al. [6], that showed Classification to Nearest Centroids (ClaNC) outperforming other methods in terms of accuracy and overall error.

In this work we propose a novel approach for class prediction from gene expression data and survival analysis of tumor and immune subtype; we are interested in finding a subset of genes which have a significant impact on survival probability. In particular, our approach relies on the definition of two groups based on the amount of certain member cell types composing the genes, and on a Kaplan-Meier survival analysis that aims at understanding which group has more chances to survive. Genes can be identified by taking into account the fact that, if the presence of some set of genes changes, the composition of the groups changes accordingly, and hence the survival probability. In our approach, we make use of machine-learning techniques in order to identify gene candidates and of CIBERSORT [16] in order to adaptively identify the induced groups. The framework is available at https://github.com/DeMaCS-UNICAL/ DLBCL-prediction.

The remainder of the paper is structured as follows: Sect. 2 reports the main methods used, including frameworks used to set up the experiments; Sect. 3 defines the experimental activities we carried out, while Sect. 4 discusses the results. Section 5 presents our conclusions and perspectives.

2   Proposed Approach

In the context of DNA microarrays, classifying and predicting the diagnostic category of a sample based on its gene expression profile constitute a challenge, as there is a large number of inputs (genes) from which to predict classes along with a relatively small number of samples. Hence, the identification of which genes contribute towards the classification is an important task.

The goal of this study is to provide a new approach for identifying a subset of genes that influence survival rate of patients having Diffuse Large B-cell Lymphoma. The proposed approach consists of three steps: *(i)* use CIBERSORT [16] to estimate the excess of certain member cell types in a mixed-cell population and subdivide the patients in different groups w.r.t. their own cell types value; *(ii)* apply Kaplan-Meier analysis to the groups, in order to estimate the survival function from lifetime data and measure the fraction of patients living for a certain amount of time after treatment; *(iii)* identify the best separating genes from the mixture that influence the survival rate of each subgroup. In particular, in order to perform an accurate prognosis prediction, we evaluate performance of three different classification algorithms: PAM, ClaNC and Proportional Overlapping Score (POS).

2.1   Subgroup Definition Using CIBERSORT

CIBERSORT [16] is a method for characterizing cell heterogeneity from nearly any tissue by using their gene expression profiles. It uses a machine learning approach called ν-Support Vector Regression ($\mathcal{V}$-SVR) and performs a deconvolution of mixtures, useful to analyze the composition of each sample in term of percentage of tumor and noise. The output of CIBERSORT is a new estimated mixtures that is expressed in the percentage relationship between genes and cell lines and, then, the composition of each gene. By using CIBERSORT, we divided the patients into two groups w.r.t. the median value computed on the B-cell proportions among all patients. In particular, let P_i be a patient, $X(P_i)$ the B-cell proportion value of that patient and M the median value, we define the *"High"* group s.t. $P_i \in High \leftrightarrow X(P_i) >= M$ and the *"Low"* group s.t. $P_i \in Low \leftrightarrow X(P_i) < M$. Basically, the *High* group contains patients, which B-cell proportion is greater or equal than the median, while the *Low* group contains patients which B-cell proportion is lower than the median. The resulting groups represent a starting point of our analysis. Indeed, for each group the Kaplan-Meier analysis is computed to obtain the overall survival of the patients.

2.2   Survival Analysis

Kaplan-Meier [17] is a method used to measure the fraction of subjects living for a certain amount of time after treatment. The Kaplan-Meier survival function is defined as the probability of surviving in a given period of time while considering time in many small intervals. Let d_i be the number of death patients at time t_i, and let n_i be the number of patients "at risk", i.e. alive patients or not censored just before t_i (a patient is censored when information is missing); at a given time t, the value of the survival function is computed as follows:

$$S(t) = \sum_{t_i \leq t} [1 - \frac{d_i}{n_i}]$$

For instance, the probability of a patient surviving two days after a chemotherapy treatment for non-Hodgkin lymphoma is computed by conditional probability [18] as follows:

$$P(t) = P(s_1 s_2 | s_1)$$

where s_1 is the probability of surviving after the first day and s_2 is the probability of surviving after the second day.

2.3   Gene Classification

In order to identify subsets of probe sets that best characterize each class with a reasonably small cross-validation error, we consider the PAM, ClaNC and POS classification techniques, widely used in problems related to cancer-gene expression studies [19]. These probe sets were removed in order to analyze the variation of survival rate by comparing the survival curves. – PAM is a statistical technique for class prediction from gene expression data using Nearest Shrunken Centroids (NSC) [20]. It is a simple, accurate and fast classifier often used to select genes directly linked with breast cancer [14]. – ClaNC is a classification algorithm based on NSC. It can be represented by the centroid components and pooled by the standard deviation of the active genes, that are most frequent genes for each class, demonstrated to be successful in selecting genes that discriminate between multiple clinical or biological classes [6]. – POS is a method based on the analysis of the overlapping regions, for each gene, yielded by the intersection between gene expression intervals of different classes with the aim to denote gene with higher discriminating power for the considered classification problem. It is able to achieve interesting results in gene selection to increase the diagnostic value of gene expression data for colorectal cancer [19].

PAM and ClaNC are based on NSC, which, in turn, is one of the most frequently used classification methods for high-dimensional data, such as microarray data [21]. NSC selects "good" genes according to two factors: *within class distance* and *between classes distance*. When expression levels of a gene for all samples in the same class are fairly consistent with a small variance, but are largely different among samples of different classes, the gene is considered a good candidate for classification because it has discriminant information for different classes. Genes whose expression levels do not significantly differ between the classes will have their centroids reduced to the overall centroids, effectively removing them from the classification procedure [20].

3   Experimental Setting

In the following we illustrate our experimental activities; in particular, we describe the dataset of use and the evaluation criteria adopted.

3.1   Dataset Description

We conduct our experiments on the publicly available dataset taken from the Gene Expression Omnibus (GEO)[1], a database consisting of microarray, next generation sequencing (NGS) and other high-throughput data. In particular, we tested our method on GSE23501 dataset composed of DNA methylation signatures define molecular subtypes of Diffuse Large B-cell Lymphoma (DLBCL), characterized by probe sets represented on GeneChip Human Genome U133 Plus 2.0 Array. The screening population consisted of 69 DLBCL cases in the 16–92 age range, subjected to the same treatment (R-CHOP); for each patient, age, overall survival, molecular subtype, gender and treatment are known. We used the LM22, signature matrix designed by Newman *et al.* [16]. LM22 contains 547 genes that distinguish 22 human hematopoietic cell phenotypes including seven T-cell types naive and memory B-cells. These cells are highly relevant since they can kill tumor cells, or in some cases promote their growth. Precisely, we focus on B-cells memory (a type of lymphocytes) that are part of the adaptive immune system, a specific defense [22]. In order to perform a foreign comparison according to [16], we converted probes of references matrix (LM22) to HUGO gene symbols [23].

3.2   Evaluation

We used *log-rank test* and *F-test* for comparing the survival distributions of two samples. The log-rank test is based on the null hypothesis that there is no difference regarding survival among two distributions. In log-rank test we calculated the expected number of events in each group, i.e. E_1 and E_2 while O_1 and O_2 are the total number of observed events in each group, respectively. The statistic test is:

$$p = \frac{(O_1 - E_1)^2}{E_1} + \frac{(O_2 - E_2)^2}{E_2}$$

Log-rank tests were computed within a level of significance of 5% [24].

Log-rank test may be invalid or less significant if the survival curves cross because of an increased probability of type II error [24]. For this reason, especially to determine whether two curves belong to different distribution, we included F-test [25] in our analysis, a statistical tool for data analysis programmed to determine whether two independent estimates of variance can be assumed to be estimates of the same variance; this allows us to perform a comparison between two treatments. Let $\bar{Y}_i$ the sample mean in the i_{th} group, n_i the number of

---

[1] https://www.ncbi.nlm.nih.gov/geo/.

observations in the i_{th} group, $\bar{Y}$ the overall mean of the data, K the number of groups, N the overall sample size; then, the formula for the F-test statistic value is:

$$F = \sum_{i=1}^{K} \frac{n_i(\bar{Y}_i - \bar{Y})^2}{K-1} / \sum_{i=1}^{K} \sum_{j=1}^{n} \frac{(\bar{Y}_i - Y_i^-)^2}{N-K}$$

F-test is performed with a significance level of 5%. It is the probability of making the wrong decision when the null hypothesis is true and it is also called α level. According to this test, the null hypothesis is rejected when both *critical f* value is smaller than *F-test* value and a p-value is smaller than α level. *Critical f* value is a cutoff value on the test distribution where the F-test value is unlikely to be wrong.

4 Experimental Analysis and Discussion

The goal of this study is to find a subset of genes that influence the survival rate and the disease. Patients are divided into two groups, and for each group the Kaplan-Meier analysis is computed in order to obtain the overall survival. Comparing performances of three classification algorithms, subsets of probe sets that best characterize each class are identified and removed in order to analyze the variation of survival rate. The first classification algorithm used is PAM. A grid search is used to estimate best score value, called *threshold*, that minimizes classification errors. The results reported in the tables are relative to the test set obtained by splitting the original dataset in training (80%) e test set (20%). In particular, the results are related to the average performance on the test set among 10-fold cross-validation. Precision, Recall, Accuracy and F-measure, derived from confusion matrix, and the overall MSE (i.e., the average of the squares of the difference between the estimated centroid and observed value) were used to assess the quality of the algorithm. In order to select the genes that best characterize each group, we tested each method by selecting sets of genes of different size (10, 50, 100, 150 and 200). For each size we performed the Kaplan-Meier analysis and we compared the overall results in order to find the best size. By increasing the size, there were no relevant changes in the survival curve. Hence, we selected only 10 genes with the additional purpose of minimizing the modifications on the genome. Table 1 reports a subset of 10 probe sets founded by PAM with an overall cross validation error of 45.5% [25 out of 38 *High* samples were correctly predicted (63%), while 18 out of 33 *Low* were misclassified (54%)].

These probe sets are removed from the original dataset and we performed the analysis over all remaining genes. Thus, Kaplan-Meier analysis is performed on the resulting new subgroups with the aim of discovering relevant correlations between genes and survival rate. After pruning (i.e. removing genes) according to the classifier, we computed the survival analysis and we noticed that some patients are automatically moved from High group to Low group or vice-versa

Table 1. Genes distinguishing best between High and Low classes, according to PAM analysis

Id	High-score	Low-score
207928_S_AT	0.0995	−0.1056
1554141_S_AT	0.0774	−0.0821
230877_AT	−0.0764	0.081
1560997_AT	−0.0651	0.0691
211821_X_AT	−0.0605	0.0642
234458_AT	−0.0581	0.0616
210607_AT	−0.0569	0.0604
240791_AT	0.0542	−0.0574
236582_AT	0.0516	−0.0547
215290_AT	−0.0473	0.0501

due to classification results. Indeed, due to removing of genes, the genome of patients and, consequently, the percentage of B-cell are changed.

Figure 1 shows a survival graph before and after removing these probe sets according to the PAM analysis, on the left and on the right, respectively. On the Y and X axes the estimated survival probability and the time of observation [26] are reported, respectively. The survival curve is drawn as a step function: the proportion surviving remains unchanged between the events, even if there are some intermediate censored observations.

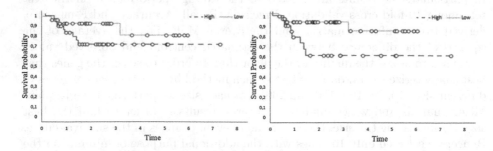

Fig. 1. Plots of Kaplan-Meier product limit estimates of survival of a group of patients (on the left), and after removing genes according the PAM analysis (on the right).

Table 2 reports the survival time for each group in which the dataset was subdivided before (on the left) and after removing probe sets according to the PAM analysis (on the right). The rows represent the number of patients belonging to the two groups, while columns represent the number of patients observed in each group, survival time and survival probability, respectively. The number of patients in the *High* class decreases as well as the average survival rate and average survival probability (i.e. 60% of survival probability), w.r.t. values obtained from original dataset (on the left) (i.e. 70% of survival probability).

Table 2. Kaplan-Meier analysis' results before (on the left) and after removing probe sets according to PAM (on the right) in terms of average survival probability and survival time for each group (±standard deviation)

Observed	Time	Survival		Observed	Time	Survival
35	3.87 (±0.36)	70%	HIGH	24	3.55 (±0.47)	60%
33	4.89 (±0.40)	90%	LOW	44	5.12 (±0.33)	90%

Table 3. Log-rank test computed before (I) and after the PAM analysis (II)

Log-rank	Observed	Critical value	P-value
I	1.93	3.84	0.17
II	4.28	3.84	0.04

In order to compare the distribution of the two obtained curves (Fig. 1), we calculated and compared the p-value according to the log-rank. Table 3 shows the comparison between log-rank test results obtained from original dataset (I) and the dataset after probe sets removed according to the PAM analysis (II). In particular, analysis (II) indicates a significant difference between the population survival curves (p-value 0.0391); analysis (I), instead, does not show a significant difference between the two curves (p-value 0.1650).

Table 4. Average and standard deviation computed before (I) and after PAM analysis (II) on survival probability (i.e Y axis) (right) and on survival time (i.e. X axis) (left)

	Average (±*stdev*)		Average (±*stdev*)
(I)	2.28 (±1.66)	(I)	0.78 (±0.08)
(II)	2.21 (±1.83)	(II)	0.73 (±0.13)

For completeness and to support our previous claims (i.e. statistical tests), we insert average comparison. As reported in Table 4, the average of survival probability and survival time decrease from analysis (I) to analysis (II).

We performed the same procedure with the other two classification algorithms. Performance of PAM, ClaNC and POS are compared to each other, as

Table 5. Comparison the PAM, the ClaNC and the POS error rate (MSE) for each class

	Class high	Class low
PAM	0.37	0.54
ClaNC	0.10	0.14
POS	0.30	0.44

illustrated in Table 5. In particular, we can observe that the overall MSE, when classifying according to ClaNC, tends to be substantially lower than the PAM error rate, in contrast to POS error.

According to each method, 10 more relevant probe sets for each class are found (see Table 6). Note that the algorithms do not find the same probe sets. Indeed, although the first two techniques are both based on NSC, they use a different approach resulting in an outcome are very different. This difference in strategy impacts on the results and the selection of the genes that best characterize each class, also taking into account that the difference between the genes is small.

Table 6. Comparison the PAM, the ClaNC and the POS top probe sets

ClaNC	PAM	POS
231192_AT	207928_S_AT	213524_S_AT
243188_AT	1554141_S_AT	201904_S_AT
227573_S_AT	230877_AT	1563203_AT
207928_S_AT	1560997_AT	241355_AT
219833_S_AT	211821_X_AT	1555801_S_AT
1554141_S_AT	234458_AT	236347_AT
221558_S_AT	210607_AT	230352_AT
215000_S_AT	240791_AT	239435_X_AT
1563127_AT	236582_AT	240529_AT
234458_AT	215290_AT	1552569_A_AT

Each subset of probe sets was removed from original dataset in order to perform Kaplan-Meier analysis and search for a correlation between these probe sets and survival rate of patients. Results of Kaplan-Meier analysis do not show a relevant change after removing each probe set selected by ClaNC and POS, as indicated in Table 7. Indeed, the survival probability is similar to the value obtained after removing probe sets according to the PAM analysis (Table 2).

Our analysis is focussed only on *High* curve, that has shown a relevant change according to the PAM analysis. Table 8 reports the comparison between log-rank

Table 7. Kaplan-Meier analysis' results after removing probe sets according to ClaNC in terms of average survival time ($\pm$ standard deviation) on the left and according to POS in terms of average survival time ($\pm$ standard deviation) on the right

	Observed	Time	Survival	Observed	Time	Survival
HIGH	21	3.80 ($\pm$0.47)	65%	37	3.87 ($\pm$0.36)	75%
LOW	47	4.78 ($\pm$0.32)	88%	31	4.89 ($\pm$0.40)	90%

test results among original dataset (I) and resulting dataset according to the PAM analysis (II), to the ClaNC analysis (III) and to the POS analysis (IV).

Table 8. Log-rank test computed among original dataset (I), analysis (II), (III) and (IV)

Log-rank	Observed	Critical value	P-value
I	1.93	3.84	0.17
II	4.28	3.84	0.04
III	1.18	3.84	0.28
IV	2.05	3.84	0.12

In particular, analysis (I), (III) and (IV) do not show a significant difference between the two curves (p-value 0.170, 0.2810 and 0.1201, respectively).

Table 9. Average and standard deviation computed before (I) and after the PAM analysis (II), after the ClaNC analysis (III), after the POS analysis (IV) on survival probability (i.e. Y axis) (right) and on survival time (i.e. X axis) (left)

	Average ($\pm stdev$)		Average ($\pm stdev$)
(I)	2.28 ($\pm$1.66)	(I)	0.78 ($\pm$0.08)
(II)	2.21 ($\pm$1.83)	(II)	0.73 ($\pm$0.13)
(III)	2.43 ($\pm$1.80)	(III)	0.76 ($\pm$0.12)
(IV)	2.26 ($\pm$1.65)	(IV)	0.79 ($\pm$0.09)

As shown in Table 9, the average value of survival probability and survival time increases from analysis (II) to analysis (III) and analysis (IV). Although the differences are small, taking into account the p-value results, we can say that the analysis (III) and (IV) do not find relevant differences between the population survival curves.

Table 10. F-test results in terms of F value, critical f value and p-value according X (Survival time) and Y (Survival probability) axes

PAM			ClaNC			POS		
	X	Y		X	Y		X	Y
F-VALUE	1.20	2.35	F-VALUE	2.12	1.97	F-VALUE	1.40	2.20
CRITICAL F	1.86	1.55	CRITICAL F	1.80	1.56	CRITICAL F	1.70	1.80
P-VALUE	0.30	0.01	P-VALUE	0.06	0.05	P-VALUE	0.40	0.06

Such result is also evident from Table 10, which reports the result of F-test computed by comparing the *High* curve between original dataset and the resulting dataset, according to each classification method used.

The F-test shows that PAM achieves good results according to survival probability (Y axis). In fact, F value is greater than critical f value and p-value is lesser that the α level (i.e. 0.05) (see Sect. 3.2).

Our findings suggest that PAM achieves the best result, implying that the distributions of the two curves (before and after the PAM analysis) are not equal.

5   Conclusion

In this work we investigated how a particular set of genes could influence the survival of two prognostic groups. In particular, we first used CIBERSORT to estimate the excess of certain member cell types in a mixed-cell population, and subdivided the patients in different groups with respect to their own cell type value. In a second phase, we performed Kaplan-Meier survival analysis in order to understand which group has more chances to survive after the same treatment. We employed different statistical techniques for class prediction from gene expression data in order to detect a set of Cells-of-Origin of disease for each prognostic subgroup. The results obtained are affected by the different probe set proportion between signature matrix and mixture. Indeed, only four probe sets over ten found according the PAM analysis is present in LM22, only one according the ClaNC analysis and no probe set according the POS analysis.

As far as future works are concerned, a new signature matrix that includes more probe sets could improve our results, and better define the correlation between genes and survival rate of patients.

References

1. National Institutes of Health: Understanding emerging and re-emerging infectious diseases. Biological Sciences Curriculum Study, Bethesda, MD, US (2007)
2. Sandlund, J.T., Martin, M.G.: Non-Hodgkin lymphoma across the pediatric and adolescent and young adult age spectrum. ASH Educ. Program Book **2016**(1), 589–597 (2016)

3. Zhao, Y., Simon, R.: Gene expression deconvolution in clinical samples. Genome Med. **2**(12), 93 (2010)
4. Van't Veer, L.J., et al.: Gene expression profiling predicts clinical outcome of breast cancer. Nature, **415**(6871), 530 (2002)
5. Ando, T., Suguro, M., Hanai, T., Kobayashi, T., Honda, H., Seto, M.: Fuzzy neural network applied to gene expression profiling for predicting the prognosis of diffuse large B-cell lymphoma. Jpn. J. Cancer Res. **93**, 1207–1212 (2002)
6. Dabney, A.R.: Classification of microarrays to nearest centroids. Bioinformatics **21**, 4148–4154 (2005)
7. Hedström, G., Hagberg, O., Jerkeman, M., Enblad, G.: The impact of age on survival of diffuse large B-cell lymphoma - a population-based study. Acta Oncol. **54**(6), 916–23 (2015)
8. Khoshhali, M., Mahjub, H., Saidijam, M., Poorolajal, J., Soltanian, A.R.: Predicting the survival time for diffuse large B-cell lymphoma using microarray data. J. Mol. Genet. Med. **6**, 287–292 (2012)
9. Lenz, G.: Novel therapeutic targets in diffuse large B-cell lymphoma. EJC Suppl. **11**, 262–263 (2013)
10. Scott, D.W., Wright, G.W., Williams, P.M., et al.: Determining cell-of-origin subtypes of diffuse large B-cell lymphoma using gene expression in formalin-fixed paraffin-embedded tissue. Blood **123**, 1214–1217 (2014)
11. Wang, S., Wang, J., Chen, H., Zhang, B.: SVM-based tumor classification with gene expression data. In: Li, X., Zaïane, O.R., Li, Z. (eds.) ADMA 2006. LNCS (LNAI), vol. 4093, pp. 864–870. Springer, Heidelberg (2006). https://doi.org/10.1007/11811305_94
12. Sharma, A., Paliwal, K.K.: Cancer classification by gradient LDA technique using microarray gene expression data. Data Knowl. Eng. **66**(2), 338–347 (2008)
13. Li, B., Zheng, C.H., Huang, D.S., Zhang, L., Han, K.: Gene expression data classification using locally linear discriminant embedding. Comput. Biol. Med. **40**(10), 802–810 (2010)
14. Orsborne, C., Byers, R.: Impact of gene expression profiling in lymphoma diagnosis and prognosis. Histopathology **58**(1), 106–127 (2011)
15. Wang, S.L., Fang, Y., Fang, J.: Diagnostic prediction of complex diseases using phase-only correlation based on virtual sample template. In: BMC Bioinformatics, vol. 14, pp. 11 (2013)
16. Newman, A.M., et al.: Robust enumeration of cell subsets from tissue expression profiles. Nat. Methods **12**(5), 453–457 (2015)
17. Altman, D.G.: Analysis of Survival times. In: Practical Statistics for Medical Research, pp. 365–93. CRC Press, Boca Raton (1990)
18. Goel, M.K., Khanna, P., Kishore, J.: Understanding survival analysis: Kaplan-Meier estimate. Int. J. Ayurveda Res. **1**(4), 274 (2010)
19. Mahmoud, O., et al.: A feature selection method for classification within functional genomics experiments based on the proportional overlapping score. BMC Bioinf. **15**(1), 274 (2014)
20. Klassen, M., Kim, N.: Nearest shrunken centroid as feature selection of microarray data. In: CATA, pp. 227–232 (2009)
21. Choi, B.Y., Bair, E., Lee, J.W.: Nearest shrunken centroids via alternative genewise shrinkages. PloS one **12**(2), e0171068 (2017)
22. Shaffer, A.L., Rosenwald, A., Staudt, L.M.: Decision making in the immune system: Lymphoid Malignancies: the dark side of B-cell differentiation. Nat. Rev. Immunol. **2**(12), 920 (2002)

23. Povey, S., Lovering, R., Bruford, E., Wright, M., Lush, M., Wain, H.: The HUGO gene nomenclature committee (HGNC). Hum. Genet. **109**(6), 678–680 (2001)
24. Bland, J.M., Altman, D.G.: The logrank test. BMJ **328**(7447), 1073 (2004)
25. Kao, L.S., Green, C.E.: Analysis of variance: is there a difference in means and What does it mean? J. Surg. Res. **144**(1), 158–170 (2008)
26. Indrayan, A., Surmukaddam, S.B.: Measurement of community health and survival analysis. Med. Biostat. **7**, 232–42 (2001)

Learning Consistent Tree-Augmented Dynamic Bayesian Networks

Margarida Sousa[1,2] and Alexandra M. Carvalho[1,2(✉)]

[1] Instituto de Telecomunicações, Lisbon, Portugal
[2] Instituto Superior Técnico, University of Lisbon, Lisbon, Portugal
{margarida.sousa,alexandra.carvalho}@tecnico.ulisboa.pt

Abstract. Dynamic Bayesian networks (DBNs) offer an approach that allows for causal and temporal dependencies between random variables repeatedly measured over time. For this reason, they have been used in several domains such as medical prognostic predictions, meteorology and econometrics. Learning the intra-slice dependencies is, however, most of the times neglected. This is due to the inherent difficulty in dealing with cyclic dependencies. We propose an algorithm for learning optimal DBNs consistent with the tree-augmented network (tDBN). This algorithm uses the topological order induced by the tDBN to increase its search space exponentially while keeping the time complexity polynomial.

1 Introduction

Bayesian networks (BN) are a powerful probabilistic representation [20] that provide interpretable models of the domain. This is achieved through the definition of a network – a directed acyclic graph (DAG) – that unravels direct conditional dependencies between random variables. This network provides nothing more than a factorization of the joint probability distribution of those variables. Learning a BN from data consists in learning this structure. Having so, it is easy to learn its parameters and make inferences over this probabilistic framework.

Dynamic Bayesian networks (DBN), on the other hand, model stochastic processes [19]. In this case, variables are measured not only once, as for the case of BNs, but repeatedly over time. The networks to be learned consist in a prior network and several transition networks. The prior network is a BN eliciting the dependencies between the random variables at their initial state. The transition network unravels the dynamic dependencies of the variables over time: from past states to current states (inter-slice dependencies); and between current states (intra-slice dependencies).

The inter-slice dependencies are easy to learn as they flow forward in time and do not create cycles [12]. On the other hand, learning the intra-slice dependencies

---

This work was supported through FCT, under contract IT (UID/EEA/50008/2013), and by projects PERSEIDS (PTDC/EMS-SIS/0642/2014), NEUROCLINOMICS2 (PTDC/EEI-SII/1937/2014), and internal IT projects QBigData and RAPID.

G. Nicosia et al. (Eds.): LOD 2018, LNCS 11331, pp. 179–190, 2019.
https://doi.org/10.1007/978-3-030-13709-0_15

suffers from the hardness of finding an acyclic graph [7,9,11]. A polynomial-time algorithm for learning optimal DBNs was proposed using the Mutual Information Tests (MIT) [22]. However, learning the inter and intra-slice networks all together is not considered. This step has been done for tree-like networks, resulting in the so-called tree-augmented DBN (tDBN) [17]. We propose to further extend this algorithm by increasing exponentially its search space to networks consistent with the topological order induced by an optimal tDBN. At the same time, we are able to maintain its time complexity polynomial in the size of the input.

The emerging availability of electronic medical records (EMR) is triggering this line of research, bringing large, feature-rich, heterogeneous, noisy, and incomplete time series. The proposed algorithm is currently being used to predict evolution of amyotrophic lateral sclerosis and treatment outcome of arthritis rheumatoid from EMR.

We start by reviewing the basic concepts of both BNs and DBNs. Then, we present the proposed learning algorithm and the experimental results. The paper concludes with a brief discussion and directions for future work.

2   Bayesian Networks

Let X denote a *discrete random variable* that takes values over a finite set $\mathcal{X}$ and $\mathbf{X} = (X_1, \ldots, X_n)$ represent an n-dimensional *random vector*, where each X_i takes values in $\mathcal{X}_i = \{x_{i1}, \ldots, x_{ir_i}\}$. Furthermore, let $P(\mathbf{x})$ denotes the probability that $\mathbf{X}$ takes the value $\mathbf{x}$. A *Bayesian network* (BN) encodes the joint probability distribution of a set of n random variables $\{X_1, \ldots, X_n\}$ [20] and it is given by a triple $B = (\mathbf{X}, G, \Theta)$, where:

- $\mathbf{X} = (X_1, \ldots, X_n)$, each random variable X_i taking values in $\{x_{i1}, \ldots, x_{ir_i}\}$, where x_{ik} denotes the k-th value X_i can take.
- $G = (\mathbf{X}, E)$ is a *directed acyclic graph* (DAG) with nodes in $\mathbf{X}$ and edges E representing direct dependencies between the nodes.
- The set Θ encodes the parameters of the network G. Each random variable X_i has an associated *conditional probability distribution* (CPD) a.k.a. local parameters: $\Theta_{ijk} = P_B(X_i = x_{ik}|\Pi_{X_i} = w_{ij})$, where Π_{X_i} denotes the set of parents of X_i in the network G and w_{ij} is the j-th *parent configuration* of Π_{X_i}, which ranges over $\{w_{i1}, \ldots, w_{iq_i}\}$, with $q_i = \prod_{X_j \in \Pi_{X_i}} r_j$.

We note that the random vector $\mathbf{X}$ coincide exactly with the set of nodes in G, and we abuse notation considering that set to be denoted by $\mathbf{X}$.

A BN B induces a unique joint probability distribution over $\mathbf{X}$ given by:

$$P_B(X_1, \ldots, X_n) = \prod_{i=1}^{n} P_B(X_i|\Pi_{X_i}). \tag{1}$$

Intuitively, the graph G of a BN can be viewed as a network structure that provides the skeleton for representing the joint probability, compactly, in a factorized way. This reduces highly the number of parameters needed to describe the full joint probability distribution over the random variables [6,16].

Learning a Bayesian network is done in two steps: first the structure is learned; having the structure fixed, the parameters are learned. This is called *structure learning* and *parameter learning*, respectively. In what follows, we assume data D is complete, i.e, each instance is fully observed, there are no missing values or hidden variables. Moreover, $D = \{\mathbf{x}_1, \ldots, \mathbf{x}_N\}$ is given by a set of N i.i.d. instances. In that case, N_{ijk} is the number of instances where X_i takes the value x_{ik} and its parents Π_{X_i} takes the configuration w_{ij}. In addition, the number of instances where Π_{X_i} takes the configuration w_{ij} is denoted by N_{ij}.

In order to learn the parameters we assume the underlying graph G is given; in this case, the goal is to estimate the parameters Θ of the network. Using general results of the maximum likelihood estimate we get the following parameters for a BN B:

$$\hat{\theta}_{ijk} = \frac{N_{ijk}}{N_{ij}}, \tag{2}$$

that is denoted by *observed frequency estimates* (OFE). When learning the structure, the aim is to find a DAG G, given D. This can be accomplished through the use of a *scoring function* $\phi : \mathcal{S} \times \mathcal{X} \to \mathbb{R}$, where $\mathcal{S}$ denotes the search space, that measures how well the BN B fits the data D; therefore, it is called *score-based learning* [2,3,5]. The main scoring criteria are Bayesian and information-theoretical [1]. We will focus only on information-theoretical ones, in particular, *log-likelihood* (LL) and *minimum description length* (MDL). The LL of a BN B is given by:

$$\mathrm{LL}(B|D) = \sum_{i=1}^{n} \sum_{j=1}^{q_i} \sum_{k=1}^{r_i} N_{ijk} \log(\theta_{ijk}). \tag{3}$$

This criterion does not generalize well as it favors complete network structures, leading to the *overfitting* of the model to the data. The MDL criterion, proposed by Rissanen [21], imposes that the parameters of the model must also be accounted, providing a penalty factor that balances between fitness and model complexity. The MDL is defined by:

$$\mathrm{MDL}(B|D) = LL(B|D) - \frac{1}{2}\ln(N)|B|, \text{ with } |B| = \sum_{i=1}^{n}(r_i - 1)q_i, \tag{4}$$

where $|B|$ corresponds to the number of parameters Θ of the network. These scoring functions have a very important property, they are *decomposable*. This means that the overall score ϕ of B can be expressed as sums of local contributions ϕ_i of each node X_i and its parents (c.f. summations in Eq. (3)). This decomposability property allows for efficient learning procedures based on local-search methods.

In light of the previous discussion, structure learning reduces to an optimization problem: given a scoring function ϕ and a data D, find the BN B that maximizes $\phi(B, D)$.

Learning general BNs is a NP-hard problem [7,9,11]. However, if we restrict the search space $\mathcal{S}$ to branchings (a.k.a. tree-like structures) [8,15] or networks

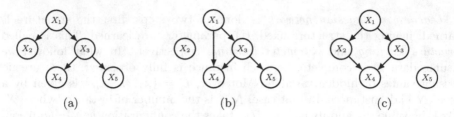

Fig. 1. Given the branching R represented in Fig. 1a, b represents a C2G w.r.t. R; Fig. 1c represents a non-consistent 2-graph w.r.t. R due to the edge from X_2 to X_4.

with bounded in-degree with a known ordering over the variables [10], it is possible to obtain global optimal solutions for this problem. A polynomial-time algorithm for learning BNs with underlying *consistent κ-graphs* (CκG) was proposed combining these ideas [4]. Therein, the authors showed that the set of networks consistent with the optimal branching is exponentially larger, in the number of variables, when comparing with branchings themselves [4]. In addition, the time-complexity of the learning procedure remained polynomial. The method we propose in this paper is an extension of the CκGs to DBNs, so in the following we further introduce notation and detail the CκG learning procedure.

A *κ-graph* is a graph where each node has in-degree at most κ. Given a branching R over a set of nodes V, a graph $G = (V, E)$ is said to be a *consistent κ-graph* (CκG) w.r.t. R if it is a κ-graph and for any edge in E from X_i to X_j the node X_i is in the path from the root of R to X_j. Intuitively, this branching R provides a topological order of the nodes from which the set of parents of each node in the network can be refined without creating cycles, avoiding the hardness of checking for cycles in the DAG. In this way, it is possible to add relevant edges, not considered previously due to the branching restriction (that allows only for one parent), and remove irrelevant ones (as branchings also requires exactly one parent per node, except from the root). For an example see Fig. 1.

The algorithm for learning CκG network structures, presented in Algorithm 1, starts by determining an optimal branching R (Step 1); for this it uses the Chow-Liu [8] or Edmond's [13] algorithm (see details in [4]). It then computes the set of candidate ancestors α_i, for each node X_i, compatible with the topological order induced by the optimal branching R (Steps 2–3). The parents of each node X_i in the network are then refined considering those in α_i (Steps 4–9). The algorithm returns a BN of in-degree κ consistent with R, augmenting the search space exponentially, in the number of variables, relatively to branchings, yet keeping a polynomial-time bound in the number of variables n.

3 Dynamic Bayesian Networks

Dynamic Bayesian networks (DBN) model the stochastic evolution of a set of random variables over time [19]. Consider the discretization of time in *time slices* $\mathcal{T} = \{0, \dots, T\}$. Let $\mathbf{X}[t] = (X_1[t], \dots, X_n[t])$ be a random vector denoting the

---

Algorithm 1. Learning CκG networks

1: Run a deterministic algorithm $\mathcal{A}_\phi$ that outputs an optimal branching R.
2: **for** each node X_i in R **do**
3:     Compute the set α_i of candidate ancestors for X_i.
4:     **for** each subset S of α_i with at most κ nodes **do**
5:         Compute $\phi_i(S, D)$.
6:         **if** $\phi_i(S, D)$ is the maximal score for X_i **then**
7:             Set $\Pi_{X_i} = S$.
8:         **end if**
9:     **end for**
10: **end for**

---

value of the set of attributes at time t. Furthermore, let $\mathbf{X}[t_1 : t_2]$ denote the set of random variables $\mathbf{X}$ for the interval $t_1 \leq t \leq t_2$. Consider a set of individuals $\mathcal{H}$ measured over T sequential instants of time. The set of observations is represented as $\{\mathbf{x}^h[t]\}_{h \in \mathcal{H}, t \in \mathcal{T}}$, where $\mathbf{x}^h[t] = (x_1^h, \ldots, x_n^h)$ is a single observation of n attributes, measured at time t and referring to individual h.

In DBNs we aim at defining a probability joint distribution over all possible *trajectories*, i.e., possible values for each attribute X_i and instant t, $X_i[t]$. Let $P(\mathbf{X}[t_1 : t_2])$ denote the joint probability distribution over the trajectory of the process from $\mathbf{X}[t_1]$ to $\mathbf{X}[t_2]$. The space of possible trajectories is enormous, therefore, it is necessary to simplify the problem and make it tractable.

In what follows, observations are viewed as i.i.d. samples of a sequence of probability distributions $\{P_{\theta[t]}\}_{t \in \mathcal{T}}$. For all individuals $h \in \mathcal{H}$, and a fixed time t, the probability distribution is considered constant, i.e., $\mathbf{x}^h[t] \sim P_{\theta[t]}, h \in \mathcal{H}$. Using the *chain rule*, the joint probability over $\mathbf{X}$ is given by:

$$P(\mathbf{X}[0 : T]) = P(\mathbf{X}[0]) \prod_{t=0}^{T-1} P(\mathbf{X}[t+1]|\mathbf{X}[0 : t]).$$

In this case the attributes in time slice $t + 1$ depend on all previous time slices t, for $t \in \{0, \ldots, T - 1\}$. Usually, not all previous time slices are considered but only a few. In that case, we say that m is the *Markov lag* of the process, also known as m^{th}-*order Markov* process, and so

$$P(\mathbf{X}[t+1]|\mathbf{X}[0 : t]) = P(\mathbf{X}[t+1]|\mathbf{X}[t - m + 1 : t]).$$

A further simpification approach is to consider that the process is *stationary*, also called *time invariant* or *homogeneous*, that is, $P(\mathbf{X}[t+1]|\mathbf{X}[t])$ is the same for all time slices $t \in \{0, \ldots, T - 1\}$. Sometimes, instead of considering the full process as stationary, we consider it *piece-wise stationary*.

In what follows we consider the stochastic process to be a first-order Markov stationary process. This eases the exposition, but its extension to a non-stationary or a mth-order Markov is straightforward. In this case, a *first-order Markov stationary dynamic Bayesian network* (DBN) consists of:

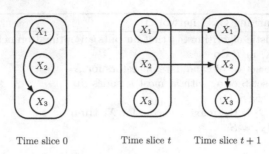

Time slice 0   Time slice t Time slice $t + 1$

Fig. 2. An example of a DBN. In the left, the prior network B_0 is depicted and in the right, the transition network B_t^{t+1} is represented. The edges $X_1[t] \rightarrow X_1[t+1]$ and $X_2[t] \rightarrow X_2[t+1]$ are the inter-slice connections and edge $X_2[t+1] \rightarrow X_3[t+1]$ represents the intra-slice connection.

- A prior network B^0, which specifies a distribution over the initial states $\mathbf{X}[0]$.
- A transition network B_t^{t+1} over the variables $\mathbf{X}[t : t+1]$, representing the state transition probabilities, for $0 \leq t \leq T - 1$.

The transition network has the additional constraint that edges between slices must flow forward in time.

We denote by G_{t+1} the subgraph of B_t^{t+1} with nodes $\mathbf{X}[t+1]$ that contains only the intra-slice dependencies. Observe that a transition network encodes the *inter-slice dependencies*, from time transitions t to $t+1$, and *intra-slice dependencies*, in time slice $t+1$ only. Figure 2 depicts an example of a DBN.

Learning dynamic Bayesian networks, considering no hidden variables or missing values, i.e., considering a fully observable process, reduces simply to learning two BNs: the initial network B_0 and the transition network B_t^{t+1}, taking into account that in B_t^{t+1} edges between slices must flow forward in time [14]. Not considering the acyclicity constraints, it was proved that learning a BN does not have to be NP-hard [12]. This result can be applied to DBNs, as the resulting *unrolled graph*, that contains a copy of each attribute in each time step, is acyclic. For this reason, several methods that consider only inter-slice dependencies appeared, as therein no cycles can arise [18, 22].

More recently, a polynomial-time algorithm was proposed that learns both the inter and intra-slice connections in a transition network; the resultant network was denoted by tree-augmented DBN (tDBN) [17]. Therein, the search space for the intra-slice networks was restricted to have a tree-like structure; each attribute in time slice $t+1$ was allowed to have at most one parent from the same time slice, and up to p parents were allowed from previous time slices; p is a user-input parameter.

We now describe the first-order Markov stationary tDBN algorithm. Let $\mathcal{P}_{\leq p}(\mathbf{X}[t])$ be the set of subsets of $\mathbf{X}[t]$ with cardinality less or equal to p. For each $X_i[t+1] \in \mathbf{X}[t+1]$, the optimal set of parents $\Pi_{X_i[t+1]} \in \mathcal{P}_{\leq p}(\mathbf{X}[t])$ yields the following score:

$$s_i = \max_{\Pi_{X_i[t+1]} \in \mathcal{P}_{\leq p}(\mathbf{X}[t])} \phi_i(\Pi_{X_i[t+1]}, D_t^{t+1}),$$

where ϕ_i is the local score of attribute $X_i[t+1]$ and D_t^{t+1} is the subset of observations for time transition $t \rightarrow t+1$. Then, allowing at most one parent $X_j[t+1]$ from the current time slice, the maximal score is defined as:

$$s_{ij} = \max_{\Pi_{X_i[t+1]} \in \mathcal{P}_{\leq p}(\mathbf{X}[t])} \phi_i(\Pi_{X_i[t+1]} \cup \{X_j[t+1]\}, D_t^{t+1}). \tag{5}$$

A complete directed graph is built such that each edge $X_j[t+1] \rightarrow X_i[t+1]$ has the following weight,

$$e_{ij} = s_{ij} - s_i, \tag{6}$$

that is, the gain in the network score of adding $X_j[t+1]$ as a parent of $X_i[t+1]$. Herein, the tDBN algorithm is able to determine the optimal set of inter and intra-slice parents of $X_i[t+1]$ in a one-step procedure.

Generally $e_{ij} \neq e_{ji}$, as the edge $X_i[t+1] \rightarrow X_j[t+1]$ may account for the contribution from the inter-slice parents and, in general, inter-slice parents of $X_i[t+1]$ and $X_j[t+1]$ are not the same. Therefore, Edmond's algorithm [13] is applied to obtain a maximum branching for the intra-slice network.

The pseudo-code of the procedure is given in Algorithm 2. A complete directed graph in $\mathbf{X}[t+1]$ is built (Step 1). Afterwards, in Step 2, the weight of all edges and the optimal set of parents for all nodes are determined according to Eq. (6) for a given scoring criterion ϕ. An optimal branching is obtained using Edmonds' algorithm [13] in Step 3. Step 4 retrieves the tree-like intra-slice transition network elicited in Step 3 with the optimal inter-slice parents determined in Step 2.

Algorithm 2. Optimal first-order Markov stationary tDBN

1: Build a complete directed graph in $\mathbf{X}[t+1]$.
2: Calculate the weight of all edges and the optimal set of parents of all nodes.
3: Apply Edmonds' algorithm to retrieve an optimal branching.
4: Extract transition network $t \rightarrow t+1$.

The tDBN algorithm has a worst-case time complexity that is linear in N (size of the input data), polynomial in n (number of variables) and r (number of values a variable can take), and exponential in p (number of parents from the previous time slice).

4   Proposed Method

Profiting from the CκG learning algorithm for BN, we propose an algorithm to learn DBN structures consistent with the tDBN. In what follows, as for tDBN, the proposed method is explained only for first-order Markov stationary DBNs; the extension to non-stationary m^{th}-order Markov, however, is straightforward.

Rigorously, a DBN is said to be a CκG, denoted by cDBN, if the intra-slice transition network G_{t+1} is a κ-graph where each edge from $X_i[t+1]$ to $X_j[t+1]$ is consistent with the intra-slice tree-network of a given tDBN. Moreover, each node $X_i[t+1]$ has at most p parents from the previous time slice. Therefore, in order to be well-defined, a cDBN needs two positive integers: κ and p. In addition, the given tDBN is an optimal tDBN computed with exactly the same number of p parents from the previous time slice.

We now describe briefly the proposed algorithm. It starts by computing an optimal tDBN. The intra-slice branching G_{t+1} is then used to refine the set of parents of each node in the network at time-slice $t+1$ so that they are consistent with the topological order induced by such branching. This is done by computing the candidate ancestors of each node $X_i[t+1]$, denoted by $\alpha_{i,t+1}$; these are exactly the set of nodes in $t+1$ connecting the root of the optimal branching given by G_{t+1} and $X_i[t+1]$. For node $X_i[t+1]$, the optimal set of past parents $\mathbf{X}_{ps}[t]$ and intra-slice parents, denoted by $\mathbf{X}_{ps}[t+1]$, are obtained in a one-step procedure by finding

$$\max_{\mathbf{X}_{ps}[t]\in\mathcal{P}_{\leq p}(\mathbf{X}[t])} \max_{\mathbf{X}_{ps}[t+1]\in\mathcal{P}_{\leq\kappa}(\alpha_{i,t+1})} \phi_i(\mathbf{X}_{ps}[t]\cup\mathbf{X}_{ps}[t+1], D_t^{t+1}), \qquad (7)$$

where $\mathcal{P}_{\leq\kappa}(\alpha_{i,t+1})$ is the set of all subsets of $\alpha_{i,t+1}$ of cardinality less than or equal to κ. Note that, if $X_i[t+1]$ is the root, $\mathcal{P}_{\leq\kappa}(\alpha_{i,t+1})=\{\emptyset\}$, so the set of intra-slice parents $\mathbf{X}_{ps}[t+1]$ of $X_i[t+1]$ is always empty.

Algorithm 3 finds an optimal first-order Markov stationary cDBN, given a decomposable scoring criterion ϕ, a set of n random variables, a maximum number of parents from the previous time slice of p, and a bounded in-degree in the intra-slice network of κ.

Algorithm 3. Learning optimal first-order Markov stationary cDBN

1: Compute an optimal tDBN with p parents with intra-slice graph given by G_{t+1}.
2: **for** each node $X_i[t+1]\in G_{t+1}$ **do**
3:     Compute the set $\alpha_{i,t+1}$ of ancestors of $X_i[t+1]$.
4:     **for** each subset P in $\mathcal{P}_{\leq p}(\mathbf{X}[t])$ **do**
5:         **for** each subset S in $\mathcal{P}_{\leq\kappa}(\alpha_{i,t+1})$ **do**
6:             Compute $\phi_i(P\cup S, D_t^{t+1})$.
7:             **if** $\phi_i(P\cup S, D_t^{t+1})$ is the maximal score for $X_i[t+1]$ **then**
8:                 Set $\Pi_{X_i[t+1]}=P\cup S$.
9:             **end if**
10:         **end for**
11:     **end for**
12: **end for**

The proposed algorithm increases exponentially the search space of the intra-slice transition network. Indeed, in the context of BNs, it was proved that the class of CκGs is exponentially larger, in the number of variables, when compared to tree-network structures [4], result which is straightforwardly extended to cDBNs. In Fig. 3 the search-space classes relating DBNs, namely tDBNs and cDBNs, are presented.

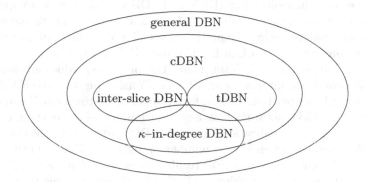

Fig. 3. Search-space classes of first-order Markov DBNs discussed in this paper. The class of inter-slice DBN contains all DBNs with no intra-slice dependencies. The class tDBN contains tree-augmented DBNs for all p parents from the previous time slice. The cDBN class contains all $(\kappa + p)$–in-degree cDBNs for all p and κ. The class of κ–in-degree DBN contains DBNs with in-degree at most $\kappa < 2n$, where n is the number of variables per time slice. This class does not include the tDBN as κ may be smaller than p. The general DBN class coincides with the $(2n - 1)$–in-degree DBNs.

In terms of worst-time complexity, when comparing with the tDBN algorithm, Algorithm 3 is linear in N (size of the input data) and T (number of time slices), polynomial in n (number of variables) and r (number of values a variable can take), and exponential in p (number of parents from the previous time slice t) and κ (number of parents in current time slice $t + 1$).

5 Experimental Results

We evaluate the proposed algorithm comparing it with the tDBN learning algorithm [17]. Our algorithm was implemented in Java and was released under a free software license.[1] The experiments were run on an Intel Core i5-3320M CPU @ 2.60GHz×4 machine.

We analyze the performance of the proposed algorithm for synthetic data generated from first-order Markov stationary cDBNs. Four cDBN structures and parameters were determined, and observations were sampled from the generated networks, for a given number of observations N. The parameters p and κ were taken to be the maximum in-degree of the inter and intra-slice network, respectively, of the transition network considered. The four transition networks

---

[1] https://margaridanarsousa.github.io/learn_cDBN/.

considered included: (i) one incomplete cDBN with $n = 5$, $\kappa = 2$ and at most $p = 1$ parents from the previous time slice; (ii) one complete cDBN with $n = 5$, $\kappa = 4$ and at most $p = 1$ parents from the previous time slice; (iii) one incomplete cDBN with with $n = 10$, $\kappa = 6$ and at most $p = 1$ parents from the previous time slice; (iv) one incomplete cDBN with $n = 10$, $\kappa = 4$ and at most $p = 1$ parents from the previous time slice. The tDBN and cDBN algorithms were applied to the resultant data sets, and the ability to learn and recover the original network structure was measured using the precision, recall and F_1-measure metrics. Two scoring functions were used: LL in Eq. (3) and MDL in Eq. (4).

The results are depicted in Table 1 and the presented values are annotated with a 95% confidence interval, over 5 trials. Considering LL, the cDBN algorithm consistently outperforms tDBN, for all number of instances N considered. As for MDL, the cDBN networks have a greater number of parameters, therefore the model complexity penalization factor of MDL leads to the selection of simple networks when considering a low number of instances. Hence, in these cases, the tDBN+MDL gives raise to better results. Generally, considering $N \geq 1000$ instances for the networks considered, cDBN+MDL outperforms tDBN+MDL. Comparing the results for networks 1 and 2, we observe that LL gives raise to better results when considering complete networks, whereas considering less complex structures, the MDL has better results. On the other hand, when comparing the results for networks 2:4 and 3:4, we conclude that considering a higher number of nodes and intra-slice in-degree κ, respectively, a higher number of instances is necessary to achieve similar recalls.

In Fig. 4, an example of the cDBN+MDL learning algorithm's ability to recover a known network is shown. The original cDBN network has $n = 5$ attributes, each taking $r = 2$ different values, having up to one parent from the previous time slice and two from the current time slice. Varying the number of input observations N, five recovered networks are shown. As N increases, the recovered network structures become more similar to the original, converging to the original for $N = 1800$.

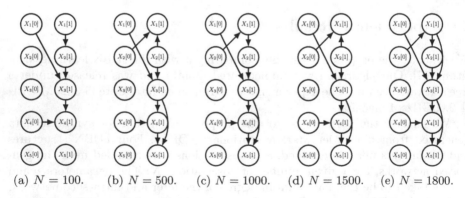

(a) $N = 100$.   (b) $N = 500$.   (c) $N = 1000$.   (d) $N = 1500$.   (e) $N = 1800$.

Fig. 4. Reconstructed networks for cDBN algorithm, where N is the number of instances used to learn. The true network was recovered when $N = 1800$.

Table 1. Comparative structure recovery results for tDBN and cDBN on simulated data. The tDBN+LL and tDBN+MDL denote, respectively, the tDBN learning algorithm with LL and MDL criteria. Similarly, for cDBN+LL and cDBN+MDL. For each network, n is the number of variables, p is the maximum inter-slice in-degree, κ is the maximum intra-slice in-degree, and r is the number of values of all attributes. On the left, N is the number of observations. Precision (Pre), recall (Rec) and F_1-measure (F_1) values are presented as percentages, running time is in seconds.

N	tDBN+LL				tDBN+MDL				cDBN+LL				cDBN+MDL			
	Pre	Rec	F_1	Time	Pre	Rec	F_1	Time	Pre	Rec	F_1	Time	Pre	Rec	F_1	Time
Network 1 ($n=5, p=1, \kappa=2, r=3$)																
100	60±5	60±5	60±5	0	92±14	51±8	66±10	0	58±5	76±7	65±6	0	100±0	20±4	33±6	0
500	78±0	78±0	78±0	0	86±8	64±4	74±5	0	73±3	98±4	84±3	0	98±4	84±5	90±4	0
1000	78±0	78±0	78±0	0	88±0	78±0	82±0	0	75±0	100±0	86±0	0	100±0	100±0	100±0	0
2000	78±0	78±0	78±0	0	88±0	78±0	82±0	0	75±0	100±0	86±0	0	100±0	100±0	100±0	0
Network 2 ($n=5, p=1, \kappa=4, r=3$)																
100	71±10	43±6	53±7	0	62±13	19±6	29±8	0	71±3	56±3	63±3	0	0±0	4±3	0±0	0
500	96±5	57±3	72±4	0	96±7	41±7	58±8	0	98±3	77±3	87±3	0	90±18	28±9	42±13	0
1000	98±4	59±2	73±3	0	100±0	47±0	64±0	0	100±0	80±0	89±0	0	100±0	44±3	61±3	0
2000	100±0	60±0	75±0	0	100±0	52±2	68±2	0	100±0	80±0	89±0	0	100±0	64±5	78±3	0
Network 3 ($n=10, p=1, \kappa=6, r=3$)																
100	53±5	33±3	41±4	0	66±8	23±4	34±5	0	36±9	38±7	37±8	2	83±18	7±2	13±4	4
500	72±5	45±3	56±4	0	88±5	40±2	55±3	0	53±2	68±7	60±4	1	100±0	33±2	50±2	1
1000	77±2	49±1	60±2	0	92±5	46±1	61±2	0	59±2	75±4	66±2	2	100±0	47±0	64±0	7
2000	78±2	49±1	60±1	0	92±2	48±1	63±2	0	60±1	78±2	68±2	10	100±0	58±3	73±2	8
Network 4 ($n=10, p=1, \kappa=4, r=3$)																
100	29±9	23±7	26±8	0	36±17	13±6	19±9	0	24±5	33±7	28±6	0	40±33	3±2	0±0	0
500	58±3	46±2	51±3	0	80±10	33±4	47±6	0	43±7	61±12	50±8	0	73±14	31±8	43±10	3
1000	60±5	48±4	53±4	0	80±8	38±3	51±5	0	41±6	69±9	51±7	4	86±6	48±4	62±5	24
2000	65±2	52±2	58±2	0	86±9	48±4	62±5	0	50±3	74±7	59±1	17	85±10	68±9	76±9	17

6 Conclusions

We conclude that the proposed algorithm allows to learn efficiently DBNs consistent with the topological order induced by the transition network of an optimal tDBN as far as the in-degree bounds p and κ are kept low. Notwithstanding, it is well known that in most practical scenarios BNs behave well with small in-degree network structures.

The resulting method is scalable (in the number of instances N, number of time slices T and number of variables n) and therefore suitable for the increasing amount of temporal data arising from medicine (and also other fields). We are currently using cDBN to predict the class of evolution of Amyotrophic Lateral Sclerosis (ALS) patients and the treatment outcome of rheumatoid arthritis (RA). These ALS and RA data is collected as a multivariate time series with heterogeneous values, which can be addressed effectively by cDBN. DBNs play the unique role of not only being able to model evolution in time of several autocorrelated variables but also provide models that are human interpretable.

Further improvements of the algorithm may include using a total order, instead of a partial one (as the topological order), and extend the learning procedure to allow hidden variables.

References

1. Carvalho, A.M.: Scoring functions for learning Bayesian networks. INESC-ID Technical Report, 12 (2009)
2. Carvalho, A.M., Adão, P., Mateus, P.: Efficient approximation of the conditional relative entropy with applications to discriminative learning of Bayesian network classifiers. Entropy 15(7), 2716–2735 (2013)
3. Carvalho, A.M., Adão, P., Mateus, P.: Hybrid learning of Bayesian multinets for binary classification. Pattern Recogn. 47(10), 3438–3450 (2014)
4. Carvalho, A.M., Oliveira, A.L.: Learning Bayesian networks consistent with the optimal branching. In: Proceedings of the ICMLA 2007, pp. 369–374. IEEE (2007)
5. Carvalho, A.M., Roos, T., Oliveira, A.L., Myllymäki, P.: Discriminative learning of Bayesian networks via factorized conditional log-likelihood. J. Mach. Learn. Res. 12, 2181–2210 (2011)
6. Charniak, E.: Bayesian networks without tears. AI Mag. 12(4), 50–63 (1991)
7. Chickering, D.M.: Learning Bayesian networks is NP-complete. In: Fisher, D., Lenz, H.J. (eds.) Learning from Data: Artificial Intelligence and Statistics, vol. 112, pp. 121–130. Springer, New York (1996). https://doi.org/10.1007/978-1-4612-2404-4_12
8. Chow, C., Liu, C.: Approximating discrete probability distributions with dependence trees. IEEE Trans. Inf. Theor. 14(3), 462–467 (1968)
9. Cooper, G.F.: The computational complexity of probabilistic inference using Bayesian belief networks. Artif. Intell. 42(2–3), 393–405 (1990)
10. Cooper, G.F., Herskovits, E.: A Bayesian method for the induction of probabilistic networks from data. Mach. Learn. 9(4), 309–347 (1992)
11. Dagum, P., Luby, M.: Approximating probabilistic inference in Bayesian belief networks is NP-hard. Artif. Intell. 60(1), 141–153 (1993)
12. Dojer, N.: Learning Bayesian networks does not have to be NP-hard. In: Královič, R., Urzyczyn, P. (eds.) MFCS 2006. LNCS, vol. 4162, pp. 305–314. Springer, Heidelberg (2006). https://doi.org/10.1007/11821069_27
13. Edmonds, J.: Optimum branchings. Math. Decis. Sci. Part 1, 335–345 (1968)
14. Friedman, N., Murphy, K., Russell, S.: Learning the structure of dynamic probabilistic networks. In: Proceedings UAI 1998, pp. 139–147 (1998)
15. Heckerman, D., Geiger, D., Chickering, D.M.: Learning Bayesian networks: the combination of knowledge and statistical data. Mach. Learn. 20(3), 197–243 (1995)
16. Koller, D., Friedman, N.: Probabilistic Graphical Models: Principles and Techniques. MIT press, Cambridge (2009)
17. Monteiro, J.L., Vinga, S., Carvalho, A.M.: Polynomial-time algorithm for learning optimal tree-augmented dynamic Bayesian networks. In: Proceedings of the UAI 2015, pp. 622–631 (2015)
18. Murphy, K.P.: The Bayes net toolbox for MATLAB. Comput. Sci. Stat. 33, 2001 (2001)
19. Murphy, K.P., Russell, S.: Dynamic Bayesian Networks: Representation, Inference and Learning (2002)
20. Pearl, J.: Probabilistic Reasoning in Intelligent Systems: Networks of Plausible Inference. Morgan Kaufmann, Burlington (2014)
21. Rissanen, J.: Minimum Description Length Principle. Wiley, Hoboken (1985)
22. Vinh, N.X., Chetty, M., Coppel, R., Wangikar, P.P.: Polynomial time algorithm for learning globally optimal dynamic Bayesian network. In: Lu, B.-L., Zhang, L., Kwok, J. (eds.) ICONIP 2011. LNCS, vol. 7064, pp. 719–729. Springer, Heidelberg (2011). https://doi.org/10.1007/978-3-642-24965-5_81

Designing Ships Using Constrained Multi-objective Efficient Global Optimization

Roy de Winter[1]($\boxtimes$), Bas van Stein[1], Matthys Dijkman[2], and Thomas Bäck[1]

[1] Leiden Institute of Advanced Computer Science, Leiden University,
Leiden, The Netherlands
r.de.winter.2@umail.leidenuniv.nl,
{b.van.stein,t.h.w.baeck}@liacs.leidenuniv.nl
[2] Research & Development, C-Job Naval Architects, Hoofddorp, The Netherlands
m.dijkman@c-job.com

Abstract. A modern ship design process is subject to a wide variety of constraints such as safety constraints, regulations, and physical constraints. Traditionally, ship designs are optimized in an iterative design process. However, this approach is very time consuming and is likely to get stuck in local optima. Not only does this optimization problem have complex constraints, it also consists of multiple objectives like resistance, stability and cost.

This constrained multi-objective optimization problem can be dealt with much more efficiently than through the traditional approach. In this paper, we propose a novel global optimization algorithm that explores the design space with the help of integrated software tools that are capable of simultaneous evaluation of the ship objectives and constraints. The optimization algorithm proposed uses the S-Metric-Selection-based Efficient Global Optimization (SMS-EGO) in combination with constraint handling techniques from an algorithm called Self-Adjusting Constrained Optimization by Radial Basis Function Approximation (SACOBRA). Since the evaluation of these ship designs is expensive in terms of computational effort, it is crucial for the algorithm to find feasible near-optimal solutions in as few evaluations as possible.

In this paper, it is shown that the proposed Constrained Efficient Global Optimization (CEGO) algorithm can significantly improve ship designs by automatic optimization using a small evaluation budget.

Keywords: Efficient Global Optimization ·
Multi-objective optimization · Constrained Optimization ·
Real-world applications

1 Introduction

The International Maritime Organization (IMO) responsible for regulating the shipping industry announced that by 2050 the greenhouse gas emissions should

© Springer Nature Switzerland AG 2019
G. Nicosia et al. (Eds.): LOD 2018, LNCS 11331, pp. 191–203, 2019.
https://doi.org/10.1007/978-3-030-13709-0_16

be reduced by 50% compared to 2008 [18]. To achieve this goal, the new ships that are currently being engineered will have to be optimized for minimum environmental impact. Of course, the environmental impact is not the only objective to consider while optimizing a ship. The ship owners also want their ship to be operationally efficient and to have the lowest building cost as possible. Additionally, safety and comfort of crew and/or passengers should meet the criteria given by the regulating authorities.

To achieve an optimal solution where all stakeholders are satisfied, typically different experts work together to optimize the ship. These experts, traditionally, optimize using the classical design spiral [9] and heuristics learned over the course of years, derived from knowledge and gained through a process of trial and error. For a single naval architect or a group of experts, it is impossible to consider the whole design space and all the relationships and dependencies between the variables, constraints and objectives [19]. Furthermore and most importantly, the traditional, expert driven, iterative approach used to design a ship can cause the design process to get stuck in local optima.

To make better design decisions in the future, the ship optimization processes such as proposed by Papanikolau [19] could be used. This integrated design approach brings together all key design aspects at the same time. In this paper it is shown that the combination of an integrated design approach and our proposed optimization algorithm results in significantly improved ship designs.

This paper is organized as follows: First, related research and algorithms are described and discussed in Sect. 2. The problem is described by giving an example ship design optimization problem in Sect. 3. The proposed algorithm is discussed into detail in Sect. 4. Next, it is shown empirically that the proposed algorithm is efficient and is able to find a good approximation of the Pareto front using a limited evaluation budget in Sect. 5. Finally, the results are discussed and conclusions are drawn in Sect. 6.

2   Related Work

Quite some work has been done in the domain of multi-objective optimization and constraint handling. The state of the art algorithms in constraint handling and multi-objective optimization together with the most relevant algorithms are listed below. Other algorithms without constraint handling (e.g. [13,17,21]) are not further considered in this research.

SACOBRA [1] In efficient constraint handling a recent model assisted optimization algorithm offers a promising efficiency in terms of the evaluation budget. This single objective optimization technique, SACOBRA, uses a *Self Adjusting parameter control in Constrained Optimization by use of Radial Basis function Approximation*. Because of self adjusting parameters and the Radial Basis function approximation of the constrained and objective space, SACOBRA is able to find high-quality results using only few function evaluations without having to spend evaluations on tuning the parameters [1].

SMS-EGO [22] *SMS-EGO* is an efficient multi-objective optimization algorithm that uses a Design and Analysis of Computer Experiments (DACE) to train Kriging [12] surrogate models in order to efficiently optimize the objective functions. Furthermore, SMS-EGO uses the S-metric or (hyper)volume contribution [2] to optimize the (hyper)volume between the current Pareto front and a reference point. This optimization algorithm, however, does not offer a constraint handling technique.

NSGA-II [6] *NSGA-II Non-dominated Sorting Genetic Algorithm, version II* is a classic multi-objective optimization algorithm. NSGA-II uses a non-dominated sorting-based selection operator. This operator creates a mating pool by combining the parent and child population to select the best N feasible solutions for the next generation. This selection operator makes sure that the mating pool is well spread and that the solutions in the pool have a high fitness.

NSGA-III [11] The adaptive *NSGA-III* algorithm is a many-objective optimization algorithm based on NSGA-II [6] and the original NSGA-III algorithm [7]. It emphasizes certain individuals in the population which are both non-dominant and close to a set of reference points which are generated on the fly. The algorithm can both be used for constrained and unconstrained problems since in every iteration the non-useful reference points are re-allocated around the useful reference points [11].

SPEA2 [29] The second *Strength Pareto Evolutionary Algorithm* (SPEA2) is an evolutionary algorithm that uses a fine-grained fitness assignment strategy that is based on how many feasible individuals each feasible individual dominates and is dominated by. Furthermore a nearest neighbor density estimation technique is incorporated which takes care of a more precise guidance of the search process. The algorithm also makes sure that the boundaries are guaranteed by truncation of the solutions that fall outside of the boundary.

MOGA This algorithm is currently a component of the widely used ship design software NAPA[1]. It is a so-called *Multi-objective Genetic Algorithm* (MOGA), which is based on the first version of the SPEA algorithm [30], where the fitness value is again based on the number of dominated feasible individuals. The selection of the parents is done by tournament selection and the children are generated by single-point crossover. Furthermore, the children have a chance to get mutated by the creep mutation operator.

3   Ship Design Optimization Example

Every ship design process starts with an initial idea from a client. After the objectives and physical constraints are known, the concept design process can begin. In the concept design phase, the naval architects translate the initial idea into the concept design of the ship. In the resulting concept design, the following components get defined and parameterized: the general arrangement, first estimations regarding stability, strength, and the main cross section.

---

[1] NAPA Oy, Release 2017.3-3 (2018), NAPA software, http://www.NAPA.fi/.

These components will define the ship's future performance, safety and cost. In this stage of the design process, all different components need to be optimized and designed in such a manner that they meet all regulations and safety criteria. This is not trivial for the following three reasons: (1) The objectives are typically conflicting. (2) Computing the constraints and objectives is very time consuming due to the required simulation time. (3) Only little parallelism is possible due to a typical limited number of commercial licences available to the ship design company. After the concept design phase, a ship yard can make an estimation of how long it will take, and how expensive it will be to build the ship.

As a real world application a dredger from C-Job Naval Architects[2] is optimized. The details about the decision variables, constraints, and objectives are given in the following subsections.

3.1   Decision Variables

The decision variables of a ship design problem are the numerical quantities for which values can be varied in the optimization process [3]. These quantities are denoted as $x = [x_1, \ldots, x_n]$, where x_j represents one decision variable.

The dredger (Fig. 1) has the following decision variables: $\Delta_{breadth}$, Δ_{length}, foreship length, hopper length extension, hopper breadth, hopper height. Here Δ means a change opposed to the original design. All the possible combinations in between a defined lower and upper bound of x together is called the design space Ω.

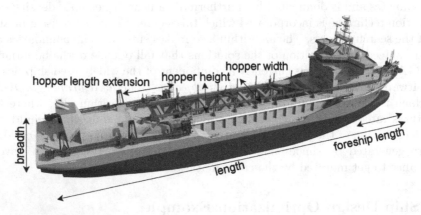

Fig. 1. Trailer suction Hopper Dredger designed by C-Job Naval Architects, with the design variables annotated.

The overall length and breadth of the hull can be transformed with the help of Free Form Deformation (FFD) [25]. For this transformation a box is drawn around the hull. Any point on the box can be moved in all directions and the

---

[2] C-Job Naval Architects, Ship Design and Engineering (2018), https://c-job.com/.

parent surface that is inside this box will be transformed accordingly. This FFD can be achieved by changing the $\Delta_{breadth}$ and Δ_{length} parameter which then applies the FFD on the original concept design.

The part of the ship from the most forward bulkhead to the front is called the foreship. The location of this last bulkhead can be changed by varying the *foreship length* decision variable.

The cargo space, where the dredged material is dumped in, is called the hopper. Changes can be made to the *height*, the *breadth*, and to the *length extension* of the hopper.

3.2   Constraints

The constraints can be expressed in terms of function inequalities; $g_i(\boldsymbol{x}) \leq 0$ where one function inequality $g_i(\boldsymbol{x})$ represents one of the m constraints. When equality constraints are present, we can simply rewrite them to two inequality constraints without loss of generality $g(\boldsymbol{x}) \leq c + \epsilon$ and $g(\boldsymbol{x}) \geq c - \epsilon$. In practice, ϵ can be neglected because it is chosen very small: $\epsilon = 0.000001$.

In the dredger case, the design has two categories of constraints: practical constraints and domain constraints. The constraints mainly make sure that everything fits in the hull and that the safety constraints are taken into account.

The practical constraints mainly consider the space reservation for: payload, fuel tank, engine, pump, and the accommodation. Every design variation is checked to see if it at least meets the minimum space required.

The domain constraints: steel arrangement, hull formation, double bottom check, location of foremost bulkhead, intact stability, draft when fully loaded, trim, and heel are checked to see if the ship meets the recommended stability criteria, and to see if it at least meets the other prescribed safety regulations.

In total, the dredger case has sixteen constraints, which are computed by subtracting the obtained constraint value from the required minimum value. When all values are negative the ship design variation is feasible.

3.3   Objectives

The objective functions are typically conflicting, as a consequence there is usually not one perfect solution but a set of alternative, so called non-dominated solutions. This non-dominated solution set contains good compromises between the objective functions: $f_j(\boldsymbol{x}), j = 1, \ldots, k$. The feasible Pareto optimal set of solutions together form the Pareto front where Pareto optimality is defined in *Coello et al.* [3].

The dredger case has two objectives: maximizing the performance and minimizing the building cost. This can be achieved by minimizing the resistance and the steel weight. This sounds trivial, but the objectives are a classical example of conflicting ones. A long and slender ship will lead to less hull resistance and a higher steel weight while a wide shorter ship will have a higher hull resistance and a lower steel weight.

The resistance of the design variation can be estimated with a Computational Fluid Dynamics (CFD) simulation. There are different types of CFD simulation methods. In the concept phase of the dredger, a relatively simple potential flow solver [26] is used. This approach does not take everything into account but it is very suitable for comparing the resistance between different design variations.

In the concept phase, an indication of the steel weight is calculated by first creating the main frame scantlings. This main frame is made strong enough so it does not exceed the maximum stress limit. This way the maximum bending moments can never be exceeded. The surface of the scantlings multiplied by the length can then be used to give an indication of the steel weight of the ship.

4   CEGO: Constrained Efficient Global Optimization

Here we propose the *Constrained Efficient Global Optimization* (CEGO) algorithm, combining the strengths of both the S-metric multi-objective optimization techniques from *SMS-EGO* and the constraint handling techniques from *SACO-BRA*. These two techniques are chosen because they showed to be very efficient in constraint handling [1] and finding a good approximation of the Pareto front [22]. The implementation of the proposed algorithm can be found on Github [4].

The proposed algorithm needs little to no parameter tuning and starts with an initial sampling of the decision variables using *Latin Hypercube Sampling* (LHS) [15]. The LSH samples then get evaluated by the evaluation function. The corresponding objective values are used to train the objective surrogate models. The objective surrogate models used are *Kriging* [14] (often also called Gaussian Process Regression models). For every objective dimension a separate Kriging model is fitted. Kriging treats every unknown objective function f as the combination of a centered Gaussian Process $\epsilon(x)$ of zero mean with an unknown constant trend μ. The advantage of using Kriging is that in addition to the predicted mean $y(x)$, the predicted uncertainty, called the Kriging variance $\sigma(x)$, is provided. The Kriging variance can be exploited in the optimization procedure.

The corresponding constraint values are used to train the constraint surrogate models. For the constraint surrogate models, *Cubic Radial Basis Functions* [1] (CRBF) are used. For every constraint function a CRBF model is fitted. The steps taken to model the constraint functions are the same as the ones in SACO-BRA [1]:

1. Rescale the decision space to an interval of $[-1, 1]$,
2. Normalize the constraint functions so that they are equally important,
3. Define the distance requirement factor (DRC) that defines how close the solutions are allowed to be to each other, and alter it at every iteration,
4. Adjust the margin (ϵ) of allowed violation of the CRBF model at every iteration.

In the first few iterations, the CRBF model might not fit the constraint function very well. Therefore, a violation of the constraints is allowed. The magnitude of the allowed violation decreases as more feasible solutions are found. In the

experiments reported in this paper, the ϵ-value used starts at 0.01. When three
feasible solutions are found, ϵ decreases by 50%. Alternatively, ϵ increases by
100% when three infeasible solutions are found.

After training the surrogate models, the feasible Pareto front approximation
is determined (denoted as Λ). To improve the Pareto front approximation, CEGO
uses the idea of Emmerich et al. to use S-metric or (hyper)volume contribution [2]
extended as an infill criterion [22]. The infill criterion function computes for
a given input vector $\boldsymbol{x}$, the predicted objective scores $\hat{y}$ and their estimated
uncertainties $\hat{s}$. If the 95% lower confidence bound of the potential solution
$\hat{y}_{pot} = \hat{y} - \alpha \cdot \hat{s}$ is still ϵ-dominant we compute the additional (hyper)volume it
adds to the Pareto front. ϵ-dominance as described in [2] is applied to support a
good distribution over the Pareto front. The size of ϵ is set every iteration:

$$\epsilon = \frac{\max(\Lambda) - \min(\Lambda)}{1+ \mid \Lambda \mid -\frac{1}{2^k} \cdot (maxEval - eval)}. \tag{1}$$

Here $max(\Lambda)/min(\Lambda)$ is the maximum/minimum value per objective on the
Pareto front, k is the number of objectives, $maxEval$ the maximum number
of allowed iterations, and $eval$ the number of evaluations executed so far. The
final (hyper)volume that $\hat{y}_{pot}$ adds to the Pareto front is the score the S-metric
criterion will return. If $\hat{y}_{pot}$ does not contribute anything, the infill criterion
will return zero. The S-metric infill criterion therefore gives the highest score to
solutions that potentially contributes the most to the Pareto front while it gives
a low score to solutions that does not contribute to the potential Pareto front.

This infill criterion is optimized using the Constrained Optimization by Linear Approximation (COBYLA) algorithm [23]. COBYLA optimizes the infill
criterion under the condition that the constraints, which are modeled with the
CRBF functions, are satisfied. The vector $\boldsymbol{x}$ that is predicted feasible and is
expected to contribute the most to the Pareto front approximation is proposed
as new solution. If no feasible solution can be found, the vector $\boldsymbol{x}$ with the
smallest expected constraint violation according to the CRBF models is chosen.

The proposed solution $\boldsymbol{x}$ then gets evaluated by the actual evaluation functions that are being optimized. This evaluation of $\boldsymbol{x}$ gives a new individual that
can be added to the population. In the next iteration, the surrogate models are
re-trained so that a new solution $\boldsymbol{x}$ can found and evaluated. This optimization
process goes on until the evaluation budget is exhausted.

5 Experiments and Results

To evaluate the performance of the proposed algorithm, three different experimental setups are used. In the first setup, seven artificially designed functions
are optimized. In the second setup, seven Real World Like Problems (RWLP)
are optimized. Finally, in the third setup, the dredger ship design is optimized.
All experiments are conducted with CEGO, NSGA-II, NSGA-III, SPEA2, and
MOGA with the default parameters, and a limited function evaluation budget

of 200 evaluations per run. In the experiments we used the two most cited performance metrics [24] to compare the diversity and the accuracy of the solutions obtained by the different multi-objective optimization algorithms. The first metric is the (hyper)volume (HV) metric that represents the HV between a fixed reference point and the Pareto front [2]. The second metric is the Generational Distance (GD), which represents how "far" the normalized obtained Pareto front is from the true normalized Pareto front [28]. Each algorithm is executed between 5 and 100 times per test function, depending on the time-complexity of the algorithm and the evaluation function.

5.1   Artificially Designed Functions

Inspired by previous studies on multi-objective optimization algorithms, seven widely used artificially designed functions are selected to experiment with: BNH [3], C3-DTLZ4 [27], OSY [3,8], SRN [8], TNK [8], CEXP1 [5], and CTP1 [5]. In Table 1 the number of objectives (k), number of variables (n), number of constraints (m), Lower Bound (LB), Upper Bound (UB) of the variables and the reference point (ref) are given for each function. To get some insight into the severity of the constraints, the percentage of feasible solutions (F(%)) is approximated by the evaluation of 1 million random samples.

Table 1. Artificially designed test problems and the corresponding dimensions.

Problem	k	n	m	LB	UB	ref	F (%)
BNH	2	2	2	[0, 0]	[5, 3]	[140, 50]	96.92
CEXP	2	2	2	[0.1, 0]	[1, 5]	[1, 9]	57.14
C3-DTLZ4	2	6	2	[0, 0, 0, 0, 0 ,0]	[1, 1, 1, 1, 1, 1]	[3, 3]	22.22
SRN	2	2	2	[−20, −20]	[20, 20]	[301, 72]	16.18
TNK	2	2	2	[1e−5, 1e−5]	[π, π]	[2, 2]	5.05
OSY	2	6	6	[0, 0, 1, 0, 1, 0]	[10, 10, 5, 6, 5, 10]	[0, 386]	2.78
CTP1	2	2	2	[0, 0]	[1, 1]	[1, 2]	92.67

5.2   Real World Like Problems

The RWLP are real world like problems which are believed to be very difficult because they have many complex constraints [11]. The following seven RWLP have been used in the experiments: Two-Bar Truss Design problem (TBTD) [10], Welded Beam problem (WB) [10], Disc Brake Design problem (DBD) [10], Speed Reducer Design problem (SRD) [16], Ship Parametric Design problem (SPD) [20], Car Side Impact problem (CSI) [11], and the Water Problem

(WP) [11]. Details about the RWLP are given in Table 2. Note that if a function was to be maximized it is transformed into a minimization problem.

5.3   Dredger Ship Design

Finally, the dredger case as described in the problem definition (Sect. 3) is optimized. The limits used for the dredger parameters are: $\Delta_{breadth} \in [-1.6, 3.4]$, $\Delta_{length} \in [-2.8, 9.8]$, foreship length $\in [16, 22]$, hopper length extension $\in [5, 9]$, hopper breadth $\in [5, 9]$, and hopper height $\in [12, 16]$. The reference point is set to $[5000, 2]$. This is the case because we are not interested in design variations with a larger resistance coefficient than 2, or design variations with a larger steel weight than 5000 tonnes. Furthermore, based on 200 random samples, approximately 24% of the design space is feasible. The original dredger designed by human experts has an approximated steel weight of 2039 tonnes and an estimated resistance coefficient of 1.08.

Table 2. Real world like problems and the corresponding dimensions.

Problem	k	n	m	LB	UB	ref	F(%)
TBTD	2	3	2	[1, 0.0005, 0.0005]	[3, 0.05, 0.05]	[0.1, 100 000]	19.46
WB	2	4	5	[0.125, 0.1, 0.1, 0.125]	[5, 10, 10, 5]	[350, 0.1]	35.28
DBD	2	4	5	[55, 75, 1 000, 2]	[80, 110, 3 000, 20]	[5, 50]	28.55
SRD	2	7	11	[2.6, 0.7, 17, 7.3, 7.3, 2.9, 5]	[3.6, 0.8, 28, 8.3, 8.3, 3.9, 5.5]	[7 000, 1 700]	96.92
SPD	3	6	9	[150, 25, 12, 8, 14, 0.63]	[274.32, 32.31, 22, 11.71, 18, 0.75]	[16, 19 000, −260 000]	3.27
CSI	3	7	10	[0.5, 0.45, 0.5, 0.5, 0.875, 0.4, 0.4]	[1.5, 1.35, 1.5, 1.5, 2.625, 1.2, 1.2]	[42, 4.5, 13]	18.17
WP	5	3	7	[0.01, 0.01, 0.01]	[0.45, 0.1, 0.1]	[83 000, 1 350, 2.85, 15 989 825, 25 000]	92.06

5.4   Results

In Table 3 it is shown that CEGO outperforms NSGA-II, NSGA-III, SPEA2 and MOGA in terms of the HV and the GD measure for all problems experimented with, except for the C3-DTLZ4 artificially designed test problem. Additionally, In the Figs. 2, 3, 4 and 5 the non-dominated solutions of a few typical test functions and the dredger case are visualized. From these figures it can clearly be seen that the approximation of the Pareto front and the spread of the CEGO algorithm is better, compared to the other algorithms.

Table 3. Mean HV and mean GD score for the obtained Pareto front by the different algorithms. Bold face denotes the method that outperforms the other methods according to a paired Welchs t-test (Welchs t-test is used because of unequal variances and unequal sample sizes, a significance level of 5% is used.).

Problem	NSGA-II		NSGA-III		SPEA2		MOGA		CEGO	
Criterion	GD	HV	GD	HV	GD	HV	GD	HV	GD	HV
BNH	0.005	5 187	0.015	4 965	0.007	5 137	0.007	4 993	**0.003**	**5 254**
CEXP	0.025	3.414	0.018	3.162	0.083	3.141	0.032	2.950	**0.002**	**3.788**
C3-DTLZ4	0.010	5.198	0.005	4.605	0.016	5.058	**0.004**	4.662	0.014	**6.098**
SRN	0.021	$5.82 \cdot 10^4$	0.035	$5.71 \cdot 10^4$	0.078	$4.88 \cdot 10^4$	0.056	$5.19 \cdot 10^4$	**0.005**	**$6.26 \cdot 10^4$**
TNK	0.025	7.247	0.007	6.763	0.045	6.449	0.011	6.074	**0.001**	**8.058**
OSY	0.136	$3.66 \cdot 10^4$	0.108	$3.92 \cdot 10^4$	0.157	$2.17 \cdot 10^4$	0.098	$4.71 \cdot 10^4$	**0.014**	**$1.00 \cdot 10^5$**
CTP1	0.037	1.248	0.022	1.218	0.055	1.221	0.042	0.661	**0.002**	**1.303**
TBTD	0.026	7 868	0.026	7 736	0.031	7 060	0.868	608.8	**0.003**	**8 805**
WB	0.028	34.07	0.058	33.74	0.054	33.67	0.019	33.93	**0.015**	**34.52**
DBD	0.041	219.4	0.031	214.8	0.050	214.6	0.016	221.4	**0.006**	**227.9**
SRD	0.118	$1.99 \cdot 10^6$	0.090	$1.81 \cdot 10^6$	0.156	$1.50 \cdot 10^6$	0.321	$1.66 \cdot 10^6$	**0.002**	**$4.16 \cdot 10^6$**
SPD	0.055	$2.45 \cdot 10^{10}$	0.047	$1.93 \cdot 10^{10}$	0.057	$2.09 \cdot 10^{10}$	0.041	$1.94 \cdot 10^{10}$	**0.026**	**$3.24 \cdot 10^{10}$**
CSI	0.032	15.34	0.034	12.77	0.032	13.95	0.026	17.13	**0.017**	**23.21**
WP	0.094	$1.28 \cdot 10^{19}$	0.100	$1.22 \cdot 10^{19}$	0.118	$1.13 \cdot 10^{19}$	0.071	$1.27 \cdot 10^{19}$	**0.053**	**$1.57 \cdot 10^{19}$**
Dredger	-	3529	-	3507	-	3579	-	3602	-	**3819**

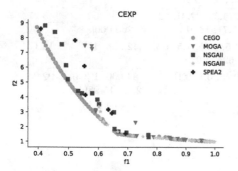

Fig. 2. Pareto front obtained by the five algorithms on CEXP problem.

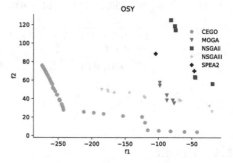

Fig. 3. Pareto front obtained by the five algorithms on OSY problem.

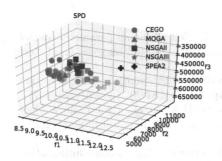

Fig. 4. Pareto front obtained by the five algorithms on SPD sproblem.

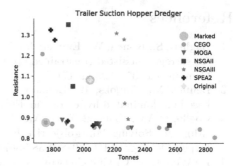

Fig. 5. Original design and Pareto front obtained by the five algorithms on dredger case.

6 Conclusions

An algorithm, *Constrained Efficient Global Optimization* (CEGO), is proposed and it is shown that CEGO is efficient in finding a Pareto front approximation using limited evaluation budgets for both Real-World Like Problems and artificially designed test functions. In case of the dredger design optimization task, ten unique non-dominated solutions are found within 200 function evaluations. The most interesting solution (marked in Fig. 5) has a resistance factor of 0.87 and a steel weight of 1748 tonnes. This means that compared to the original design, the improved design has a 19% smaller resistance coefficient and 14% less steel weight. As a post processing step, a naval architect inspected the design. After a few, very small, practical changes the ship was good to go to the next phase in the design process.

CEGO also outperforms state-of-the-art alternatives on all of the fourteen test problems used in the experimental setup. The novel proposed CEGO algorithm shows great potential and can be used to optimize ships that are more energy efficient while maintaining or even improving all other objectives. Of course the CEGO algorithm could also be used for any other application with expensive function evaluations with or without constraints.

For future work, the proposed algorithm could be improved by taking the CRBF constraint surrogate models into account when defining a new infill-criterion instead of using them as a constraint when minimizing the S-metric infill-criterion. It would also be beneficial to parallelize the CEGO algorithm such that multiple evaluations can be run at the same time. For practical ship design purposes, it would also be interesting to extend the algorithm in such a manner that constrained multi-objective mixed integer problems can be solved.

References

1. Bagheri, S., Konen, W., Emmerich, M., Bäck, T.: Self-adjusting parameter control for surrogate-assisted constrained optimization under limited budgets. Appl. Soft Comput. **61**, 377–393 (2017)
2. Beume, N., Naujoks, B., Emmerich, M.: SMS-EMOA: multiobjective selection based on dominated hypervolume. Eur. J. Oper. Res. **181**(3), 1653–1669 (2007)
3. Coello, C.A.C., Lamont, G.B., Van Veldhuizen, D.A., et al.: Evolutionary Algorithms for Solving Multi-objective Problems, vol. 5. Springer, New York (2007). https://doi.org/10.1007/978-0-387-36797-2
4. De Winter, R.: CEGO: Constrained Multi-Objective Efficient Global Optimization (2018). https://github.com/RoydeZomer/CEGO/
5. Deb, K.: Multi-objective Optimization Using Evolutionary Algorithms, vol. 16. Wiley, Hoboken (2001)
6. Deb, K., Agrawal, S., Pratap, A., Meyarivan, T.: A fast elitist non-dominated sorting genetic algorithm for multi-objective optimization: NSGA-II. In: Schoenauer, M., et al. (eds.) PPSN 2000. LNCS, vol. 1917, pp. 849–858. Springer, Heidelberg (2000). https://doi.org/10.1007/3-540-45356-3_83
7. Deb, K., Jain, H.: An evolutionary many-objective optimization algorithm using reference-point-based nondominated sorting approach, part i: Solving problems with box constraints. IEEE Trans. Evol. Comput. **18**(4), 577–601 (2014)
8. Deb, K., Pratap, A., Meyarivan, T.: Constrained test problems for multi-objective evolutionary optimization. In: Zitzler, E., Thiele, L., Deb, K., Coello Coello, C.A., Corne, D. (eds.) EMO 2001. LNCS, vol. 1993, pp. 284–298. Springer, Heidelberg (2001). https://doi.org/10.1007/3-540-44719-9_20
9. Evans, J.H.: Basic design concepts. J. Am. Soc. NavalEngineers **71**(4), 671–678 (1959). https://doi.org/10.1111/j.1559-3584.1959.tb01836.x
10. Gong, W., Cai, Z., Zhu, L.: An efficient multiobjective differential evolution algorithm for engineering design. Struct. Multidiscip. Optim. **38**(2), 137–157 (2009)
11. Jain, H., Deb, K.: An evolutionary many-objective optimization algorithm using reference-point based nondominated sorting approach, part ii: Handling constraints and extending to an adaptive approach. IEEE Trans. Evol. Comput. **18**(4), 602–622 (2014)
12. Jones, D.R., Schonlau, M., Welch, W.J.: Efficient global optimization of expensive black-box functions. J. Glob. Optim. **13**(4), 455–492 (1998)
13. Knowles, J.: ParEGO: a hybrid algorithm with on-line landscape approximation for expensive multiobjective optimization problems. IEEE Trans. Evol. Comput. **10**(1), 50–66 (2006)
14. Krige, D.G.: A statistical approach to some basic mine valuation problems on the witwatersrand. J. Chem. Metall. Min. Soc. S. Afr. **52**(6), 119–139 (1951)
15. McKay, M.D., Beckman, R.J., Conover, W.J.: A comparison of three methods for selecting values of input variables in the analysis of output from a computer code. Technometrics **21**(2), 239–245 (1979). http://www.jstor.org/stable/1268522
16. Mirjalili, S., Jangir, P., Saremi, S.: Multi-objective ant lion optimizer: a multi-objective optimization algorithm for solving engineering problems. Appl. Intell. **46**(1), 79–95 (2017)
17. Müller, J.: Socemo: surrogate optimization of computationally expensive multiobjective problems. INFORMS J. Comput. **29**(4), 581–596 (2017)
18. International Maritime Organization: Adoption of the inital IMO strategy on reduction of GHG emissions from ships. Note by the IMO to the 48 session of subsidiary body of scientific and technological advice, Bonn, Germany (2018)

19. Papanikolaou, A., Harries, S., Wilken, M., Zaraphonitis, G.: Integrated design and multiobjective optimization approach to ship design. In: Proceedings of International Conference on Computer Application in Shipbuilding, vol. 3 (2011)
20. Parsons, M.G., Scott, R.L.: Formulation of multicriterion design optimization problems for solution with scalar numerical optimization methods. J. Ship Res. 48(1), 61–76 (2004)
21. Picheny, V.: Multiobjective optimization using Gaussian process emulators via stepwise uncertainty reduction. Stat. Comput. 25(6), 1265–1280 (2015)
22. Ponweiser, W., Wagner, T., Biermann, D., Vincze, M.: Multiobjective optimization on a limited budget of evaluations using model-assisted S-metric selection. In: Rudolph, G., Jansen, T., Beume, N., Lucas, S., Poloni, C. (eds.) PPSN 2008. LNCS, vol. 5199, pp. 784–794. Springer, Heidelberg (2008). https://doi.org/10.1007/978-3-540-87700-4_78
23. Powell, M.J.: A direct search optimization method that models the objective and constraint functions by linear interpolation. In: Gomez, S., Hennart, J.P. (eds.) Advances in Optimization and Numerical Analysis. Mathematics and Its Applications, pp. 51–67. Springer, Heidelberg (1994). https://doi.org/10.1007/978-94-015-8330-5_4
24. Riquelme, N., Von Lücken, C., Baran, B.: Performance metrics in multi-objective optimization. In: 2015 Latin American Computing Conference (CLEI), pp. 1–11. IEEE (2015)
25. Sederberg, T.W., Parry, S.R.: Free-form deformation of solid geometric models. ACM SIGGRAPH Comput. Graph. 20(4), 151–160 (1986)
26. Tahara, Y., Stern, F., Himeno, Y.: Computational fluid dynamics-based optimization of a surface combatant. J. Ship Res. 48(4), 273–287 (2004)
27. Tanabe, R., Oyama, A.: A note on constrained multi-objective optimization benchmark problems. In: 2017 IEEE Congress on Evolutionary Computation (CEC), pp. 1127–1134. IEEE (2017)
28. Van Veldhuizen, D.A., Lamont, G.B.: Evolutionary computation and convergence to a pareto front. In: Late Breaking Papers at the Genetic Programming 1998 Conference, pp. 221–228 (1998)
29. Zitzler, E., Laumanns, M., Thiele, L.: SPEA2: improving the strength pareto evolutionary algorithm. In: EUROGEN 2001. Evolutionary Methods for Design, Optimization and Control with Applications to Industrial Problems, Athens, Greece, pp. 95–100 (2001)
30. Zitzler, E., Thiele, L.: Multiobjective evolutionary algorithms: a comparative case study and the strength pareto approach. IEEE Trans. Evol. Comput. 3(4), 257–271 (1999)

A New Approach to Measuring Distances in Dense Graphs

Fatimah A. Almulhim[1,2(✉)], Peter A. Thwaites[1], and Charles C. Taylor[1]

[1] University of Leeds, Leeds LS2 9JT, UK
mmfa@leeds.ac.uk
[2] Princess Nourah Bint Abdulrahman University, Riyadh, Saudi Arabia

Abstract. The problem of computing distances and shortest paths between vertices in graphs is one of the fundamental issues in graph theory. It is of great importance in many different applications, for example, transportation, and social network analysis. However, efficient shortest distance algorithms are still desired in many disciplines. Basically, the majority of dense graphs have ties between the shortest distances. Therefore, we consider a different approach and introduce a new measure to solve all-pairs shortest paths for undirected and unweighted graphs. This measures the shortest distance between any two vertices by considering the length and the number of all possible paths between them. The main aim of this new approach is to break the ties between equal shortest paths SP, which can be obtained by the Breadth-first search algorithm (BFS), and distinguish meaningfully between these equal distances. Moreover, using the new measure in clustering produces higher quality results compared with SP. In our study, we apply two different clustering techniques: hierarchical clustering and K-means clustering, with four different graph models, and for a various number of clusters. We compare the results using a modularity function to check the quality of our clustering results.

Keywords: Network · Adjacency matrix · K-means clustering · Hierarchical clustering · Modularity function

1 Introduction

The problem of computing distances and shortest paths between vertices in graphs is one of the most fundamental and well-studied problems in graph theory. The shortest path means the minimum path length between any pair of vertices in a graph, and in the case of directed graphs, there are source and destination vertices which determine the direction of the path. The shortest path is of great importance in many different applications, which has induced researchers to produce different measures to match their applications' purposes and graph types, for example, social network analysis, transportation, and computer science. And due to this variety of applications, there are many studies on

Supported by Saudi Arabian Cultural Bureau in London.

graph types. For example, the graph can be either static with a fixed number of vertices and edges or dynamic which updates in the graph's structure by adding or deleting vertices and edges or changing the position of edges. The edges of a graph can be directed or undirected and can have either a positive or negative weight. There is no one universal approach to solve the shortest path that could be suitable for all different graph models.

The survey by [9] classifies the shortest-path algorithms into two groups: (1) single source shortest-path (SSSP) algorithms, which calculate the shortest path from a source vertex to other vertices in the graph based on the adjacency list representation, and (2) all-pair shortest path (APSP) algorithms, which calculate the shortest path between all pairs of vertices in the graph based on the adjacency matrix representation. In their survey, they presented a taxonomy of multiple levels of shortest path algorithms as a useful tool for the researcher to understand the shortest path categories, and to guide them to suitable techniques, which depend on the application. They also mention some challenges and solutions in each group of algorithms.

The survey by [12] reviews the shortest path algorithms on static graphs that produce exact results for the APSP problems for both weighted and unweighted graphs as well as dense and sparse graphs. He also represented some studies on APSP for restricted families of graphs, such as interval graphs that determine an interval for each vertex with its neighbours.

[13] presented a survey of APSP and SSSP for weighted and unweighted graphs as well as directed and undirected graphs.

The different studies in the literature present different methods to capture the SP in graphs. Although SP results are consistent, the difference is in the methods or in the time taken. However, an efficient shortest distance algorithm is still desired in many disciplines, we think about it differently, and introduce a new measure to solve APSP for undirected and unweighted graphs. The new algorithm has a unique feature: the ability to distinguish between equal SP distances. More precisely, whereas SP is a positive integer, the new measure breaks up the equal integers into rational number. This idea and its framework is presented in Sect. 2. In Sect. 3, we have a short discussion about graph clustering. In particular, we explain the hierarchical clustering method (HC) and K-means clustering algorithm. We also describe the modularity function as a useful measure of clusters quality. A simulation study with four different graph models is described in Sect. 4 to compare the clustering results between the proposed distance and SP. In Sect. 5, we consider a real data example from the Facebook network and we conclude in Sect. 6 with a discussion of our approach.

2   A New Distance in Graphs (Breaking Ties Distance - BTD)

Most of the straightforward and essential approaches that solve the shortest path problem use the tree search idea. They start from the source vertex (root) and pass along the branches to reach the destination. In the case of undirected, unweighted graphs, this may be done by following the Breadth-First Search (BFS) algorithm from each vertex to the rest of vertices by counting the number of edges of the shortest path. In recent years, there have been many improvements in shortest paths algorithms, which usually focus on decreasing the time complexity of the essential algorithms [9].

In this study, we focus on undirected, unweighted dense graph models. As the majority of dense graphs have ties between the shortest distances, these distances appear as equal shortest distances. However, these ties are real obstacles in several applications, for example, in clustering of vertices; when we allocate a vertex to the nearest cluster's centroid and then find a tie in the shortest distances between the vertex and a couple of centroids. Therefore, due to this issue, we consider a new way of measuring distances in graphs, which allows for breaking of ties. Whereas the majority of shortest path measures produce integer SP lengths, our distance metric produces real values for the distances. The proposed distance measure is not a transient thought; rather, it is a result of cumulative work and trials of various updated distance metrics until the aim is achieved.

During several experiments with these measures, we determined the important and effective parameters that cause these ties in graphs: the vertex degree, graph diameter and the plurality of shortest paths between a pair of nodes. Our concept is inspired by the general principle: 'more relations, more strength'. We believe that a pair of nodes joined together by multiple equal shortest paths should be thought of as closer than a pair of nodes joined by a single path of the same length. As a simple example from social networks, any two vertices (two persons) connected by three relations (such as gender, neighbourhood, and school) appear to be more strongly linked (closer) than two vertices connected by one relation. Based on this assumption, we have carried out a lot of experiments to find a distance metric formula that combines all available paths between any pair of vertices. However, the resulting distances do not refer to existing routes in the graph, in contrast, the new distance measure produces distances between vertices which often distinguish between equal shortest paths. Our measure starts with a simple version of distance and then undergoes several updates until reaching its final version. The breaking-ties distance by BTD satisfies some conditions which make it suitable to measure distances in all graph models in this study:

Let $G(V, E)$ be an undirected, unweighted graph with vertices V (nodes) and edges E , then:

- BTD is symmetric, i.e. $d_{\mathrm{BTD}}(v_i, v_j) = d_{\mathrm{BTD}}(v_j, v_i)$ for $v_i, v_j \in V$.
- BTD preserves the shortest paths order, i.e. if $d_{\mathrm{SP}}(v_i, v_j) < d_{\mathrm{SP}}(v_k, v_s)$, then BTD satisfies $d_{\mathrm{BTD}}(v_i, v_j) < d_{\mathrm{BTD}}(v_k, v_s)$. This is evident from the examples in Fig. 1.
- BTD satisfies the triangle inequality $d_{\mathrm{BTD}}(v_i, v_j) \leq d_{\mathrm{BTD}}(v_i, v_s) + d_{\mathrm{BTD}}(v_s, v_j)$ as does SP.

We define a similarity matrix S_{ij}:

$$S_{ij} = \sum_{r=1}^{diam} \frac{(A^r)_{ij}}{(2\max(A^r))^r}, \tag{1}$$

where:

- $A_{(n \times n)}$ is the adjacency matrix, which has entries $0, 1$ with 0 in the diagonal, and n is the number of nodes.
- $diam$ is the diameter, the longest SP in the graph.
- $\max(A^r)$ is the largest element in A^r.
- A^r is the usual matrix multiplication.

Since $(A^r)_{ij} \leq \max(A^r)_{ij}^r$, with equality when $r = 1$ and in the case of $r > 1$, S_{ij} is always less that < 1.

Now define the dissimilarity matrix by:

$$D_{ij} = -\log(S_{ij}), \quad i \neq j \quad \text{and} \quad D_{ii} = 0, \tag{2}$$

We carry out many experiments which compare BTD and SP on different random graphs according to four different graph models from [4]. Model ER is the Erdős-Rényi graph model, model WS is Watts-Strogatz graph model, model BA is Barabási-Albert graph model, and model FF is Forest-fire graph model. The experiments show that *BTD* produces distances which follows the SP order without any overlap between distances.

In Fig. 1 (first row), all graph models show a spreading in the BTD in each SP group. In the BA model, the spreading is limited compared with other models due to the tree graph structure which has a lack of ties between nodes (the range of the distances which correspond to SP= 2 is equal 10^{-3}).

From this experiment, we believe that BTD succeeds in breaking ties between equal SP especially in dense graphs. In the next section, we compare the performance of BTD with SP in graph clustering and show that BTD produces higher quality results compared with SP.

3   Graph Clustering

In graph theory, the clustering idea is to divide the nodes V into different groups, each group's nodes share similar features. These groups are called clusters, and in some cases, there are some constraints on the number or size of the clusters.

In real networks, these groups are sometimes called communities and appear naturally due to the real network's structure. This is because, first, the general feature of a real graph structure is the inhomogeneities in the edges distributions, and second, the nodes degree distribution (node degree is the number of edges which connect the node with the other nodes in a graph) often follows the power law distribution. This gives the majority of the low degree nodes a higher tendency to connect with large degree nodes which lead to the creation of some

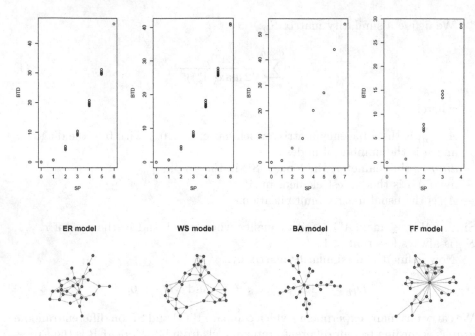

Fig. 1. Examples of graph models with $n = 30$, and the corresponding plots illustrate the breaking of ties SP distance.

communities in real networks. These communities may be visible in some small real networks.

In graph theory, there are different methods designed to cluster graphs, called graph partitioning methods. The main idea of these methods is to divide the graph's nodes into K clusters of predefined equal size and keep the number of edges between clusters minimal. These methods grew in pure mathematics among researchers who were interested in graph clustering [6]. Although these methods are simple and fast, they are still not a preferred tool to detect communities in graphs because of the preliminary assumption: equal groups size, which is considered as a drawback of this tool. Therefore, most of the traditional clustering methods which relax this condition are accepted in graph theory, such as hierarchical clustering, partitional clustering, and spectral clustering [5]. In the next sections, we apply BTD in hierarchical clustering and K-means algorithm.

3.1   Hierarchical Clustering- HC

The Hierarchical clustering (HC) method is one of the most popular methods in graph theory, it represents the clusters usually as a dendrogram. The leaves correspond to the graph nodes, and the root joins all nodes. It can be done by the agglomerative technique, which assigns each node in a separate cluster, and merges clusters to end with all nodes in one cluster. Alternatively, by the divisive technique, which initially puts all the nodes into one cluster, then divides it into sub-clusters. It continues division until finding a desirable structure.

The HC method has a special feature distinguishing it from the rest of clustering methods in that it produces multi-level clustering. Each level produces different clusters and each higher level cut produces of a subset of clusters from the lower level structure. This feature makes HC common in graph clustering of real networks as they have a hierarchical structure of communities, for example, social, biology, marketing networks, etc [1].

We use one of the HC techniques in association with our BTD metric, and as most of the work in literature is based on the agglomerative strategy, we choose it in our study. However, as the clusters merge from the lower level, there are different criteria to merge the clusters, each one estimates the similarity between clusters in a different way, for example, single, complete, and average linkage.

In this study, we apply agglomerative hierarchical clustering with complete linkage on four different graph models and show a comparison between both distances: SP and BTD. The results are discussed in Sect. 4.

3.2 K-means Clustering

An alternative clustering method is K-means clustering algorithm, which is one of the oldest methods in cluster analysis. Although its first appearance was in the 1950's, it is still one of the most commonly used methods. Also, it has a rich history in the literature as it is applied in various scientific areas. Three essential ingredients are required in K-means algorithm: the number of clusters K, initial centroids $\{c_1, c_2, \ldots, c_K\}$ and a distance metric [7]. Given a graph $G(V, E)$, Algorithm 1 below summarizes the steps:

Algorithm 1. K-means clustering $G(V, E)$

1: Compute the $(n \times n)$ distance matrix based on BTD.
2: Select K nodes randomly as initial centroids $v_1^*, \ldots, v_K^*$.
3: Allocate each node v_i to cluster C_k, where $k = arg \min_j d(v_i, v_j^*)$

4: **repeat**
5: Allocate each node v_i to C_k if $k = arg \min_l \frac{\sum_{v_j \in C_l} d(v_i, v_j)}{|C_l|}$
6: **until** convergence criterion is met no change in clusters members.

Starting the algorithm by computing the distance matrix BTD, then choose K initial centroids randomly. To form the clusters, we assign each node v_i to cluster k which satisfies the condition in 3. After the first allocation, we repeat step 5 until allocating all nodes to their fitted clusters. This differs from the original K-means algorithm where the points have coordinates and it is possible to calculate centroids in each iteration. In graphs, the nodes do not have coordinates, and in this case, we can not recompute the centroids in each iteration.

The most crucial point in K-means algorithm that can affect its performance is the selection of the initial centroids. It is noticeable that the random selection of initial starts often leads to very different clustering solutions. Therefore, K-means can only converge to a local minima, and this problem increases if the dataset structure does not have natural clusters. At most, we can repeat the algorithm for different sets of initial centroids, and either evaluate the results by subjective choice (in small data sets it may be possible to choose the solution which has obvious clusters by eye) or choose the cluster's solution which has a minimum squared error between the nodes. Given a graph of size n, $V = \{v_1, v_2, \ldots, v_n\}$, and K-means clustering result $C = \{C_1, C_2, \ldots, C_K\}$, the sum of squared errors (SSE) is given by:

$$SSE(C) = \frac{1}{2} \sum_{k=1}^{K} \sum_{v_i, v_j \in C_k} d(v_i, v_j)^2, \tag{3}$$

where K is the number of clusters.

In our study, we did a simulation study of clustering four different graph models using the K-means clustering algorithm, the goal of this simulation was to compare the performance of both distances BTD and SP in graphs. The results are presented in Sect. 4.

3.3   Modularity Function

The evaluation of the quality of a cluster is one of the most critical tasks in cluster analysis. As one of the clustering goals is exploring the latent structure of a graph, high-quality clustering result could describe the communities in the underlying graph. However, [2] argued that there is no single unique measurement to check the clusters quality, and in case of graphs which can be easily visualized by the researcher, the evaluation could be subjective.

One of the most popular quality functions for measuring the goodness of network partitions is the modularity function, introduced by [10]. The idea of this approach is that most random networks do not have clear communities in the graph structure. So, It assesses the partitions quality based on the difference between the arrangement of the edges within clusters in graphs, and the random distribution of these edges between nodes in case of no community structure. It can be either positive or negative, and we are looking for divisions with high modularity as a sign of proper partitions. It can be written as

$$Q = \frac{1}{4m} \sum_{i,j} \left(A_{ij} - \frac{k_i k_j}{2m} \right) \delta(c_i, c_j),$$

where k_i is the degree of the vertex v_i, $m = |E|$ is the total number of edges in the graph, $\frac{k_i k_j}{2m}$ is the expected number of edges between vertices i and j if edges are placed at random (null model), and the function $\delta(c_i, c_j)$ is equal to 1 if i and j are in the same group and 0 otherwise.

In our study, we use the modularity measure as an optimization function given K of the quality of our clustering results. The reasons behind our choice are that the modularity function is widely used by most of the academic researchers in cluster analysis as the best measure of the goodness of partitions [3,6,14]. Also, the modularity is quite a simple tool and is faster than most of the available quality measures even for large and sparse networks [11].

4   Simulation Study

The first simulation study is HC of four different graph models, the goal of this simulation is to compare the clustering results between the SP and BTD. For each model, we simulate 1000 graphs; each graph has size 100. Then, apply HC (complete link) with the number of clusters $K = 2, 3, \ldots, 10$. The Q results in Fig. 2 show that BTD always exceeds SP results in all graph models, and this is an evidence of the efficiency of BTD in hierarchical graph clustering.

The second simulation study is K-means clustering with the same graph models to compare the performance of BTD and SP by the modularity function Q. For each model, we simulate 50 graphs of size 200. For each graph, we choose 100 different sets of random initial centroids in each of cases $k = 5, 7$. Table 1 shows the simulation results for the four models. In each simulation, we calculate the modularity measure for each of the clustering results for all 100 random initial centroids and choose the maximum modularity over these 100 modularities. So, over 50 simulations, we obtain 50 modularity measures. In Table 1, max, min, and avg correspond to maximum, minimum and average values over the 50 modularity measures. Avg Itr is the average iteration number of the K-means algorithm over 50 simulations. The avg time is the average time taken over all 50 simulations. All these measures are calculated for both cases of distances: BTD and SP. Length 1 is the average number of BTD clustering results which have modularity measures bigger than the maximum modularity measure of SP clustering results for all 50 simulations. Length 2 is the average number of SP clustering results which have modularity measures less than the minimum modularity measure of BTD results over all 50 simulations. Table 1 shows that for the ER model, WS model, and FF model, the BTD produces slightly higher quality results when compared with SP using the modularity function. In the ER model, length 1 and length 2 are higher than the other models due to lots of ties in the graph structures. In the BA model, the results look similar between BTD and SP because the graph model has a tree structure which has a lack of the ties between nodes.

Also, we apply a paired t-test on the simulation results to check if the mean difference between BTD and SP results is zero:

$$H_0 : \max(Q_{\mathrm{BTD}}) = \max(Q_{\mathrm{SP}})$$

$$H_1 : \max(Q_{\mathrm{BTD}}) \neq \max(Q_{\mathrm{SP}}),$$

where the max is taken over the 100 random starts. The p-values of the paired test are less than 0.05 for ER, WS, and FF models, which is a significant sign

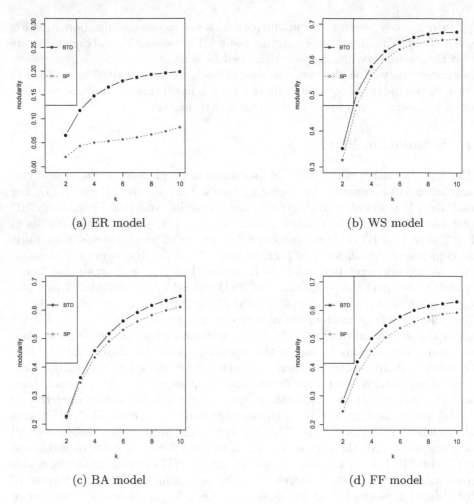

Fig. 2. Scatter plot of the mean of the modularity Q (vertical axes) of HC (complete linkage) using BTD (black line) and SP (red line) over 1000 simulations of each graph model, with different number of clusters $K = 2, 3, \ldots, 10$ (horizontal axes). (Color figure online)

of the difference between both distances results. Figure 3 illustrates the box plots of the differences between maximum/minimum modularity values between BTD and SP over 50 simulations; each figure corresponds to one model and one number of clusters. We can see that for all graph models except the BA model, most of the differences appeared higher than zero in most simulations. This is evidence that BTD generally produces higher cluster quality than SP.

From Table 1 and Fig. 3, we conclude that for dense graphs, the BTD produces higher quality clusters than the SP when assessed by the modularity function Q. This makes the BTD a more preferable distance measure in dense graphs than the SP. Note that because the difficulties in covering all parameter settings

Table 1. Table of statistics measurements to compare between the efficiency of BTD and SP over four different graph models with K-means algorithm for $K = 5, 7$.

	ER model				WS model			
	$k = 5$		$k = 7$		$k = 5$		$k = 7$	
Distance	BTD	SP	BTD	SP	BTD	SP	BTD	SP
Max	0.35	0.33	0.35	0.33	0.70	0.70	0.75	0.73
Min	0.31	0.29	0.31	0.30	0.64	0.63	0.68	0.66
Avg	0.33	0.32	0.33	0.31	0.67	0.66	0.70	0.69
Avg Itr	13.6	12	13.06	13	12.24	13.24	13.24	12.74
Avg time	13.3	12	13.11	11.91	13.81	13.14	14.39	13.28
Length 1	13.4		21.46		3.24		6.24	
Length 2	19.16		35.02		4.46		6.76	
	BA model				FF model			
	$K = 5$		$K = 7$		$K = 5$		$K = 7$	
Distance	BTD	SP	BTD	SP	BTD	SP	BTD	SP
Max	0.78	0.78	0.82	0.82	0.57	0.55	0.58	0.56
Min	0.67	0.68	0.77	0.77	0.32	0.32	0.26	0.27
Avg	0.75	0.75	0.80	0.80	0.46	0.44	0.45	0.43
Avg Itr	6.5	6.4	6.7	6.6	10.44	8.72	12.04	8.74
Avg time	7.6	7.3	8.7	8.3	1.35	1.39	1.57	1.38
Length 1	0.42		0.38		3.34		3.38	
Length 2	0.54		0.48		2.96		3.38	

and all simulation parameters: number of simulations, graph size, the number of clusters and the models parameters are subjective choices.

5 Facebook Example [8]

Our Facebook Network dataset is a combination of 10 ego-networks which consist of 4,039 users (nodes) and 88,234 edges. Each ego user has connections with all nodes in his/her network. In this example, we compare the performance of BTD with SP in hierarchical and K-means clustering algorithms on one connected component consisting of both the first and second ego networks excluding the ego nodes. It has 547 nodes and 5706 edges.

In Fig. 4, we show the proposed Facebook network with $K = 3$ clusters produced by K-means algorithm and BTD. As well as the HC results for BTD and SP for $K = 2, \ldots, 10$. The results show that BTD produces higher modularity than SP for the more important K values: in this network it appears that there are three natural communities, and for $K = 3, 4, 5$, HC produces a higher modularity score for BTD compared with SP. For $K > 8$, both distance

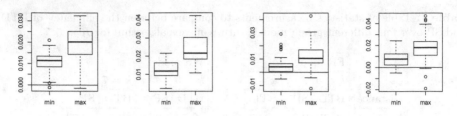

(a) ER model, $k = 5$ (b) ER model, $k = 7$ (c) WS model, $k = 5$ (d) WS model, $k = 7$

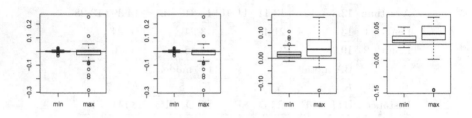

(e) BA model, $k = 5$ (f) BA model, $k = 7$ (g) FF model, $k = 5$ (h) FF model, $k = 7$

Fig. 3. Box plots of the differences of maximum/minimum modularity values between BTD and SP over 50 simulations of all proposed graph models of size $n = 200$, $K = 5, 7$ and 100 initial starts group in each simulation. The horizontal line crosses at zero to show the positive differences.

measures produce equal results, but the structural features have by this point disappeared. Even though Q reaches higher scores for large K, this is not a sign for a better number of clusters or communities, as the Q function is constructed to assess the cluster quality but not to choose the number of clusters [6]. We apply K-means algorithm with a different number of clusters $K = 3, \ldots, 10$, in each K we choose 10 different random initial centroids sets and check the maximum and minimum over these 10 sets for each K. We compare the performance of BTD with SP in this experiment by running the same 10 sets with each distance. Figure 4 illustrates the differences between BTD and SP by the minimum modularity scores over K. The results show a significant difference in favour of the BTD; this means the arbitrary choice of initial centroids always has a lower modularity score limit by SP than BTD. In the maximum comparison, both clustering methods behave similarly and produce similar modularity score. From HC and K-means algorithms experiments, we conclude that BTD produces higher quality results compared with SP in the Facebook network.

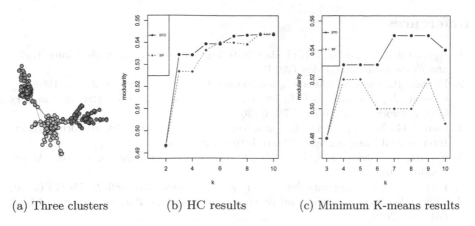

(a) Three clusters   (b) HC results   (c) Minimum K-means results

Fig. 4. Facebook graph with 3 clusters, comparison between BTD and SP in HC, and K-means clustering results.

6 Discussion

This paper presents a novel approach to measure distances in undirected, unweighted graphs. Its main idea is to break the ties between similar distances in dense graphs. In the experimental study, we examined the proposed distance BTD with four different graph models and concluded that BTD breaks the ties in SP and keeps the same SP order. Also, BTD has effective results compared with SP in graph clustering. In all simulation experiments, the results give evidence for the superiority of BTD compared with SP. This result is a significant finding in graph theory especially for graph clustering as most of the literature depends on SP.

Moreover, we have introduced a new way of assigning the nodes to the clusters in K-means algorithm based on the dissimilarity matrix, which differs from the standard K-means algorithm which is co-ordinate based.

Finally, we reaffirm our results by considering real data from the Facebook network.

Currently, we think over further study to reduce the time taken in repeating K-means algorithm for different initial centroids sets by considering deterministic choices. We intend to look for other real data which has a different structure from Facebook and compare the results. Moreover, we will check the validity of BTD in different statistical learning topics, for example, classification and regression.

Acknowledgments. The first author of this manuscript is grateful to the Saudi Arabian Cultural Bureau in London for financial support.

References

1. Aggarwal, C.C., Reddy, C.K.: Data Clustering: Algorithm and Applications. Taylor and Francis Group, London (2014)
2. Bonner, R.: On some clustering techniques. IBM J. Res. Dev. **8**, 22–32 (1964)
3. Clauset, A., Newman, M.E.J., Moore, C.: Finding community structure in very large networks. Phys. Rev. E **70**, 6 (2004)
4. Csardi, G., Nepusz, T.: The igraph software package for complex network research. InterJournal Complex Syst. (2006). http://igraph.org
5. Everitt, S., Landau, B.S., Leese, M., Stahl, D.: Cluster Analysis, 5th edn. Wiley, London (2011)
6. Fortunato, S.: Community detection in graphs. Phys. Rep. **486**(3), 75–174 (2010)
7. Jain, A.K.: Data clustering: 50 years beyond K-means. Pattern Recognit. **31**(8), 651–666 (2010)
8. Leskovec, J., Krevl, A.: SNAP datasets: Stanford large network dataset collection. (2014). http://snap.stanford.edu/data/index.html
9. Madkour, A., Aref, G., Rehman, F., Abdur Rahman, M., Basalamah, S.: A survey of shortest-path algorithm. CoRR abs/1705.02044 (2017). http://arxiv.org/abs/1705.02044
10. Newman, M.E.J.: Modularity and community structure in networks. Proc. Natl. Acad. Sci. USA **103**(23), 8577–8582 (2006)
11. Song, S., Zhao, J.: Survey of graph clustering algorithms using amazon reviews (2014). http://snap.stanford.edu/class/cs224w-2014/projects2014
12. Reddy, K.R.U.K.: A survey of the all-pairs shortest paths problem and its variants in graphs. Acta Universitatis Sapientiae, Informatica **8**(1), 16–40 (2016). https://doi.org/10.1515/ausi-2016-0002
13. Zwick, U.: Exact and approximate distances in graphs — A survey. In: auf der Heide, F.M. (ed.) ESA 2001. LNCS, vol. 2161, pp. 33–48. Springer, Heidelberg (2001). https://doi.org/10.1007/3-540-44676-1_3
14. Blondel, V.D., Guillaume, J., Lambiotte, R., Lefebvre, E.: Fast unfolding of communities in large networks. J. Stat. Mech.: Theory Exp. **2008**(10), P10008 (2008)

Ant Colony Optimization for Markov Blanket-Based Feature Selection. Application for Precision Medicine

Christine Sinoquet[(✉)] and Clément Niel

LS2N, UMR CNRS 6004, Université de Nantes, 2 rue de la Houssinière,
44322 Nantes Cedex, France
{christine.sinoquet,clement.niel}@univ-nantes.fr

Abstract. In this work, we address feature subset selection in the case when the variables exert a null or weak marginal effect on the target variable, a situation called "pure" epistasis hereafter. We explore the Markov blanket approach, to tackle epistasis detection, and we introduce SMMB-ACO. This method combines Markov blanket learning with stochastic and ensemble features and guides the stochastic sampling process by incorporating ant colony optimization. We first analyze the impact of parameter adjustment on SMMB-ACO complexity. Then using simulated and real data, we compare SMMB-ACO with four other methods, including its former version SMMB. We show that SMMB-ACO compares well with three state-of-the-art methods and that SMMB-ACO is more stable than SMMB. On the real dataset, the detection ability of SMMB-ACO is close to that of the best approach, which is a slow method, and SMMB-ACO is the fastest algorithm behind a much less performing method.

Keywords: Feature subset selection · Epistasis pattern ·
Bayesian network · Markov blanket · Metaheuristic ·
High dimensionality

1 Introduction

The intensive use of high-throughput genotyping technologies has opened the era for precision medicine, whose paradigm relies on targeted prevention and drug treatment, depending on the genetic profiles of patients. In this context, the aim of *genetic association studies* is to generate cutting-edge knowledge on the relationships between genotypes and some complex pathologies. Epistasis characterizes the situation in which a combination of variables (*i.e.*, genetic markers in interaction) is influential on a target variable (affected/unaffected status), whereas each of these variables shows a null or small individual effect. All studies involving univariate statistical tests are prone to miss the situations of epistasis. Exhaustive strategies cannot cope with high dimensionality, a characteristic of genotype data, and are limited to the exploration of two-way interactions (*e.g.*,

© Springer Nature Switzerland AG 2019
G. Nicosia et al. (Eds.): LOD 2018, LNCS 11331, pp. 217–230, 2019.
https://doi.org/10.1007/978-3-030-13709-0_18

GWIS [1]) or the examination of combinations within a user-specified size upper bound [2]. On the other hand, by definition, parametric statistical approaches suffer from limitations. A recent review has highlighted the contribution of artificial intelligence to the field of epistasis detection [3]. The proposals embrace ensemble learning strategies based on random forests (*e.g.*, Random Jungle [4]), metaheuristics designed for combinatorial optimization (*e.g.*, AntEpiSeeker [5]), and Bayesian network-based methods (*e.g.*, BEAM [6]). Other approaches combine several methods. For instance, HiSeeker incorporates a two-way interaction filtering stage prior ant colony optimization [7], KNN-MDR combines K-Nearest Neighbors and MDR methods [8]. A common drawback of all above mentioned approaches is the lack of detection power, especially in the case of pure epistasis (no marginal effect or weak marginal effect). Besides, reducing the search space is inescapable to handle a wealth of data. However, filtering strategies rely more or less on biased or incomplete knowledge. In contrast, artificial intelligence is poised to have a big impact in this regard, notably by combining novel methods with stochastic search algorithms designed to explore the combinatorial search space. In the remainder of this paper, we deal with discrete variables and a binary categorical target variable.

In this article, we state the problem of epistasis detection as a feature subset selection problem, and explore the Markov blanket (MB) approach to solve it. To this aim, we introduce SMMB-ACO (Stochastic Multiple Markov Blankets with Ant Colony Optimization), an innovative hybrid approach which combines Markov blanket construction with stochastic and ensemble features and incorporates an ant colony optimization (ACO) strategy.

Section 2 first puts forth a central property related to the Markov blanket concept. Then it briefly mentions the weaknesses of existing MB-based algorithms when addressing the epistasis detection issue. Section 3 introduces SMMB-ACO. Experimental results and discussion are presented in Sect. 4.

2   Markov Blanket Learning

In a Bayesian network built over the variables of set V, the Markov Blanket of a target variable T, $MB(T)$, is defined as a minimal set of variables that makes any variable outside $MB(T)$ statistically independent of T, conditional on $MB(T)$: $\forall\, X \in V \setminus MB(T),\ X \perp\!\!\!\perp T \mid MB(T)$.

The reference algorithm IAMB (Incremental Association Markov Blanket) chains two phases to grow an optimal MB from the empty set [9]. The forward phase successively incorporates variables into the MB under construction; the backward phase is meant to dismiss false positives. Several proposals were developed following this scheme, with variations in design, including variations in interleaving the forward and backward phases. The conditional test aforementioned is one of the essential ingredients used in the two phases. Various drawbacks are identified in these methods: (i) intractability in high-dimensional settings, (ii) requirement to verify strong assumptions, such as the faithfulness property [10] and absence of noise in the data, (iii) lack of power. The stochastic and ensemble features of SMMB-ACO were designed to alleviate issue (i) through an efficient exploration of the search space. The MB concept and optimal, principled feature subset selection are strongly connected under the faithfulness property, which ensures the unicity of the MB [10]. However, finding a best MB as the solution to feature selection from real life data is still more challenging. Again, SMMB-ACO addresses issue (ii) through an ensemble-based technique, to find multiple suboptimal MBs and enhance the construction of a consensus MB. Point (iii) is a recurring issue in feature subset selection and is highly challenging under the hypothesis of epistasis. In MB learning approaches, incorporating variables one at a time impedes the detection of pure epistasis: since the independence test achieved at first iteration is conditioned on the empty MB, a variable marginally dependent with the target variable is incorporated from the outset, which skews the whole construction. SMMB-ACO addresses this issue by including groups of variables instead. Importantly, SMMB-ACO is the enhanced version of a former method, SMMB [11], to be briefly described in Sect. 4.1. In SMMB-ACO, the stochastic sampling process is guided by incorporating ant colony optimization, for a more efficient resolution of issues (ii) and (iii).

3   The SMMB-ACO Algorithm

The input data for the SMMB-ACO algorithm consists of a matrix D (p observations $\times$ n variables), and a vector T (dimension p) to describe the target variable. The set of n variables is denoted V. For a didactical presentation, we introduce SMMB-ACO in a top-down fashion.

3.1   Top-Level Procedure

SMMB-ACO is structured into two procedures. The top-level procedure (Algorithm 1) successively runs n_{it_t} iterations (Algorithm 1, line 4). Each iteration supervises n_{ants} ants (Algorithm 1, line 7) that operate *in parallel* to build n_{ants} suboptimal MBs. Each ant a is first assigned a submatrix D_a of dimension $p \times K$, sampled from matrix D (Algorithm 1, line 8). The sampling is performed following a probability distribution $\mathbb{P}$ (Algorithm 1, line 5), to be further described. The learning of a suboptimal MB is carried out by function learnMB (Algorithm 1, line 10). Notably, this function implements conditional independence tests, whose statistics are collected in memory mem_a (Algorithm 1, lines 9 and 11). The test used is the conditional G-test. Once all ants have achieved their tasks, global memory mem collects all statistics output in the scope of ith ACO iteration (Algorithm 1, lines 13 and 16). Second, if an ant returns a non-empty suboptimal MB, then this set of variables is added to MBs (Algorithm 1, line 17).

We have adapted the standard ACO framework [12] to cope with MB learning. To further update the probability distribution over the n variables in V, we update the pheromone rates τ using global memory mem (Algorithm 1, line 19). Knowing how function learnMB proceeds is crucial to understand

Algorithm 1. SMMB-ACO

INPUT:

- D, matrix of p observations $\times$ n variables
- T, vector of dimension p, representing the target variable
- n_{it_t}, number of ACO iterations
- n_{ants}, number of ants
- K, size of the subset of variables sampled from D by each ant
- k, size of a combination of variables sampled amongst the K above variables ($k < K$)
- n_{it_n}, maximal number of iterations to coerce the exploration of the search space, in nested function learnMB, in case the MB under construction remains empty
- α', global type I error threshold
- Parameters for ACO optimization:

    - τ_0, constant to initialize the pheromone rates
    - ρ and λ, two constants used in pheromone rate updates
    - η, vector of weights (of dimension p), to account for prior knowledge on the variables in D
    - α and β, two constants used to adjust the relative importance between pheromone rate and *a priori* knowledge on the variables

OUTPUT:

- MB^*, a Markov blanket, built as the consensus of at most $n_{it_t} \times n_{ants}$ suboptimal Markov blankets

```
 1: MBs ← ∅
 2: /* τ, vector of dimension n, records the pheromone rates for the variables of D */
 3: τ ← init(τ₀)

 4: for i = 1 to n_{it_t}
 5:    /* ℙ is a probability distribution over the variables of D */
 6:    ℙ ← computeDistribution(τ, η, α, β)

 7:    for a = 1 to n_{ants}
 8:       D_a ← sample(D, K, ℙ)
 9:       mem_a ← ∅
10:       MB_a ← learnMB(D_a, T, k, n_{it_n}, α', mem_a, ℙ)
11:       /* mem_a now records the statistics for all conditional tests performed by ant a */
12:    end for

13:    mem ← ∅  /* mem will record the statistics for all conditional tests performed */
14:             /* during iteration i */
15:    for a = 1 to n_{ants}
16:       add(mem, mem_a)
17:       if not empty(MB_a) then MBs ← MBs ∪ MB_a end if
18:    end for

19:    τ ← updatePheromoneRates(τ, mem, ρ, λ)
20: end for

21: U ← ⋃_{M∈MBs} M /* consensus precursor */
22: MB* ← backwardPhase(U, T, α') /* refined consensus */
```

how pheromone rates are updated. Therefore we postpone this explanation to Sect. 3.3, after function learnMB has been described. Once the pheromone rates have been updated, the next iteration starts by computing probability distribution $\mathbb{P}$ as:

$$\forall\, X \in \boldsymbol{V},\ \mathbb{P}(X) = \frac{\tau(X)^\alpha \cdot \eta(X)^\beta}{\sum_{X' \in \boldsymbol{V}} \tau(X')^\alpha \cdot \eta(X')^\beta}, \text{(Algorithm 1, line 6),}$$

where $\tau(X)$ is the pheromone rate for variable X. In the feature selection problem, the pheromone rate $\tau(X)$ deposited by the ants indicates the significance of X to contribute to interactions with other variables, to determine the target variable. $\eta(X)$ is designed to integrate prior knowledge on variable X. Parameters α and β allow to adjust the relative weights between pheromome rate and prior knowledge.

When the last ACO iteration is completed in the top-level procedure, a MB consensus is built (Algorithm 1, lines 21 and 22). To this aim, the current version of SMMB-ACO first performs an union operation over all suboptimal MBs, followed by a backward phase. The resulting set is the output of SMMB-ACO. The backward phase used to refine the consensus is the same as in learnMB, with the exception that p-values are computed using permutations. This common process will be described in Sect. 3.2.

3.2   Learning a Suboptimal Markov Blanket

The nested procedure learnMB (Algorithm 2) driven by each ant starts with the initialization of the suboptimal MB to the empty set (Algorithm 2, line 1). Function learnMB iterates a series of forward steps each followed by a full backward phase. This process is iterated as long as the suboptimal MB can be modified, or as long as the MB remains empty and a maximal number of iterations, n_{it_n}, is not reached (Algorithm 2, line 3). Each forward step starts with the sampling of a submatrix S of dimension $p \times k$, sampled from the submatrix D_a of dimension $p \times K$ that is handled by the ant (Algorithm 2, line 6). Again, the probability distribution $\mathbb{P}$ is used for this purpose. One can construct $2^k - 1$ non-empty subsets from the variables in S. The subset s of S which maximizes a quality score is identified among all these combinations (Algorithm 2, line 7). Example 1 illustrates the computation of this score on a toy example. Once the best candidate s is identified, the p-value made available by the quality score computation (see Example 1) is examined. If the conditional dependence between s and T is statistically significant (Algorithm 2, line 8), the combination s is added to the suboptimal MB under construction (Algorithm 2, line 9) and a full backward phase is triggered (Algorithm 2, line 11). It is important to note that when we compute association scores (Algorithm 2, line 7), all the statistics generated by this calculus are stored in mem_a, the ant's memory (see Example 2 for an illustration).

The backward phase in the consensus construction (Algorithm 1, line 22) follows the same scheme as the backward phase in function learnMB (Algorithm 2, line 11). In the backward phase of the reference algorithm IAMB, the conditioning set used in all independence tests is the MB in its current state. SMMB-ACO implements a more refined strategy (Algorithm 3, line 2): to discard a false positive variable X, the independence between X and target T is tested conditional on each subset S of the current MB. Indeed, one of these subsets is necessarily the final MB to be discovered.

3.3   Back to Top-Level Procedure

Now we know that the statistics from conditional tests are stored in global memory mem (Algorithm 1, lines 11 and 16), we can explain how the current version of SMMB-ACO updates the vector τ (Algorithm 1, line 19). In standard ACO, the pheromone rate of a variable is updated as soon as the variable is selected by an ant. SMMB-ACO departs from this scheme: τ is updated at the end of an ACO iteration, when all the statistics provided by the ants are available. This choice is motivated by the parallelization of the ants' tasks. To update $\tau(X)$, SMMB-ACO takes into account the statistics output from all the conditional independence tests that involved X (across all ants). The update involves ρ ($0 \leq \rho \leq 1$), the pheromone evaporation rate, and a constant λ ($0 \leq \lambda \leq 1$). For instance, if $mem(X) = \{t_1, t_2, t_3, t_4\}$, the principle is to iterate operation $\tau(X) \leftarrow (1 - \rho)\, \tau(X) + \lambda\, t_i$, through the elements t_i of $mem(X)$.

Algorithm 2. learnMB

INPUT: see Algorithm 1 for input parameter description

OUTPUT:
- MB_a, a Markov blanket of T (possibly empty), learnt by current ant a.

1: $\mathbf{MB_a} \leftarrow \emptyset$
2: $MB_a\_modified \leftarrow true$; $j \leftarrow 0$

3: **while** ($MB_a\_modified$ or $(empty(\mathbf{MB_a})$ and $j < n_{it_n}))$
4:     $MB_a\_modified \leftarrow false$
5:     /* forward step */
6:     $\mathbf{S} \leftarrow sample(D_a, k, \mathbb{P})$
7:     $\mathbf{s} \leftarrow argmax_{\mathbf{s'} \subseteq \mathbf{S}}\{assocScore(s', T, MB_a, mem_a)\}$
8:     **if** ($p\text{-}value(\mathbf{s}) < \alpha'$) **then**
9:         $\mathbf{MB_a} \leftarrow \mathbf{MB_a} \cup \mathbf{s}$; $MB_a\_modified \leftarrow true$
10:        /* interleaved backward phase */
11:        $backwardPhase(\mathbf{MB_a}, \mathbf{T}, \alpha')$
12:    **end if**
13:    $incr(j)$
14: **end while**

15: **return $\mathbf{MB_a}$**

Example 1 (Computation of assocScore). We wish to identify the subset s of $S = \{X_1, X_2, X_3\}$, eligible to inclusion in the current MB (Algorithm 2, line 7). For this purpose, we rely on the conditional independence G-test $ind(X; T \mid Z)$, with X a variable, T the target variable, and Z a set of variables. This test outputs the statistic of the G-test of independence between variables X and T, conditional on set Z.

We successively compute:

subset: assocScore

① $\{X_1\}$: $\mathbf{t_1} = ind(X_1; T \mid MB)$
     $\{X_2\}$: $\mathbf{t_2} = ind(X_2; T \mid MB)$
     $\{X_3\}$: $\mathbf{t_3} = ind(X_3; T \mid MB)$

② $\{X_1, X_2\}$: $\max(\mathbf{t_4}, \mathbf{t_5})$
     $t_4 = ind(X_1; T \mid MB \cup \{X_2\})$
     $t_5 = ind(X_2; T \mid MB \cup \{X_1\})$

③ $\{X_1, X_3\}$: $\max(\mathbf{t_6}, \mathbf{t_7})$
     $t_6 = ind(X_1; T \mid MB \cup \{X_3\})$
     $t_7 = ind(X_3; T \mid MB \cup \{X_1\})$

④ $\{X_2, X_3\}$: $\max(\mathbf{t_8}, \mathbf{t_9})$
     $t_8 = ind(X_2; T \mid MB \cup \{X_3\})$
     $t_9 = ind(X_3; T \mid MB \cup \{X_2\})$

⑤ $\{X_1, X_2, X_3\}$: $\max(\mathbf{t_{10}}, \mathbf{t_{11}}, \mathbf{t_{12}})$
     $t_{10} = ind(X_1; T \mid MB \cup \{X_2, X_3\})$
     $t_{11} = ind(X_2; T \mid MB \cup \{X_1, X_3\})$
     $t_{12} = ind(X_3; T \mid MB \cup \{X_1, X_2\})$

If *assocScore* is the highest for $\{X_1, X2\}$, thanks to, say, t_5, then $\{X_1, X2\}$ is the candidate identified for inclusion in the current MB, and the p-value relative to statistic t_5 is memorized to be further used in Algorithm 2, line 8.

Example 2 (Update of ant memory). In the computation shown in Example 1, mem_a would be updated as follows:

$$mem_a(X_1) = mem_a(X_1) \cup \{t_1, \ t_4, \ t_6, \ t_{10}\},$$
$$mem_a(X_2) = mem_a(X_2) \cup \{t_2, \ t_5, \ t_8, \ t_{11}\},$$
$$mem_a(X_3) = mem_a(X_3) \cup \{t_3, \ t_7, \ t_9, \ t_{12}\}.$$

Algorithm 3. backwardPhase

INPUT:
- M, a Markov blanket whose false positive variables must be discarded from
- T, the target variable, $\mathbf{T} \notin \mathbf{M}$
- α', global type I error threshold

OUTPUT:
- M, possibly modified

1: **for each** $\mathbf{X} \in \mathbf{M}$ /* Is **X** a false positive? */
2:    **for each** $\mathbf{S} \subseteq \mathbf{M} \setminus \{\mathbf{X}\}, \mathbf{S} \neq \emptyset$ /* We try any **S** because the "true" MB is some $\mathbf{S} \subseteq \mathbf{M}$ */
3:      **if**($significant\_conditional\_independence(\mathbf{X}, \mathbf{T}, \mathbf{S}, \alpha')$) **then**
4:        $\mathbf{M} \leftarrow \mathbf{M} \setminus \{\mathbf{X}\}$; **break** /* X is a false positive and is therefore discarded from **M** */
5:      **end if**
6:    **end for**
7: **end for**

---

4   Experiments

SMMB-ACO was implemented in C++, using OpenMP to parallelize the iteration on the ants [13]. SMMB-ACO and four other methods were run using six cores composed of biprocessors XEON 5462 2.66 GHz. This section first describes the experimental protocol. Then we present a study relative to the impact of parameter adjustment on SMMB-ACO complexity, and its former version, SMMB. Finally, we compare SMMB-ACO with SMMB and three state-of-the-art methods of the domain, both on simulated and real data.

4.1   Experimental Road Map

Datasets. Our proof-of-concept framework considers epistasis detection in genetic association studies. Therein, we handle n genetic markers, taking their values in $\{0, 1, 2\}$, and a binary target variable (0:unaffected/1:affected), to describe $p/2$ unaffected subjects and $p/2$ patients, respectively. We simulated data following three interaction models described in the literature. The 2-way epistatic models 1 and 2 [14] each describe the interaction of two influential genetic markers. The 3-way epistatic model 3 was also used in previous works [6]. First, for each model, we generated 100 datasets (4,000 observations × 100 variables), using GAMETES software simulator (version 2.1) [15]. For each genetic marker simulated *via* GAMETES, a characteristic called "minor allele frequency" (MAF) must be specified. For each non-causal genetic marker, the MAF was uniformly drawn from [0.05, 0.5]. In our protocol, the simulated influential variables were specified to share the same MAF. Across our experiments, we varied this parameter in $\{0.05, 0.1, 0.2, 0.5\}$. Therefore, we simulated 12 conditions in total [16]. Second, to study the impact of parameter adjustment, we also simulated 100 datasets (4,000 observations × 5,000 variables) using model 2 (MAF 0.10). Third, we simulated data under the null hypothesis (no epistasis).

Finally, we used the genome-wide Rheumatoid Arthritis (RA) dataset provided by the Wellcome Trust Case Control Consortium [17]. This dataset describes 23 human chromosomes showing between 5,754 and 38,867 variables (20,236 variables on average; 469,616 in total). The numbers of unaffected and affected subjects are 2,938 and 1,860, respectively.

Methods Compared. SMMB-ACO was compared to BEAM [6], DASSO-MB [18] and AntEpiSeeker [5], all three dedicated to epistasis detection. We also compared SMMB-ACO to SMMB [11], our former brute-force proposal. The parameter corresponding to $n_{it_t} \times n_{ants}$ is n_{MBs} in SMMB. SMMB successively runs function learnMB (n_{MBs} times). SMMB and SMMB-ACO share K and k, the parameters involved in variable sampling. The naive approach SMMB samples variables using a uniform distribution, whereas ant colony optimization is used to explore the search space more efficiently in SMMB-ACO. Each of the first three methods shares a feature with SMMB-ACO. DASSO-MB implements the deterministic IAMB algorithm modified with interleaved forward and backward phases. AntEpiSeeker relies on ant colony optimization. As SMMB-ACO, BEAM relies on a Bayesian framework, this time using Monte-Carlo Markov Chains. For reasons of computational burden in the simulations, we limited the number of methods compared with SMMB-ACO.

Besides, to ensure scalability at the genome scale, we adapted our implementation of SMMB and SMMB-ACO. Since the bottleneck was the size of the consensus precursor U (see Algorithm 1, line 21), we modified SMMB et SMMB-ACO as follows: a first pass builds MBs and stops after it has produced U; in a second pass, SMMB (respectively SMMB-ACO) is run on the dataset consisting in U.

Criteria Used for Method Comparisons. For each of the 100 datasets simulated under the same condition, we ran each method 100 times and computed its F-Measure as $2/(1/recall + 1/precision)$, with $recall = TP/(TP + FN)$ and $precision = TP/(TP + FP)$.

We ran each stochastic method 10 times, and deterministic DASSO-MB one time, on each of the 23 human chromosome datasets and on the genome-wide dataset as well. At this scale, the consensus MB^* (Algorithm 1, line 22) output by SMMB or SMMB-ACO is still large. We decided to define the output of SMMB or SMMB-ACO as the set of the suboptimal MBs generated throughout a whole run (see Algorithm 1, line 17), whose variables belong to the refined consensus MB^*. For instance, if $MB^* = \{X_6, X_{43}, X_{51}, X_{77}\}$, the suboptimal MB $\{X_6, X_{51}\}$ is reported as a solution, whereas the suboptimal MB $\{X_{33}, X_{51}\}$ is not.

4.2 Results and Discussion

Impact of Parameter Adjustment. Most machine learning methods require the adjustment of a number of parameters. In all experiments presented in this paper, we easily adjusted 9 of the 12 parameters of SMMB-ACO (τ_0, ρ, λ, η, α, β, α', k, n_{it_n}; see Table 1). The three remaining parameters, n_{it_t}, n_{ants} and K, are crucial to control SMMB-ACO's complexity. Figure 1 provides insights on the impact of these parameters. Figure 1 (A) shows a plateau for consensus precursor size (≈ 9) when the total number of MBs ($n_{it_t} \times n_{ants}$) is increased beyond 100 MBs. This suggests that learning a number of MBs greater than this threshold is unuseful. Besides, the consensus precursor size corresponding to the plateau is larger for SMMB (≈ 14), which suggests a lower number of false positives for SMMB-ACO. Moreover, Fig. 1 (B) shows that learning more MBs than necessary is detrimental to time complexity. Figure 1 (B) also shows that SMMB-ACO is faster than SMMB. Interestingly, Fig. 1 (C) highlights the existence of a plateau for consensus precursor size (≈ 9 for $K = 100$) when K is increased. The plateau value is around 14 for SMMB. In a more thorough study, we inspected the consensus precursors obtained for K larger than the threshold evidenced. As from this threshold, the rate of consensus precursors containing the variables in interaction was constant (results not shown). Finally, Fig. 1 (D) shows how increasing the number of ACO iterations (n_{it_t}) impacts the running time.

Performances on Simulated Data. Figure 2 shows the comparison across 5 methods, 3 models and 4 MAFs. SMMB-ACO, SMMB, AntEpiSeeker and BEAM respectively require 12, 6, 12 and 11 parameters. To identify correct

Table 1. Parameter adjustment common to all experiments run on simulated data.

SMMB-ACO	Adjustment of the 6 parameters related to the ACO feature, following [5]: $\tau_0 = 100$, $\rho = 0.05$, $\lambda = 0.1$, $\eta = \alpha = \beta = 1$
SMMB-ACO & SMMB	Standard statistical significance threshold of independence tests: $\alpha' = 0.05$
	$n_{it_n} = 30$ (SMMB-ACO); $n_{MBs} = 100$ (SMMB); $k = 3$ (both) (empirical feedback)

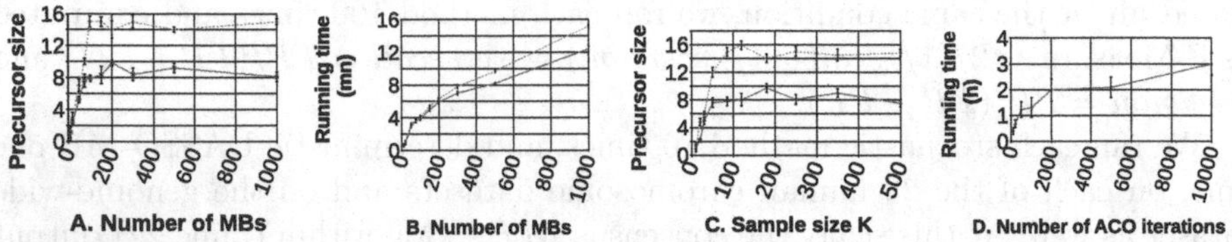

A. Number of MBs   B. Number of MBs   C. Sample size K   D. Number of ACO iterations

Fig. 1. Impact of parameter adjustment on SMMB-ACO and SMMB complexities. Continuous line: SMMB-ACO; dotted line: SMMB. Precursor size denotes the size of the consensus Markov blanket U, prior refinement (see Algorithm 1, line 21). One hundred datasets were generated using model 2 (MAF 0.10) (see Sect. 4.1/Datasets), with various n values. Total number of MBs: $n_{it_t} \times n_{ants}$ (SMMB-ACO) or n_{MBs} (SMMB). (A) and (B): $p = 4,000$; $n = 100$ (see Algorithm 1, INPUT); $n_{it_t} = 34$; $K = 10$. (C) and (D): $p = 4,000$; $n = 5,000$. (C): $n_{it_t} = 34$; $n_{ants} = 3$; $n_{MBs} = 100$. (D): $n_{ants} = 3$; $K = 100$.

orders of magnitude for n_{it_t}, n_{ants} and K, we ran tests as in Fig. 1. In [5], minimal thresholds are given for the parameters corresponding to n_{it_t} and n_{ants}, depending on the dataset size processed by AntEpiSeeker. Default values are indicated for the other parameters. The unique parameter of DASSO-MB was set to the default value. Orders of magnitude for the number of MCMC iterations in BEAM are provided in [6].

We report a significant discrepancy of SMMB or SMMB-ACO with another method if the two standard deviations are lower than the difference between the performance averages. A salient feature is the relatively large variability of AntEpiseeker and SMMB. In contrast, SMMB-ACO and BEAM show the lowest magnitudes of variation. SMMB-ACO apart, SMMB performs slightly better than the other methods under 3 conditions. SMMB ranks second behind BEAM for model 3 and MAF equal to 0.50. In the 8 remaining conditions, two or three methods, including SMMB, cannot be distinguished. No case arises for which SMMB is the worst method. In sharp contrast explained by its low variability, SMMB-ACO ranks first under 6 conditions and it shares the first rank with a single method under 5 other conditions (SMMB apart). We conclude that SMMB-ACO ranks first in half of the 12 simulated conditions studied, with regards to (i) the high number of datasets simulated under the same condition (100), (ii) the high number of runs operated on the same dataset (100) to compute the dataset's F-measure, and (iii) the identical processings of the four stochastic methods compared. Besides, fastest to slowest, on simulations, we always find DASSO-MB, SMMB-ACO/SMMB, BEAM and AntEpiSeeker (results not shown).

For datasets of dimension $p = 4,000 \times n = 100$ simulated under the null hypothesis, we observed that SMMB-ACO ranks second with a false positive rate of 12.2%, in comparison of BEAM (0%), SMMB (16%), DASSO-MB (20.5%), and AntEpiSeeker (100%).

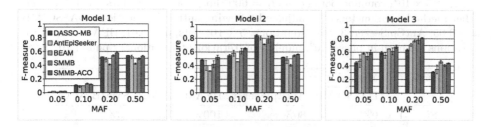

Fig. 2. Comparison of performances for the five methods, on simulated data. $p = 4,000$; $n = 100$ (see Algorithm 1, INPUT). Parameters for SMMB-ACO: $n_{it_t} = 34$, $n_{ants} = 3$; parameter for SMMB: $n_{MBs} = 100$; common parameter: $K = 10$; for other parameters, see Table 1; for other methods, see Sect. 4.2/Performances on simulated data.

Performances on Real Data. Except for DASSO-MB, all methods discovered 7 2-way patterns through their 10 runs. For instance, SMMB-ACO discovered 7 epistasis patterns in 5 runs, 6 patterns in 4 runs and 4 patterns in 1 run. DASSO-MB identified only 5 of the 7 former patterns in its single run. None of the methods yielded results outside this set of 7 patterns. The target variable was regressed against the set of variables corresponding to each 2-way pattern. The p-values for these 7 logistic regressions are all smaller than 10^{-16}. We also computed the odds ratio relative to the logistic regression of the target variable (i) against each variable identified in a pattern, and (ii) against each pattern. In all seven cases, the marginal effect of each variable on the target variable is smaller than the effect of the variables in interaction (results not shown).

For any stochastic method, each of the 10 runs output between 4 and 7 interactions. For insights on the stability of each stochastic method, we computed the percentage of runs which output at least 6 of the 7 interactions. We set 6 as an arbitrary threshold. Table 2 highlights the discrepancies between the five methods. On the whole genome set, 90% of the runs of SMMB-ACO output at least 6 of the 7 interactions (versus 80% for AntEpiSeeker which is 5.3 times as slow as SMMB-ACO). Given this performance of 90%, and since SMMB-ACO is also 4.5 times as fast as BEAM, it is affordable to launch several runs of SMMB-ACO.

Table 2. Comparison of the five methods on the real RA dataset. Separate chromosomes (SC) *versus* whole genome (WG). Each stochastic method discovered 7 two-way interactions through its 10 runs and all stochastic methods discovered the same 7 interactions. The P6 rate measures the percentage of runs over 10 runs (1 run for deterministic DASSO-MB) which identified at least 6 of the 7 interactions. RT measures the average running time. Parameters for runs on separate chromosomes: SMMB-ACO: $n_{it_t} = 10^4$, $n_{ants} = 4$; SMMB: $n_{MBs} = 4 \times 10^4$; common parameter: $K = 180$. Parameters for runs on the whole genome: SMMB-ACO: $n_{it_t} = 8,333$, $n_{ants} = 6$; SMMB: $n_{MBs} = 5 \times 10^4$; common parameter: $K = 600$. For other parameters see Table 1. For other methods, parameters were adjusted following authors' indications: DASSO-MB: $\alpha = 0.05$; AntEpiSeeker: 3×10^4 iterations, 5×10^4 ants, $\alpha = 0.01$; BEAM: number of iterations for burn-in phase: 10^6, number of iterations for stationary phase: 10^7.

Data		Method				
		DASSO-MB	AntEpiSeeker	BEAM	SMMB	SMMB-ACO
SC	P6	0%	80%	100%	70%	90%
	RT	12 h	69 h	59 h	34 h	13 h
WG	P6	0%	80%	100%	70%	90%
	RT	17 h	47 h	53 h	23 h	19 h

4.3 Discussion

Comparison with SMMB. When SMMB ranked first on simulated data, it could not be demarcated from 2 or 3 other methods; and it was never the worst method. Its performance is mitigated for the RA dataset. The SMMB sampling strategy enhanced with ACO allowed to improve SMMB's performances, both on simulated and real datasets. Meanwhile, the time complexity was improved. The comparison of SMMB and SMMB-ACO's behaviors under the same parameter adjustment allowed to shade light on the improvement brought by the ACO technique. A consensus precursor obtained from SMMB-ACO is likely to contain less false positives than a consensus precursor obtained from SMMB. This fact translates into the decrease of the running time in SMMB-ACO, as verified in Table 2. The explanation lies in the fact that a lower number of false positives allows to decrease the complexity of the backward phase that is applied to the consensus precursor. The above explanation allows to emphasize that the effectiveness brought by memory management (mem_a and mem, see Algorithm 1), together with updating procedures (pheromone rates, probability distribution) does not come at the cost of higher running times.

Comparison with BEAM. BEAM is more successful than the other methods on the RA dataset. However, on simulations, BEAM does not outperform the other methods. Besides, it is not advisable to rely on the single run of a stochastic method, BEAM included, *a fortiori* on large datasets. BEAM is not parallelized, whereas the iteration on the ants is, in SMMB-ACO. However, the backward phase used to refine the consensus precursor is intrinsincally not parallelisable.

Moreover, in this backward phase, type I error is controled with computationally expensive permutations. The construction of the MB consensus consumes between 50% and 60% of the total running time. In spite of this additional computational burden, SMMB-ACO runs faster than BEAM. The complexity of SMMB-ACO is an advantage over BEAM, to launch several runs.

Complexities. Because of its stochastic feature, it is impossible to assess the theoretical overall time complexity of SMMB-ACO. The complexity of the backward phase is at worst $O(q\, 2^q)$, q being the size of the set processed by procedure backwardPhase. Experimental feedback indicates that q is lower than 4–5 for the simulated datasets and the real datasets as well. Besides, we also observed that the size of the conditioning set (*i.e.*, the MB) rapidly decreases: generally, the first conditional test performed for a variable indicates that this variable must be discarded from the MB (Algorithm 3, lines 2 to 4). The complexity of the backward phase is therefore not an issue in function learnMB and is expected to be close to $O(q)$. In contrast, the final backward phase handles the union of all Markov blankets that have been generated (Algorithm 1, lines 21 and 22). In this case, q may rise to a few hundreds for a chromosome. Only a few variables will require that all conditional tests are performed (3 if there is a 3-way epistatic pattern in the data). However, the computational burden is increased by the use of permutations at this stage. Thus the complexity of the backward phase is expected to be close to $O(q\, e)$, where e denotes the number of permutations (*e.g.*, 1000).

5   Conclusions and Future Work

In this work, we have tackled the problem of feature selection under the epistasis assumption. The performances obtained by SMMB-ACO are very promising but there is still room for improvement. First, on large datasets, the bottleneck is the computational burden entailed by the size of the MB consensus, prior refinement. Our future directions of research will explore various strategies to either diminish the size of this set or discard this stage. Second, a brute-force method is currently used to update the pheromone rates in SMMB-ACO. Further work will examine various leads to update these rates. Third, we are currently investigating alternatives to the ACO optimization technique. In particular, the "less is more approach" (LIMA) represents a recent line of inquiry to increase the effectiveness and efficiency of a heuristic customized to solve a given optimization problem [19]. Finally, beyond this present proof-of-concept in the challenging domain of precision medicine, the further optimized version of SMMB will be developed and tested as a tool for feature selection under epistasis assumption, but in a more general framework (*e.g.*, with no constraint on the cardinalities of the variables, with a continuous target variable ...).

References

1. Goudey, B., et al.: GWIS - model-free, fast and exhaustive search for epistatic interactions in case-control GWAS. BMC Genomics **14**(Suppl. 3), S10 (2013)
2. Hahn, L.W., Ritchie, M.D., Moore, J.H.: Multifactor dimensionality reduction software for detecting gene-gene and gene-environment interactions. Bioinformatics **19**, 376–382 (2003)
3. Niel, C., Sinoquet, C., Dina, C., Rocheleau, G.: A survey about methods dedicated to epistasis detection. Front. Genet. **6**, 285 (2015)
4. Schwarz, D.F., König, I.R., Ziegler, A.: On safari to Random Jungle: a fast implementation of random forests for high-dimensional data. Bioinformatics **26**(14), 1752–1758 (2010)
5. Wang, Y., Liu, X., Robbins, K., Rekaya, R.: AntEpiSeeker: detecting epistatic interactions for case-control studies using a two-stage ant colony optimization algorithm. BMC Res. Note **3**, 117 (2010)
6. Zhang, Y., Liu, J.S.: Bayesian inference of epistatic interactions in case-control studies. Nat. Genet. **39**, 1167–1173 (2007)
7. Liu, J., Yu, G., Jiang, Y., Wang, J.: HiSeeker: detecting high-order SNP interactions based on pairwise SNP combinations. Genes (Basel) **8**(6), 153 (2017)
8. Abo Alchamlat, S., Farnir, F.: KNN-MDR: a learning approach for improving interactions mapping performances in genome wide association studies. BMC Bioinform. **18**(1), 184 (2017)
9. Tsamardinos, I., Aliferis, C.F., Statnikov, A.: Algorithms for large scale Markov blanket discovery. In: 16th International Florida Artificial Intelligence Research Society (FLAIRS) Conference, pp. 376–380. AAAI Press (2003)
10. Tsamardinos, I., Aliferis, C.F.: Towards principled feature selection: relevancy, filters, and wrappers. In: 9th International Workshop on Artificial Intelligence and Statistics (2003)
11. Niel, C., Sinoquet, C., Dina, C., Rocheleau, G.: SMMB - A stochastic Markov-blanket framework strategy for epistasis detection in GWAS. Bioinformatics **34**(16), 2773–2780 (2018)
12. Dorigo, M., Stützle, T.: Ant Colony Optimization. The MIT Press, Cambridge (2004)
13. SMMB-ACO. https://ls2n.fr/listelogicielsequipe/DUKe/130/
14. Marchini, J., Donnelly, P., Cardon, L.R.: Genome-wide strategies for detecting multiple loci that influence complex diseases. Nat. Genet. **37**, 413–417 (2005)
15. Urbanowicz, R.J., Kiralis, J., Sinnott-Armstrong, N.A., Heberling, T., Fisher, J.M., Moore, J.H.: GAMETES: a fast, direct algorithm for generating pure, strict, epistatic models with random architectures. BioData Min. **5**, 16 (2012)
16. Simulated data repository. https://uncloud.univ-nantes.fr/index.php/s/rLG7QA JQaxjZ7ef
17. WTCCC. https://www.wtccc.org.uk/
18. Han, B., Park, M., Chen, X.-W.: A Markov blanket-based method for detecting causal SNPs in GWAS. BMC Bioinform. **11**(Suppl. 3), S5 (2010)
19. Mladenovic, N., Todosijevic, R., Urosevic, D.: Less is more: basic variable neighborhood search for minimum differential dispersion problem. Inf. Sci. **326**, 160–171 (2016)

Average Performance Analysis
of the Stochastic Gradient Method
for Online PCA

Stéphane Chrétien[1(✉)], Christophe Guyeux[2], and Zhen-Wai Olivier Ho[3]

[1] National Physical Laboratory, Teddington, UK
stephane.chretien@npl.co.uk
[2] FEMTO-ST Institute, UMR 6174 CNRS, Besançon, France
christophe.guyeux@univ-fcomte.fr
[3] LMB Université de Bourgogne Franche-Comté,
16 route de Gray, 25030 Besançon, France
zhen-wai-olivier@univ-fcomte.fr
https://sites.google.com/view/stephanechretien

Abstract. This paper studies the complexity of the stochastic gradient algorithm for PCA when the data are observed in a streaming setting. We also propose an online approach for selecting the learning rate. Simulation experiments confirm the practical relevance of the plain stochastic gradient approach and that drastic improvements can be achieved by learning the learning rate.

Keywords: Stochastic gradient · Online PCA ·
Non-convex optimisation · Average case analysis

1 Introduction

1.1 Background

Principal Component Analysis (PCA) is a paramount tool in an amazingly wide scope of applications. PCA belongs to the small list of algorithms which are extensively used in data science, medicine, finance, machine learning, etc. and the list is almost infinite. PCA is one of the basic blocks in Data Analytics. Computing singular/eigenvectors also appears key to discovering nonlinear embeddings of the data such as Laplacian eigenmaps [2].

In the era of Big Data, computing a set of singular vectors might turn to be a computationally difficult task to achieve. In practice the data matrix itself cannot be imported into the RAM and the data can only be accessed in small samples. In face of such hard memory management problems, Online Convex Optimisation often provides efficient alternatives to standard computations in machine learning [6,10,13]. On the other hand, computing eigen/singular vectors is not a convex optimisation problem. Instead, PCA can be seen as an optimisation problem over the sphere and as such, requires a different type of analysis.

© Springer Nature Switzerland AG 2019
G. Nicosia et al. (Eds.): LOD 2018, LNCS 11331, pp. 231–242, 2019.
https://doi.org/10.1007/978-3-030-13709-0_19

Online or stochastic versions of PCA have been extensively studied lately; see in particular the review [3]. On the theoretical side, [11] proposed a very clear analysis of the stochastic gradient algorithm for PCA which does not require information about the gap between successive eigenvalues. Better convergence rates were subsequently obtained in [1,7,12] using more advanced algorithms. All these previous works rely on the assumption that the data arrive sequentially and are i.i.d., and their objective is to compute their common covariance matrix.

Our contribution explores a different set up. In the present work, we assume that the entrees of the covariance matrix are revealed one at a time in a sequential fashion. In such a set up, only some correlations between certain components of the data vectors, supposed to be chosen uniformly at random, are assumed to be available at each round, and not the data themselves. Therefore, our set up pertains to the activity around the important problem of Positive Semi-Definite matrix completion [5,8,9].

Our first main contribution is a mathematical proof that the method of [11] extends to the online matrix completion problem. Our theoretical findings also include a formula for the learning rate which can be optimised depending on the problem at hand. Practical optimisation of the learning rate is our second contribution. Our tuning algorithm is an adaptation of Freund and Shapire's online Hedge algorithm and is shown to provide substantial improvement of the practical convergence speed of the online gradient scheme for PCA.

1.2   Organisation of the Paper

Our main results are presented in Sect. 2 where the algorithm is described and our main theorem is given. The proof of our main theorem is exposed in Sect. 3. Implementation and numerical experiments are given in Sect. 4. In particular, a simple method for choosing the learning rate is described in Sect. 4.1. The technical lemmæ which are used in the proof of Sect. 3 are gathered in Sect. A at the end of the paper.

2   Main Results

2.1   Presentation of the Problem and Prior Result

We use bold-faced letters to denote vectors, and capital letters to denote matrices unless specified otherwise. Given a matrix A, we denote by $A^\top$ its transpose matrix, $\|A\|$ its spectral norm and $\|A\|_{1\to 2} = \max \|\mathbf{A}_j\|_2$ the maximum ℓ_2 norm of its column. For a vector $\mathbf{v}$, we denote by $\mathbf{v}^\top$ its transpose. Moreover $(\mathbf{e}_i)_i$ denote the canonical basis of $\mathbb{R}^d$. The optimisation problem can be written

$$\min_{\mathbf{w}:\|\mathbf{w}\|=1} -\mathbf{w}^\top A\mathbf{w}, \tag{1}$$

where $d > 1$ and A is a symmetric positive semi-definite matrix supposed unknown. We suppose that we have access to a stream of i.i.d. matrices A_t defined as

$$A_t = d^2 \, A_{i_t,j_t} \, \mathbf{e}_{i_t}\mathbf{e}_{j_t}^\top \tag{2}$$

and (i_t, j_t) is drawn uniformly at random from $\{1, \ldots, n\}^2$. It is easily seen that $\mathbb{E}[A_t] = A$, therefore each matrix A_t can be seen as a properly rescaled noiseless random component of A. It can be readily seen that any leading eigenvector of A is a solution of the optimisation problem.

2.2 The Stochastic Projected Gradient Algorithm

Given a symmetric matrix $A \in \mathbb{R}^{d \times d}$, the projected gradient algorithm writes

$$\mathbf{w}_{t+1} = (I + \eta A)\mathbf{w}_t / \|(I + \eta A)\mathbf{w}_t\|_2 \tag{3}$$

where η is a step-size parameter and $\mathbf{w}_0$ is the initial estimate for a leading eigenvector of A. This algorithm correspond to initialising at $\mathbf{w}_0$ then make a gradient step at each iteration followed by a projection into the unit sphere. However, since A is unknown, the stochastic gradient we will study in this paper is simply defined as

$$\mathbf{w}_{t+1} = (I + \eta A_t)\mathbf{w}_t / \|(I + \eta A_t)\mathbf{w}_t\|_2 \tag{4}$$

obtained by replacing A with the random matrix A_t. Since the projection on the unit sphere is a rescaling operation which is commutative with respect to the matrix product, we can leave the projection operation to the end. That is, for our analysis, it is enough to consider the equivalent algorithm which only performs projection at the end:

- Initialise $\mathbf{w}_0$ on a unit sphere,
- Perform $T > 0$ stochastic gradient step: $\mathbf{w}_{t+1} = (I + \eta A_t)\mathbf{w}_t$
- Return $\mathbf{w}_T / \|\mathbf{w}_T\|_2$.

In [12], the stream of i.i.d. matrices A_t are also assumed positive semidefinite. The main result in [12] is the following theorem.

Theorem 1. *Suppose that the matrices $(A_t)_{t \in \mathbb{N}}$ are positive semi-definite, real i.i.d for some leading eigenvector $\mathbf{v}$ of A, $\frac{1}{p} < \langle \mathbf{w}_0, \mathbf{v} \rangle^2$ for some $p > 0$ and that for some $b \geq 1$, both $\|A_t\|/\|A\|$ and $\|A_t - A\|/\|A\|$ are at most b with probability 1. Then, after T iterations of (4) with $\eta = \frac{1}{b\sqrt{pT}}$, then with probability at least $\frac{1}{cp}$, the return $\mathbf{w}_T$ satisfies*

$$1 - \frac{\mathbf{w}_T^\top A \mathbf{w}_T}{\|A\|} \leq c' \frac{\log(T)b\sqrt{p}}{\sqrt{T}}, \tag{5}$$

where c and c' are positive constants.

Note that our online positive semidefinite matrix completion framework is not compatible with the assumptions required for Theorem 1 to apply. In our problem, the matrices A_t are not themself positive semidefinite.

2.3   Main Theorem

Without loss of generality, we will throughout assume that $\|A\| = 1$. Our goal is to show that, for $\varepsilon > 0$, the vector $\mathbf{w}_T$ obtained after T iterations of the stochastic gradient method, satisfies

$$1 - \mathbf{w}_T A \mathbf{w}_T \le \varepsilon \tag{6}$$

in expectation for T a sufficiently large integer and η tuned accordingly. Since $\|\mathbf{w}_T\|_2 = 1$, this is equivalent to showing that

$$\mathbf{w}_T^\top((1 - \varepsilon)I - A)\mathbf{w}_T \le 0. \tag{7}$$

The next theorem summarizes our main findings.

Theorem 2. *Let $\varepsilon > 0$ and assume that $0 < \frac{1}{p} < \langle \mathbf{w}_0, \mathbf{v} \rangle^2$ for a leading eigenvector $\mathbf{v}$ of A. Define*

$$V_T = \mathbf{w}_0^\top \prod_{i=T}^{1}(I + \eta A_i)^\top((1 - \epsilon)I - A)\prod_{i=1}^{T}(I + \eta A_i)\mathbf{w}_0. \tag{8}$$

Then for T satisfying

$$T > \max\left(\frac{4p^2d^2}{\varepsilon}, \frac{\log 4p\varepsilon^{-1}}{\log\left(1 + \frac{\varepsilon}{pd^2}\right)} \right), \tag{9}$$

and $\eta = \frac{\varepsilon}{4pd^2}$, it holds that

$$\mathbb{E}[V_T] \le -\frac{\varepsilon}{4p}(1 + 2\eta)^T. \tag{10}$$

Since $V_T = \|\mathbf{w}_T\|_2^2 \mathbf{w}_T^\top((1 - \varepsilon)I - A)\mathbf{w}_T$, the theorem implies the desired result.

3   Proof of the Theorem 2

In this section, we prove our main result, namely Theorem 2. Define

$$\mathbf{B}_T = \prod_{i=T}^{1}(I + \eta A_i)^\top((1 - \epsilon)I - A)\prod_{i=1}^{T}(I + \eta A_i) \tag{11}$$

so that $V_T = \mathbf{w}_0^\top B_T \mathbf{w}_0$.

Lemma 1. *We have that*

$$\mathbb{E}[B_T] = \mathbb{E}[B_{T-1}] + \eta\left(A^\top \mathbb{E}[B_{T-1}] + \mathbb{E}[B_{T-1}]A\right)$$

$$+ \eta^2 d^2 \mathrm{diag}\left(A^\top \mathrm{diag}(\mathbb{E}[B_{T-1}])A\right). \tag{12}$$

Proof. Expand the recurrence relationship and take the expectation. Finally use Lemma 2 to obtain the last term of the inequality.

Expanding the recurrence in Lemma 1, we have

$$\mathbb{E}[V_T] \leq \mathbf{w}_0^\top (I + 2\eta A)^\top ((1 - \varepsilon)I - A)\mathbf{w}_0$$
$$+ \eta^2 d^2 \sum_{i=1}^{T} (1 + 2\eta)^{T-i} \|\text{diag}(\mathbb{E}[B_{i-1}])\| \|\mathbf{w}_0\|_2^2. \tag{13}$$

where the last term was obtained by using inequality (28) and $\|A\|_{1 \to 2} \leq 1$. Using an eigendecomposition of A and $\|\mathbf{w}_0\|_2^2 = 1$ gives

$$\mathbb{E}[V_T] \leq \sum_{j=1}^{d} (1 + 2\eta s_j)^T (1 - \varepsilon - s_j) w_{0,j}^2 + \eta^2 d^2 \sum_{i=1}^{T} (1 + 2\eta)^{T-i} \|\text{diag}(\mathbb{E}[B_{i-1}])\|. \tag{14}$$

where $s_1 > \cdots > s_d$ denote the eigenvalues of A and $w_{0,j} = \langle \mathbf{w}_0, \mathbf{v}_j \rangle$ denotes the *j-th* component of $\mathbf{w}_0$ in the basis of the eigenvectors of A. Since $s_1 = 1$, this inequality rewrites

$$\mathbb{E}[V_T] \leq -\varepsilon(1 + 2\eta)^T w_{0,1}^2 + \sum_{j=2}^{d} (1 + 2\eta s_j)^T (1 - \varepsilon - s_j) w_{0,j}^2$$
$$+ \eta^2 d^2 \sum_{i=1}^{T} (1 + 2\eta)^{T-i} \|\text{diag}(\mathbb{E}[B_{i-1}])\|. \tag{15}$$

In the remainder of the proof, we prove that the negative term $-\varepsilon(1 + 2\eta)^T w_{0,1}^2$ dominates the positive terms. The terms $w_{0,j}^2$ sum to $1 - w_{0,1}^2$. Therefore the sum $\sum_{j=2}^{d}(1 + 2\eta s_j)^T(1 - \varepsilon - s_j)w_{0,j}^2$ is less than $\max_{s \in [0,1]}(1 + 2\eta s)^T(1 - \varepsilon - s)$, which can be bounded from above using Lemma 7. Therefore, we get the following inequality

$$\mathbb{E}[V_T] \leq -\varepsilon(1 + 2\eta)^T w_{0,1}^2 + (1 + \frac{(1 + 2\eta(1 - \varepsilon))^T}{\eta(T + 1)})$$
$$+ \eta^2 d^2 \sum_{i=1}^{T} (1 + 2\eta)^{T-i} \|\text{diag}(\mathbb{E}[B_{i-1}])\|. \tag{16}$$

Factoring out $(1 + 2\eta)^T$, the inequality now writes

$$\mathbb{E}[V_T] \leq (1 + 2\eta)^T \Big(-\varepsilon w_{0,1}^2 + \frac{1}{(1 + 2\eta)^T} + \frac{(1 + 2\eta(1 - \varepsilon))^T}{(1 + 2\eta)^T \eta(T + 1)}$$
$$+ \eta^2 d^2 \sum_{i=1}^{T} (1 + 2\eta)^{-i} \|\text{diag}(\mathbb{E}[B_{i-1}])\| \Big) \tag{17}$$

For the sake of simplifying the analysis, we will use a uniform bound on the spectral norm of $\mathrm{diag}(\mathbb{E}[B_k])$. More precisely, Lemma 6 implies that

$$
\begin{aligned}
\|\mathrm{diag}(\mathbb{E}[B_k])\| &\leq 2\frac{\eta}{\eta d^2+1}\left(\frac{1}{1-\eta(\eta d^2+2)}-\frac{1}{1-\eta}\right)(1-\varepsilon) \\
&\quad + 2\frac{\eta}{\eta d^2+1}\left(\eta d^2\frac{1}{1-\eta(\eta d^2+2)}\right. \\
&\qquad \left.+\frac{1}{1-\eta}\right)(2-\varepsilon)+\left(1+\frac{\eta^2 d^2}{1-\eta(\eta d^2+2)}\right)(1-\varepsilon)
\end{aligned}
$$
(18)

$$
\leq 2\frac{\eta}{\eta d^2+1}\left(\frac{1-\varepsilon+(2-\varepsilon)\eta d^2}{1-\eta(\eta d^2+2)}+\frac{1}{1-\eta}\right)+\left(1+\frac{\eta^2 d^2}{1-\eta(\eta d^2+2)}\right)(1-\varepsilon)
$$

$$
\leq 2\frac{\eta}{\eta d^2+1}\frac{2-\varepsilon+(2-\varepsilon)\eta d^2}{1-\eta(\eta d^2+2)}+1+\frac{\eta^2 d^2}{1-\eta(\eta d^2+2)}
$$
(19)

for all k. This simplifies into

$$
\|\mathrm{diag}(\mathbb{E}[B_k])\| \leq 1+\frac{\eta^2 d^2+4\eta}{1-\eta(\eta d^2+2)}.
$$
(20)

Thus we obtain

$$
\begin{aligned}
\mathbb{E}[V_T] &\leq (1+2\eta)^T\left(-\varepsilon w_{0,1}^2+\frac{1}{(1+2\eta)^T}+\frac{(1+2\eta(1-\varepsilon))^T}{(1+2\eta)^T\eta(T+1)}\right. \\
&\quad \left.+\eta^2 d^2\left(1+\frac{\eta^2 d^2+4\eta}{1-\eta(\eta d^2+2)}\right)\sum_{i=1}^{T}(1+2\eta)^{-i}\right)
\end{aligned}
$$
(21)

Bounding $\sum_{i=1}^{T}(1+2\eta)^{-i}$ by its infinite series $\sum_{i=1}^{\infty}(1+2\eta)^{-i}=(2\eta)^{-1}$ yields

$$
\mathbb{E}[V_T] \leq (1+2\eta)^T\left(-\varepsilon w_{0,1}^2+\frac{1}{(1+2\eta)^T}+\frac{(1+2\eta(1-\varepsilon))^T}{(1+2\eta)^T\eta(T+1)}\right.
$$
(22)

$$
\left.+\eta/2d^2\left(1+\frac{\eta^2 d^2+4\eta}{1-\eta(\eta d^2+2)}\right)\right).
$$
(23)

We can show that, for well chosen values of η and T, the term between parenthesis can be made to be less that $-\varepsilon/4p$. Taking for example $\eta=\frac{\varepsilon}{4Cpd^2}$ for some constant C such that $\left(1+\frac{\eta^2 d^2+4\eta}{1-\eta(\eta d^2+2)}\right)\leq 2$ and $T>\max(4p^2 d^2 C/\varepsilon,\log(4p\varepsilon^{-1})/\log(1+\varepsilon/(Cpd^2)))$ is consistent with the constraints. Notice further that for ε sufficiently small, this can be simplified further by taking $C=1$. One of the benefits of using this approach over standard methods from the literature, is that it is all at the same time elementary, intuitive and it can easily be checked to enjoy the same theoretical guarantees as the original method devised in [4]. Full details will be provided in a longer version of the paper.

4 Implementation

4.1 Choosing the Learning Rate

In this section, we address the question of choosing the learning rate, i.e. the step-size η in iterations (4). Tuning the learning rate is essential in practice as it is well known to have a huge impact on the convergence speed of the method. Our idea to tune the learning rate is as follows:

- Choose the tolerance $\epsilon \in (0,1)$, and the algorithm's parameters R, $K \in \mathbb{N}_*$, $\rho \in (0,1)$ and $\beta > 0$.
- *Burn-in period:*
    - For $\eta \in \{\rho^k\}_{k=1:K}$, run R gradient iterations in parallel whose iterates are denoted by $\mathbf{w}_t^{(k,r)}$, $t = 1,\ldots, B$.
    - Define $\pi_0^{(k)} = 1/K$, $k = 1,\ldots, K$. For $t = 1,\ldots, B$, let

$$L_t^{(k)} = \frac{2}{R(R-2)} \sum_{r<r'=2,\ldots,R} \langle \mathbf{w}_t^{(k,r)}, \mathbf{w}_t^{(k,r')} \rangle, \tag{24}$$

    and for $k = 1,\ldots, K$, define $\pi_{t+1}^{(k)} = \pi_t^{(k)} \exp\left(\beta\, L_t^{(k)}\right)$.
    - Stop when $\max_{k=1,\ldots,K}\, L_t^{(k)} \geq 1 - 10\,\epsilon$.
- *After burn-in:*
    - Reset R to 1 and K to 1.
    - Normalise π.
    - At each step $t = B+1, \ldots$, choose the stepsize with probability π_B.
    - Stop when $L_t^{(1)} \geq 1 - \epsilon$.

Choosing the parameter β is more robust than choosing the learning rate. Moreover, a reasonably effective value for β is given by (see [4]):

$$\beta = \sqrt{\frac{\log(K)}{B}}. \tag{25}$$

4.2 Numerical Experiment

In this section, we present a simple numerical experiment which shows that

- The stochastic gradient method actually works in practice
- The adaptive selection of the learning rate/step-size described in the previous subsection actually accelerates the method's convergence drastically.

We run a simple experiment on a random i.i.d. Gaussian matrix of size 10000×10000. The convergence of $(L_t^{(1)})_{t \in \mathbb{N}}$ to 1 of the plain stochastic gradient method is shown in Fig. 1a below. The accelerated version's convergence for the same experiment is shown in Fig. 1b below. These results show that the method of the previous Section actually provides a substantial acceleration. We carefully checked that the selected learning rate is not equal to the smallest nor the largest value on the proposed grid of values between $2^{-3}, 2^{-2}, \ldots 2^{17}$. The observed gain in convergence speed was by a factor of 8.75. Extensive numerical experiment demonstrating this behaviour at larger scales will be included in an expanded version of this work.

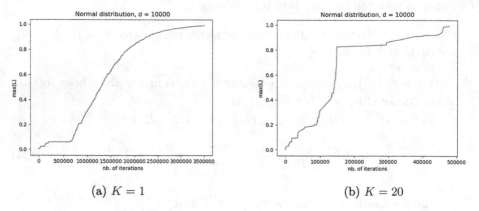

(a) $K = 1$ (b) $K = 20$

Fig. 1. Convergence of $(L_t^{(1)})_{t \in \mathbb{N}}$ as a function of the iteration index: (a) is for the case of the arbitrary choice of learning rate equal to 2^{-4} and (b) shows the behaviour of the method using the learning procedure of Sect. 4.1 for values of the learning rate equal to $2^{-3}, 2^{-2}, 2^{-1}, 1, 2, \ldots, 2^{17}$.

5   Conclusion

In the present paper, we have studied the average behaviour of the stochastic gradient for the computation of the principal eigen-vector of positive semi-definite matrices, in the setting where the entrees are revealed one at a time. The analysis provides the first complexity analysis in this online setting. A preliminary computer experiment integrating a novel learning rate optimisation procedure is included.

A   Technical lemmæ

Recall that

$$B_T = \prod_{t=T}^{1} (I + \eta A_t)^\top ((1 - \varepsilon)I - A) \prod_{t=1}^{T} (I + \eta A_t). \tag{26}$$

Lemma 2. *In the case of matrix completion, given a matrix X, we have*

$$\mathbb{E}[A_t^\top X A_t] = d^2 \ \mathrm{diag}(A \ \mathrm{diag}(X)A).$$

Proof. The resulting matrix writes

$$A_t^\top X A_t = d^4 A_{ij} A_{ji} \mathbf{e}_{j_t} \mathbf{e}_{i_t}^\top X \mathbf{e}_{i_t} \mathbf{e}_{j_t}^\top$$
$$= d^4 A_{ij} A_{ji} X_{ii} \mathbf{e}_{j_t} \mathbf{e}_{j_t}^\top.$$

Therefore the expected matrix writes

$$\mathbb{E}[A_t^\top X A_t] = d^2 \sum_{i,j}^{d} A_{ij} A_{ji} X_{ii} \mathbf{e}_j \mathbf{e}_j^\top$$

Using the symmetry of A gives the result.

Now our next goal is to see how diag $\left(A^\top \mathrm{diag}(\mathbb{E}[B_{T-1}])A\right)$ evolves with the iterations. For this purpose, take the diagonal of (12), multiply from the left by $A^\top$ and from the right by A and take the diagonal of the resulting expression.

Lemma 3. *We have that*

$$\|\mathrm{diag}\left(\mathbb{E}[B_T]\right)\| \le 2\eta \,\|\mathbb{E}[B_{T-1}]\|_{1\to 2} + (1 + \eta^2 d^2) \,\|\mathrm{diag}(\mathbb{E}[B_{T-1}])\| \qquad (27)$$

Proof. Expanding the recurrence relationship (12) gives

$$\mathrm{diag}(\mathbb{E}[B_T]) = \mathrm{diag}(\mathbb{E}[B_{T-1}]) + \eta \left(\mathrm{diag}\left(A^\top \mathbb{E}[B_{T-1}] + \mathbb{E}[B_{T-1}]A\right)\right)$$
$$+ \eta^2 d^2 \mathrm{diag}\left(A^\top \mathrm{diag}(\mathbb{E}[B_{T-1}])A\right).$$

For any diagonal matrix Δ and symmetric matrix A, we have

$$\|\mathrm{diag}(A^\top \Delta A)\| \le \|A\|_{1\to 2}^2 \|\Delta\|. \qquad (28)$$

Therefore, by taking the operator norm on both sides of the equality, we have

$$\|\mathrm{diag}(\mathbb{E}[B_T])\| \le (1 + \eta^2 d^2 \|A\|_{1\to 2}^2)\|\mathrm{diag}(\mathbb{E}[B_{T-1}])\| + 2\eta\|\mathrm{diag}(A^\top \mathbb{E}[B_{T-1}])\| \qquad (29)$$

We conclude using $\|\mathrm{diag}(A^\top E[B_{T-1}])\| \le \|A\|_{1\to 2}\|\mathbb{E}[B_{T-1}]\|_{1\to 2}$ and $\|A\|_{1\to 2} \le 1$.

We also have to understand how the $\ell_{1\to 2}$ norm evolves.

Lemma 4. *We have*

$$\|\mathbb{E}[B_T]\|_{1\to 2} \le \eta \,\|\mathbb{E}[B_{T-1}]\| + (1 + \eta) \,\|\mathbb{E}[B_{T-1}]\|_{1\to 2} + \eta^2 d^2 \,\|\mathrm{diag}(\mathbb{E}[B_{T-1}])\|. \qquad (30)$$

Proof. Expanding the recurrence relationship gives

$$\|\mathbb{E}[B_T]\|_{1\to 2} = \|\mathbb{E}[B_{T-1}]\|_{1\to 2} + \eta \left(\|A^\top \mathbb{E}[B_{T-1}]\|_{1\to 2} + \|\mathbb{E}[B_{T-1}]^\top A\|_{1\to 2}\right)$$
$$+ \eta^2 d^2 \|\mathrm{diag}(A^\top \mathrm{diag}(\mathbb{E}[B_{T-1}])A)\|_{1\to 2}.$$

For a diagonal matrix Δ, we have $\|\Delta\|_{1\to 2} = \|\Delta\|$. This leads to

$$\|\mathbb{E}[B_T]\|_{1\to 2} = \|\mathbb{E}[B_{T-1}]\|_{1\to 2} + \eta \left(\|A\|\|\mathbb{E}[B_{T-1}]\|_{1\to 2} + \|\mathbb{E}[B_{T-1}]\|\|A\|_{1\to 2}\right)$$
$$+ \eta^2 d^2 \|A\|_{1\to 2}^2 \|\mathrm{diag}(\mathbb{E}[B_{T-1}])\|.$$

Finally, using $\|A\|_{1\to 2} \le 1$ concludes the proof.

We then have to understand how the operator norm of $\mathbb{E}[B_T]$ evolves

Lemma 5. *We have*

$$\|\mathbb{E}[B_T]\| \leq (1+2\eta)\|\mathbb{E}[B_{T-1}]\| + \eta^2 d^2 \,\|\text{diag}(\mathbb{E}[B_{T-1}])\|. \tag{31}$$

Proof. Expanding the recurrence relationship (12) return

$$\|\mathbb{E}[B_T]\| = \mathbb{E}[B_{T-1}] + \eta(\|A^\top \mathbb{E}[B_{T-1}]\| + \|\mathbb{E}[B_{T-1}]A\|)$$
$$+\eta^2 d^2\|\text{diag}(A^\top \text{diag}(\mathbb{E}[B_{T-1}])A)\|.$$

Then using similar inequalities as in the proof of the lemmas above, we have the result.

Lemma 6. *Let* $\|A\| = 1$, *then we have*

$$\|\text{diag}(\mathbb{E}[B_T])\| \leq \alpha \max_j(1 - \varepsilon - s_j) + \beta\|(1-\varepsilon)I - A\|_{1\to 2} + \gamma \max_j(1 - \varepsilon - A_{jj}) \tag{32}$$

where

$$\alpha = 2\frac{\eta}{\eta d^2 + 1}\left(\frac{1 - \eta^{T-2}(\eta d^2 + 2)^{T-2}}{1 - \eta(\eta d^2 + 2)} - \frac{1 - \eta^{T-2}}{1 - \eta}\right)$$

$$\beta = 2\frac{\eta}{\eta d^2 + 1}\left(\eta d^2\frac{1 - \eta^{T-2}(\eta d^2 + 2)^{T-2}}{1 - \eta(\eta d^2 + 2)} + \frac{1 - \eta^{T-2}}{1 - \eta}\right)$$

$$\gamma = 1 + \eta^2 d^2\frac{1 - \eta^{T-2}(\eta d^2 + 2)^{T-2}}{1 - \eta(\eta d^2 + 2)}$$

Proof. Expanding the recurrence and using Eqs. (27), (30), and (31) yields the following system

$$\begin{bmatrix} \|\mathbb{E}[B_T]\| \\ \|\mathbb{E}[B_T]\|_{1\to 2} \\ \|\text{diag}(\mathbb{E}[B_T])\| \end{bmatrix} \leq \left(I + \eta \begin{bmatrix} 2 & 0 & \eta d^2 \\ 1 & 1 & \eta d^2 \\ 0 & 2 & \eta d^2 \end{bmatrix}\right) \begin{bmatrix} \|\mathbb{E}[B_{T-1}]\| \\ \|\mathbb{E}[B_{T-1}]\|_{1\to 2} \\ \|\text{diag}(\mathbb{E}[B_{T-1}])\| \end{bmatrix} \tag{33}$$

To obtain the result, we expand the inequality by recurrence. Therefore, we are interested in computing the T-th power of the matrix in inequality (33). We have

$$\left(I + \eta \begin{bmatrix} 2 & 0 & \eta d^2 \\ 1 & 1 & \eta d^2 \\ 0 & 2 & \eta d^2 \end{bmatrix}\right)^T = I + \sum_{i=1}^T \eta^i \begin{bmatrix} 2 & 0 & \eta d^2 \\ 1 & 1 & \eta d^2 \\ 0 & 2 & \eta d^2 \end{bmatrix}^i. \tag{34}$$

After computing the power matrices, it result that

$$\|\text{diag}(\mathbb{E}[B_T])\| \leq \sum_{i=1}^T \left(\eta^i \frac{2(\eta d^2 + 2)^{i-1} - 1}{\eta d^2 + 1}\right) \|\mathbb{E}[B_0]\|$$

$$+ \sum_{i=1}^T \left(\eta^i \frac{2\eta d^2(\eta d^2 + 2)^{i-1} + 1}{\eta d^2 + 1}\right) \|\mathbb{E}[B_0]\|_{1\to 2}$$

$$+ \left(1 + \eta^2 d^2 \sum_{i=1}^T (\eta^2 d^2 + 2\eta)^{i-1}\right) \|\text{diag}(\mathbb{E}[B_0])\|. \tag{35}$$

We conclude after computing the sums and bounding from above $\|\mathbb{E}[B_0]\|$ by $\max_j(1 - \varepsilon - s_j)$.

Lemma 7. *For $\eta < 1$ and $\varepsilon > 0$, we have*

$$\max_{s \in [0,1]} (1 + 2\eta \, s)^T (1 - \varepsilon - s) \leq 1 + \frac{(1 + 2\eta(1 - \varepsilon))^T}{\eta(T + 1)} \tag{36}$$

Proof. Denote $f(s) = (1 + 2\eta \, s)^T (1 - \varepsilon - s)$. Differentiating f and setting to zero, we obtain

$$2\eta T(1 + 2\eta \, s)^{T-1}(1 - \varepsilon - s) - (1 + 2\eta \, s)^T = 0$$
$$\iff 2\eta T(1 - \varepsilon - s) - (1 + 2\eta \, s) = 0$$
$$\iff \frac{T(1 - \varepsilon) - 1/2\eta}{T + 1} = s$$

Let $s_c = \frac{T - \varepsilon - 1/2\eta}{T+1}$ denote this critical point. Consider the two following cases:

– if $s_c \notin [0,1]$, then f has no critical point in the domain and therefore is maximised at either domain endpoint, i.e.

$$\max_{s \in [0,1]} f(s) = \max\{f(0) = 1 - \varepsilon, f(1) = -\varepsilon(1 + 2\eta)^T\} \leq 1$$

– if $s_c \in [0,1]$, then f is maximised at s_c and the value of f at s_c is

$$\left(1 + 2\eta \frac{T(1 - \varepsilon) - 1/2\eta}{T + 1}\right)^T \left(1 - \varepsilon - \frac{T(1 - \varepsilon) - 1/2\eta}{T + 1}\right)$$
$$= \left(1 + \frac{2\eta T(1 - \varepsilon) - 1}{T + 1}\right)^T \left(\frac{1 - \varepsilon + 1/2\eta}{T + 1}\right)$$
$$\leq (1 + 2\eta(1 - \varepsilon))^T \left(\frac{1 + 1/2\eta}{T + 1}\right) \leq \frac{(1 + 2\eta(1 - \varepsilon))^T}{\eta(T + 1)}.$$

This analysis proves that the maximum value f can achieve is less than $\max\{1, \frac{(1+2\eta(1-\varepsilon))^T}{\eta(T+1)}\} \leq 1 + \frac{(1+2\eta(1-\varepsilon))^T}{\eta(T+1)}\}$. Hence the result.

References

1. Allen-Zhu, Z., Li, Y.: LazySVD: even faster SVD decomposition yet without agonizing pain. In: Advances in Neural Information Processing Systems, pp. 974–982 (2016)
2. Bandeira, A.S.: Ten lectures and forty-two open problems in the mathematics of data science (2015)
3. Cardot, H., Degras, D.: Online principal component analysis in high dimension: which algorithm to choose? arXiv preprint arXiv:1511.03688 (2015)

4. Freund, Y., Schapire, R.E.: A decision-theoretic generalization of on-line learning and an application to boosting. J. Comput. Syst. Sci. **55**(1), 119–139 (1997)
5. Grone, R., Johnson, C.R., Sá, E.M., Wolkowicz, H.: Positive definite completions of partial Hermitian matrices. Linear Algebra Appl. **58**, 109–124 (1984)
6. Hazan, E., et al.: Introduction to online convex optimization. Found. Trends® Optim. **2**(3–4), 157–325 (2016)
7. Jin, C., Kakade, S.M., Musco, C., Netrapalli, P., Sidford, A.: Robust shift-and-invert preconditioning: faster and more sample efficient algorithms for eigenvector computation, arXiv preprint arXiv:1510.08896 (2015)
8. Laurent, M.: A tour d'horizon on positive semidefinite and euclidean distance matrix completion problems. Top. Semidefinite Inter.-Point Methods **18**, 51–76 (1998)
9. Laurent, M.: Matrix completion problems. In: Floudas, C.A., Pardalos, P.M. (eds.) Encyclopedia of Optimization, pp. 1311–1319. Springer, Boston (2001). https://doi.org/10.1007/0-306-48332-7_271
10. Shalev-Shwartz, S., et al.: Online learning and online convex optimization. Found. Trends®Mach. Learn. **4**(2), 107–194 (2012)
11. Shamir, O.: A stochastic PCA and SVD algorithm with an exponential convergence rate. In: ICML, pp. 144–152 (2015)
12. Shamir, O.: Convergence of stochastic gradient descent for PCA. In: Proceedings of the 33rd International Conference on International Conference on Machine Learning, ICML 2016, pp. 257–265, vol. 48. JMLR.org (2016)
13. Sra, S., Nowozin, S., Wright, S.J.: Optimization for Machine Learning. MIT Press, Cambridge (2012)

Improving Traditional Dual Ascent Algorithm for the Uncapacitated Multiple Allocation Hub Location Problem: A RAMP Approach

Telmo Matos[(✉)] [iD], Fábio Maia, and Dorabela Gamboa [iD]

School of Technology and Management, Polytechnic of Porto,
CIICESI – Center for Research and Innovation in Business Sciences
and Information Systems, Porto, Portugal
{tsm, dgamboa}@estg.ipp.pt, fabio7maia@gmail.com

Abstract. Hub Location Problems are complex combinatorial optimization problems that raised a lot of interest in the literature and have a huge number of practical applications, going from the telecommunications, airline transportation among others. In this paper we propose a primal-dual algorithm to solve the Uncapacitated Multiple Allocation Hub Location Problem (UMAHLP). RAMP algorithm combines information of traditional Dual Ascent procedure on the dual side with an improvement method on the primal side, together with adaptive memory structures. The overall performance of the proposed algorithm was tested on standard Australian Post (AP) and Civil Aeronautics Boarding (CAB) instances, comprising 192 test instances. The effectiveness of our approach has been proven by comparing with other state-of-the-art algorithms.

Keywords: Hub Location Problem · Primal-dual algorithm ·
Dual ascent procedure · RAMP algorithm

1 Introduction

The Hub Location Problem (HLP) is one of the most studied problem by the scientific community, and in recent years has assumed a great focus giving rise to many algorithms for different variants (some surveys on HLP can be found in [1, 11, 12]). We tackle the Uncapacitated Multiple Allocation Hub Location Problem (UMAHLP) in which the objective is to choose the nodes that will act as hubs and the optimal assignment of other nodes to the selected hubs.

Campbell [6] proposed the first linear programming formulation for the HLP, with and without capacity constraints, and addressed the single and multiple allocation. Klincewicz [16] presented an algorithm that applies the dual ascent and dual adjustment procedure together with a branch-and-bound method for UMAHLP. Dual ascent procedure was based on the work of Erlenkotter [9] to solve the Uncapacitated Facility Location Problem. Later, Mayer and Wagner [19] implemented some improvements to the dual ascent procedure, considering the aggregated (considering the combination of $i - j - k - m$) model formulation despite the disaggregated one (replacing the combination of node pair (i,j) and hub pair (k,m) by single variables h and l, respectively)

© Springer Nature Switzerland AG 2019
G. Nicosia et al. (Eds.): LOD 2018, LNCS 11331, pp. 243–253, 2019.
https://doi.org/10.1007/978-3-030-13709-0_20

leading to the reduce number of constraints in the constraint set and reducing the computation time of branch-and-bound procedure. Boland et al. [4] presented some formulations and solution approaches for three variants of the HLP, implementing preprocessing techniques in order to decreased the constraints of the problem. A computational study is presented with different formulations using two different commercial solvers. Kratica et al. [17] developed a genetic based algorithm using the cache technique. The main goal of this technique is to store the "genetic code" of the visited solutions in order to avoid returning to the same solution.

Later in 2007, Cánovas et al. [7] proposed a new heuristic based on dual ascent, also using a branch-and-bound technique. The dual ascent procedure is initiated with initial preprocessing to improve the solution, following the feasibility of each solution. Next, the dual adjustment takes place and the algorithm continues in primal side with an exact method. The algorithm's performance was tested over the standard benchmark providing results to instances up to 120 nodes. In the next year, de Camargo et al. [5] presented an algorithm based on benders decomposition [3] to solve the problem. This approach divides the problem into two simple problems: a higher-level problem (or master problem), which determines the chosen hubs and a lower-level problem, (known as subproblem) that defines the allocation of nodes to the chosen hubs. Contreras et al. [8] proposed an exact algorithm to solve large instances for this problem, performing tests with instances up to 500 nodes. The algorithm is based on benders decomposition with the inclusion of several features such as reduction tests, and a heuristic procedure incorporating two distinct phases (namely the estimation and intensification phase), which aims to construct an initial solution and to generates feasible solutions covering the sets of open hubs obtained in the previous phase, respectively.

More recently, Mokhtar et al. [20] also used benders procedure for the UMAHLP. A modified version of benders procedure is presented with different parameters tuning and with subproblems reformulation. These subproblems are then solved using a minimum cost network flow algorithm resulting in a more effective benders algorithm with less number of benders iterations and therefore better running times. The performance of the algorithm is compared only with other benders algorithms [3, 8] and for AP and CAB instances, obtaining in average around two thirds less computational time with same solutions quality then the other algorithms with the same technique.

2   Problem Description

UMAHLP is a well-known combinatorial optimization problem belonging to the class of the NP-Hard problems [21]. This problem can be described as follows. Consider the complete graph $G = (N, A)$, where N is the set of nodes $N = \{1, 2, \ldots, n\}$, that correspond to origins/destinations as well as potential hub locations. Let w_{ij} be the flow between i and j. For each node $i \in N$, let f_i the fixed set-up cost of hub i. The distance between nodes i and j is assumed to satisfy the triangle inequality and is denoted by d_{ij}. We will use these distances as a measure of the per unit flow transportation costs along the links of the graph. These distances are weighted by some discount factors, denoted χ, α and δ, to represent the collection, transfer and distribution costs per unit of flow, respectively. The objective consists in choosing the set of nodes to be established as hubs, while minimizing the total cost of assigning all the non-hubs to the chosen hubs. The total cost of routing the flow along the path $i - j - k - m$ (these are the paths between origin destination pairs, where i and j represents the origin and destination, respectively, and k and m are the hubs to which i and j are allocated, respectively) is given by:

$$F_{ijkm} = w_{ij}\left(\chi d_{ik} + \alpha d_{km} + \delta d_{mj}\right) \tag{1}$$

and for each pair $i, k \in N$ the following sets of binary decision variables are defined by:

$$Z_{ik} = \begin{cases} 1 \; if \; node \; i \; is \; assigned \; to \; hub \; k; \\ 0 \; otherwise. \end{cases} \tag{2}$$

Variable Z_{kk} denotes the establishment or not of a hub at node k, when $i = k$. An additional set of binary variables will be defined. These variables indicate if there is flow through each link of the graph. For each $i, j, k, m \in N$ the fraction of flow is defined by X_{ijkm}.

The mathematical formulation for UMAHLP is:

$$\text{UMAHLP} = min \sum_{k \in N} f_k Z_{kk} + \sum_{i \in N} \sum_{j \in N} \sum_{k \in N} \sum_{m \in N} F_{ijkm} X_{ijkm} \tag{3}$$

$$s.t. \quad \sum_{k \in N} \sum_{m \in N} X_{ijkm} = 1 \, \forall \, i, j \in N \tag{4}$$

$$\sum_{m \in N} X_{ijkm} \leq Z_{ik} \, \forall \, i, j, k \in N \tag{5}$$

$$\sum_{k \in N} X_{ijkm} \leq Z_{jm} \, \forall \, i, j, m \in N \tag{6}$$

$$Z_{ik} \in \{0, 1\} \, \forall \, i, k \in N \tag{7}$$

$$X_{ijkm} \geq 0 \, \forall \, i, j, k, m \in N \tag{8}$$

Constraints (4) assure that every single node is assigned to one hub. Constraints (5) insure that if node i is assigned to hub k then all the flow from this node to any other (fixed) node j must go through some other hub m. Constraints (6) state a similar interpretation regarding to the flow sent to a node j, assigned to hub m, from some node i. Constraints (7) and (8) are integrality and non-negativity constraints. The objective function (3) represents the total cost of hub establishments plus total cost of routing the flow along the path $i - j - k - m$. Being uncapacitated and multiple allocation, means that the hubs do not have capacity limits and each node can be assigned to more than one hub, respectively. Unlike the single allocation, where a node is assigned to only a hub, this version is more difficult to solve, involving much more variables in its formulation. Below (Fig. 1), an image is presented showing the difference between the two versions.

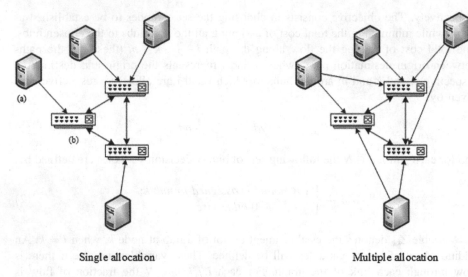

Single allocation   Multiple allocation

Fig. 1. Difference between single and multiple allocation, where switches (a) represents hubs and workstations (b) the nodes.

A special feature of the uncapacitated multiple allocation hub location problems, is that once the hubs are defined, the allocations of nodes to hubs are immediately assigned, that is, each pair of nodes (i, j) is immediately allocated to the pair of hubs (k, m), where the cost along the path is minimum.

3   RAMP Algorithm for the UMAHLP

The Relaxation Adaptive Memory Programming (RAMP) emerged in 2005 by Rego [22], combining fundamental principles of mathematical relaxation with concepts of adaptive memory programming techniques, with the objective of incorporating information obtained by primal and dual solutions spaces. RAMP comprises two levels of sophistication, namely Dual-RAMP and Primal-Dual RAMP. At the first level of sophistication (Dual-RAMP or simply RAMP), this framework explores more intensively the dual side, restricting the primal side interaction to the projection of dual solutions to the primal solutions space and to the improvement of these solutions. Higher levels of sophistication (Primal-Dual RAMP or simply PD-RAMP) allow a more intensive exploration of the primal side, incorporating the simple level, the Dual-RAMP, with more complex memory structures. Several combinatorial optimization problems have already been solved by RAMP applications, producing excellent results, in some cases with new best-known solutions. Some examples of RAMP approaches with different levels of sophistication are the capacitated minimum spanning tree [23], the linear ordering problem [13], the resource constrained project scheduling problem [24], or the capacitated single allocation hub location problem [18], among others.

The proposed algorithm embraces the simplest level of RAMP framework, the Dual-RAMP and explores the dual side of the problem through a dual ascent procedure. The primal side is explored by a simple method based on tabu search procedure. Below, (Fig. 2) the RAMP model for UMAHLP is presented.

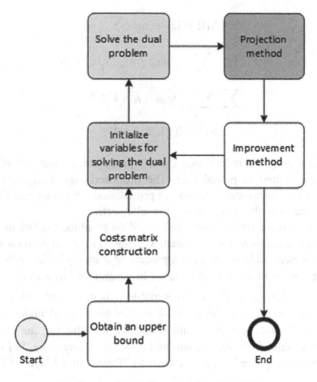

Fig. 2. Dual-RAMP model for UMAHLP.

The algorithm employs a local search procedure to construct a feasible solution to get a good starting solution to dual ascent (DA) procedure. Then, for each iteration of DA, it increases the dual variables till no more improvement can be made. When DA procedure ends, the dual solution is projected to the primal solution space by the projection method and the remaining solution is improved by an improvement method. After the solution is improved, the algorithm alternates to dual solution space, completing one iteration of the DA. The algorithm stops when it reaches the maximum number of predefined iterations.

3.1   The Dual Method

The algorithm relies on DA to explore the dual side of the problem. Following the dualization approach proposed by Wagner and Mayer [19], the condensed formulation of dual problem ($UMAHLP(D)$) is presented as follows:

Let v_h and w_{khl} be the associating dual variables, we obtain the following formulation of the dual problem. Variable P_k contains all pairs of hubs l, where the hub k appears. The dual objective function only depends on the variable v_h. Thus, if these variables are known, the variable w_{khl} can be assigned to any value, since it respects the constraints (10) and (11). The objective is to maximize the value v_h, respecting all the constraints, as it can be seen below.

$$\text{UMAHLP(D)} = \max \sum_h v_h \tag{9}$$

$$s.t. \quad v_h - \sum_{k \in H_l} w_{khl} \leq C'_{hl} \; \forall h, l \in N \tag{10}$$

$$\sum_l \sum_{l \in P_k} w_{khl} \leq f_k \; \forall k \in N \tag{11}$$

$$w_{khl} \geq 0 \; \forall k, h, l \in N \tag{12}$$

The algorithm proposed in [19] uses the DA procedure in dual side of the problem and a branch-and-bound in primal side. The proposed algorithm enhances the DA algorithm, using an improvement method in primal side, obtaining excellent results as will be demonstrated in the computational results section.

Initially, the costs matrix is constructed based on combined solutions of all pair of hubs l, with pairs of nodes h. This cost matrix includes all possible routes between the chosen pairs of nodes and hubs. Next, DA procedure starts by building the matrix costs C'^q_h, that is, the values of C'_{hl} for all h sorted in ascending order to $q = 1, \ldots, n^2$ and setting $C'^q_h = +\infty$ when $q = n^2 + 1$. The variable v_h is initialized with $min \; C'^{1\,T}_h$. Next, the procedure tries to increase the value of v_h (for all pairs of nodes h) to the next highest value of C'^1_{hl}. Note that, only pairs of nodes h for which v_h can be increased are covered. The procedure ends when v_h cannot be increased any more. For more detailed information about this procedure, please refer to Wagner and Mayer [19].

3.2   The Projection Method

The DA procedure provides a dual solution from which it's possible build a primal feasible solution, through a projection method. This method is very simple and takes in consideration constraint (11) and specifically, the variable z_k:

$$z_k \left(f_k - \sum_l \sum_{l \in P_k} w_{khl} \right) = 0 \; \forall k \in N \tag{13}$$

If the sum of allocations (w_{khl}, where it represents the quantity of each hub k in the pairs h and l) of the assigned nodes is greater or equal to the opening cost (f_k, the opening cost of hub k), then node k is chosen to be a hub.

3.3   The Primal Method

Before the RAMP algorithm begins the primal-dual procedure, a solution (upper bound) is obtained to start the DA procedure. This procedure needs a good starting solution to start maximizing the dual function (formulation from 9 to 12). To achieve a good solution a greedy heuristic was used. This heuristic is known to be very robust by finding good solutions in reduced computational time. Basically, this procedure starts with an initial primal solution $S = \emptyset$, and the value of its objective function vs is a very large number $(+\infty)$. For each iteration of the greedy algorithm, a search in the neighborhood is made to accomplish if any move improves the current solution. If so, then the move is made, and the solution is improved. The algorithm ends when there are no more moves that improves the current solution.

In each iteration of DA procedure, the projection method projects dual solution onto primal solution space, and the resulting solution is improved through an improvement method (IM) based on Tabu Search [14, 15]. The IM starts with an admissible initial solution S_0, with the objective function value vs_0. The initial size of the tabu list $ts = p/2$, and p is the number of hubs in the solution returned by the greedy method. The maximum number of iterations without improve the best solution found so far depends on the number of nodes of the problem. After several tests performed and taking in consideration the computational time, $\sqrt{numNodes}$ was used as the maximum number of iterations.

For each iteration, this improvement method checks if there is any movement which improves the current solution. If so, the movement is made, and the current solution is updated. If there are no movements that improves the current solution, a hub is randomly chosen to be closed. The size of tabu list is dynamic, that is, will be changed according to whether find a current better solution or the best so far. Its size can be between 1 and $p + 2$. The tabu list influences the choice of the next possible move, since only movements outside the tabu list are permitted. If the movement is tabu, it only be considered if improves the best solution found so far (the common aspiration criterion). As said, the algorithm stops at $\sqrt{numNodes}$ consecutive iterations without improving the best solution found. In both improvement methods (the greedy and the IM) described above, the same neighborhood structure is used, that is, defining a node as hub or defining a hub as node.

4   Computational Results

The performance of the RAMP algorithm was evaluated on a standard AP (Australia Post) dataset introduced by Ernst and Krishnamoorthy [10] and obtained in Beasley's OR-Library [2] and CAB (Civil Aeronautics Boarding) data benchmark proposed by O'Kelly [21]. The AP benchmark includes 184 instances, of which 28 are standard instances available by Beasley [2], 84 were proposed by Cánovas et al. [7] (where the author has 3 sets of 28 asymmetric instances) and 72 were proposed by Contreras et al. [8]. The CAB benchmark consists of 8 instances, being two instances proposed by O'Kelly [21] and the remaining six are variants of standard instances, proposed by Cánovas et al. [7].

The algorithm was coded in C programming language and run on an Intel Pentium I7 2.40 GHz (only one processor was used) with 8 GB RAM under Ubuntu operating system. The Dual-RAMP algorithm was compared with the state of the art algorithms for the solution of the UMAHLP. The best-known approaches are the dual ascent combined with the branch and bound method proposed by Cánovas et al. [7] (DA-BB), genetic algorithm proposed by Katrica et al. [17] (GA) and benders decomposition heuristic proposed by Contreras et al. [8] (BD).

For all results tables, "instances" stands for the designation of the dataset and "OF" is the number of optimal/best-known solutions found. The value of the column "gap" was computed as $(UB - Z*)/UB*100$ ($Z*$ is the optimal/best-known solution and UB is the value of the upper bound obtained). The "cpubs" and "cputot" columns are the computational time (in seconds) needed to achieve the best value found and the total running time, respectively. The "b-k" is the best-known solution present in literature and "Nb-k" is the new best-known solution found by RAMP algorithm.

Table 1 shows the results for AP benchmark proposed by Beasley [2] and Cánovas et al. [7]. Analyzing the table, the RAMP algorithm achieved the optimal solutions for all instances, while the GA algorithm cannot reach the optimal solution for four instances. Although computational time is not comparable, we choose to present this result.

Table 1. Aggregated results for the AP data instances (Beasley and Cánovas) and comparing RAMP vs GA algorithm.

Instances	GA				RAMP			
	of	gap	cpubs	cputot	of	gap	cpubs	cputot
Beasley	27/28	0.030	2.470	11.610	28/28	0.000	3.100	5.480
Cánovas-1	28/28	0.000	1.310	9.370	28/28	0.000	7.700	10.540
Cánovas-2	26/28	0.010	0.930	8.860	28/28	0.000	0.540	10.250
Cánovas-3	27/28	0.000	1.080	9.090	28/28	0.000	0.040	8.600
Average	108/112	0.010	1.448	9.733	112/112	0.000	2.845	8.718

The results presented in Table 2 concern the AP dataset proposed by Contreras et al. [8], namely the 72 instances. The RAMP algorithm achieved 70 out of 72 optimal/best-known solutions in a short computational. Comparing with BD algorithm,

Table 2. Aggregated results for the AP data instances (Contreras) and comparing the RAMP vs BD algorithm.

Instances	BD			RAMP		
	of	gap	cputot	of	gap	cputot
25	9/9	0.000	0.060	9/9	0.000	0.044
50	9/9	0.000	0.443	9/9	0.000	0.392
75	9/9	0.000	1.478	9/9	0.000	1.210
100	9/9	0.000	3.566	8/9	0.004	3.208
125	7/9	0.017	6.113	8/9	0.002	6.525
150	8/9	0.019	14.477	9/9	0.000	12.812
175	9/9	0.000	16.404	9/9	0.000	20.102
200	8/9	0.031	33.441	9/9	0.000	30.507
Average	68/72	0.008	9.498	70/72	0.001	9.350

RAMP obtained best quality results (BD got 68 out of 72 optimal/best-known). As well as the previous table, we choose to show the computational time for this group of instances. We can say that the proposed algorithm achieved excellent result in low computational time.

For the last table, Table 3, we can see that once again that RAMP algorithm found the best results in the literature for the CAB instances. Our algorithm was able to find all optimal/best-known solutions. Comparing with DA-BB algorithm, that uses the same dualization as the proposed algorithm, RAMP achieved higher quality results due to its exploration on primal solutions space.

Table 3. Results for the CAB data instances and comparing RAMP vs DA-BB algorithm.

Instances	b-k	DA-BB		RAMP		Nb-k
		gap	cputot	gap	cputot	
25La	390369	2.020	0.000	0.000	0.056	-
25L1	196903	0.000	0.000	0.000	0.041	-
25L5	234523	2.600	0.000	0.000	0.034	-
25L9	256691	0.380	0.000	0.000	0.034	-
25Ta	484591	1.950	0.000	0.000	0.041	-
25T1	258577	0.420	1.000	−0.008	0.042	258557
25T5	279144	0.000	0.000	−0.001	0.034	279141
25T9	284852	0.000	0.000	0.000	0.033	-
Average		0.921	0.125	−0.001	0.039	-

For the CAB instances, RAMP algorithm managed to improve the best-known solutions for two instances, the 25T1 and the 25T5. For these two instances, 25T1 and 25T5 found new solutions with values 258557 and 279141, respectively, as indicated in the table below in column "Nb-k".

5 Conclusions

The Hub Location Problem (HLP) have been extremely studied by the scientific community due to its complexity and the vastly real-world applications, motivating many authors to present state of the art algorithms. A Dual-RAMP algorithm to solve the UMAHLP is proposed, combining dual ascent procedure in the dual side with an improvement method on the primal side. The proposed algorithm managed to improve the traditional dual ascent algorithm, where a dual solution projected to the primal space is improved and subjected to an improvement method based on tabu search metaheuristic.

Numerous tests were performed to access to effectiveness of our algorithm. For the 192 standard instances used, the RAMP algorithm successful achieved excellent results in very reduced time outperforming all other best approaches in literature. In fact, for all instances shown, only two of them the proposed algorithm could not reach the best-known solution but could get new best-known solutions for two instances of the AP dataset.

Once again, the RAMP approach was able to solve efficiently a complex optimization problem. The use of primal-dual exploration techniques with the use of adaptive memory in metaheuristics such as tabu search are extremely efficient in such problems. It is estimated that the application of this technique to other complex problems of difficult resolution obtain results of the same quality as we obtained for the UMAHLP.

References

1. Alumur, S., Kara, B.Y.: Network hub location problems: the state of the art. Eur. J. Oper. Res. **190**(1), 1–21 (2008)
2. Beasley, J.: OR-Library: distributing test problems by electronic mail. J. Oper. Res. Soc. **65**, 1069–1072 (1990)
3. Benders, J.F.: Partitioning procedures for solving mixed-variables programming problems. Numer. Math. **4**(1), 238–252 (1962)
4. Boland, N., et al.: Preprocessing and cutting for multiple allocation hub location problems. Eur. J. Oper. Res. **155**(3), 638–653 (2004)
5. de Camargo, R.S., et al.: Benders decomposition for the uncapacitated multiple allocation hub location problem. Comput. Oper. Res. **35**(4), 1047–1064 (2008)
6. Campbell, J.F.: Integer programming formulations of discrete hub location problems. Eur. J. Oper. Res. **72**(2), 387–405 (1994)
7. Cánovas, L., et al.: Solving the uncapacitated multiple allocation hub location problem by means of a dual-ascent technique. Eur. J. Oper. Res. **179**(3), 990–1007 (2007)
8. Contreras, I., et al.: Benders decomposition for large-scale uncapacitated hub location. Oper. Res. **59**(6), 1477–1490 (2011)
9. Erlenkotter, D.: A dual-based procedure for uncapacitated facility location. Oper. Res. **26**(6), 992–1009 (1978)
10. Ernst, A.T., Krishnamoorthy, M.: Solution algorithms for the capacitated single allocation hub location problem. Ann. Oper. Res. **86**, 141–159 (1999)
11. Farahani, R.Z., et al.: Hub location problems: a review of models, classification, solution techniques, and applications. Comput. Ind. Eng. **64**(4), 1096–1109 (2013)
12. Fernandez, E.: Locating hubs: an overview of models and potential applications (2013)
13. Gamboa, D.: Adaptive memory algorithms for the solution of large scale combinatorial optimization problems. Ph.D. thesis (in Portuguese), Instituto Superior Técnico, Universidade Técnica de Lisboa (2008)
14. Glover, F.: Tabu search—part I. ORSA J. Comput. **1**(3), 190–206 (1989)
15. Glover, F.: Tabu search—part II. ORSA J. Comput. **2**(1), 4–32 (1990)
16. Klincewicz, J.G.: A dual algorithm for the uncapacitated hub location problem. Locat. Sci. **4**(3), 173–184 (1996)
17. Kratica, J., et al.: Genetic algorithm for solving uncapacitated multiple allocation hub location problem. Comput. Inform. **24**(4), 415–426 (2005)
18. Matos, T., Gamboa, D.: Dual-RAMP for the capacitated single allocation hub location problem. In: Gervasi, O., et al. (eds.) ICCSA 2017. LNCS, vol. 10405, pp. 696–708. Springer, Cham (2017). https://doi.org/10.1007/978-3-319-62395-5_48
19. Mayer, G., Wagner, B.: HubLocator: an exact solution method for the multiple allocation hub location problem. Comput. Oper. Res. **29**(6), 715–739 (2002)

20. Mokhtar, H., et al.: A new Benders decomposition acceleration procedure for large scale multiple allocation hub location problems. In: International Congress on Modelling and Simulation, pp. 340–346 (2017)
21. O'Kelly, M.E.: A quadratic integer program for the location of interacting hub facilities. Eur. J. Oper. Res. **32**(3), 393–404 (1987)
22. Rego, C.: RAMP: a new metaheuristic framework for combinatorial optimization. In: Rego, C., Alidaee, B. (eds.) Metaheuristic Optimization via Memory and Evolution: Tabu Search and Scatter Search, pp. 441–460. Kluwer Academic Publishers, Boston (2005)
23. Rego, C., et al.: RAMP for the capacitated minimum spanning tree problem. Ann. Oper. Res. **181**(1), 661–681 (2010)
24. Riley, C., et al.: A simple dual-RAMP algorithm for resource constraint project scheduling. In: Proceedings of the 48th Annual Southeast Regional Conference on - ACM SE 2010, p. 1 ACM Press, New York (2010)

Supervised Learning Approach for Surface-Mount Device Production

Eva Jabbar[1,2]([✉]), Philippe Besse[1], Jean-Michel Loubes[1],
Nathalie Barbosa Roa[2], Christophe Merle[2], and Rémi Dettai[2]

[1] IMT - Institut de Mathématiques, Université de Toulouse, CNRS UMR 5219,
Toulouse, France
[2] Continental Automotive France SAS, Toulouse, France
eva.jabbar1@gmail.com

Abstract. In this paper, we propose a decision-making tool based on supervised learning techniques that detects defects and proposes to the Surface-Mount Technology (SMT) operator a probability of being a false call. In this work, we compare four tree-based learning methods. The result of our experiments shows that a XGBoost model trained with our real-world dataset can accurately classify most real defects and false calls with an accuracy score of about 99.4% and a recall of about 98.6%. Moreover, we investigated the computing time of our prediction model and concluded that integration of our classification tool based on the XGBoost algorithm is realistic and feasible in the SMT production line. We believe that our tool will significantly improve the daily work of the SMT verify operator.

Keywords: Supervised learning · Industry 4.0 ·
Decision-making tool · Big data analytics · Surface-Mount Technology

1 Introduction

Nowadays, the manufacturing of Printed Circuit Board (PCB) uses mostly Surface Mount Technology (SMT). A Surface-Mount Device (SMD) assembly line of PBCs consists of several operations among which solder paste printing, component placing and reflow soldering are critical processes. There are various inspection and test methods used in the SMT line; the optical inspection system is the most common, which includes Solder Paste Inspection (SPI) and Automated Optical inspection (AOI). In the first process, a solder paste layer is printed on the surface of the board: this process is known as Solder Paste Printing (SPP). In the second, components are picked from the equipment feeders and placed on the board. In the third process, the solder joints take shape by the reflowing of the solder paste. A schematic view of a SMT line is presented in Fig. 1. Heterogeneous data are generated within these steps, in terms of size, format and frequency. These data are processed in real time and pushed to the cloud.

Each inspection process consists of two stages: an AOI and its verify station. In the first step the AOI machine assigns a tag to the PCB. This tag is either

© Springer Nature Switzerland AG 2019
G. Nicosia et al. (Eds.): LOD 2018, LNCS 11331, pp. 254–263, 2019.
https://doi.org/10.1007/978-3-030-13709-0_21

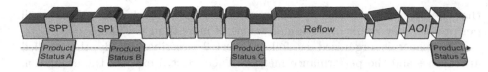

Fig. 1. This figure presents a schematic view of a SMT line.

good or fail. In case the PCB is considered as a fail, a manual inspection is then performed by an operator to double check and judge whether the defect is real or not (false call). Since the production is operating continuously 24/7, the verify operator plays a crucial role in a AOI system. It is clear that the inspection process becomes rapidly laborious regarding the capacity of human operators [1].

Soukup [1] proposed a preventive methodology for optimizing the false call rate in AOI based on experience of AOI experts and performed analysis. Regarding the machine learning application in production line, the studies focused on automatically detecting defects using inspection images and training deep learning.

Richter et al. [2] propose to integrate deep learning in the AOI system by using convolutional networks in order to automatically detect defects without manual user interactions. Acciani et al. [3] focused on detecting solder joint defects by applying feature extraction techniques to extract regions of interests from images and then used neural networks to classify defects achieving an accuracy of 98%.

These techniques frequently related to training images are used offline and are unlikely to be successful in case new products are introduced into production. In fact, most defects occur due to physical deviations in SMD [1]. Tavakolizadeh et al. [4] used simulated production data of multiple SMT stations to build a binary classification model based on random forests to detect defective products with a Matthews correlation coefficient (MCC) score of 0.96. In our use case, we use real mass production datasets with 3 classes: good, real defect and false call which positions this study as multiclass and in a big data configuration. Considering false calls as a separated class came after some exploratory analyses and particularly after discussions with experts who insisted that false calls were a population with its proper characteristics (can be due to values approaching thresholds, dimensions of components...).

To the best of our knowledge, there is no application of data-driven supervised learning methods in a real-world production line focused on reducing the time-consuming task of checking false calls. In this work, we are specially concerned by the SMT assembly line problems with a focus on defects caused by paste printing and reflow. Predictive models studied under this work have been trained by combining available data. The tree-based model will provide the operator a tag (good, false call or real defect) with the corresponding confidence score. The operator can then make his or her decision. As calculated in [5] the cost of a single false call is estimated at around 0.65 eurocents. Beside the convenience that the approach provides to the operator, the fact of reducing one false call per product may result in an annual saving of more than € 6500.

The paper is organized as follows: Sect. 2 describes briefly the studied approaches and the performance metrics. Then, Sect. 3 exposes the evaluation procedure. Finally, conclusions and an outlook of future work are provided.

2 Methodology

The aim of this work is to propose a new approach based on supervised learning techniques in order to reduce the time-consuming task of detecting AOI false calls.

In this case study, we focus on six types of PCB $P_1, \cdots, P_6$. Each product P_i is associated to a $n_i \times f$ matrix X_i of n_i pad measurements and $f = 187$ features presented as follows:

$$
X_i = \begin{pmatrix}
m_{11}^{(i)} & m_{12}^{(i)} & \cdots & m_{1f}^{(i)} \\
m_{21}^{(i)} & m_{22}^{(i)} & \cdots & m_{2f}^{(i)} \\
\vdots & \vdots & \ddots & \vdots \\
m_{n_i1}^{(i)} & m_{n_i2}^{(i)} & \cdots & m_{n_if}^{(i)}
\end{pmatrix},
\tag{1}
$$

where the number of pads n_i varies depending on the PCB type. Each feature vector $m_{k.}^{(i)}$ is made of SPI and reflow measurements. It is associated with a label Y set by AOI verify which belongs to the discrete set:

$$\{good, false\ call, real\ defect\}.$$

As previously mentioned, among existing algorithms, we focus on a tree-based learning method CART [6], and several improvements of this method using aggregation: Random Forest [7], using boosting methodology AdaBoost [8] and finally one of the latest learning algorithm in statistical learning XGBoost [9]. These algorithms are described in Sect. 2.1, then Sect. 2.2 introduces the selected performance metrics. We used the scikit-learn implementation of studied algorithms.

2.1 Tree-Based Learning Techniques

On the one hand, by extracting the set of rules depicting normal zones in CART algorithm, we can easily detect which variables play an important role in anomaly detection and thus may lead to process improvements. On the other hand, ensemble learning methods such as RF, Adaboost and XGboost extract only the most important features.

CART. [6] is a decision tree (DT) method that stands for Classification And Regression Tree. The principle of DT is based on a series of if-then rules that form the tree branches. DT uses nodes and leaves to split the instance space into classification or regression case. The affiliation of a point to a branch and therefore to a class is assigned according to the rule followed by this point. Detailed descriptions are available in [6].

Random Forest (RF). [7] classifies data points by aggregating the predictions given by multiple DT during the training phase. Random Forest generates a fixed number k of decision tree classifiers. For the classification, classes are given to each tree in RF, which then returns its prediction about the given classes. The decision of each tree is considered as a vote for obtaining the final decision which is based on the majority rule. Actually, RF is expected to perform better in terms of prediction accuracy than single base learners. Indeed, using multiple DTs reduces the variance by aggregating classifiers. This makes random forests less prone to overfitting and more robust on imbalanced datasets.

AdaBoost. [8] stands for Adaptive Boosting and is used to boost the performance of any machine learning algorithm for two-class classification problems. The boosting approach is based on the idea of creating a highly accurate prediction rule by combining relatively weak rules.

The AdaBoost algorithm proceeds by applying, in the first step, the chosen weak classifier to the training samples and then produces class labels. In our case the weak learner is DT. Then, a higher weight is attributed to all the misclassified points (boosted). A second classifier is built based on these new weights and the procedure is repeated. The final classifier is defined as the linear combination of classifiers from previous stages.

The multi-class classification version of the Adaboost algorithm, which is used in this paper, is the one proposed by Zhu et al. [10].

XGBoost. [9] stands for eXtreme Gradient Boosting and is a scalable machine learning method for tree boosting. It is called gradient boosting because it uses a gradient descent algorithm to minimize the loss when adding these new models. By adding models on top of each other iteratively, the errors of the previous model are corrected by the next predictor, until the training data is accurately predicted or reproduced by the model.

Instead of assigning different weights to the misclassified points after every iteration (as AdaBoost), this method fits the new model to new residuals of the previous prediction and then minimizes the loss when adding the latest prediction. XGBoost is specifically developed with an additional custom regularisation term in objective function and a penalization term at every iteration over the DTs.

2.2 Performance Metrics

Machine learning algorithm evaluation has been achieved by comparing:

- the quality prediction measured according to different metrics (accuracy, Hamming loss, precision, recall and f1-score),
- the computation time C_t: need to satisfy the condition $C_t < 180(s)$

In order to have a better overview of what the model is correctly predicting and what types of errors it is making, a confusion matrix can be calculated to give the full picture. We define below in Table 1 the confusion matrix of a predictive model results in the case of three class problem with the classes A, B and C.

Table 1. Three-class confusion matrix

	A	B	C
A	TP_A	E_{AB}	E_{AC}
B	E_{BA}	TP_B	E_{BC}
C	E_{CA}	E_{CB}	TP_C

The confusion matrix shows how the predictions are made by the model and gives the full picture through each class performance. The rows correspond to the known class of the data and the columns correspond to the predictions made by the model. The diagonal elements show the number of correct classifications made for each class, and the off-diagonal elements show the errors made [11].

We use the accuracy score rather than the Matthews correlation coefficient (MCC) [12] because we have decided to balance our datasets by down-sampling good records, while MCC considers the weight of classes.

Hamming loss [13] computes the average loss between two sets of samples. Considering $\widehat{y}_j$ as the predicted value for the j-th label of a given sample, y_j as the corresponding true value, and n_{labels} as the number of classes, then the Hamming loss $L_{Hamming}$ between two samples is defined as:

$$L_{Hamming}(y, \widehat{y}) = \frac{1}{n_{labels}} \sum_{j=0}^{n_{labels}-1} 1_{(\widehat{y}_j \neq y_j)}. \qquad (2)$$

The generalization of multi-class problems is to sum over rows/columns of the confusion matrix. Given that the matrix is oriented as explained before, i.e. rows of the matrix correspond with the "truth" value, we have:

$$Precision_i = \frac{TP_i}{TP_i + \sum_{j \neq i} E_{ji}}, \qquad (3)$$

$$Recall_i = \frac{TP_i}{TP_i + \sum_{j \neq i} E_{jij}}, \qquad (4)$$

$$F_i = \frac{2}{\frac{1}{Recall_i} + \frac{1}{Precision_i}} = 2 \cdot \frac{Precision_i . Recall_i}{Precision_i + Recall_i} \qquad (5)$$

with $i, j \in \{A, B, C\}$.

3   Evaluation and Analysis

This work analyses a strongly imbalanced dataset that includes some hundreds of observed features in which defects are very rare. In this section the evaluation procedure is exposed by first, describing the multiclass datasets, and then showing our choice of the optimal algorithm configurations and finally comparing the performance metrics defined above for the studied algorithms.

3.1   Dataset Description

In this paper, we propose a decision-aiding tool to discern between real defects and false calls in a SMT production line. With that objective in mind, we assembled and analysed a multi-source dataset composed of data from three different stations in the line. This dataset $(X = (X_1, ..X_i., X_N))$ is composed of more than $N = 150,000$ products of multiple surface-mount device types, where X_i is the matrix of the product P_i as introduced before. Each product contains multiple components having a number of pads going from 2 to 100, which results in a dataset of more than 378 million records.

A cloud cluster was used to combine the data sources. For each pad k, we extracted a feature vector $M_k^{(i)} \in \mathbb{R}^{187}$. The features contain measurements from SPI such as pads position and size, area, height, volume and offset of the solder paste and measurements from reflow soldering of multiple individual temperature-controlled zones. After filtering the constant features, a final feature vector $M_k^{(i)} \in \mathbb{R}^{117}$ with three labels $\{good, false\ call, real\ defect\}$ was obtained.

As the 117 features have different scales, we standardize them by removing the mean and scaling to unit variance.

Table 2. Frequency of classes

	Frequency	Percent
Good	378,544,970	99.930
False call	240,000	0.063
Real defect	15,030	0.007
TOTAL	378,800,000	100.0

Performance of the classifier is seriously affected by the highly imbalanced dataset Table 2, which impacts the performance of the classifier during the training phase. To handle this problem, external techniques are required during the pre-processing phase. Two strategies are mainly used in the literature: data level strategy and algorithm level strategy [14]. Data level methods adjust data sets by adding or deleting records to minimize class differences, whereas algorithm level strategy focuses on adjusting classifier algorithms to improve the learning process relative to the minority class. Batista et al. [15] show that Some of the most commonly used classification algorithm currently, for example, decision tree are not adaptive to dealing with unbalanced data. Given that our approach is using tree-based methods, we therefore opted for data level strategy. We created a balanced learning dataset by down-sampling the original dataset taking randomly an ratio of good and false calls records Table 3. We use an 75-15-10 ratio to retain more of the good class and reduce any loss of information. Specifically, we retain all real defect records and randomly sample without replacement from false call and good records. This configuration has provided the best results.

Table 3. Frequency of classes after down-sampling

	Frequency	Percent
Good	112,725	75
False call	22,545	15
Real defect	15,030	10
TOTAL	150,300	100

Four supervised learning algorithms were evaluated on Standardized real-world data and their performance is presented in Sect. 3.2. Further details about features and exploratory analysis are not reported in this work due to Continental AG confidentiality restrictions.

3.2 Performance Comparison

We used the scikit-learn machine learning library for python. To avoid overly complex trees, the implementation of decision trees in this library imposes to takes into account the maximum depth of the tree and the minimum number of samples as hyper-parameters rather than mechanisms such as pruning (rpart package in R).

For each algorithm we tried various combinations of hyper-parameters using the grid search method (as shown below) to finally choose the optimal configuration that gives a high accuracy and minimises the training computing cost. The accuracy score is measured using a 10-fold cross-score unbiased estimate.

- CART
  - max_depth (the maximum depth of the tree)

    {1,10,20,30,40,50}
- Random Forest
  - max_depth (the maximum depth of the tree)

    {1,10,20,30,40,50}
  - max_features (features to consider for the split

    {1,10,20,30,40,50,60,70,80,90,100}
  - n_estimators (number of trees in the forest)

    {10,20,30,40,50,60}

– AdaBoost
- base_estimator

(default=DecisionTreeClassifier)
- n_estimators

{10,20,30,40,50,60}
– XGBoost
- base_estimator

(default=DecisionTreeClassifier)
- n_estimators

{10,20,30,40,50,60}
- max_depth (Maximum tree depth for base learners)

{1,10,20,30,40,50}
- learning_rate (Boosting learning rate (xgb's "eta"))

{0,10,0,20}

For CART, we found that the optimal max_depth (maximum tree depth) is 30. For RF, we noticed that with n_estimators equal to 60 (number of trees in the forest) and max_features equal to 65 (features to consider for the split), we got a higher accuracy score.

For AdaBoost, we used a decision tree classifier as the base estimator with a max_depth of 30, while for XGBoost, we used a max_depth of 30 with learning_rate of 0.20.

To evaluate the efficiency of the ensemble with the different settings, we trained a model with 80% of the dataset selected and 20% for test. The result for each algorithm is represented in Table 4. As we can see, all algorithms give a high accuracy score at around 98%. The difference is seen rather in the test time which remains more important for XGBoost without being too constrained for our environment. The validation step with a product gives especially good results for XGboost with Computation time (C_t) of 840 ms and an accuracy score of 98.3%.

Table 4. Performance of algorithms on SMT datasets

	Acc	HLoss	Ttime (s)	C_t	Recall
CART	97.4%	0.026	2.84	0.015	97%
Random forest	98.5%	0.014	12.20	0.046	97%
AdaBoost	98.0%	0.019	3.85	0.015	97%
XGBoost	98.6%	0.005	211.58	0.840	98.3%

As our focus is to propose a classification prediction to the operator, we have also evaluated the efficiency of the selected algorithms regarding each class as show in Table 5.

It can be seen from Table 5 that XGBoost outperforms the other tree-based algorithms.

Table 5. Performance of algorithms regarding each class

	CART			Random forest			AdaBoost			XGBoost		
	Preci.	Re.	F1	Preci.	Re.	F1	Preci.	Re.	F1	Preci.	Re.	F1
False call	0.92	0.96	0.94	0.96	0.97	0.97	0.95	0.95	0.95	0.98	0.99	0.99
Good	0.99	0.98	0.98	0.99	0.99	0.99	0.99	0.99	0.99	1.00	0.98	0.99
Real defect	0.94	0.97	0.95	1.00	0.95	0.98	1.0	0.97	0.98	0.99	0.99	0.99

3.3   Results Interpretation

We showed that a XGBoost model trained with a large dataset can accurately classify most real defects and false calls. The fraction of wrongly classified items (Hamming Loss) goes below 0.005 with more than 366,590 observations. It is worth noting that, in this use case, the model prediction will have an actual added value, if and only if the product classification tag is generated before the visual diagnostic stage. In this case, the process operator could use the provided class information as an aid in its diagnosis. Note that today false calls represent about 95% of AOI fails. We tested our final algorithm using a new dataset consisting of one product observation of about 2000 pads, and obtained an accuracy of about 98.3% in less than 840 ms. The goal was to verify the correctness of decision made in case the algorithm is implemented in the production. The big data architecture defined by Continental Automotive Midi-Pyrénées allows a real time transfer of data in the cloud. According to the process experts, the product reaches AOI verify about 3 min after leaving the reflow soldering process. As the computing time of our prediction method is much lower than these 3 min, it is therefore concluded that the integration of our classification tool based on the XGBoost algorithm is realistic and feasible in the SMT production line.

3.4   Conclusions and Future Work

In this paper, the performance of several tree-based machine learning methods was evaluated through an experimental application on real-world production data. Our aim is to implement the best chosen algorithm in production, which is why we also considered the computing cost and the scalability/complexity of the algorithm to be implemented. The best result is achieved when applying XGBoost, which results in a prediction accuracy of 99.4% with a recall of 98.6% and a validation time C_t near to 840 ms when considering as a dataset one product with about 2000 pads.

Further investigations are necessary for the implementation of this decision-making tool in production. Going further with this approach, we assume to be able to facilitate the hard task of the verify operators and to reduce human classification errors. Furthermore, we want to analyse feature importance as computed by XGBoost to improve the production process.

Acknowledgements. This work is supported by Continental Automotive FRANCE.

References

1. Soukup, R.: A methodology for optimization of false call rate in automated optical inspection post reflow. In: 33rd International Spring Seminar on Electronics Technology, ISSE 2010, pp. 263–267, May 2010
2. Richter, J., Streitferdt, D., Rozova, E.: On the development of intelligent optical inspections. In: 2017 IEEE 7th Annual Computing and Communication Workshop and Conference (CCWC), pp. 1–6, January 2017
3. Acciani, G., Brunetti, G., Fornarelli, G.: Application of neural networks in optical inspection and classification of solder joints in surface mount technology. IEEE Trans. Ind. Inform. **2**(3), 200–209 (2006)
4. Tavakolizadeh, F., Soto, J.Á.C., Gyulai, D., Beecks, C.: Industry 4.0: mining physical defects in production of surface-mount devices. In: 17th Industrial Conference on Data Mining ICDM, pp. 146–151, July 2017
5. Ellenbogen, R.: Cutting down on false alarms. OnBoard Technology, September 2006. www.Onboard-Technology.com
6. Breiman, L., Friedman, J., Olshen, R., Stone, C.: Classification and Regression Trees. Wadsworth and Brooks, Monterey (1984)
7. Breiman, L.: Random forests. Mach. Learn. **45**(1), 5–32 (2001)
8. Yoav, F., Robert, S.: A decision-theoretic generalization of on-line learning and an application to boosting. J. Comput. Syst. Sci. **55**(1), 119–139 (1997)
9. Chen, T., Guestrin, C.: XGBoost: a scalable tree boosting system. In: Proceedings of the 22nd ACM SIGKDD International Conference on Knowledge Discovery and Data Mining, KDD 2016, pp. 785–794. ACM, New York (2016)
10. Zhu, J., Zou, H., Rosset, S., Hastie, T.: Multi-class AdaBoost (2009)
11. Sokolova, M., Lapalme, G.: A systematic analysis of performance measures for classification tasks. Inf. Process. Manag. **45**(4), 427–437 (2009)
12. Matthews, B.W.: Comparison of the predicted and observed secondary structure of T4 phage lysozyme. Biochimica et Biophysica Acta (BBA) - Protein Struct. **405**, 442–451 (1975)
13. Tsoumakas, G., Katakis, I.: Multi-label classification: an overview. Int. J. Data Warehous. Min. **1–13**, 2007 (2007)
14. Li, Y., Sun, G., Zhu, Y.: Data imbalance problem in text classification. In: 2010 Third International Symposium on Information Processing, pp. 301–305, October 2010
15. Batista, G.E.A.P.A., Prati, R.C., Monard, M.C.: A study of the behavior of several methods for balancing machine learning training data. SIGKDD Explor. Newsl. **6**(1), 20–29 (2004)

Crawling in Rogue's Dungeons
with (Partitioned) A3C

Andrea Asperti[(✉)], Daniele Cortesi, and Francesco Sovrano

Department of Informatics: Science and Engineering (DISI), University of Bologna,
Mura Anteo Zamboni 7, 40127 Bologna, Italy
andrea.asperti@unibo.it

Abstract. Rogue is a famous dungeon-crawling video-game of the 80ies,
the ancestor of its gender. Rogue-like games are known for the necessity
to explore partially observable and always different randomly-generated
labyrinths, preventing any form of level replay. As such, they serve as
a very natural and challenging task for reinforcement learning, requiring the acquisition of complex, non-reactive behaviors involving memory
and planning. In this article we show how, exploiting a version of Asynchronous Advantage Actor-Critic (A3C) partitioned on different situations, the agent is able to reach the stairs and descend to the next level
in 98% of cases.

Keywords: Deep reinforcement learning ·
Asynchronous actor-critic advantage ·
Partially observable Markov decision process · Multi-task learning

1 Introduction

In recent years, there has been a huge amount of work on the application of
deep learning techniques in combination with reinforcement learning (the so
called *deep reinforcement learning*) for the development of automatic agents for
different kind of games. Game-like environments provide realistic abstractions of
real-life situations, creating new and innovative dimensions in the kind of problems that can be addressed by means of neural networks. Since the seminal work
by Mnih et al. [16] exploiting a combination of Qlearning and neural networks
(Deep Q-Networks, DQN) in application to Atari games [5], the field has rapidly
evolved, offering several improvements such as Double Qlearning [8] (correcting
overestimations in the action value of the original version) to the recent breack-
through provided by the introduction of asynchronous methods, the so called
A3C model [15].

In this work, we apply a version of A3C to automatically move a player
in the dungeons of the famous Rogue video game. Rogue was the ancestor of
this gender of games, and the first application exploiting a procedural, random
creation of its levels; we use it precisely in this way: as a generator of different
kind of labyrinths, with a reasonable level of complexity. Of course, the full game

© Springer Nature Switzerland AG 2019
G. Nicosia et al. (Eds.): LOD 2018, LNCS 11331, pp. 264–275, 2019.
https://doi.org/10.1007/978-3-030-13709-0_22

offers many other challenges, comprising collecting objects, evolving the rogue, and fighting with monsters of increasing power, but, at least for the moment, we are not addressing these aspects (although they may provide interesting cues for future developments).

We largely based this work on the learning environment that was previously created to this aim in [3,4], and that allows a simple interaction with Rogue. At the same time, the extension to A3C forced a major revision of the environment, that will be discussed in Sect. 6.

The reasons for addressing Rogue, apart from the fascination of this vintage game, have been extensively discussed in [3,4] (see also [6]), and we just recall here the main motivations. In particular, Rogue's dungeons are a classical example of Partially Observable Markov Decision Problem (POMDP), since each level is initially unknown and not entirely visible. Solving this kind of task is notoriously difficult and challenging [20], since it requires an important amount of *exploration*.

The other important characteristic that differentiates it from other, more modern, 3D dungeons-based games such as ViZDoom [11] or the Labyrinth in [15] is precisely the graphical interface, that in the case of Rogue is ASCII-based. Our claim is that, at the current state of knowledge, decoupling vision from more intelligent activities such as planning can only be beneficial, allowing to focus the attention on the really challenging aspects of the player behavior.

1.1 Achievements Overview

Rogue is a complex game, where the player (the "rogue") is supposed to roam through many different levels of a dungeon trying to retrieve the amulet of Yendor. In his quest, the player must be able to: 1. explore the dungeon (partially visible, when you enter a new level); 2. defend himself from enemies, using the items scattered through the dungeon; 3. avoid traps; 4. avoid starvation, looking for and eating food inside the dungeon.

Currently, we are merely focusing on exploring the maze: as explained in Sect. 6 monsters and traps may be easily disabled in the game (Fig. 1).

The dungeon consists of 26 floors (configurable) and each floor consists of up to 9 rooms of varying size and location, randomly connected through non linear corridors, and small mazes. To reach a new floor the agent needs to find and to go down the stairs, whose position is likely hidden from sight, located in a yet unexplored room and in a different spot at each new level. Finding and taking the stairs are the main ingredients governing the agent movement: the

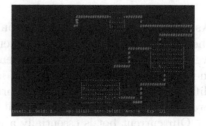

Fig. 1. The two dimensional ASCII-based interface of Rogue.

only differences between the first floors and the subsequent ones are related to the frequency of meeting enemies, dark rooms, mazes or hidden doors. As a consequence, we organized the training process on the base of a single level, terminating the episode as soon as the rogue takes the stairs. In the rest of the work, when we talk about the *performance* of an agent, we refer to the probability that it correctly completes a *single* level, finding and taking the stairs within a maximum of 500 moves[1]. The performance is measured on a set of 200 consecutive (i.e. random) games and we show a comparison with previous work in Table 1. The results are not conclusive, partly because the approaches rely on vastly different models.

Table 1.

Agent	Random	DQN [4]	This work
Performance	7%[a]	23%	98%

[a]The mobility resulting from brownian motion is always impressive.

There are essentially three ingredients behind this achievement:

1. the adoption of A3C as a base learning algorithm, in substitution of DQN; we shall diffusely talk about A3C in Sect. 3.2
2. an agent-centered, cropped representation of the state
3. a supervised partition of the problem in a predefined set of *situations*, each one delegated to a different A3C agent, sharing nevertheless a common value function (i.e. a common evaluation of the state).[2] We shall talk about situations in Sect. 4.1.

While the adoption of A3C and the idea of experimenting with *situations* was a planned activity [3], the shift to an agent-centered view, as well as the choice of the agent situations have been mostly the result of trial-and-error, through an extremely long and painful experimentation process.

2   Related Work

As we mentioned in the introduction, there is a *huge* amount of research around the application of deep reinforcement learning to video games. In this section we shall merely mention some recent works that, in addition to those already mentioned, have been a source of inspiration for our work, or the subject of different experimentations we performed. A few more works that seems to offer promising developments [18,22] will be discussed in the conclusions.

Our current bot is essentially a partitioned multi-task agent in the sense of [19]. Its tree-like structure may be reminiscent of Hierarchical models [7,13,21], but they are in fact distinct notions. In Hierarchical models a Master cooperates with one or more Workers, by dictating them macro actions (e.g. "reach the next

---

[1] For a good agent, in average, little more than one hundred move are typically enough.
[2] Source code and weights are publicly available at [2].

room"), that are taken by Workers as their objectives. The Master typically gets rewards from the environment and gives ad hoc, possibly *intrinsic* bonuses to Workers. The hope is to let top-level agents focus on planning while sub-parts of the hierarchy manage simple atomic actions, improving the learning process.

In our case, we simply split the task according to different situations the rogue may be faced with: a room, a corridor, the proximity to stairs/walls, etc. (see Sect. 4.1 for details). We did several experiments with hierarchical structures, but so far none of them gave satisfactory result.

We also experimented with several forms of *intrinsic* rewards [17], especially after passing to a rogue-centered view. Intrinsic motivations are stimuli received form the surrounding environment different from explicit, extrinsic rewards, and that could be used by the agent for alternative form of training, learning to do a particular action because *inherently enjoyable*. Examples are *empowerment* [12] or *auxiliary tasks* [9]. In this case too, we have not been able to obtain interesting results.

3   Reinforcement Learning Background

A Reinforcement Learning problem is usually formalized as a Markov Decision Process (MDP). In this setting, an agent interacts at discrete timesteps with an external environment. At each time step t, the agent observes a state s_t and choose an action a_t according to some policy π, that is a mapping (or more generally a probability distribution) from states to actions. As a result of its action, the environment change to a new state $s' = s_{t+1}$; moreover the agent obtains a reward r_t (see Fig. 2). The process is then iterated until a terminal state is reached.

The future cumulative reward $R_t = \sum_{k=0}^{\infty} \gamma^k r_{t+k}$ is the total accumulated reward from time starting at t. $\gamma \in [0,1]$ is the so called *discount factor*: it represents the difference in importance between present and future rewards.

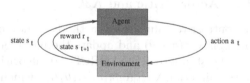

Fig. 2. Basic operations of a Markov Decision Process

The goal of the agent is to maximize the expected return starting from an initial state $s = s_t$.

The *action value* $Q^\pi(s,a) = \mathbb{E}^\pi[R_t | s = s_t, a = a_t]$ is the expected return for selecting action a in state s_t and prosecuting with strategy π.

Given a state s and an action a, the optimal action value function $Q^*(s,a) = \max_\pi Q^\pi(s,a)$ is the best possible action value achievable by any policy.

Similarly, the *value* of state s given a policy π is $V^\pi(s) = \mathbb{E}^\pi[R_t | s = s_t]$ and the optimal value function is $V^*(s) = \max_\pi V^\pi(s)$.

3.1   Q-learning and DQN

The Q-function, similarly to the V-function can be represented by suitable function approximators, e.g. neural networks. We shall use the notation $Q(s, a; \theta)$ to denote an approximate action-value function with parameters θ.

In (one-step) Q-learning, we try to approximate the optimal action value function: $Q^{(}s, a) \approx Q(s, a; \theta)$ by learning the parameters via backpropagation according to a sequence of loss function functions defined as follows:

$$L_i(\theta_i) = \mathbb{E}_{(s,a,r,s') \sim U(D)} \left[(r + \gamma \max_{a'} Q(s', a', \theta_{i-1}) - Q(s, a, \theta_i))^2 \right]$$

where s' is the new state reached from s taking action a and $U(D)$ is the uniform distribution on stored transitions for experience replay.

The previous loss function is motivated by the well know Bellman equation, that must be satisfied by the optimal Q^* function:

$$Q^*(s, a) = \mathbb{E}_{s'}[r_0 + \gamma max_{a'} Q^*(s', a')]$$

Indeed, if we know the optimal state-action values $Q^*(s', a')$ for next states, the optimal strategy is to take the action that maximizes $r_0 + \gamma max_{a'} Q^*(s', a')$.

Q-learning is an *off-policy* reinforcement learning algorithm. The main drawback of this method is that a reward only directly affects the value of the state action pair s,a that led to the reward. The values of other state action pairs are affected only indirectly through the updated value $Q(s, a)$. The back propagation to relevant preceding states and actions may require several updates, slowing down the learning process.

3.2   Actor-Critic and A3C

In contrast to value-based methods, policy-based methods directly parameterize the policy $\pi(a|s; \theta)$ and update the parameters θ by gradient ascent on $\mathbb{E}[R_t]$.

The standard REINFORCE [20] algorithm updates the policy parameters θ in the direction $\nabla_\theta \mathbb{E}[log\pi(a_t|s_t; \theta)R_t]$, which is an unbiased estimate of $\nabla_\theta \mathbb{E}[R_t]$.

It is possible to reduce the variance of this estimate while keeping it unbiased by subtracting a learned function of the state $b_t(s_t)$ known as a baseline. The gradient is then $\nabla_\theta \mathbb{E}[log\pi(a_t|s_t; \theta)(R_t - b_t)]$.

A learned estimate of the value function is commonly used as the baseline $b_t(s_t) \approx V^\pi(s_t)$. In this case, the quantity $R_t - b_t$ can be seen as an estimate of the *advantage* of action a_t in state s_t for policy π, defined as $A^\pi(a_t|s_t) = Q^\pi(s_t, a_t) - V^\pi(s_t)$, just because R_t is an estimate of $Q^\pi(s_t, a_t)$ and b_t is an estimate of $V^\pi(s_t)$.

This approach can be viewed as an actor-critic architecture where the policy π is the actor and the baseline b_t is the critic.

A3C [15] is a particular implementation of this technique based on the asynchronous interaction of several parallel couples of Actor and Critic. The experience of each agent is independent from that of the other agents, which stabilizes learning without the need for experience replay as in DQN.

4 Neural Network Architecture

Our implementation is essentially based on A3C. In this section we describe a novel technique that partitions the sample space into a predefined set of *situations*, each one addressed by a different A3C agent. All of these agents contributes to build a common cumulative reward without sharing any other information, and for this reason they are said to be highly independent. Each agent employs the same architecture, state representation and reward function. In this section we discuss: the situations (Sect. 4.1), the state representation (Sect. 4.2), how we shaped the reward function (Sect. 4.3), the neural network (Sect. 4.4), hyperparameters tuning (Sect. 4.5).

4.1 Situations

In our work, with the term *situation* we mean the environment state used to discriminate which situational agent should perform the next action. We experimented the four situations listed below, from higher to lower priority:

1. The rogue (the agent) stands on a corridor
2. The stairs are visible
3. The rogue is next to a wall
4. Any other case

The situations are determined programmatically and are not learned. This is a simplistic choice, mostly dictated by frustration: in future work we plan to learn them in an end-to-end way. When multiple conditions in the above list are met, the one with higher priority will be selected. For example, if the stairs are visible but the rogue is walking on a corridor, the situation is determined to be 1 rather than 2, because the former has higher priority.
We define:

- $s4$ as the configuration made of all the aforementioned situations
- $s2$ as the configuration made of situations 2 and 4
- $s1$ as the configuration with no situations at all.

We believe that situations may be seen as a way to simplify the overall problem, breaking it down into easier sub-problems.

4.2 State Representation

The state is a 17×17 matrix corresponding to a cropped view of the map centered on the rogue (i.e. the rogue position is always on the center of the matrix). This representation has the advantage to be sufficiently small to be fed to dense layers (possibly after convolutions); moreover, it does not require to represent the rogue into the map. In our experiments we adopted two variations of the above matrix. The first (called $c1$) has a single channel, resulting in a $17 \times 17 \times 1$ shape, and it is filled with the following values:

4   for stairs
8   for walls
16   for doors and corridors
0   everywhere else

The second (called $c2$) is made of two channels (the stairs channel and the environment channel) and thus has shape $17 \times 17 \times 2$. The values used for $c2$ are the same of $c1$.

4.3   Reward Shaping

We designed the following reward function:

1. a positive reward $(+1)$ is given when using a door never used before
2. a positive reward $(+1)$ is given when, after an action, one or more new doors are found
3. a huge positive reward $(+10)$ is given when descending the stairs
4. a small negative reward (-0.01) is given when taking an action that does not change the state (eg.: try to cross a wall).

The chosen reward values are not random. In fact each floor contains at most 9 rooms and each room has maximum 4 doors, thus on each floor the cumulative reward of the rewards of type 1 and 2 can not exceed $9 \cdot (4 + 2) = 54$. But what normally happens, in the episodes with the best return, is that only about $\frac{2}{3}$ of the cumulative reward is given by finding new rooms. This is true because negative rewards are enough to teach the agent not to take useless actions and, in the meantime, they do not significantly affect the balance between room exploration and stair descent.

The result is that the agent is encouraged both to descend the stairs and to explore the floor, and this impacts positively and significantly on its performance. In future work we plan to employ sparser reward functions that are not as problem specific.

4.4   Neural Network

The neural network architecture we used is shown in Fig. 3. This network consists of two convolutional layers followed by a dense layer to process spatial dependencies and a LSTM layer to process temporal dependencies, and finally, value and policy output layers. The convolutions have a ReLU activation, a 3×3 kernel with unitary stride and respectively 16 and 32 filters. Their output is flattened and fed to a FC with ReLU and 256 units. We call this structure: tower.

The tower input is the state representation described in Sect. 4.2 and its output is concatenated with a numerical "one hot" representation of the action taken in the previous state and the obtained reward. This concatenation is fed into an LSTM composed of 256 units. The idea of concatenating previous actions and rewards to the LSTM input comes from [9].

The output of the LSTM is then the input for the value and policy layers.

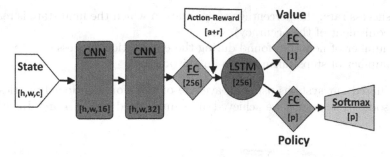

Fig. 3. The neural networks architecture

A network with the aforementioned structure implements an agent for each situation described in Sect. 4.1. The loss is computed separately for each network, and corresponds to the A3C loss computed in [9].

4.5 Hyper-parameters Tuning

Each episode lasts at most 500 steps/actions, and it may end either achieving success (i.e. descending the stairs), or reaching the steps limit. Thus, the death state is impossible for the agent, since in our experiments monsters and traps have been disabled and 500 steps are not enough to die for starvation.

Most of the remaining hyper-parameters values we adopted (for example the entropy $\beta = 0.001$) came from [14], an Open-Source implementation of [9], except the following:

discount factor γ	0.95
batch size t_{max}	60

We employed the same Tensorflow's RMSprop optimizer [1] available in [14], with parameters:

decay	0.99
momentum	0
epsilon	0.1
clip norm	40

The learning rate is annealed over time according to the following equation: $\alpha = \eta \cdot \frac{T_{max} - T}{T_{max}}$, where T_{max} is the maximum global step, and T is the current global step.

The initial learning rate is approximatively $\eta = 0.0007$.

5 Evaluation

For evaluation purposes we want to measure how often the agent is able to descend the stairs and to explore the floor. In our experiments, the final state is reached when the agent descend the stairs. For this reason, a good evaluation metric for a Rogue-like exploration-only system should be based at least on:

- the success rate: the percentage of episodes in which the final state is reached (an equivalent of the accuracy)
- the number of new tiles found during the exploration process
- the number of steps taken to win an episode

We evaluated our systems using an average of the aforementioned metrics over 200 episodes. The results we achieved are summarized in Fig. 4 and Table 2.[3]

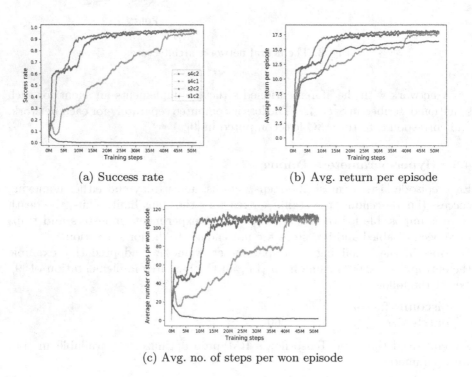

(a) Success rate   (b) Avg. return per episode

(c) Avg. no. of steps per won episode

Fig. 4. Results comparison. In the legend the labels sX denote the use of X situations, while cY a state representation with Y channels. Please see Sects. 4.1 and 4.2 for details.

Our best agent[4] shows remarkable skills in exploring the dungeon, searching for the stairs.

Using four situations instead of just two did not prove to be beneficial, however adopting a separate *situation* (and hence a separate neural network) for the case when the stairs are visible was fundamental. In fact, as can be seen in Fig. 4, the policy learned by *s1-c2* completely ignored the stairs, thus achieving a very low success rate.

The experiment with 4 situations resulted in the development of the peculiar inclination for the agent of walking alongside walls.

---

[3] Source code and weights are publicly available at [2].

[4] A video of our agent playing is available at https://youtu.be/1j6_165Q46w.

Table 2. Learned policies evaluation. With sX we denote the use of X situations and with cY a state representation with Y channels. Please see Sects. 4.1 and 4.2 for details.

Agent	s1-c2	s2-c2	s4-c1	s4-c2
Success rate	0.03%	98%	96.5%	97.6%
Avg return	16.16	17.97	17.66	17.99
Avg number of seen tiles	655.02	386.46	365.88	389.27
Avg number of steps to succeed	2.11	111.48	108.22	110.26

Finally, state representation $c2$ induced faster learning, but only a slight increase in the resulting success rate.

6   Refactoring the *Rogue in a Box* library

With this article, we release a new version [2] of the *Rogue In A Box* library [3,4] that improves modularity, efficiency and usability with respect to the previous version. In particular, the old library was mainly centered around DQN-agents, that at the time looked as the most promising approach for the application of deep reinforcement learning to this kind of games. With the advent of A3C and other techniques, we restructured the learning environment, neatly decoupling the interface with the game, supported by a suitable API, from the design of the agents.

Other innovative features comprise:

1. Screen parser and frames memory
2. Communication between Rogue and the library
3. Enabling or disabling monsters and traps
4. Evaluation module

Of particular note is the evaluation module, which provides statistics on the history of environment interactions, allowing to properly compare the policies of different agents.

7   Conclusions

In this article, we have shown how we can address the Partially Observable Markov Decision Problem behind the exploration of Rogue's dungeons, achieving a success rate of 98%, with a simple technique that partitions the sample space into situations. Each situation is handled by a different A3C agent, all of them sharing a common value function. The interest of Rogue is that the planar, ASCII-based, bi-dimensional interface permits to decouple vision from more intelligent activities such as planning: in this way we may better investigate and understand the most challenging aspects of the player's behavior.

The current version of the agent works very well, but still has some problems in cul-de-sac situations, where the agent should trace-back his path. Moreover, to completely solve the Rogue's exploration problem, *dark rooms* and *hidden doors* are also required to be handled. We predict that the main challenge is going to be provided by *hidden doors*, since they are almost completely unpredictable and hard to detect even for a human. Different aspects of the game, such as collecting objects and fighting could also be taken into account, possibly delegating them to ad-hoc situations.

In spite of the fact that the overall performance of our agent is really good, its design is not yet entirely satisfactory. In fact, too much intelligence about the game is built in, both in the design of situations, and especially in their identification and attribution to specific networks. Also the rogue-centered, cropped view introduces a major simplification of the problem, completely by-passing the *attention* problem (see e.g. [10]) that, as discussed in [3], was one of the interesting aspects of Rogue.

Currently, our efforts are going in the direction of designing an unsupervised version of the work described in this paper, where the agent is able to autonomously detect interesting situations, delegating them to specific subnets. As additional research topics, we are

- exploring the role of sample-efficiency in our context, along the lines of [22].
- looking Multi-Task Adaptive Networks, following the ideas in [18].

References

1. RMSPropOptimizer. https://www.tensorflow.org/api_docs/python/tf/train/RMS PropOptimizer
2. Asperti, A., Cortesi, D., Sovrano, F.: Partitioned A3C for rogueinabox. https://github.com/Francesco-Sovrano/Partitioned-A3C-for-RogueInABox
3. Asperti, A., Pieri, C.D., Maldini, M., Pedrini, G., Sovrano, F.: A modular deep-learning environment for rogue. WSEAS Trans. Syst. Control **12**, 362–373 (2017). http://www.wseas.org/multimedia/journals/control/2017/a785903-070.php
4. Asperti, A., Pieri, C.D., Pedrini, G.: Rogueinabox: an environment for rogue like learning. Int. J. Comput. **2**, 146–154 (2017). http://www.iaras.org/iaras/filedownloads/ijc/2017/006-0022(2017).pdf
5. Bellemare, M.G., Naddaf, Y., Veness, J., Bowling, M.: The arcade learning environment: an evaluation platform for general agents. J. Artif. Intell. Res. (JAIR) **47**, 253–279 (2013). https://doi.org/10.1613/jair.3912
6. Cerny, V., Dechterenko, F.: Rogue-like games as a playground for artificial intelligence – evolutionary approach. In: Chorianopoulos, K., Divitini, M., Hauge, J.B., Jaccheri, L., Malaka, R. (eds.) ICEC 2015. LNCS, vol. 9353, pp. 261–271. Springer, Cham (2015). https://doi.org/10.1007/978-3-319-24589-8_20
7. Dilokthanakul, N., Kaplanis, C., Pawlowski, N., Shanahan, M.: Feature control as intrinsic motivation for hierarchical reinforcement learning. CoRR abs/1705.06769 (2017). http://arxiv.org/abs/1705.06769
8. van Hasselt, H., Guez, A., Silver, D.: Deep reinforcement learning with double Q-learning. CoRR abs/1509.06461 (2015). http://arxiv.org/abs/1509.06461

9. Jaderberg, M., et al.: Reinforcement learning with unsupervised auxiliary tasks. CoRR abs/1611.05397 (2016). http://arxiv.org/abs/1611.05397

10. Jaderberg, M., Simonyan, K., Zisserman, A., Kavukcuoglu, K.: Spatial transformer networks. In: Cortes, C., Lawrence, N.D., Lee, D.D., Sugiyama, M., Garnett, R. (eds.) Advances in Neural Information Processing Systems: Annual Conference on Neural Information Processing Systems 2015, Montreal, Quebec, Canada, 7–12 December 2015, vol. 28, pp. 2017–2025 (2015). http://papers.nips.cc/paper/5854-spatial-transformer-networks

11. Kempka, M., Wydmuch, M., Runc, G., Toczek, J., Jaskowski, W.: ViZDoom: a doom-based AI research platform for visual reinforcement learning. CoRR abs/1605.02097 (2016). http://arxiv.org/abs/1605.02097

12. Klyubin, A.S., Polani, D., Nehaniv, C.L.: Empowerment: a universal agent-centric measure of control. In: Proceedings of the IEEE Congress on Evolutionary Computation, CEC 2005, Edinburgh, UK, 2–4 September 2005, pp. 128–135 (2005). https://doi.org/10.1109/CEC.2005.1554676

13. Kulkarni, T.D., Narasimhan, K., Saeedi, A., Tenenbaum, J.B.: Hierarchical deep reinforcement learning: integrating temporal abstraction and intrinsic motivation. CoRR abs/1604.06057 (2016). http://arxiv.org/abs/1604.06057

14. Miyoshi, K.: Unreal implementation. https://github.com/miyosuda/unreal

15. Mnih, V., et al.: Asynchronous methods for deep reinforcement learning. CoRR abs/1602.01783 (2016). http://arxiv.org/abs/1602.01783

16. Mnih, V., et al.: Human-level control through deep reinforcement learning. Nature **518**(7540), 529–533 (2015). https://doi.org/10.1038/nature14236

17. Singh, S.P., Barto, A.G., Chentanez, N.: Intrinsically motivated reinforcement learning. In: Advances in Neural Information Processing Systems: Neural Information Processing Systems, NIPS 2004, Vancouver, British Columbia, Canada, 13–18 December 2004, vol. 17, pp. 1281–1288 (2004). http://papers.nips.cc/paper/2552-intrinsically-motivated-reinforcement-learning

18. Song, Y., Xu, M., Zhang, S., Huo, L.: Generalization tower network: a novel deep neural network architecture for multi-task learning. CoRR abs/1710.10036 (2017). http://arxiv.org/abs/1710.10036

19. Sun, R., Peterson, T.: Multi-agent reinforcement learning: weighting and partitioning. Neural Netw. **12**(4–5), 727–753 (1999). https://doi.org/10.1016/S0893-6080(99)00024-6

20. Sutton, R.S., Barto, A.G.: Introduction to Reinforcement Learning, 1st edn. MIT Press, Cambridge (1998)

21. Vezhnevets, A.S., et al.: Feudal networks for hierarchical reinforcement learning. CoRR abs/1703.01161 (2017). http://arxiv.org/abs/1703.01161

22. Wang, Z.: Sample efficient actor-critic with experience replay (2016)

Decision of Neural Networks Hyperparameters with a Population-Based Algorithm

Yağız Nalçakan[1]([⊠]) and Tolga Ensari[2]

[1] Altınbaş University, Istanbul, Turkey
yagiz.nalcakan@altinbas.edu.tr
[2] Istanbul University, Istanbul, Turkey
ensari@istanbul.edu.tr

Abstract. This paper proposes a method named Population-based Algorithm (PBA) to decide the best hyperparameters for a neural network (NN). The study focuses on which type of hyperparameters achieve better results in neural network problems. Population-based algorithm inspired from evolutionary algorithms and uses basic steps of genetic algorithms. The distinctive feature of our algorithm from genetic algorithms is fitness evaluation of individuals. To test our approach, we implemented our algorithm to a handwritten digits recognition problem to find the best hyperparameters for a simple neural network and we reached 98.66 accuracy score. Finally, we conclude, how PBA used in neural networks for the best way.

Keywords: Neural networks · Population-based Algorithm ·
Hyperparameter optimization · Character recognition

1 Introduction

Since the beginning of humankind, nature has always been a source of inspiration for people to solve their problems. In computer science, we call that problem-solving algorithms Nature-inspired Algorithms. Population-based algorithm (PBA) is one of these algorithms and PBA is very similar to genetic algorithm (GA). To understand how PBA works, understanding of GA will be needed [1]. GAs are optimization methods that inspired by the genetic processes of biological organisms. Simulating this natural process on computers result in genetic algorithms that can often out-perform classical optimization methods when applied to difficult real-world problems in engineering. Their evolutionary process can be applied to problems where heuristic solutions generally lead to unsatisfactory results. In addition, neural networks (NN), which are the most popular artificial intelligence algorithms of recent times, are methods that are based on working principle of neurons in brain. Simple neural networks consist of input, output and hidden layers. Although more complex neural networks can solve more complex problems, simple neural networks are also very successful in solving many of today's problems.

The variation of the methods in the different layers of neural networks depends on the problem, and the success rate of NN directly affected by selection of these methods. In our work, we applied a PBA to decide the best hyperparameters for a NN.

© Springer Nature Switzerland AG 2019
G. Nicosia et al. (Eds.): LOD 2018, LNCS 11331, pp. 276–281, 2019.
https://doi.org/10.1007/978-3-030-13709-0_23

This paper consists of 6 sections: Sect. 2 explains neural networks and their parameters and hyperparameters, Sect. 3 brief description of other hyperparameter selection methods, Sect. 4 about an explanation of Population-based Algorithm (PBA), Sect. 5 has experimental results of PBA in a character recognition problem, and Sect. 6 concludes the paper and references for this paper.

2 Neural Networks and Hyperparameters'

Neural Networks (NN) are an attempt to mimic the way biological brain works. NN can work on complex non-linear problems without understanding the underlying relation completely, this is possible due to the large number of neurons connected with individual reconfigurable weights. It is analogous to the parallel computing system comprising of huge number simple processors. NN is being used in almost every aspect of computer science problems with complexity where input data is fed to the input nodes of network and output is taken after numerous hidden layers. Due to self-learning ability of neural networks the optimum weight assignment, selection of the correct number of hidden layers and the correct activation function leads to satisfactory results in problem and that's why it is important to decide useful and effective parameters and hyperparameters for that NN.

Neural networks have many types of parameters and hyperparameters. Parameters are weights(W) and bias(b); and learning rate, number of hidden layers, number of neurons in hidden layers, activation function, optimizer, iteration count, etc. are hyperparameters. Every parameter and hyperparameter will affect the result of NN [2–4].

Different kind of activation functions and optimizers using for different problem types. ReLu, eLu, tanh, sigmoid, etc. are kinds of activation functions and the optimizer can be select as RMSProp, AdaDelta, AdaGrad, etc. These number of selections gives us many possibilities that we can't decide which one is the best for the problem. With no explicit formula for choosing a correct set of hyperparameters, their selection often depends on a combination of previous experience and trial-and-error. Therefore, we create a population-based algorithm (PBA) to adjust four of that hyperparameters to accomplish better results.

3 Hyperparameter Selection Methods and Related Work

There are many different types of hyperparameter selection methods in the literature. Some of the most widely used methods for hyperparameter selection in neural networks are manual search, grid search and random search [5]. Manual search refers to the process of a researcher manually selecting hyperparameter sets. This is often chosen for its ease and can reach a reasonable solution quickly. However, it is difficult to reuse these sets in a new dataset and find an expert on that specific data. Grid search allows for reproducible results, but it is not efficient in searching a high dimensional hyperparameter space. However, it is widely used because of its easy to implement architecture but, grid search wastes resources exploring what could be unimportant dimensions of the hyperparameter space while holding other values constant, some of which could be more important. Thus, random search is much more effective at searching the hyperparameter space since it avoids this problem of under-sampling important dimensions, all while being just as easy to implement and just easy as to parallelize as grid search [5]. Random search suffers from being non-adaptive and can be outperformed with combination of manual and grid search.

On the other hand, Evolutionary algorithms provide opportunity to both search hyperparameter space randomly also utilize previous results to direct the search [5]. They are frequently used in the hyperparameter optimization field currently. Especially in neural networks.

After the rise of deep learning (DL), neural networks started to be more complex. As the complexity of neural networks increases, the hyperparameters that need to be determined increase also. To deal with this optimization problem, Evolutionary Algorithms like Genetic Algorithm (GA) widely used in the literature [10, 11]. In [10], Suganuma et al., attempted to automatically construct Convolutional Neural Networks (CNN) architectures for an image classification task with Cartesian genetic programming (CGP). The CNN architecture is trained on a learning task and assigned the validation accuracy of the trained model as the fitness. The evolutionary algorithm that they created, searches for the better CNN architectures. To test their approach, they applied their approach to ResNet and VGG, and their CGP-ResNet achieved 23.47 error rate and CGP-VGG 23.48. Also, in [11], they used an evolutionary approach to decide which hyperparameters of the architecture are getting the better result. Their algorithm, EDEN, creates an initial population with different simple neural networks with random hyperparameters then classic GA steps applied to the population. But in mutation step, to create diversity, randomly selected hidden layers randomly changed or deleted. They achieved 98.4 accuracy score in MNIST dataset.

4   Population-Based Algorithm

The effort to find fittest individuals in a population always led that population to create more powerful, more healthy individuals. This process of evolution has also inspired the computer science community. Evolutionary algorithms come from this inspiration. Also, our Population-based Algorithm approach uses this evolution principle.

In our approach, we created a population of individuals that have hyperparameters as their genes. Every individual has four hyperparameter types in their allele. Number of hidden layers, number of neurons in hidden layers, activation function and optimizer selected as hyperparameters for PBA. Selected hyperparameter types and their possible contents are given in Table 1.

Since the Population-based Algorithm is inspired by evolutionary algorithms, we used the selection, crossover, mutation steps of genetic algorithm (GA). Each individual was defined to include four hyperparameters in Table 1. To create diversity, all hyperparameters of all individuals in the population were randomly selected. This randomization process done by brute force algorithm. Brute force algorithm creates every possible individual from these hyperparameter types. After all possible individuals created twenty individuals selected as initial population.

Table 1. Selected hyperparameter types for individuals.

Number of hidden layers	Number of neurons in hidden layers	Activation function	Optimizer
1	64	ReLu	RMSProp
2	128	eLu	Adam
3	256	Sigmoid	SGD
4	512	Tanh	AdaGrad
	768		AdaDelta
	1024		AdaMax
			NAdam

Main difference of PBA from genetic algorithms [7–9] calculation of the fittest. Fitness calculation done for all hyperparameters separately. Every individuals' accuracy score assigned as each hyperparameter type's fitness score and then two individuals selected from the hyperparameters having the best average fitness score. The crossover and mutation steps in GA work the same in the PBA. We used single point crossover and bit-flip mutation.

Bit-flip mutation done slightly different from normal one. One of the hyperparameter selected randomly and randomly changed to another type of that hyperparameter. Block diagram of PBA can be seen on Fig. 1.

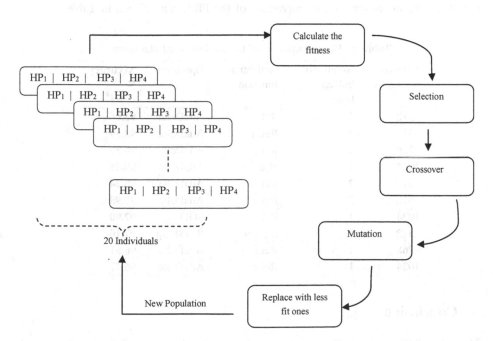

Fig. 1. The block diagram of population-based algorithm. HP_α represent the hyperparameter types as genes of the individual.

These types of algorithms need a stop criterion. In PBA, to reach the tenth population selected as stopping criterion due to long computation time. It can be thought that this can prevent the achievement of the desired success rate, but also a neural network that has been optimized with PBA has very good results in the handwritten digits recognition problem which is selected as our testing problem.

5   Experimental Results of PBA Optimized NN

We simulated these ideas in a project. In this project, we used our PBA to find the best hyperparameters for a simple neural network that build for recognition of handwritten digits with MNIST handwritten digits database [6]. MNIST dataset has 50000 train and 10000 test images in 28 × 28 pixels format. PBA used for optimizing four hyperparameter types and every optimized NN individual gets better and better in every population. In this NN, output layer defined with Softmax activation function and applied 0.2 dropout to the hidden layers. To avoid looping over same hyperparameters and to tune our epochs we used a callback called *early stopping*.

Early stopping is basically stopping the training when your loss starts to increase, or in other words validation accuracy starts to decrease. This approach is one of the best stop criterions when optimization of NN done by a genetic algorithm like optimization.

Each PBA optimized NN trained an average of 22.8 min. Best NN reached 98.66% accuracy with 512 neurons per layers, 2 hidden layers, with ReLu activation function and AdaDelta optimizer. Other generations of the PBA can be seen in Table 2.

Table 2. Best generations of population-based algorithm.

Neurons per layers	Number of hidden layers	Activation function	Optimizer	Accuracy %
512	2	ReLu	AdaDelta	**98.66**
512	2	ReLu	AdaDelta	**98.59**
768	2	ReLu	AdaDelta	**98.55**
1024	1	eLu	AdaDelta	**98.45**
512	2	ReLu	AdaDelta	**98.44**
512	2	ReLu	AdaDelta	**97.98**
1024	1	ReLu	SGD	**97.90**
512	1	Sigmoid	RMSProp	**97.10**
768	1	ReLu	AdaDelta	**96.81**
1024	1	ReLu	AdaDelta	**96.77**

6   Conclusion

The goal of this paper is to propose a method to decide hyperparameters of a neural network. This method named Population-based Algorithm which inspired by evolutionary algorithms. We implemented basic genetic algorithm steps to an algorithm with a different fitness evaluation. Each hyperparameter type evaluated separately and the best ones selected.

To test our approach MNIST handwritten digits database used. PBA used to optimize, activation functions (Sigmoid, tanh, Softmax, ReLU), optimizers (RMSProp, Adam, SGD, AdaGrad, AdaDelta, AdaMax), hidden layer count and neuron count in hidden layers. The best optimized NN's hyperparameters are ReLu activation function, AdaDelta optimizer with two hidden layers and 512 neurons per layer which reached the accuracy score of 98.66.

PBA can be used in many types of NN like deep neural networks. As a future work, we intend to use our PBA for some Deep Learning problems.

References

1. Fiszelew, A., et al.: Finding optimal neural network architecture using genetic algorithms. Adv. Comput. Sci. Eng. Res. Comput. Sci. **27**, 15–24 (2007)
2. Bergstra, J.S., et al.: Algorithms for hyper-parameter optimization. In: Advances in Neural Information Processing Systems (2011)
3. Vose, A., et al.: Deep neural network hyperparameter optimization with genetic algorithms (2017)
4. Fridrich, M.: Hyperparameter optimization of artificial neural network in customer churn prediction using genetic algorithm. Trendy Ekon. Manag. **11**(28), 9–21 (2017)
5. Bergstra, J., Bengio, Y.: Random search for hyperparameter optimization. J. Mach. Learn. Res. **13**(1), 281–305 (2012)
6. LeCun, Y., Cortes, C., Burges, C.J.C.: The MNIST handwritten digits database. http://yann. lecun.com/exdb/mnist
7. Patil, S.S, Bhalchandra, A.S: pattern recognition using genetic algorithm (2017)
8. Valdez, F., Melin, P., Parra, H.: Parallel genetic algorithms for optimization of modular neural networks in pattern recognition (2011)
9. Jaderberg, M., et al.: Population based training of neural networks. arXiv:1711.09846 (2017)
10. Suganuma, M., Shirakawa, S., Nagao, T.: A genetic programming approach to designing convolutional neural network architectures. In: Proceedings of the Genetic and Evolutionary Computation Conference. ACM (2017)
11. Dufourq, E., Bassett, B.A.: EDEN: evolutionary deep networks for efficient machine learning. In: Pattern Recognition Association of South Africa and Robotics and Mechatronics (PRASA-RobMech, 2017). IEEE (2017)

Strong Duality of the Kantorovich-Rubinstein Mass Transshipment Problem in Metric Spaces

José Rigoberto Gabriel-Argüelles[1], Martha Lorena Avendaño-Garrido[1]([✉]) [iD],
Luis Antonio Montero[1], and Juan González-Hernández[2]

[1] Facultad de Matemáticas, Universidad Veracruzana, Xalapa, Veracruz, Mexico
maravendano@uv.mx
[2] Departamento de Probabilidad y Estadística,
Instituto de Investigaciones en Matemáticas Aplicadas y Sistemas,
Universidad Nacional Autónoma de México, Mexico City, Mexico

Abstract. This paper studies the Kantorovich-Rubinstein mass transshipment (KR) problem on metric spaces and with an unbounded cost function. Some assumptions are given under which the strong duality condition holds; that is, the KR problem and its dual are both solvable and their optimal values coincide.

Keywords: Strong duality · Kantorovich-Rubinstein problem · Infinite linear program

1 Introduction

The mass transshipment problem or Kantorovich-Rubinstein (KR) problem (see [11,14,17,18]) is an optimization problem in spaces of measures. It is well known that the KR problem has relation with the mass transfer problem ([17, Chaps. 4 and 6]). In both problems it is required to transfer an initial mass ν_1 to a final mass ν_2 with minimal cost, However, while the mass transfer problem is a generalization of the transportation problem, the KR problem is a generalization of the network flow problem, this is, in the first one the movement is made in "one trip" and in the second one this movement can be made in several "shorter trips" ([17, Chap. 5]). If the cost function c is a metric or with similar properties to a metric, the equivalency of both problems has been proved in [12] and [17, Chap. 4], but if the cost function c is not a metric, as in this paper, then the problems can be quite different.

Regarding duality, relevant works on this topic are as following: Rachev studied the duality of a class of mass transshipment problems on $\mathbb{R}^n$, with the Euclidian distance as the cost function, and he worked with differential distributions [14]. Hanin and Rachev also studied the KR problem on $\mathbb{R}^n$; they found a dual representation for it and defined a class of ideal metrics, the cost function being quite similar to a distance [6]. Moreover, they worked with a generalization of

© Springer Nature Switzerland AG 2019
G. Nicosia et al. (Eds.): LOD 2018, LNCS 11331, pp. 282–292, 2019.
https://doi.org/10.1007/978-3-030-13709-0_24

this problem, where the feasible measures satisfied a condition of moments [5]. Dedecker, Prieur and De Fitte worked with a completely regular pre-Radon space and gave an hypothesis for getting that the absence of duality gap condition holds, that is, the primal and dual values of the mass transshipment problem coincide, this condition is also known as the Kantorovich-Rubinstein's Theorem or the duality condition. However, the optimal measure only can be found if we already know the optimal solution of the dual problem [2]. An extension to previous work is given in [3]. Also Rachev and Shortt proved a duality theorem for a cost function similar to a metric in one separable metric space [18].

Among the applications of the KR problem we can mention: applications of Fortet-Mourier metrics for studying the quantitative stability of two-stage models [8], probability metrics [15], control of cancer radiotherapy [7], image registration and warping [4], limit theorems and recursive stochastic equations [17].

In this paper we shall study the KR problem by using infinite linear programming under standard conditions. The main contributions are: the solvability of primal and dual problems and we prove the no duality gap condition.

The strong duality is proved in cases when X is a general metric space and c is a moment or inf-compact function and when X is a σ-compact space and c is a lower semicontinuous cost function.

Unlike this article, most of the KR problem literature assumes that the cost function is a metric (or similar to a metric); therefore the solvability and no duality gap follows from mass transfer problem, which is already known, e.g., for Polish spaces [9,17].

When c is not metric, the strong duality is very important since it includes Kantorovich-Rubinstein's Theorem. Moreover there are some researches on the mass transfer problem with quadratic costs; therefore it is quite natural to study the KR problem for quadratic costs or even for more general cost functions (see [13]).

There are just a few works on the KR problem, because when the cost function is a metric (or quite similar to) then the KR problem and MT problem are equivalent, in the sense that their optimal values and solutions are the same. However, when the cost function is not a metric the results for the KR problem are not consequences of results for the MT problem. Even more, the demonstrations are quite different, as we show in this paper.

The remainder of the paper is organized as follows. In Sect. 2, we present the KR problem. In Sect. 3, we prove the theorem of solvability of the KR problem. In Sect. 4, we prove that the strong duality condition holds for the KR problem. Section 5 are conclusions and future work and finally, Sect. 6 is an appendix.

Throughout this article we use following notation: If Z is a metric space, then $\mathbb{B}(Z)$ is its Borel σ-algebra, $M(Z)$ is the set of all signed finite measures on the measurable space $(Z, \mathbb{B}(Z))$, $M^+(Z)$ is the set of all nonnegative finite measures and $P(Z)$ is the set of all probability measures.

2   The KR Problem

In the Kantorovich-Rubinstein mass transshipment problem that we are concerned with we are given the following data:

(i) A metric space X, endowed with the corresponding Borel σ-algebra.
(ii) A nonnegative measurable function $c : X \times X \to \mathbb{R}$.
(iii) Two probability measures (p.m.) ν_1, ν_2 in $P(X)$.

For any measure μ in $M(X \times X)$, we denote by $\Pi_1 \mu$ and $\Pi_2 \mu$ the marginals (or projections) of μ on X, that is, for all $A, B \in \mathbb{B}(X)$

$$\Pi_1 \mu(A) := \mu(A \times X), \quad \text{and} \quad \Pi_2 \mu(B) := \mu(X \times B).$$

Then, with $\langle \mu, c \rangle := \int c d\mu$, the KR problem can be stated as follows:

KR   minimize: $\langle \mu, c \rangle$
          subject to: $\Pi_1 \mu - \Pi_2 \mu = \nu_1 - \nu_2, \quad \mu \in M^+(X \times X).$ (1)

A measure μ on $M^+(X \times X)$ is a feasible solution for the KR problem if it satisfies (1) and $\int c d\mu$ is finite. The KR problem is called consistent if the set of feasible solutions is nonempty, in which case its optimum value is defined as

$$\inf(\text{KR}) := \inf\{\langle \mu, c \rangle \mid \mu \in \mathcal{F}\}, \tag{2}$$

where $\mathcal{F}$ denotes the class of all feasible solutions for the KR problem. It is said that the KR problem is solvable if there is a feasibe solution μ^* that attains its optimum value. In this case, μ^* is called an optimal solution for the KR problem, and the value $\inf(\text{KR})$ is written as $\min(\text{KR}) = \langle \mu^*, c \rangle$.

3   Solvability of the KR Problem

In this article we work on the space $M_r^+(X \times X)$ that denotes the family of nonnegative measures such that $\mu(X \times X) \leq r$ for some $r > 0$. To prove the solvability of the KR problem, we need either one of the following assumptions.

Assumption 1

(a) *The KR problem is consistent.*
(b) *The cost function $c(x, y)$ is inf-compact or is a moment.*

Assumption 2

(a) *The KR problem is consistent.*
(b) *X is a metric σ-compact space.*
(c) *The "cost" function $c(x, y)$ is lower semi-continuous (l.s.c.).*

Theorem 3. *If either Assumption 1 or Assumption 2 holds, then KR is solvable.*

Proof. By Assumption 1 (a) or 2 (a), we have that there exists μ_0 in $\mathcal{F}$ such that $0 \leq \inf(\mathrm{KR}) \leq \langle \mu_0, c \rangle < \infty$; therefore there exists $\{\mu_n\}$ in $\mathcal{F}$, a minimizing sequence, such that $\langle \mu_n, c \rangle \downarrow \inf(\mathrm{KR})$.

Given $\varepsilon > 0$, there exists a positive integer N such that

$$\inf(\mathrm{KR}) \leq \langle \mu_n, c \rangle \leq \inf(\mathrm{KR}) + \varepsilon \quad \forall n \geq N.$$

Finally, by Assumption 1 (b) or Assumption 2 (b)–(c) and by taking $b := \inf(\mathrm{KR}) + \varepsilon$ in Proposition 1 (see Appendix), we have that the measure set

$$\Gamma := \{\mu_n\}_{n \geq N}$$

is tight; then by extended Prohorov Theorem, Γ is relatively compact and so there are a subsequence $\{\mu_m\}$ in Γ and a measure μ^* in $M_r^+(X \times X)$, such that $\{\mu_m\}$ weakly converges to μ^*.

By Lemma in [9] we have that

$$\langle \mu^*, c \rangle = \inf(\mathrm{KR}).$$

Finally to prove that μ^* is an optimal solution for the KR problem, it suffices to show that μ^* is a feasible solution, that is, μ^* satisfies the equality (1). To do this, we shall prove that the marginals $\Pi_i \mu_m$, for $i = 1, 2$, converge weakly to the marginals $\Pi_i \mu^*$. Take $i = 1$, by Portmanteau's Theorem it suffices to show that

$$\liminf_{m \to \infty} \Pi_1 \mu_m(G) \geq \Pi_1 \mu(G),$$

for any open $G \subset X$. Let G be an open subset of X, then $G \times X$ is open in $X \times X$ and so, by weak convergence,

$$\liminf_{m \to \infty} \Pi_1 \mu_m(G) = \liminf_{m \to \infty} \mu_m(G \times X) \geq \mu(G \times X) = \Pi_1 \mu(G).$$

Finally we have that $\nu_1 - \nu_2 = \Pi_1 \mu_n - \Pi_2 \mu_n$, for all n in $\mathbb{N}$ then

$$\nu_1 - \nu_2 = \lim_{n \to \infty} (\Pi_1 \mu_n - \Pi_2 \mu_n) = \Pi_1 \mu^* - \Pi_2 \mu^*.$$

This shows that μ^* is in $\mathcal{F}$ and, therefore, μ^* is an optimal solution for the KR problem. $\qquad\square$

4   Strong Duality of KR Problem

In this section we reformulate the KR problem as an infinite dimensional linear program. We use a similar approach to that of Anderson and Nash [1], but for cost functions c possibly unbounded and general metric spaces. With this in mind, we introduce suitable linear spaces with weighted norms.

Definition 1. Let $w(x,y) := 1 + c(x,y)$. For each μ in $M(X \times X)$ and f a function on $X \times X$, we define

$$\|\mu\|_w := \int_{X \times X} w \, d|\mu| \text{ and } \|f\|_w := \sup_{(x,y)} \frac{|f(x,y)|}{w(x,y)}, \tag{3}$$

where $|\mu|$ denotes the total variation of μ, $M^w(X \times X)$ denotes the (normed) linear space of finite signed measures μ on $X \times X$ such that $\|\mu\|_w < \infty$ and $F^w(X \times X)$ stands for the (normed) linear space of measurable function f on $X \times X$ such that $\|f\|_w < \infty$.

Observe that the cost function c belongs to $F^w(X \times X)$. Furthermore, the pair $(M^w(X \times X), F^w(X \times X))$ is a dual pair of vector spaces with respect to the bilinear form

$$\langle \mu, f \rangle := \int_{X \times X} f \, d\mu, \text{ for } \mu \in M^w(X \times X) \text{ and } f \in F^w(X \times X). \tag{4}$$

Let $M^{w+}(X \times X) := \{\mu \in M^w(X \times X) \mid \mu \geq 0\}$ the cone of nonnegative measures in $M^w(X \times X)$. We observe that as $w(x,y) \geq 1$, then $\mu(X \times X) < \infty$ for all μ in $M^{w+}(X \times X)$. Now let us consider the weight functions

$$w_1(x) := \inf_{y \in X} w(x,y) \text{ and } w_2(y) := \inf_{x \in X} w(x,y).$$

Assumption 4. *The functions w_1 and w_2 are measurable.*

We define $\overline{w}(x) := \min\{w_1(x), w_2(x)\}$, then $\overline{w}$ is a measurable function on X.

We define the dual pair $(M^{\overline{w}}(X), F^{\overline{w}}(X))$, where $M^{\overline{w}}(X)$ is the linear space of finite signed measures η on X such that

$$\|\eta\|_{\overline{w}} := \int_X \overline{w} \, d|\eta| < \infty,$$

and $F^{\overline{w}}(X)$ is the linear space of measurable functions $f : X \to \mathbb{R}$, such that

$$\|f\|_{\overline{w}} := \sup_x \frac{|f(x)|}{\overline{w}(x)} < \infty,$$

the corresponding bilinear form is $\langle \eta, f \rangle := \int_X f \, d\eta$.

Assumption 5. *The measure $\hat{\mu} := \nu_1 \times \nu_2$ is in $M^w(X \times X)$.*

Hence, we have.

Remark 1. The marginal measures ν_1, ν_2 are in $M^{\overline{w}}(X)$.

Remark 2. (See [10]) For any μ in $M^{w+}(X \times X)$ there exists a stochastic kernel ϕ, such that

$$\int_{X \times X} f(x)\mu(d(x,y)) = \int_X \int_X f(x)\phi(dy|x)(\Pi_1\mu)(d(x)) = \int_X f(x)(\Pi_1\mu)(d(x))$$

and similarly

$$\int_{X \times X} f(y)\mu(d(x,y)) = \int_X f(y)(\Pi_2\mu)(d(y)).$$

Hence,

$$\int_{X \times X} (f(x) - f(y))\mu(d(x,y)) = \int_X f(x)(\Pi_1\mu)(d(x)) - \int_X f(y)(\Pi_2\mu)(d(y)).$$

By the last remark and under Assumptions 4 and 5, we can now rewrite the KR problem as a linear program in the following way

$$\text{KR} \quad \text{minimize} \quad \langle \mu, c \rangle \tag{5}$$

$$\text{subject to:} \quad A\mu = \nu_1 - \nu_2, \quad \mu \in M^{w+}(X \times X), \tag{6}$$

where $A : M^w(X \times X) \to M^{\overline{w}}(X)$ is the linear map given by

$$A\mu := \Pi_1\mu - \Pi_2\mu \tag{7}$$

and the adjoint is given by

$$A^*(f)(x,y) = f(x) - f(y). \tag{8}$$

Moreover $A^*(F^{\overline{w}}(X)) \subset F^w(X \times X)$, in fact, let $f \in F^w(X)$, by definition of $\overline{w}$, we have that

$$\| A^*(f) \|_w = \sup_{(x,y)} \frac{|f(x) - f(y)|}{w(x,y)}$$

$$\leq \sup_{(x,y)} \frac{|f(x)|}{w(x,y)} + \sup_{(x,y)} \frac{|f(y)|}{w(x,y)}$$

$$\leq \sup_x \frac{|f(x)|}{\overline{w}(x)} + \sup_y \frac{|f(y)|}{\overline{w}(y)} = 2 \| f \|_{\overline{w}} < \infty.$$

This implies, in particular, that the linear map A is weakly continuous with respect to the weak topologies $\sigma(M^w(X \times X), F^w(X \times X))$ and $\sigma(M^{\overline{w}}(X), F^{\overline{w}}(X))$ on $M^w(X \times X)$ and $M^{\overline{w}}(X)$ respectively (see [1], Proposition 4 p. 37).

Then, we have that the dual of KR problem is

$$\textbf{KR}^* \quad \text{maximize} \quad \langle \nu_1 - \nu_2, f \rangle \tag{9}$$

$$\text{subject to:} \quad A^*(f) \leq c, \quad f \in F^{\overline{w}}(X). \tag{10}$$

As c is nonnegative, the dual KR* is consistent because $f = 0$ satisfies (10). Therefore, we can define the value of the dual problem as

$$\sup(\text{KR}^*) = \sup\{\langle \nu_1 - \nu_2, f \rangle \mid f \, satisfies \, (10)\}$$

and it fulfills the weak duality condition

$$\sup(KR^*) \leq \inf(\text{KR}).$$

We now wish to show that KR* is solvable, in which case we write max(KR*) in lieu of sup(KR*). Moreover we wish to prove

$$\sup(\text{KR}^*) = \inf(\text{KR}). \tag{11}$$

We need the following Assumption.

Assumption 6. *There exists a maximizing sequence $\{f_n\}$ for KR*, which is bounded in $F^{\overline{w}}(X)$.*

Theorem 7. *Under Assumption 1 (a) or (b) of Sect. 3 and Assumptions 4, 5 and 6.*

(a) The KR* problem is solvable, and
(b) there is no duality gap for the KR problem.

Proof.(a) Let $\{f_n\}$ be as in Assumption 6, that is, the functions f_n are in $F^{\overline{w}}(X)$ for all n in $\mathbb{N}$, and

$$f_n(x) - f_n(y) \leq c(x,y) \, \forall n \in \mathbb{N}, \tag{12}$$

$$\langle \nu_1 - \nu_2, f_n \rangle \uparrow \sup \text{KR}^* \tag{13}$$

and there is a constant m, such that

$$\|f_n\|_{\overline{w}} \leq m, \text{ for all } n \in \mathbb{N}. \tag{14}$$

We take

$$f^* := \limsup f_n. \tag{15}$$

By Eq. (15) we have

$$\mid f_n(x) \mid \leq m\overline{w}(x) \quad \forall x \in X \text{ and } \forall n \in \mathbb{N},$$

$$-m\overline{w}(x) \leq f_n(x) \leq m\overline{w}(x) \quad \forall x \in X \text{ and } \forall n \in \mathbb{N},$$

hence

$$-m\overline{w}(x) \leq \limsup f_n(x) \leq m\overline{w}(x) \quad \forall x \in X \text{ and } \forall n \in \mathbb{N}.$$

Therefore

$$\mid f^*(x) \mid = \mid \limsup f_n(x) \mid \leq m\overline{w}(x) \quad \forall x \in X,$$

this implies that $f^*(x)$ is in $F^{\overline{w}}(X)$. On the other hand, by Exercise 16, page 39 in [19], we have

$$f^*(x) - f^*(y) = \limsup f_n(x) - \limsup f_n(y) = \limsup f_n(x) + \liminf(-f_n(y))$$

$$\leq \limsup(f_n(x) - f_n(y)) \leq c(x, y) \quad \forall x, y \in X,$$

that is, f^* is a feasible solution to KR*. Finally, by Fatou's Lemma (see, for instance, [19] p. 264) we get that

$$\langle \nu_1 - \nu_2, f^* \rangle = \sup(\text{KR}^*),$$

that is, f^* is an optimal solution to KR*.

(b) To prove this, consider the subset H of $M_{\overline{w}}(X) \times \mathbb{R}$ defined as

$$H := \{(A\mu, \langle \mu, c \rangle + r) \mid \mu \in M_w^+(X \times X), r \in \mathbb{R}_+\}.$$

Then, according to Theorem 3.9 from [1], we have that, Eq. (11) will follow if we can show that

$$H \text{ is weakly closed,} \tag{16}$$

that is, closed in the weak topology $\sigma(M(X) \times \mathbb{R}, F(X) \times \mathbb{R})$. To see this, let $(N, \leq)$ be a directed set and let $\{(\mu_\alpha, r_\alpha), \alpha \in N\}$ be a net in $M_+(X \times X) \times \mathbb{R}_+$ such that $(A\mu_\alpha, \langle \mu_\alpha, c \rangle + r_\alpha)$ converges weakly to (λ, ρ) in $M_{\overline{w}}(X) \times \mathbb{R}$, i.e.,

$$\langle A\mu_\alpha, f \rangle \to \langle \lambda, f \rangle \quad \forall f \in F(X) \tag{17}$$

and

$$\langle \mu_\alpha, c \rangle + r_\alpha \to \rho. \tag{18}$$

We wish to show that (λ, ρ) is in H, that is, there exist $\widehat{\mu} \in M_+(X \times X)$ and $\widehat{r} \in \mathbb{R}_+$, with

$$(i) \ A\widehat{\mu} = \lambda \quad \text{and} \quad (ii) \ \langle \widehat{\mu}, c \rangle + \widehat{r} = \rho. \tag{19}$$

To prove (19), first note that all the terms in (18) are nonnegative, and so we may consider two cases, $\rho = 0$ and $\rho > 0$. Observe that if $A\mu_\alpha \to \lambda$, then $A(\min\{\mu_{\alpha_0}, \mu_\alpha\}) \to \lambda$. Let us first consider the case of $\rho > 0$. If $\rho > 0$, then for any given $\varepsilon > 0$, Eq. (19) implies the existence of $\alpha_0 \in N$, such that

$$\langle \mu_\alpha, c \rangle \leq \rho + \varepsilon \quad \forall \alpha \geq \alpha_0$$

Thus by Prohorov extended Theorem, the net $\Gamma := \{\mu_\alpha, \alpha \geq \alpha_0\}$ is relatively compact, and, therefore, there exists a subnet $\{\mu_{\alpha(i)}\}$ of Γ and a measure $\widehat{\mu}$ in $M_+(X \times X)$ such that $\mu_{\alpha(i)}$ converges to $\widehat{\mu}$ in the weak topology $\sigma(M(X \times X), C_b(X \times X))$, i.e.,

$$\langle \mu_{\alpha(i)}, v \rangle \to \langle \widehat{\mu}, v \rangle \quad \forall v \in C_b(X \times X). \tag{20}$$

On the other hand, the adjoint A^* maps $C_b(X)$ into $C_b(X \times X)$, which together with expression (20) yields

$$\lim \langle A\mu_{\alpha(i)}, f \rangle = \lim \langle \mu_{\alpha(i)}, A^* f \rangle$$
$$= \langle \widehat{\mu}, A^* f \rangle$$
$$= \langle A\widehat{\mu}, f \rangle \quad \forall f \in C_b(X).$$

From this fact and expression (17), we conclude that

$$\langle A\widehat{\mu}, f \rangle = \langle \lambda, f \rangle \quad \forall f \in C_b(X)$$

and it follows that $\widehat{\mu}$ satisfies (19)(i).

Furthermore, by expression (20), (18) and Lemma 2.6 in [9] we have

$$\rho \geq \liminf \langle \mu_{\alpha(i)}, c \rangle \geq \langle \widehat{\mu}, c \rangle.$$

Hence, by taking $\widehat{r} := \rho - \langle \widehat{\mu}, c \rangle$ we obtain (19)(ii). This completes the proof of (19) in the case $\rho > 0$.

Finally, if $\rho = 0$, the same arguments used in the previous paragraph show that there exists a measure $\widehat{\mu}$ in $M_+(X \times X)$ such that $\widehat{\mu}$ satisfies (19)(i) and $\langle \widehat{\mu}, c \rangle = 0$. Therefore, (19)(ii) holds with $\widehat{r} = 0$.   □

5 Conclusions and Future Work

Firstly, we emphasize the differences between mass transfer problem and KR problem. That is to say, the feasible solutions of the mass transfer problem are measures of probability and it does not happen in the KR problem, its feasible solutions belong to a set of more general measures: non-negative measures, which leads us to deal with a not bounded set of feasible solutions. In this work, the solvability of primal and dual problems of the KR problem are studied and we proved the no duality gap condition. Moreover, the strong duality is proved when X is a metric space and c is a moment or inf-compact function and when X is a σ-compact space and c is a lower semicontinuous cost function.

The applications of the KR problem are the Fortet-Mourier metrics to study the quantitative stability of two-stage models (see [8]), probability metrics (see [15] and [16]), control of cancer radiotherapy (see [7]), image registration and warping (see [4]), limit theorems and recursive stochastic equations (see [17]), among others. For all of them, it is important to have an efficient approximation scheme, in which we are working.

6 Appendix

Proposition 1. Let f be a nonnegative function on Z. Then, if one of the following assumptions holds

(a) f is an inf-compact.
(b) f is a moment.
(c) f is a lower semicontinuous function and unbounded above and Z is a σ-compact space,

then $\Gamma = \{\mu \in M_+(Z) \mid \langle \mu, f \rangle \leq b\}$ (where b is a fixed constant) is tight.

Proof.(a) As f is an inf-compact function, then $K_n := \{x \in Z \mid f(x) \le n\}$ is a compact set. Hence, for any measure $\mu \in \Gamma$

$$b \ge \langle \mu, f \rangle \ge \int_{K_n^c} f \, d\mu \ge n\mu(K_n^c),$$

that is, $\mu(K_n^c) \le \frac{b}{n}$ for all n. Therefore Γ is tight.

(b) As f is a moment, there exists a sequence $\{F_n\}$ of compact sets $F_n \uparrow X$, such that $\lim_{n\to\infty} \inf_{x \in F_n^c} f(x) = +\infty$.

Let $\varepsilon > 0$ then there exists $M > 0$, such that $\frac{b}{M} < \varepsilon$. Moreover, there is N in $\mathbb{N}$ such that

$$M \le \inf_{x \in F_n^c} f(x) \quad \forall n \ge N.$$

Hence, for all $n \ge N$ we have that $M\chi_{F_n^c} \le f(x)\chi_{F_n^c}$, where χ_A is the characteristic function of the set A, and

$$M\mu(F_n^c) \le \int_{F_n^c} f \, d\mu \le \int f \, d\mu \le b,$$

therefore

$$\mu(F_N^c) \le \frac{b}{M} < \varepsilon$$

and Γ is tight.

(c) Let f be a lower semicontinuous function and unbounded above and let Z be a σ-compact set, then, there exists a sequence $\{F_n\}$ of compact sets with $F_n \uparrow Z$. Let

$$Z_k := \{x \in Z \mid f(x) \le k\},$$

then, Z_k is a closed set. Let $G_k := F_k \cap Z_k$, G_k is a compact set and $G_k \uparrow Z$. Let

$$h(x) := \begin{cases} 0 \text{ if } x \in G_1; \\ k \text{ if } x \in G_{k+1} \backslash G_k; \end{cases}$$

hence, h is an inf-compact function. Moreover $h(x) \le f(x)$ and for any $\mu \in \Gamma$, we find that $\int h \, d\mu \le \int f \, d\mu \le b$. By (a) we conclude that Γ is tight. □

References

1. Anderson, E.J., Nash, P.: Linear Programming in Infinite-Dimensional Spaces. Wiley, Chichester (1987)
2. Dedecker, J., Prieur, C., de Fitte, R.P.: Parametrized Kantorovich-Rubinstein theorem and application to coupling of random variables. In: Bertail, P., Soulier, P., Doukhan, P. (eds.) Lecture Notes in Statistics, vol. 187, pp. 105–121. Springer, New York (2006). https://doi.org/10.1007/0-387-36062-X_5
3. Edward, D.A.: on the Kantorovich-Rubinstein theorem. Expo. Math. **29**, 387–398 (2011)

4. Haker, S., Zhu, L., Tannenbaum, A., Angenent, S.: Optimal mass transport for registration and warping. Int. J. Comput. Vis. **63**, 225–240 (2004)
5. Hanin, L., Rachev, S.T.: An extension of the Kantorovich-Rubinstein mass transshipment problem. Num. Funct. Anal. Optimiz. **16**, 701–735 (1995)
6. Hanin, L.G., Rachev, S.T.: Mass transshipment problems and ideal metrics. J. Comp. Appl. Math. **56**, 183–196 (1994)
7. Hanin, L., Rachev, S.T., Yakovlev, A.Y.: On the optimal control of cancer radiotherapy for non-homogeneous cell populations. Adv. Appl. Prob. **25**, 1–23 (1993)
8. Heitsch, H., Romisch, W.: A note on scenario reduction for two-stage stochastic programs. Oper. Res. Lett. **35**, 731–738 (2007)
9. Hernandez-Lerma, O., Gabriel, J.R.: Strong duality of the Monge-Kantorovich mass transfer problem in metric spaces. Math. Z. **239**, 579–591 (2002)
10. Hernandez-Lerma, O., Lasserre, J.B.: Approximation schemes for infinite linear programs. SIOPT **8**, 973–988 (1998)
11. Levin, V.L.: On the mass transfer problem. Soviet Math. Dokl. **16**, 1349–1353 (1975)
12. Levin, V.L.: General Monge-Kantorovich problem and its applications in measure theory and mathematical economics. In: Leifman, L.J. (ed.) Functional Analysis, Optimization, and Mathematical Economics: A collection of papers dedicated to the memory of Leonid VitaFevich Kantorovich, pp. 141–176. Oxford University Press, USA (1990)
13. Mikami, T.: Monge's problem with a quadratic cost by the zero-noise limit of h-path processes. Probab. Theory Relat. Fields **129**, 245–260 (2004)
14. Rachev, S.T.: Mass transshipment problems and ideal metrics. Numer. Funct. Anal. Optimiz. **12**, 563–573 (1991)
15. Rachev, S.T.: Probability Metrics and the Stability of Stochastic Models. Wiley, New York (1991)
16. Rachev, S.T., Klebanov, L., Stoyanov, S.V., Fabozzi, F.: The Methods of Distances in the Theory of Probability and Statistics. Springer, New York (2013). https://doi.org/10.1007/978-1-4614-4869-3
17. Rachev, S.T., Rüschendorf, R.L.: Mass Transportation Problems, vol. 1, 2. Springer, New York (1998)
18. Rachev, S.T., Shortt, R.M.: Duality theorems for Kantorovich-Rubinstein and Wassertein functionals. Diss. Math. **299**, 647–676 (1990)
19. Royden, H.L.: Real Analysis. Prentice Hall, New York (1988)

Evolutionary Construction
of Convolutional Neural Networks

Marijn van Knippenberg[1(✉)], Vlado Menkovski[1], and Sergio Consoli[2]

[1] Eindhoven University of Technology (TU/e), Eindhoven, The Netherlands
{m.s.v.knippenberg,v.menkovski}@tue.nl
[2] Philips Research Eindhoven, Eindhoven, The Netherlands
sergio.consoli@philips.com

Abstract. Neuro-Evolution is a field of study that has recently gained significantly increased traction in the deep learning community. It combines deep neural networks and evolutionary algorithms to improve and/or automate the construction of neural networks. Recent Neuro-Evolution approaches have shown promising results, rivaling hand-crafted neural networks in terms of accuracy.

A two-step approach is introduced where a convolutional autoencoder is created that efficiently compresses the input data in the first step, and a convolutional neural network is created to classify the compressed data in the second step. The creation of networks in both steps is guided by an evolutionary process, where new networks are constantly being generated by mutating members of a collection of existing networks. Additionally, a method is introduced that considers the trade-off between compression and information loss of different convolutional autoencoders. This is used to select the optimal convolutional autoencoder from among those evolved to compress the data for the second step.

The complete framework is implemented, tested on the popular CIFAR-10 data set, and the results are discussed. Finally, a number of possible directions for future work with this particular framework in mind are considered, including opportunities to improve its efficiency and its application in particular areas.

Keywords: Neuro-evolution · Genetic algorithms ·
Convolutional autoencoders · Convolutional neural networks

1 Introduction

Neural networks have become a popular data analysis tool in both academia and industry, especially when tasks like image classification, natural language processing, and speech recognition need to be addressed. They have shown to be very adept at these tasks, which are classic problems in which we like the computer to show "human-like" behavior. One of the challenges surrounding neural networks which has kept researchers around the world occupied is the matter of their design, and the possible automation of this task. Constructing

© Springer Nature Switzerland AG 2019
G. Nicosia et al. (Eds.): LOD 2018, LNCS 11331, pp. 293–304, 2019.
https://doi.org/10.1007/978-3-030-13709-0_25

a neural network is still often seen as a somewhat "magic" skill by many: a combination of knowledge, past experience, and intuition. Network performance is greatly affected by its size and structure, the type and order of layers, the choice of loss function and the way data is presented to the network. These are all decisions that a designer has to make in what is often a lengthy continuous cycle of re-design, re-training, and re-evaluation. This makes it an attractive target for automation [7].

In this paper we introduce a neuro-evolution approach to the above challenge that is more efficient than current solutions in terms of computational resource consumption [7]. This will hopefully make this type of approach more viable in real-life settings. It takes inspiration from a specific, recently published study, which establishes a kind of baseline for the application of evolutionary algorithms in this setting [10]. Implementation of the approach also requires a look at solution efficiency and multi-criteria decision making (MCDM), which helps reasoning about the relation between the performance metrics of different types of neural networks [9, 14].

Both the baseline and the proposed approach are implemented and tested in order to confirm their behavior and to draw comparisons. The results are then inspected to establish whether the new framework has significant impact in terms of final accuracy and consumed computational resources. These two will form a trade-off that is controlled by the MCDM process. Finally, a number of alternatives and future study topics are presented and considered.

2   Related Work

Neuro-evolution is the field of study that combines neural networks with evolutionary algorithms in the search for innovative training methods [16]. The field has recently gained increased interest with the sharp increase of popularity of deep neural networks. Likely the most well-known algorithm in the field of neuro-evolution is the neuro-evolution of augmented topologies algorithm (NEAT) [13]. Recent studies have applied evolutionary strategies to deep neural networks in a variety of manners, such as evolving network weights via genetic programming [9,15], differential evolution [11], pitting networks against each other in a tournament selection environment [10], more recent extensions of NEAT [3,7], and by evolving a network's activation functions [4]. The general "mood" of the intersection of these two fields is very much an exploratory one. There is no clear precedent on how to apply techniques from the meta-heuristics field to deep learning [4]. Another interesting direction in neuro-evolution that has recently been taken under the growing complexity of deep neural networks is that of network reduction. Evolutionary strategies can be used to simplify networks, in the hope of reducing future training and operating cost [12].

This work is based in large parts on a recent Neuro-Evolution study [10]. The goal of this study was to establish the viability of genetic programming in replacing human input during the design phase of neural networks. In the remainder of this paper, their approach will be referred to as the "baseline framework".

The approach is based on multiple worker processes operating in parallel, each applying a genetic programming algorithm that continually evolves a population of neural networks. In genetic programming, a collection of candidate solutions, called the *population*, is updated in steps in the hopes of finding better candidate solutions. A candidate solution is also known as an *individual*. An individual is considered to have "DNA"; some simplified representation from which the solution can be reconstructed. By mutating this DNA in different ways, an individual can be evolved into new individual, one which hopefully has better *fitness*, which is some measure of the quality of the solution. In this case, each candidate solution is a convolutional neural network, and the DNA is simply the structure of the network. In order to apply genetic programming, a *selection method* is required. This is the method that is used to determine which individuals in the population are chosen for mutation. The approach is based on the basic process of tournament selection [8]. It is a simple and direct selection method that is based on "rounds". Each round, k individuals are selected from the population according to some probability distribution, and their fitness is compared. The individual with the best fitness is copied and its copy is mutated before inserting it into the population, while the one with the worst fitness is removed from the population. This method of selecting individuals is intuitive, easy to implement, and easy to adjust. An added benefit is that it is quite easy to parallelize. In the case of the study, $k = 2$ individuals are selected uniformly at random. In order to establish a population to select from, each worker process generates some initial individuals. In this case, the initial individuals are very simple networks with very low fitness scores.

The major issue of this approach is the computational resource required. To arrive at a competitive result, many networks have to be generated, trained, and evaluated. The authors' main goal was to establish viability in terms of accuracy, so their approach is fairly basic and direct. This leaves many opportunities for improvements and further study, which is the point where this work steps in.

3 Method

3.1 Evolving Autoencoders

The attempt of this work to reduce computation time is based on reducing the size of the input samples that are used for training classification networks. Reducing the input sample size has a double positive effect on the evolutionary process: training each individual network takes less effort, since there are less values to process for each sample, and networks can be expected to be shallower since the input is smaller, reducing the average training time per network in the overall evolutionary process.

Autoencoders are well-suited tools for reducing sample sizes, and in the case of image data, a convolutional autoencoder can be used to maintain the spatial relations of the input data in the encoded data. Since convolutional autoencoders are simply convolutional neural networks with some extra restraints and a different error metric, the baseline framework can also be applied to obtain the most suitable autoencoder. The overall process is then as follows:

1. Perform an evolutionary process on a population of convolutional autoencoders. Training is based on the original input and the reconstruction error of the decoder.
2. Pick the best autoencoder and encode the entire original input data set.
3. Perform an evolutionary process on a population of convolutional neural networks that classify input samples. Training is based on the encoded input and classification error.
4. Pick the classifier with the highest validation accuracy and append it to the encoder used to encode the data in order to obtain the best overall network.

In order to apply the baseline framework to convolutional autoencoders, a new selection process needs to be refined. In the baseline framework, selection was purely based on accuracy. In the case of autoencoders, there are now two selection criteria: reconstruction accuracy and compression ratio. We like both of these to be as high as possible, but it is important to consider the case where one autoencoder has a higher compression ratio, while another one has a higher accuracy. Taking inspiration from the Non-dominated Sorting Genetic Algorithm (NSGA)[2], autoencoders are grouped into Pareto fronts based on dominance. If two sampled individuals are members of different fronts, the one that is a member of the dominating front "wins". If both individuals are part of the same front, neither is strictly better than the other in terms of Pareto efficiency. In this case, the individual that is furthest away the other individuals in that front is chosen as winner. In other words, a higher probability of innovation is rewarded if both individuals come from the same front. The final adjustment to make to the evolutionary process is to define a new set of mutations so actual convolutional autoencoders are evolved.

Below are listed the mutations that are made available to the evolutionary framework for the purpose of evolving convolutional autoencoders. These are based on the baseline framework, with some small changes, such as the inclusion of pooling-related mutations. Note that only the encoder is mutated, since the decoder is directly derived from it by mirroring it and replacing each pooling layer with an up-sampling layer (see Fig. 1).

- IDENTITY: The network structure is not changed in any way. In practice this means the same network trained longer, since weights are maintained when copying networks.
- INSERT CONVOLUTION: This mutation inserts a convolutional layer at a random location in the network in the encoder.
- REMOVE CONVOLUTION: This mutation removes a random convolutional layer from the encoder.
- ALTER STRIDE: The stride length of a random convolutional layer is incremented or decremented by 1 at random.

- INSERT POOL: This mutation inserts a max-pooling layer at a random location in the encoder. The initial pool size is 2×2.
- REMOVE POOL: In this case a random max-pooling layer from the encoder is removed.
- ALTER FILTER NUMBER: The number of filters of a random convolutional layer is adjusted.
- ALTER FILTER SIZE: One of the dimensions of the filters of a random convolutional layer is incremented or decremented at random
- ALTER POOL SIZE: The pool size of a random max-pooling layer is incremented or decremented at random.

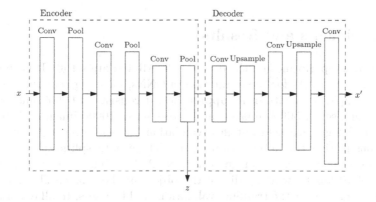

Fig. 1. Convolutional autoencoder structure. For some input x, the encoder produces an encoding z, and the decoder produces a reconstruction x'. The decoder mirrors the encoder, replacing pooling layers with up-sampling layers.

On top of the mutations themselves, there is an additional constraint on each mutation. For a mutation to be applied successfully, the output size of the encoder must be smaller than input size of the encoder. This forces the evolutionary process to evolve the encoder that actually compresses the data, and helps preventing the autoencoder from learning a dictionary of the input.

At the end of the evolutionary process, one of the autoencoders has to be chosen from the Pareto front to process the original data set to obtain a compressed data set that is used in the second evolutionary process. A method is needed that considers the trade-off between the compression ratio and the accuracy of the autoencoders. This relies on MCDM, since the best autoencoder choice is based on multiple, possibly conflicting, criteria (accuracy and compression) to be optimized. There are many different existing algorithms that help in picking a best solution based on multiple criteria. In this study, the preferred method should be able to process any collection of solutions, as it is not known beforehand what kind of autoencoders will be involved. TOPSIS is a straight-forward and intuitive algorithm that can do this [5].

User-provided weights are applied to each of the criteria of each solution, allowing users to place emphasis on certain criteria. Then, the best and worst possible solutions, called the positive ideal alternative and the negative ideal alternative, are determined. In the context of convolutional autoencoders, the positive ideal has an accuracy and compression ratio of 1.0, while the negative ideal has an accuracy and compression ratio of 0.0. Finally, the ratio of L^2 distance between solutions and the ideal alternatives determines which solution is chosen.

The matter left open by this algorithm is that of the weights of criteria. Since there are no previous results to draw conclusion from, the weights will initially be set to equal values. The weights will depend on the relation between the autoencoder compression ratio and the accuracy of the classification networks.

4   Experiments and Results

Experiments are performed on the popular CIFAR-10 data set [6]. It is one of the most popular data sets for the evaluation of CNNs, meaning that its use makes for easier comparisons with other approaches. It consists of 60000 images, split into a training set of 50000 images, and a test set of 10000 images. 5000 of the training set images are held out in a validation set, with the remaining 45000 images constituting the actual training set. These data splits are randomized between experiments. Each image is a 32×32 RGB image, and is associated with one of 10 labels, which indicate the object in the image. The classes are evenly distributed over the training, validation, and test sets. In all experiments, stochastic gradient descent with a momentum of 0.9 is used to train networks. Training data is divided into batches of 50 samples, making for 1000 batches per epoch. Each network is trained for 25 epochs. This setup is copied from the baseline framework [10] in order to help facilitate comparisons. Each worker process has access to an Nvidia Tesla K80 GPU. Concurrent access to the population of networks is facilitated purely by a shared file system; worker processes to not directly communicate with each other. By taking advantage of POSIX standards[1], minimal parallel programming has been required [1]. The source code and accompanying documentation is available online.[2]

4.1   Baseline Framework

Initially, the baseline framework is run to confirm its performance, and to inspect its behavior in a more resource-restricted setting. The results of this experiment are listed in Table 1. Perhaps the most surprising result here is that after 12 h, 5 worker processes have reached a much higher best classification accuracy than 250 worker processes. This may be partly because of the more limited available mutations, which "focuses" the evolutionary process better in the early stages

---

[1] http://standards.ieee.org/develop/wg/POSIX.html.
[2] https://github.com/marijnvk/LargeScaleEvolution.

in the case of five worker processes. At the 24 h point, the baseline has best accuracy. Nonetheless, this shows that even a very small number of workers can be very effective at finding good neural networks. Figure 2 illustrates the evolutionary process.

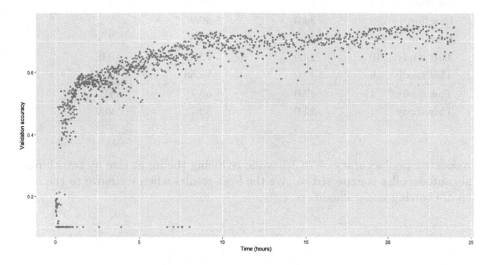

Fig. 2. Results of the reproduction of the baseline framework, run with five worker processes. Each point indicates a completely trained CNN, color-coded by worker process. (Color figure online)

4.2   Evolving Convolutional Autoencoders

The second experimental setting is the evolution of convolutional autoencoders. After evolving over 900 autoencoders in a span of two days on a single worker process, the status of the Pareto front is as in Fig. 3. The most telling structure in the population is the appearance of vertical groups of autoencoders. These groups share the same compression ratio. Generally those low in reconstruction accuracy tend to have fewer convolutional layers compared to those with a higher reconstruction accuracy. Their compression instead comes mostly from pooling layers. Evolution quickly pushes networks towards very high compression ratios, mostly due to the insertion of multiple pooling layers. Intuitively, however, it is the autoencoders that have a more even trade-off between compression ratio and reconstruction accuracy that are of most interest. These all have a fairly simple structure, usually being not more than five layers in depth. This is due to the nature of the training data. Images in the CIFAR-10 data set are already in compressed form when they are presented to the autoencoders. This undoubtedly reduces the effectiveness of the process step of evolving convolutional autoencoders.

At generation 507, an autoencoder is generated that clearly deviates from the otherwise quite linear shape of the final Pareto front. This individual attains

Table 1. Performance results for the baseline setup of the framework. Accuracy values from the baseline study for the 12 and 24 h interval are derived from a figure and thus are not precise. The baseline study did not report the number of generated networks.

Study	# Workers	Time elapsed (h)	Best accuracy (%)	# Networks
Baseline	250	12.0	~56.6	?
Baseline	250	24.0	~86.9	?
Baseline	250	256.2	94.6	?
This study	10	12.0	72.32	848
This study	10	24.0	75.68	1197
This study	1	12.0	58.7	190
This study	1	24.0	59.7	299

a reconstruction accuracy of 70.3% while reducing the input size by two-thirds. This autoencoder is expected to give the best results when we move to the last step of evolving image classifiers.

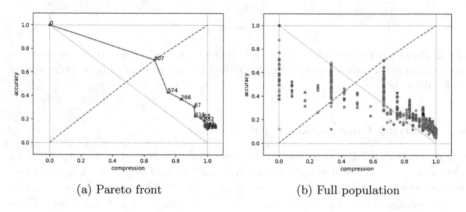

(a) Pareto front   (b) Full population

Fig. 3. Results of the evolutionary process after generation 915. Each point illustrates a convolutional autoencoder, plotted by encoder compression and decoder accuracy. Fronts are marked by color. A compression of 1 means that samples are reduced to a single value. The individual chosen by TOPSIS is indicated by the dashed line. In the left plot, the number indicates the generation of that individual (i.e. the number of individuals that were generated before it). (Color figure online)

4.3   Complete Framework

The final setting combines both versions of the framework into a full approach that is supposed to outperform the baseline framework in terms of computation time at the cost of some accuracy. A number of representative autoencoders are picked from the Pareto font of evolved autoencoders and used to encode the

CIFAR-10 data set. Table 2 shows the chosen autoencoders and their characteristics. As expected, the original data set results in the eventual highest classification accuracy. It is the data coming from autoencoder 507, the one which shows most promise and that progresses most similarly to the baseline. While the results are quite similar in terms of accuracy, the smaller, encoded data from autoencoder 507 means that approximately 20% more networks are generated, trained, and evaluated. For more heavily compressed data sets, this increased number of generated networks is even higher, up to nearly 50%. As the number of generated and evaluated networks in the same time frame has increased, we have been able to increase the efficiency of the method. But on the other hand, in all other cases the encoding of data causes a significant loss of accuracy and early plateauing of evolutionary progress. It is clear that in the encoded setting, the evolutionary process has a significantly faster turn-over of networks. Note that this effect can be magnified if mutations are chosen in a more efficient way, which gives each new network a higher probability of being an improvement in terms of accuracy.

While the gains in this setting appear rather small, if applied to the baseline study, they would likely result in time savings on the scale of days. As the evolutionary process on classifiers is run longer, gains increase. After all, smaller sample sizes means that each batch is processed faster, and reduces the need for larger networks, which pushes down the average network size. Importantly, these results also show that the initial fair weighing of compression ratio and accuracy of the autoencoders for constructing the Pareto front is successful in determining

Table 2. Convolutional autoencoders (CAEs) sampled from the Pareto front. For each CAE, its generation ID, compression ratio, and mean classification accuracy over 5 runs are listed (variance in parentheses). Also listed is the average number of networks generated when evolving classifiers with training data encoded by this CAE. The last two CAEs do not come from the Pareto front, but share the same compression rate with the best CAE.

CAE generation	Compression rate	Accuracy (Var)	#Generated
0	1.0	0.0 (0.0)	105.8
507	0.66	0.7028 (0.0004)	121.6
574	0.75	0.4289 (0.0004)	130.4
266	0.83	0.3710 (0.0003)	136.9
67	0.91	0.3054 (0.0003)	141.2
611	0.92	0.2255 (0.0003)	140.9
355	0.95	0.2123 (0.0003)	145.0
882	0.97	0.1885 (0.0003)	144.2
841	0.98	0.1493 (0.0002)	146.1
130	0.66	0.5650 (0.0004)	122.1
725	0.66	0.4271 (0.0003)	118.4

the best autoencoder. Based on the results though, it may be advisable to weigh the autoencoder accuracy somewhat heavier since high compression rates seem to cause heavy plateauing. Note that TOPSIS can easily be run again on the results of the evolution of autoencoders to obtain a differently-weighed result quickly.

5   Conclusion

A framework for evolving the structure of convolutional neural networks is introduced. Its main weakness is clearly the large amount of computational resources that are required to obtain networks with competitive accuracy. A method to remedy this weakness is proposed in this paper in the form of a two-step process. It consists of applying the baseline framework to the evolution of convolutional autoencoders in the hope of reducing the sample size of the input data. This makes any subsequent training of networks cheaper and helps in limiting their size.

A significant result is the viability of much smaller number of worker processes than previous studies demonstrated. This gives confidence in the other results of this study, and in future work with this framework. Experiments that investigate the impact of the new framework show the trade-off between compressing the input data and maintaining classification accuracy. The initial weighing of autoencoder accuracy and compression rate is deemed successful, while some tuning may help if the framework has to be used for other data sets. Although the experiments were necessarily limited in scope due to the available computational resources, the results indicate that this extended approach would result in significant running time reductions when applied at larger scales, both in terms of input size and computational time spent.

5.1   Future Work

The fundamental nature of the proposed work means that there are many more directions to explore. Mentioned here are some the more interesting ones.

More Complex Data Sets: As already mentioned before, the CIFAR-10 data set used in this study consists of images that are already compressed significantly. This naturally decreases the effectiveness of any convolutional autoencoder that attempts to reduce the image size further. Running the extended framework on a data set of larger images can realistically be expected to result in a much larger impact of the step that evolves the autoencoders.

Beating Humans at Their Own Game: The initial population for the evolutionary process does not necessarily have to consist of very simple networks. An interesting application is to take state-of-the-art, human-designed networks as the individuals for the initial population.

Adaptive Mutations: The availability of mutations affects the efficiency of the overall evolutionary process. Different kinds of mutations will be more or less likely to improve performance at different points in that process. The process would benefit from some sort of adaptive mutation choosing process, whereby mutations are sampled in a weighed fashion, perhaps even completely excluding some mutations at certain points in the process.

Learning How to Evolve: Given similar tasks, image classification for example, the evolutionary process can be expected to behave somewhat similarly in terms of which mutations are most effective at what points in the process. By adding a layer of abstraction on top of the evolutionary process which keeps track of this, it may be possible to learn how to best evolve networks given a task.

Constrained Evolution for Constrained Networks: There are various situations, most notably in embedded systems, where compact networks are very desirable. The framework can easily be adapted to only consider a smaller solution space that can be limited in a variety of ways by placing additional constraints on the success of a mutation.

Acknowledgements. Research leading to these results has received funding from the EU ECSEL Joint Undertaking under grant agreement no. 737459 (project Productive4.0) and from Philips Research.

References

1. Breuel, T., Shafait, F.: AutoMLP: simple, effective, fully automated learning rate and size adjustment. In: The Learning Workshop, vol. 4, p. 51, Utah (2010)
2. Deb, K., Agrawal, S., Pratap, A., Meyarivan, T.: A fast and elitist multiobjective genetic algorithm: NSGA-II. IEEE Trans. Evol. Comput. **6**(2), 182–197 (2002). https://doi.org/10.1109/4235.996017
3. Desell, T.: Large scale evolution of convolutional neural networks using volunteer computing. In: Genetic and Evolutionary Computation Conference, Berlin, Germany, 15–19 July 2017, Companion Material Proceedings, pp. 127–128 (2017). https://doi.org/10.1145/3067695.3076002
4. Hagg, A., Mensing, M., Asteroth, A.: Evolving parsimonious networks by mixing activation functions. In: Proceedings of the Genetic and Evolutionary Computation Conference, GECCO 2017, Berlin, Germany, 15–19 July 2017, pp. 425–432 (2017). https://doi.org/10.1145/3071178.3071275
5. Hwang, C., Lai, Y., Liu, T.: A new approach for multiple objective decision making. Comput. OR **20**(8), 889–899 (1993)
6. Krizhevsky, A., Hinton, G.: Learning multiple layers of features from tiny images (2009)
7. Miikkulainen, R., et al.: Evolving deep neural networks. CoRR abs/1703.00548 (2017). http://arxiv.org/abs/1703.00548
8. Miller, B.L., Goldberg, D.E.: Genetic algorithms, tournament selection, and the effects of noise. Complex Syst. 9(3) (1995). http://www.complex-systems.com/abstracts/v09_i03_a02.html
9. Morse, G., Stanley, K.O.: Simple evolutionary optimization can rival stochastic gradient descent in neural networks. In: Proceedings of the 2016 on Genetic and Evolutionary Computation Conference, Denver, CO, USA, 20–24 July 2016, pp. 477–484 (2016). https://doi.org/10.1145/2908812.2908916

10. Real, E., et al.: Large-scale evolution of image classifiers. In: Proceedings of the 34th International Conference on Machine Learning, ICML 2017, Sydney, NSW, Australia, 6–11 August 2017, pp. 2902–2911 (2017). http://proceedings.mlr.press/v70/real17a.html
11. Rere, L.M.R., Fanany, M.I., Arymurthy, A.M.: Metaheuristic algorithms for convolution neural network. Comp. Int. Neurosc. 1537325:1–1537325:13 (2016). https://doi.org/10.1155/2016/1537325
12. Shafiee, M.J., Barshan, E., Wong, A.: Evolution in groups: a deeper look at synaptic cluster driven evolution of deep neural networks. CoRR abs/1704.02081 (2017). http://arxiv.org/abs/1704.02081
13. Stanley, K.O., Miikkulainen, R.: Evolving neural networks through augmenting topologies. Evol. Comput. 10(2), 99–127 (2002)
14. Such, F.P., Madhavan, V., Conti, E., Lehman, J., Stanley, K.O., Clune, J.: Deep neuroevolution: genetic algorithms are a competitive alternative for training deep neural networks for reinforcement learning. CoRR abs/1712.06567 (2017). http://arxiv.org/abs/1712.06567
15. Suganuma, M., Shirakawa, S., Nagao, T.: A genetic programming approach to designing convolutional neural network architectures. In: Proceedings of the Genetic and Evolutionary Computation Conference, GECCO 2017, Berlin, Germany, 15–19 July 2017, pp. 497–504 (2017). https://doi.org/10.1145/3071178.3071229
16. Turner, A.J., Miller, J.F.: Neuroevolution: evolving heterogeneous artificial neural networks. Evol. Comput. 7(3), 135–154 (2014). https://doi.org/10.1007/s12065-014-0115-5

Improving Clinical Subjects Clustering by Learning and Optimizing Feature Weights

Sergio Consoli[✉], Monique Hendriks, Pieter Vos, Jacek Kustra,
Dimitrios Mavroeidis, and Ralf Hoffmann

Philips Research, High Tech Campus 34, 5656 Eindhoven, AE, The Netherlands
{sergio.consoli,monique.hendriks,pieter.vos,jacek.kustra,
dimitrios.mavroeidis,ralf.hoffmann}@philips.com

Abstract. Data analytics methods in the clinical domain are challenging to put into practice. Unsupervised learning provides opportunity for giving the level of personalization in evidence based decision-making that can otherwise only be achieved through the use of prediction models, by helping doctors gaining insights from data. In this context, grouping of clinical subjects, in terms of biomedical information of patients, is an important task for patient cohort identification for comparative effectiveness studies and clinical decision-support applications. It allows the decision-making process to leverage not only on data but also on doctors' domain knowledge. However, one of the issues that needs to be addressed for a focused and realist unsupervised clustering of clinical subjects, is the fact that in the majority of the cases patients datasets are heterogeneous, i.e. their data features belong to several different feature spaces, e.g. nominal, ordinal, interval or rational, with completely different variation ranges and statistical distributions, affecting clustering quality and performance. In order to use these data measurements properly in an unsupervised manner, their corresponding weights need to be modeled. In this paper, we present a method for learning feature weights on clinical data. We show that learning feature weights is necessary in order to generate meaningful separation of data in high dimensional space. The method is based on silhouette score and principal component analysis, demonstrating its performance on a clinical test dataset.

Keywords: Patients data clustering · Features engineering ·
Principal component analysis · Optimization · Local search ·
Parameters tuning

1 Introduction

With the widespread adoption of electronic health records (EHRs), patient data storage is becoming digital and standardized, paving the way for data analytics. Data analytics is a broad concept encompassing methods for gathering, storing,

© Springer Nature Switzerland AG 2019
G. Nicosia et al. (Eds.): LOD 2018, LNCS 11331, pp. 305–316, 2019.
https://doi.org/10.1007/978-3-030-13709-0_26

and cleaning data, as well as describing, modeling and interpreting information, which is needed for exploring possible interrelations, and for confirming, or disproving, a hypothesis.

As data collection in clinical practice is also becoming digital and standardized, it becomes possible to extract additional knowledge by performing data analysis techniques. This allows for types of explorative analysis where it is not necessary to define a hypothesis and/or the type of data that needs to be collected to test the hypothesis beforehand, as is the case with clinical trials. In this way it is possible get an earlier insight generation from new data arising, e.g. from new treatments, improvements on devices for imaging, better image analysis techniques, or new diagnostic tests.

In this context, clustering of clinical subjects (e.g. biomedical information of patients) has been shown to be a useful data analysis technique. It is indeed a regular activity in clinical practice to define the best treatment options for a given patient [8]. Although this is widely an empirical process using the physician's experience, automatic systems can enhance the doctor ability to choose a treatment by analysing a much larger amount of patients even outside of the doctors network, which is of high importance in the context of modern clinical decision support systems.

Clustering of similar patients is therefore an important task, e.g., in the context of patient cohort identification for comparative effectiveness studies and clinical decision-support applications. In this context it is not given a-priori an output to tune the patients grouping or to make predictions and classification, but it is necessary to rely only on unsupervised clustering. The goal is to derive clinically meaningful distance metrics to measure the similarity among a given patients dataset. Unsupervised patients clustering still remains a complex and daunting task. As proven in [15,16], popular clustering metrics, like the average Silhouette width [11], do not usually reach high levels on unsupervised clustering tasks related to patients data, since it is always very hard to derive an appropriate and consistent measure of similarity, or dissimilarity, among individuals. This is due to the excessive requirements of domain knowledge and the multiple variables involved: the question whether two patients are similar or not is always very hard to answer, even for experienced medical experts [15,16]. In the real-world how to come up with consistent patients similarity measures and the related classifications remain critical open issues.

The fundamental starting point for any clustering approach is the assumption that a data object can be represented as a high-dimensional feature vector. In many traditional applications, all the features (i.e. variables or characteristics) are essentially of the same "type", and determination of appropriate feature spaces is often clear. If two features can be treated as homogeneous, we should only reluctantly treat them otherwise. However, many emerging real-life datasets often consist of several different feature spaces, like it is in the case in the biomedical domain for patients clustering. Here, indeed, it is needed to deal with both clinical and pathology features, with quantitative measurements belonging to completely different domains, and with completely different variation ranges and statistical distributions.

A necessary condition for a set of scalar features to be homogeneous is whether an intuitively meaningful symmetric distortion can be defined on them. A distortion between two data objects is defined as a weighted sum of suitable distortion measures on individual component feature vectors, where the distortions on individual components are allowed to be different [12]. Instead, a sufficient condition for two sets of features to be considered heterogeneous is whether, after clustering the dataset, it is possible to interpret the obtained clusters along one set of features independently of the clusters along the other set of features or, conversely, whether it is possible to study "association" or "causality" between clusters along different feature spaces.

When tackling the problem of clustering of a heterogeneous clinical subjects dataset, there is then the need of a measure of distortion between each of the features of the patients data. Since different types of features may have radically different statistical distributions, in general, it is unnatural to disregard fundamental differences between various different types of features and to impose a uniform, unweighted distortion measure across disparate feature spaces. Therefore, a practical issue when clustering heterogeneous patient datasets is how to determine the various feature spaces. In general, optimal feature weightings cannot be empirically computed using classic existing clustering approaches since it is a form of gradient-descent heuristic, susceptible to local minima. It is common therefore to end on a resulting clustering with a very low number of clusters, misclassifying patients that are radically different as belonging to the same group. Figure 1 shows an extreme example of such a situation, where a single cluster containing all the patients data is produced.

In this paper we propose a method for grouping similar patients which is able to learn patient feature weights so that they can be used on top of classic existing clustering algorithms, like, e.g., k-means [18] or hierarchical clustering [2], to derive a calculated weighted distortion measure among the input patients dataset. The method tackles the issue of weighting appropriately features of clinical subjects data coming from a multidimensional space, allowing to better separate those clinical subjects data and to improve the overall performance of clustering algorithms applied on the given patients data.

The rest of the paper is organized as follows. In the following section we provide some background material related to the influence of biases on clinical data modelling, and on the use of unsupervised machine learning as a potential solution (Sect. 2). Section 3 presents the proposed method of clustering or grouping subjects that are similar to one another. The method is based on silhouette score and principal component analysis, and its performance is demonstrated on a clinical test dataset (Sect. 4). The paper ends with conclusions and directions of future research in Sect. 5.

2   Background on the Influence of Biases on Clinical Data Modelling

While there is a great focus in the domain of statistical analysis on identification and prevention of bias, the risk of introducing bias into a clinical model remains

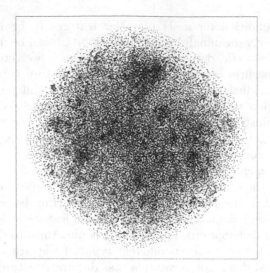

Fig. 1. Example of a single cluster containing all the patients.

high, especially in the domain of clinical data analysis, where there are often many features which may influence outcome and features are sparsely collected. For example, in oncology, survival is one of the main outcomes that is measured in order to determine successfulness of a treatment. However, as cancer patients are often older, survival is also influenced largely by comorbidities which can have a wide range. As comorbidities can also influence eligibility for certain treatment types, not taking into account comorbdities may therefore confound the results of a model towards suggesting that one treatment is better than the other, while the first treatment was given to healthier patients than the second.

This bias can take extreme forms, where trends found in the data may be reversed when the confounding variable is included in the model. Consider for example two hospitals A and B performing surgery. Looking at the outcome in terms of survival, hospital A may seem to be outperforming hospital B. However, if we include the type of surgery, heavy or light surgery, in the model, we may find that hospital B is performing more heavy surgeries and given the proportion of heavy surgeries, hospital B may be outperforming hospital A.

On the other hand, the inclusion of more features is not without problems. For example, in oncology, new developments in immunotherapy require assessment of genetic mutations in the tumor. Genetic data is highly dimensional and not measured for all patients, leading to sparse data. Highly dimensional data is subject to the curse of dimensionality. The curse of dimensionality means that predictive power of models decreases rapidly as the number of dimensions increases, due to the fact that it becomes easier to find coincidental relations.

Sparseness of data requires thorough thought on how to deal with missing values. Simply removing incomplete cases will lead to an extreme reduction of the data set and in many cases there might not even be a single complete case. So some form of imputation is necessary, taking into account the fact that missing values are most probably not missing at random but due to the fact that a certain measurement is not done for a patient for a reason (e.g. genetic testing is not done for early stage cancers). Dimensionality might even cause problems in outcome data, as treatment efficacy measures are moving more towards quality of life as a combination of outcomes such as survival, side effects and patient reported outcomes through questionnaires.

Another risk of bias in clinical data modeling is that conventional methods assume that all variables or characteristics are statistically similar, for example, having a similar range or having similar variation ranges and statistical distributions. However, subject data is usually heterogeneous, and different data therefore tends to belong in different domains with a high degree of statistically dissimilarity. This leads to poor performance on conventional clinical data modeling methods. Clinicians have therefore started to employ automated or unsupervised "machine learning" methods for classical data modeling tasks, like, e.g., for grouping clinical subjects. Indeed, due to the typically large number of variables or possibly influential characteristics of a subject, it is difficult for even the experienced clinician to determine whether two subjects are similar or not.

Unsupervised machine learning is the machine learning task of inferring a function to describe hidden structure from "unlabeled" data (a classification or categorization is not included in the observations). Since the examples given to the learner are unlabeled, there is no evaluation of the accuracy of the structure that is output by the relevant algorithm, which is one way of distinguishing unsupervised learning from supervised learning and reinforcement learning. A central case of unsupervised learning is the problem of density estimation in statistics [7], though unsupervised learning encompasses many other problems (and solutions) involving summarizing and explaining key features of the data. As stated in [4], we need to solve the unsupervised learning problem before we can even think of getting to true Artificial Intelligence.

3   Optimization Approach for Unsupervised Features Engineering of Clinical Subjects

In this section we describe the technical features of the proposed method for similar patients clustering, which is able to learn patient feature weights so that they can be used on top of classic existing clustering algorithms to derive a calculated weighted distortion measure among the input patients dataset. The proposed method consists of the following main components (see also Fig. 2):

1. A clustering algorithm;
2. An optimization strategy, able to adjust iteratively the weights of the patients features so that the clustering quality of the patients is improved;
3. An evaluation mechanism of the quality of the patients clustering by means of appropriate quantitative clustering metrics;
4. A procedure able to learn the feature weights at each iteration in order to guide the optimization strategy towards improved patients grouping.

These components, which are described in the following, represent new understandings on the features engineering task for improved patients clustering and similar patients concept.

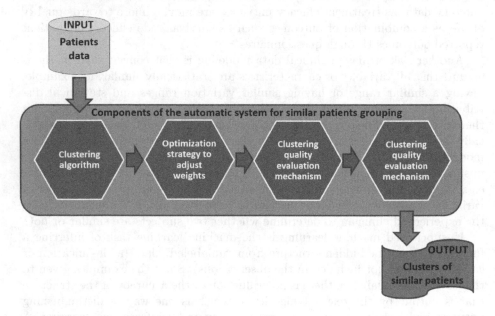

Fig. 2. Diagram of the main components of the proposed method for similar patients grouping.

We are given in input a dataset of patients data, including clinical and pathology features belonging to completely different domains, and with completely different variation ranges and statistical distributions. Each such feature, j, is referred to as F_j.

Consider now a generic patient i, referred to as:

$$P_i = [F_{i1}; F_{i2}; ...; F_{ij}; ...F_{in}].$$

The whole matrix of patients is therefore:

$$P = [F_1; F_2; ...; F_j; ...F_n].$$

The aim of the system is to find the set of feature weights, w, that provide an improved clustering, i.e.

$$P = [w_1 F_1; w_2 F_2; ...; w_j F_j; ...w_n F_n] = w \cdot F$$

so that to have a better representation of the data.

Although the proposed method supports any kind of clustering approach sensitive to perturbation [12], we consider the classic k-means [18] as an example.

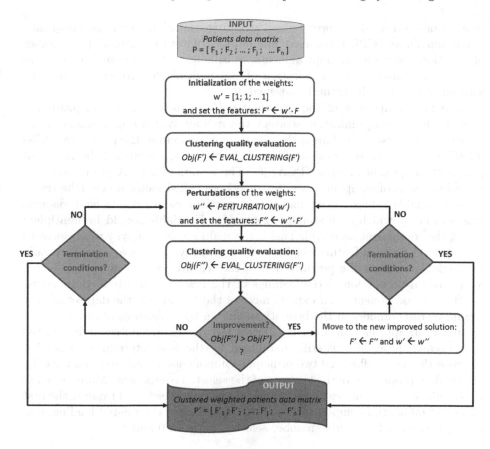

Fig. 3. Flow-chart of the clustering system for similar patients grouping.

The optimization strategy in (2) can be any approximate optimization approach producing high-quality results, ranging from basic routines to more advanced metaheuristics [3], like, e.g. Genetic Algorithms [5], Particle Swarm Optimization [14], or Simulated Annealing [1]. In our case it is enough to use a classic local search heuristic [10,13] as the optimization routine to achieve already good performance.

In Fig. 3 it is provided a flowchart of the algorithm. At the initial step, all the weights are set equal to one, i.e. $w = [1; 1; ...1]$. Then we iteratively perform a perturbation [12] of these weights (procedure $PERTURBATION\,(w')$), obtaining a new set of weights, w''. If the new set of weights leads to a distortion, F'', which improves the separability of the dataset with respect to the objective function value, then the perturbation is accepted and the new solution F'' becomes the new incumbent one, otherwise the move is rejected.

At each iteration, we evaluate the quality of the patients clustering (i.e., Component (3) of the system) using, for example, but not limited to, the average Silhouette width and the Dunn Index [11]. The iterations of the optimization routine proceed until the user-defined stopping conditions, such as, classically, maximum allowed CPU time, maximum number of iterations, maximum number of iterations between two improvements, or Silhouette width of at least 0.5, are satisfied. In this way, the integration of the multiple feature spaces into the k-means clustering algorithm is obtained [9].

The last component of the system, i.e. Component (4), is the procedure used to guide the optimization strategy towards an overall improvement of the patients data clustering. This is the crucial routine behind the procedure *PER-TURBATION(·)* in the local search and, in our case, we guide it by means of principal component analysis (PCA) [6]. The results of a PCA are usually discussed in terms of component scores, sometimes called factor scores (the transformed variable values corresponding to a particular data point), and loadings (the weight by which each standardized original variable should be multiplied to get the component score) [6]. This is the main motivation why we propose to employ PCA to learn feature weights as an alternative simply to a random walk. In particular, we set the perturbation of the feature weights at each iteration as a random proportion of the loadings of the first two principal components, i.e. the two components that explain most of the variance in the dataset at that iteration. Randomness in the heuristic search, a basic features of any heuristic optimization approach in order to allow a fair diversification capability in the whole search procedure, is still guaranteed by the non-deterministic selection between the most influential two principal components at each step, and also by the random proportion of the loadings of the selected component. More formally: $c_{i,j}/r$, where $i = 1, 2$ indicates the principal component selected (among the first two most influential components), $j = 1, .., n$ indicates the related loading, and $r \in [0, 1]$ is a random double number selected between 0 and 1.

4   An Experiment Showing the Potential of the Method

In the following we report the results of an experiment concerning the application of the described procedure on an artificial dataset of patients with risk of prostate cancer tumors who underwent radical prostatectomy [17]. This dataset includes the following clinical and pathology features: age at surgery, prostate specific antigen (PSA) density, percentage of positive biopsy cores, primary and second biopsy Gleason scores, and clinical stage, among others.

When we apply the k-means clustering at the first iteration, we have not a clear separation of data, as shown by the Bivariate Clusters plot, also known as Clusplot, and Clusters Silhouette plot shown in Figs. 4 and 5, respectively. Clusplot is a graphical display method in which the objects are represented as points in a bivariate plot and the clusters as ellipses of various sizes and shapes. All observation are represented by points in the plot, using principal components.

Quantitatively a clustering is commonly accepted as a satisfying one if the obtained Silhouette score is above 0.5. As shown in the figure, the Silhouette score is much below this threshold (i.e. 0.2 on average), which suggests a poor separation of data for the produced three clusters.

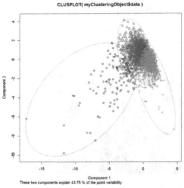

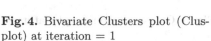

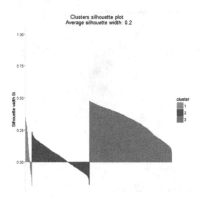

Fig. 4. Bivariate Clusters plot (Clusplot) at iteration = 1

Fig. 5. Clusters Silhouette plot at iteration = 1

Iterating our procedure, the clustering is improved. Figures 6 and 7 show the Clusplot and Clusters Silhouette plot at twelfth iteration. Now we obtain twelve different clusters with a better average Silhouette width, which is equal to 0.4; data are gaining a better separation. Continuing the iterations, the algorithm stops after few minutes at iteration 21, giving finally the Clusplot and Clusters Silhouette plot showed in Figs. 8 and 9.

We may see how clustering is evidently improved, producing now twelve well-separated clusters with an average Silhouette width equal to 0.51. The reported experiment shows a quantitative evaluation of the proposed features weighting k-means clustering. The method is able to obtain clusters of data with similar mean features, e.g. clinical stage, PSA density, prostate volume, etc., that are then representative of data with similar characteristics and more easily interpretable too.

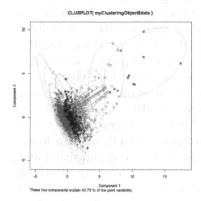

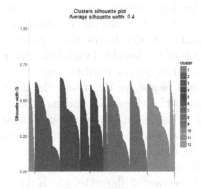

Fig. 6. Bivariate Clusters plot (Clusplot) at iteration = 12

Fig. 7. Clusters Silhouette plot at iteration = 12

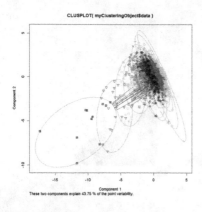

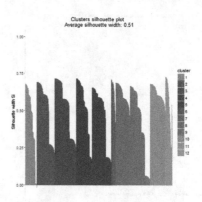

Fig. 8. Bivariate Clusters plot (Clusplot) at Iteration = 21

Fig. 9. Clusters Silhouette plot at iteration = 21

The obtained results demonstrates the feasibility of the method and its capability of producing meaningful groups of similar patients. These resulting groups of patients are well-separated each other instead of being misclassified in large agglomerations. The patients in each group are spreading within pairwise similarity boundaries, guarantying to place patients that are radically different in other, well-separated groups.

5   Conclusions

When tackling the problem of grouping heterogeneous clinical subjects containing biomedical information of patients, there is the need to measure the distortion between each of the features of the patients data. Popular clustering metrics, like the average Silhouette width, do not usually reach high levels on unsupervised clustering tasks related to patients data, since it is always very hard to derive an appropriate and consistent measure of similarity, or dissimilarity, among individuals. This is due to the excessive requirements of domain knowledge and the multitude of the involved variables: the question whether two patients are similar or not is always very hard to answer, even for experienced medical experts.

In this paper we described the details of an optimization algorithm which generalizes the classical k-means algorithm to derive a calculated weighted distortion measure for clustering clinical subjects. This optimization routine is a classic local search that, by using principal component analysis at each iterative step, determines the optimal feature weighting, within computational and heuristic constraints, to be the one that yields the "best" clustering according to both average Silhouette width and Dunn index. In this way, the gathered clustering simultaneously minimizes the average within-cluster dispersion and maximizes the average between-cluster dispersion along all the feature spaces.

The proposed method can be used in the context of any unsupervised clinical decision support system where physicians analyse similar patient groups to make a more precise and accurate diagnosis or treatment selection. The prototype presented in this paper has demonstrated its high potential for improving prostate cancer patient clustering.

References

1. Aarts, E., Korst, J., Michiels, W.: Simulated annealing. In: Burke, E.K., Kendall, G. (eds.) Search Methodologies: Introductory Tutorials in Optimization and Decision Support Techniques, pp. 187–210. Springer, Heidelberg (2005). https://doi.org/10.1007/0-387-28356-0_7
2. Dhillon, I.S., Modha, D.S.: Concept decompositions for large sparse text data using clustering. Mach. Learn. **42**, 143–175 (2001)
3. Gendreau, M., Potvin, J.-Y.: Metaheuristics in combinatorial optimization. Ann. Oper. Res. **140**, 189–213 (2005)
4. Goroshin, R., Bruna, J., Tompson, J., Eigen, D., LeCun, Y.: Unsupervised learning of spatiotemporally coherent metrics. In: ICCV 2015, pp. 4086–4093 (2015)
5. Holland, J.H.: Adaptation in Natural and Artificial Systems: An Introductory Analysis with Applications to Biology, Control, and Artificial Intelligence. The MIT Press, Cambridge (1992)
6. Jolliffe, I.T.: Principal Component Analysis, 2nd edn. Springer, New York (2002). https://doi.org/10.1007/b98835
7. Jordan, M.I., Bishop, C.M.: Neural networks. In: Tucker, A.B. (ed.) Computer Science Handbook (Section VII: Intelligent Systems), 2nd edn, pp. 137–142. Chapman & Hall/CRC Press LLC, Boca Raton (2004)
8. Kang, J., Schwartz, R., Flickinger, J., Beriwal, S.: Machine learning approaches for predicting radiation therapy outcomes: a clinician's perspective. Int. J. Radiat. Oncol. Biol. Phys. **93**(5), 1127–1135 (2015)
9. Lameski, P., Zdravevski, E., Mingov, R., Kulakov, A.: SVM parameter tuning with grid search and its impact on reduction of model over-fitting. In: Yao, Y., Hu, Q., Yu, H., Grzymala-Busse, J.W. (eds.) RSFDGrC 2015. LNCS (LNAI), vol. 9437, pp. 464–474. Springer, Cham (2015). https://doi.org/10.1007/978-3-319-25783-9_41
10. Lourenço, H.R., Martin, O.C., Stützle, T.: Iterated local search: framework and applications. In: Gendreau, M., Potvin, J.-Y. (eds.) Handbook of Metaheuristics, vol. 146, pp. 363–397. Springer, USA (2010). https://doi.org/10.1007/978-1-4419-1665-5_12
11. Modha, D.S., Scott Spangler, W.: Feature weighting in k-means clustering. J. Mach. Learn. **52**, 217–237 (2001)
12. Moore, J., Ackerman, M.: Foundations of perturbation robust clustering. In: Proceedings of the IEEE 16th International Conference on Data Mining (ICDM), pp. 1089–1094 (2016)
13. Pardalos, P.M., Resende, M.G.C.: Handbook of Applied Optimization. Oxford University Press, Oxford (2002)
14. Pugh, J., Martinoli, A.: Discrete multi-valued particle swarm optimization. In: Proceedings of IEEE Swarm Intelligence Symposium, vol. 1, pp. 103–110 (2006)
15. Qian, B., Wang, X., Cao, N., Li, H., Jiang, Y.-G.: A relative similarity based method for interactive patient risk prediction. Data Min. Knowl. Discov. **29**(4), 1070–1093 (2015)

16. Wang, F., Sun, J., Ebadollahi, S.: Composite distance metric integration by leveraging multiple experts' inputs and its application in patient similarity assessment. Stat. Anal. Data Min. **5**(1), 54–69 (2012)
17. Weinstein, J.N., et al.: The cancer genome atlas pan-cancer analysis project. Nat. Genet. **45**(10), 1113–1120 (2013)
18. Xiao, Y., Yu, J.: Partitive clustering (k-means family). Wiley Interdiscip. Rev.: Data Min. Knowl. Discov. **2**(3), 209–225 (2012)

A Framework to Automatically Extract Funding Information from Text

Subhradeep Kayal[✉], Zubair Afzal, George Tsatsaronis, Marius Doornenbal,
Sophia Katrenko, and Michelle Gregory

Content and Innovation Group, Elsevier B.V., Amsterdam, Netherlands
d.kayal@elsevier.com

Abstract. Many would argue that the currency of research is citations;
however, researchers and funding organizations alike are lacking tools
with which they can explore how this currency translates to funding
opportunities. Motivated by this need, in this paper we address one of
the fundamental problems facing the development of such a tool, namely
the problem of automatically extracting funding information from scientific articles. For this purpose, we experiment with a two-stage framework
which ingests text, filters paragraphs which contain funding information,
and then combines sequential learning methods to detect named entities in a novel ensemble approach. We present a comparative analysis of
each independent component of this pipeline, named *FundingFinder*, the
results of which indicate that the said pipeline can extract the funding
organizations and the associated grants, from scientific articles, accurately and efficiently.

1 Introduction

The *US* Government's policy says that *all federal funding agencies must ensure
public access to all articles and data which result from federally-funded research*,
as a result of which, institutions and researchers are required to report on funded
research outcomes, and acknowledge the funding source and grants. This information, if captured effectively from scientific text, will enable funding organizations to be in a position to trace back these acknowledgements and justify
the impact of their allocated research funds to their stakeholders and tax-payers
alike. At the same time, this information will also help researchers discover appropriate funding opportunities for their scientific interests.

In this paper we address the problem of automating the extraction of funding
information from text, using natural language processing and machine learning
techniques. We present *FundingFinder*, a pipeline that is engineered to accept
a scientific article as input, and provide the detected funding organizations and
associated grants as output annotations. All of the experiments done for the
purposes of evaluation were performed on a benchmark dataset that was created
exclusively for this purpose, a part of which we release publicly[1].

---

[1] https://drive.google.com/file/d/0B2RjZ7vHfzMldDVZUzQ4eUNkbkU/view?
usp=sharing.

© Springer Nature Switzerland AG 2019
G. Nicosia et al. (Eds.): LOD 2018, LNCS 11331, pp. 317–328, 2019.
https://doi.org/10.1007/978-3-030-13709-0_27

2   Background

2.1   Problem Definition

Given a scientific article as raw text input, the automated extraction of funding information from text performs two separate tasks: (1) identify all text segments which contain funding information, and (2) process all the funding text segments in order to detect the set of the funding bodies, denoted as *FB*, and the set of grants, denoted as *GR* that appear in the text. Provided that there is training data available, the former task can be seen as a binary text classification problem, whereas, the latter task can be seen as a named entity recognition (*NER*) problem. In the following section we give a small overview on existing state-of-the-art *NER* methods, which form the background for the current work.

2.2   Named Entity Recognition

Named entity recognition (*NER*) extracts information, known as *named entities*, from unstructured text; for example, the names of persons, locations and organizations. In this work, entities fall into the category of either *Funding Bodies* (*FB*) or *Grants* (*GR*). As an example, given a text of the form: "*The authors would like to thank the National Funding Foundation for grant number FF-1234*", we wish to label "*National Funding Foundation*" as a *FB* and "*FF-1234*" as *GR*. In literature, *NER* systems have been found to employ rule-based, gazetteer and machine learning approaches, a detailed survey of which can be found in the work of Nadeau et al. [8]. In this work, we near-exhaustively utilize several sequential learning approaches for *NER*, as discussed next.

Sequential Learning Approaches. Sequential learning approaches model the relationships between nearby data points and their class labels, and can be classified into *generative* or *discriminative*. In the context of *NER*, *Hidden Markov Models* (*HMMs*) are popular generative models that learn the joint distribution between words and their labels [11]. A *HMM* is a *Markov chain* with hidden states, and in *NER* the observed states are words while the hidden states are their labels. Given labelled sentences as training examples, *NER HMMs* find the maximum likelihood estimate of the parameters of the joint distribution, a problem for which many algorithmic solutions are known. *Conditional Random Fields* (*CRFs*) are discriminative, in contrast to *HMMs*, and find the most likely sequence of labels or entities given a sequence of words. The relationship between the labels is modelled by a *Markov Random Field*. *Linear chain CRFs* are well suited to sequence analysis and have been applied successfully in the past in *NER* [6]. Finally, another way of modelling data for *NER*, although not sequential, are *Maximum Entropy* (*MaxEnt*) models, which select the probability distribution that maximizes entropy, thereby making as little assumptions about the data as possible. Maximum entropy estimation has also been successfully applied to *NER* in works such as [1].

State-of-the-Art Toolkits for Information Extraction. Several open-source toolkits implement one or more of the learning approaches mentioned in the previous paragraph. The *Stanford CoreNLP toolkit*[2], for example, has a *CRF* implementation, enhanced with long-distance features to capture more of the structure in text. An important feature of the toolkit is the ability to use distributional similarity measures, which assume that similar words appear in similar contexts [3]. Additionally, the toolkit also has pre-trained models for recognizing *persons, locations* and *organizations*. *LingPipe*[3] is another *NLP* toolkit, whose efficient *HMM* implementation includes n-gram features. Finally, in this work we also use the *Apache OpenNLP*[4] toolkit, which has a *MaxEnt* implementation for *NER*.

Apart from the aforementioned open-source tools, this work makes use of *Elsevier's Fingerprint Engine (FPE)*[5], which is an industrial solution for annotating text with ontological concepts, given a vocabulary.

2.3 Related Work

Not much literature exists that is aimed at systematically exploring the concept of extracting funding information from the full text of scientific articles. A close category of related published research aims at extracting names of organizations from affiliation strings, e.g., the works of Jonnalagadda et al. [5], and Yu et al. [10], both of which aim at extracting names of organizations from the metadata of published scientific articles. The work by Giles et al. [4] is also noteworthy, as it aims at automatically tagging acknowledgment sections from text, in order to combine acknowledgment analysis with citation indexing. All of these works use regular expressions to extract the relevant entities from text.

Apart from the aforementioned published works, there are also several initiatives that started recently and are aiming at a similar direction to the current work, such as the *ERC* project *"Extracting funding statements from full text research articles in the life sciences"*[6].

3 Methodology

3.1 Overview

Our approach receives the raw full text of a scientific article as an input, and annotates the text with entities corresponding to *Funding Bodies (FBs)* and *Grants (GRs)*, where present. A two-stage search strategy for finding *FB* and *GR* entities in text has been devised: (1) The first step starts by splitting the input text into paragraphs and feeding them sequentially to a binary text classifier that

---

[2] http://stanfordnlp.github.io/CoreNLP/.
[3] http://alias-i.com/lingpipe/demos/tutorial/read-me.html.
[4] https://opennlp.apache.org/.
[5] https://www.elsevier.com/solutions/elsevier-fingerprint-engine.
[6] http://cordis.europa.eu/result/rcn/186297_en.html.

detects paragraphs that may contain any funding information. (2) The second step involves performing *NER* only on the filtered text paragraphs, to annotate them with *FB* and *GR* labels.

This two-step design has the following benefits: (1) It minimizes the execution time of the approach as the costliest component, namely *NER*, can now be executed on only a small selection of paragraphs, in which the binary text classifier has detected evidence of funding information. (2) It reduces the number of false positives, as there are many text segments in a scientific full text article that contain strings which a *NER* component could potentially annotate falsely as *FB*, e.g., the organisation names in the affiliation information of the authors, or *GR*, e.g., scientific formulae in text.

3.2   Data Collection

In this work, supervised *NER* algorithms are used for entity extraction. Since they are supervised, labeled data is needed to train the algorithms. The process of gathering data is described next.

The *"Silver"* Set. *NER* requires manually annotated training data, which is expensive and time-consuming to collect. Though we have created such a *"gold"* dataset, as explained next, we have also explored the creation of a *"silver"* dataset by collecting unannotated data for training, in order to examine whether for this application the costs for collecting additional training data could be minimized. The process is as follows:

1. 100,000 full text articles were randomly selected, from various journals published in the last 10 years, from *ScienceDirect*[7], a large scientific article database.
2. The articles were analyzed for the presence of an acknowledgement section. If present, such a section was extracted under the assumption that often the information about the funding of the research work is mentioned here. A total of 60,271 acknowledgement sections were extracted.
3. *Elsevier's Fingerprint Engine (FPE)* was used on these sections to annotate *FBs* using the *CrossRef's Open Funder Registry*[8] as the vocabulary, which contains 12,928 funding organizations, both active and defunct.
4. As it contains abbreviations to several funding organizations, many false positives can be found in the results of this process, e.g., funding organisations that carry person's names were often mixed with real persons mentioned in the acknowledgements. To clean the dataset, a pre-trained 4-class *NER* model, provided with the *Stanford CoreNLP* toolkit, was used to detect person names from the same sections, and the overlapping annotations between the two aforementioned detections were removed as noise.
5. At the end of this step, the number of retained sections with at least one annotated *FB* resulted in 44,660, which constitute the *"silver"* dataset.

---

[7] http://www.sciencedirect.com/.

[8] http://www.crossref.org/fundingdata/registry.html.

The set was used for two purposes: firstly, learning the word clusters for the distributional similarity measure that can be employed within the *Stanford CoreNLP* toolkit; this was done by using the popular *Word2Vec* algorithm by Mikolov et al. [7], followed by k-means clustering to create the clusters, based on cosine-similarity. Secondly, it was used to train *NER* models for detecting *FB* labelled entities.

The *"Gold"* Set. A *"Gold"* set was also created for benchmarking, i.e., a manually curated and annotated set of scientific articles with *FB* and *GR* labels, the details of which were as follows:

1. 1,950 journal articles were picked randomly from a large number of scientific publishers, and sent to be annotated by three different professional annotators.
2. The annotators were provided with comprehensive guidelines, explaining the process and the entities, and examples.
3. A harmonization process took place, merging the annotations of the three experts; when all three agreed, annotations were automatically harmonized, whilst the disagreements between the annotators were resolved manually by a subject matter expert. From the 1,950 articles, 1,682 contained at least one funding-related annotation. As for the individual entities, a total of 4,537 *FB* and 3,156 *GR* annotations exist in the set.
4. In order to check the quality of the prepared dataset, pair-wise averaged *Cohen's kappa* [2] was used to calculate the inter-annotators agreement, which for this set was measured at 0.89, suggesting a high-quality dataset.

The *"gold"* set was used for two purposes: (1) to train the binary text classifier that detects the paragraphs of text which contain funding information, and (2) to train the *NER* components that detect *FB* and *GR* entities.

A summary of the *"gold"* set is provided in Table 1.

Table 1. A summary of the *"gold"* set.

Property	Numeric value
Documents	1950
Documents with funding annotation	1682
Paragraphs with funding information	1682
Paragraphs without funding annotation	47565
Funding bodies (*FB*)	4537
Grant IDs (*GR*)	3156

3.3 Detecting Text with Funding Information

As the first step of the proposed annotation framework, the text segments which contain funding information are to be separated from the rest. As explained in

Sect. 2.1, this problem can be formulated as a binary text classification problem. To address this problem, we have adopted in this work the usage of *Support Vector Machines (SVMs)*, which are known to perform favourably on text classification problems. More precisely, an *L2-regularized linear SVM* has been used, operating on *TF-IDF* vectors extracted from the segments of each input text, based on a bigram bag-of-words text. As text segments, the input text paragraphs were used. The *SVM* was trained on the examples of positive and negative segments, i.e., paragraphs with and without funding information, which could be found in the "gold" set described in the previous section.

Data: Trained annotators A, Set of annotated training text sections S
Result: Trained logistic regression R
Let us initialize an empty set of features F and labels L;
do
    do
        Use annotator A_j to annotate S_i, producing a set of annotations N_j;
        Let the unique pool of all such annotations be U_i;
    while $A_j \in A$;
    do
        Initialize a vector V of length $|A|$ filled with 0s;
        If $U_k \in N_j$, then $V_j = 1$. Do this for all j;
        Append V to F;
        If U_k is a true annotation for S_i, append "positive" to L, else "negative";
    while $U_k \in U_i$;
while $S_i \in S$;
Train R using F and L;

Algorithm 1. FundingFinder: Ensemble learning of base approaches for text annotation

---

3.4   Extracting Funding Information

Base Models. In order to annotate a piece of text with *FB* and *GR* labels, a variety of models were used: trained on the "gold" set, the "silver" set, or pre-trained models available in the toolkits. For *FB* labels the following models were used as base models: (1) pre-trained models packaged as part of the *Stanford CoreNLP* and *LingPipe* suites; in this work they were used to identify the "Organization" labels in the text, which were then stored as *FB*, (2) *Stanford CRF*, *LingPipe HMM* and *OpenNLP MaxEnt* models trained on the "silver set, and on folds of the "gold set, as a part of the 10-fold cross validation process we adopted for the experimental analysis, and, (3) *Stanford CRF* classifiers using distributional similarity features based on the word clusters created from external data, as described in Sect. 3.2, trained on the "gold" set, using the same folds as described in the previous point. For the *GR* labels the following base models were used: (1) a rule-based approach, considering every word inside the funding section with at least a digit, as a grant ID, and (2) *Stanford CRF*, *LingPipe HMM* and *OpenNLP MaxEnt* models trained on the "silver" set and on folds of the "gold" set, as was the case with the *FB* labels.

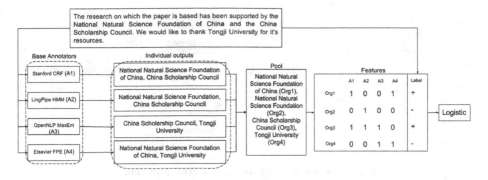

Fig. 1. An example of the *ensemble* approach for extracting funding information from text.

Ensembling. The base annotators described in the previous section are all examples of supervised learning algorithms or rule-based approaches. *Ensemble learning* combines the hypotheses from the different base approaches, in order to learn how to formulate a better hypothesis which takes the best of all (underlying) worlds, improving performance [9]. In this work we use a novel supervised ensembling mechanism for text annotation, using stacked generalization, such that the final classifier is a logistic regression model. In this approach, all of the annotated entities that are returned by the base annotators are first put in a pool, constructed by keeping all the unique entities across the annotators. Next, for each of the entities in the pool, a binary feature vector is constructed, based on whether it was positively identified by the respective annotators. The class label for the feature vector is positive if it is found to be a true positive from the training data, and negative if it's a false negative. The exact steps followed to train the ensemble mechanism are described in Algorithm 1. To give an example, let us consider the following piece of text:

The research on which this paper is based has been financially supported by the National Natural Science Foundation of China, and the China Scholarship Council. We would like to thank Tongji University for its resources.

In this example, the manually labelled terms are *National Natural Science Foundation of China* and *China Scholarship Council* as *FB*, but not the term *Tongji University* which is also an organization name, but in this case it is not acting explicitly as a *FB* according to the text. The ensemble's training process begins by annotating this piece of text using the base annotators. In this real example, the *Stanford CRF* base model annotates both entities correctly, the *HMM* model annotates the *China Scholarship Council* but misses the *National Natural Science Foundation of China* by wrongly annotating *National Natural Science Foundation* as *FB*, the *OpenNLP* implementation of the *MaxEnt* model also annotates the *China Scholarship Council* but confuses *Tongji University* as *FB*, and finally, the *Fingerprint Engine* misses *China Scholarship Council*,

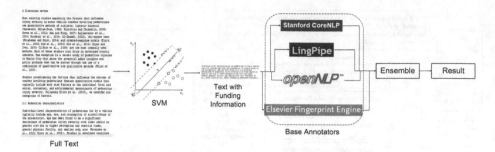

Fig. 2. Schematic showing the overall pipeline.

possibly because it is not part of the vocabulary used by the engine. Thus, we have four unique labels from all the annotators: *National Natural Science Foundation of China* and *China Scholarship Council*, which are both correct, and *National Natural Science Foundation* and *Tongji University*, which are wrong.

For each of the unique annotation, a vector with binary values is produced. In this example case, such a vector, for a single annotation, would have four components. Each component corresponds to a base annotator has a value of 1, if this annotation has been performed correctly by the respective base annotator, and 0 otherwise. In this specific example, the vector corresponding to the label *National Natural Science Foundation of China* will be: $[1, 0, 0, 1]$, since the *Stanford CRF model* and the *FPE* correctly identified it, but the *HMM* and the *MaxEnt* models did not. The class label of this data point is *positive*, as it has been manually labelled so. On repeating this process for all of the labels in the pool, we get a matrix of values where each row corresponds to the vector of values for an annotation label in the pool, and the features are binary values and correspond to whether a particular base annotator correctly identified that specific annotation label in the text. This data matrix is used for training a logistic regression model, which acts as the *ensemble* model. The process is illustrated in Fig. 1, while Fig. 2 shows the end-to-end pipeline.

4   Experiments and Results

In the following we present the experimental results on evaluating the different components of the pipeline. For the comparison of the alternative approaches, the *micro-averaged* measures of *Precision* (P), *Recall* (R), and *F1-score* ($F1$) were used.

4.1   Detecting Text with Funding Information

The results shown in Table 2 are obtained using a 10-fold cross-validation scheme, on the "*gold*" set data, described in Sect. 3.2. The results of two *SVM* setups are presented on the positive class: a linear *SVM*, and an *L2*-regularized *SVM* with the regularization (C) parameter set to 2, which makes the *SVM* robust

Table 2. Results for the identification of text with funding information using *SVM*.

Section	P	R	F1
SVM	99	5	9
L2-SVM (C = 2)	95	85	90

against overfitting. The difference in the $F1$ score is very big, indicating the problems faced in this case in the application of *SVM* in highly imbalanced sets. The *"gold"* set is highly skewed towards the negative examples of text containing funding information, leading the baseline linear *SVM* to very low recall levels. When switching to the *L2-SVM*, the model is now not affected by this skewness, which is also depicted by the much higher recall, at a very small precision cost. Therefore, these results indicate that the detection of text with funding information is possible, with an $F1$ score at 90%, using an *L2-SVM* as a trained model.

4.2 Extracting Funding Information

Tables 3 and 4 present the results of the annotation of *FB* and *GR* respectively, on the text segments which contain funding information, and which were identified by the previous step of the process. The tables present the results from all approaches discussed in this paper, and compare them against the suggested *FundingFinder* approach, which ensembles the base annotators in a novel manner. Besides two approaches listed as *FPE* and *Rule-based*, the tables list the compared approaches by using the naming convention: <*approach_name*>-{*Pre,S,G*}, where *Pre* indicates that the approach is used with one of the pretrained models in the original toolkit, *S* indicates that the approach was trained on the *"silver"* set, and *"G"* indicates that the approach was trained on the *"gold"* set. For example, *CRF-S* refers to the *Stanford Core NLP CRF* approach, as discussed in Sect. 2.2, and, more precisely, that in this case it was tested with a model trained on the *"silver"* set. The *FPE* and the *Rule-based* approaches do not require such a clarification, as they do not use any of these sets as training. All of the presented approaches were tested on the *"gold"* set, using 10–fold cross validation.

From Table 3 one can make the following observations:

- Amongst the pretrained models, the Fingerprint Engine performs most effectively, which can be attributed to it having a task-specific vocabulary. This may be regarded as the baseline and is a measure of what is possible in the absence of any training data.
- When trained with even a noisy (*"silver"*) dataset, all of the models perform at par and reasonably.
- The *CRF* models consistently perform more favourably than the other models in this task, which is not surprising, with an observed increased overall performance when the distributional similarity features are used.

Table 3. NER Results for Funding Body (FB) annotation label. Best performing model is highlighted in **bold** while the second best is in *italics*.

Method	P	R	F1
HMM-Pre	18($\pm$0)	31($\pm$0)	23($\pm$0)
CRF-Pre	35($\pm$0)	54($\pm$0)	42($\pm$0)
FPE	48($\pm$0)	46($\pm$0)	47($\pm$0)
CRF-S	49($\pm$0)	43($\pm$0)	46($\pm$0)
HMM-S	36($\pm$0)	48($\pm$0)	41($\pm$0)
MaxEnt-S	50($\pm$0)	39($\pm$0)	44($\pm$0)
CRF-G	64($\pm$.2)	58($\pm$.2)	61($\pm$.2)
CRF-dsim-G	*66($\pm$.2)*	*61($\pm$.3)*	*63($\pm$.2)*
HMM-G	49($\pm$.3)	54($\pm$.2)	52($\pm$.2)
MaxEnt-G	64($\pm$.4)	54($\pm$.2)	59($\pm$.3)
FundingFinder	**72($\pm$.3)**	**63($\pm$.2)**	**68($\pm$.3)**

Table 4. NER Results for Grant (GR) annotation label. Best performing model is highlighted in **bold** while the second best is in *italics*.

Method	P	R	F1
Rule-based	78($\pm$0)	89($\pm$0)	83($\pm$0)
CRF-G	*91($\pm$.1)*	*91($\pm$.08)*	*91($\pm$.1)*
HMM-G	76($\pm$.2)	77($\pm$.2)	76($\pm$.2)
MaxEnt-G	87($\pm$.2)	89($\pm$.1)	88($\pm$.2)
FundingFinder	**92($\pm$.1)**	**91($\pm$.1)**	**92($\pm$.1)**

– *FundingFinder*, which ensembles the base annotators, outperforms the next best model, namely the *CRF* with distributional similarity features, by 6 percentage points (*p.p.*) in precision, 2 p.p. in recall and 5 in $F1$ score, all of them being statistically significant with a p-value of less than 0.01%. This signifies that all of the base models perform differently on the task to finally form a positive symbiosis.

Similar findings can be drawn from Table 4 on the annotation of *GR*, with the main difference being that the absolute scores reported in this table are significantly higher from the ones reported in Table 3, for all approaches. This suggests that the annotation of *FB* is much harder than the annotation of *GR*, and also implies that the room for improvement by applying ensemble (*FundingFinder*) is smaller.

5 Conclusions

In this paper we have formally tackled the problem of funding information extraction from scientific articles. A benchmark dataset was created for comparative evaluation, a part of which is released in public. The results on this benchmark set indicate that the annotation of funding bodies (*FB*) is a significantly more difficult problem than the annotation of grants (*GR*). Additionally, by creating an ensemble of a number of base annotators, *FundingFinder* could perform the two tasks with $F1$ scores of 68% and 92% respectively, which is significantly more than the best-performing base annotator.

To conclude, the main contributions of the paper can be listed as follows:

1. We have discussed on the practically important problem of extracting funding information from text, and have experimentally provided a comprehensive overview of the state-of-the-art methods that could be used for the same. This may prove to be a significant head-start for researchers delving into the same problem for further research.
2. Empirically, we have shown that a small and high quality dataset is more suitable for this *NER* task than a larger, but noisier, dataset.
3. We have suggested an efficient two-stage pipeline for the task of funding information extraction.
4. A learning mechanism, based on an ensemble of state-of-the-art base annotators, was suggested, which should be easily extensible to any *NER* task.

References

1. Chieu, H.L.: Named entity recognition: a maximum entropy approach using global information. In: Proceedings of the 2002 International Conference on Computational Linguistics, pp. 190–196 (2002)
2. Cohen, J.: A coefficient of agreement for nominal scales. Educ. Psychol. Measur. **20**(1), 37 (1960)
3. Curran, J.R.: From distributional to semantic similarity. Ph.D. thesis, University of Edinburgh (2003)
4. Giles, C.L., Councill, I.G.: Who gets acknowledged: measuring scientific contributions through automatic acknowledgment indexing. Proc. Natl. Acad. Sci. U.S.A. **101**, 17599–17604 (2004)
5. Jonnalagadda, S., Topham, P.: NEMO: extraction and normalization of organization names from pubmed affiliation strings. J. Biomed. Discov. Collab. **5**, 50–75 (2010)
6. McCallum, A., Li, W.: Early results for named entity recognition with conditional random fields, feature induction and web-enhanced lexicons. In: Proceedings of the Seventh Conference on Natural Language Learning at HLT-NAACL 2003, vol. 4, pp. 188–191 (2003)
7. Mikolov, T., Sutskever, I., Chen, K., Corrado, G., Dean, J.: Distributed representations of words and phrases and their compositionality. In: Proceedings of the 26th International Conference on Neural Information Processing Systems, pp. 3111–3119 (2013)

328   S. Kayal et al.

8. Nadeau, D., Sekine, S.: A survey of named entity recognition and classification. Linguisticae Investig. **30**(1), 3–26 (2007)
9. Rokach, L.: Ensemble-based classifiers. Artif. Intell. Rev. **33**, 1–39 (2010)
10. Yu, W., Yesupriya, A., Wulf, A., Qu, J., Gwinn, M., Khoury, M.J.: An automatic method to generate domain-specific investigator networks using pubmed abstracts. BMC Med. Inf. Decis. Making **7**(1), 17 (2007)
11. Zhou, G., Su, J.: Named entity recognition using an HMM-based chunk tagger. In: Proceedings of the 40th Annual Meeting on Association for Computational Linguistics, pp. 473–480 (2002)

Speeding Up Budgeted Stochastic Gradient Descent SVM Training with Precomputed Golden Section Search

Tobias Glasmachers[✉] and Sahar Qaadan

Institute for Neural Computation, Ruhr University Bochum,
44801 Bochum, NRW, Germany
{tobias.glasmachers,sahar.qaadan}@ini.rub.de

Abstract. Limiting the model size of a kernel support vector machine to a pre-defined budget is a well-established technique that allows to scale SVM learning and prediction to large-scale data. Its core addition to simple stochastic gradient training is budget maintenance through merging of support vectors. This requires solving an inner optimization problem with an iterative method many times per gradient step. In this paper we replace the iterative procedure with a fast lookup. We manage to reduce the merging time by up to 65% and the total training time by 44% without any loss of accuracy.

1 Introduction

The Support Vector Machine (SVM; [5]) is a widespread standard machine learning method, in particular for binary classification problems. Being a kernel method, it employs a linear algorithm in an implicitly defined kernel-induced feature space [24]. SVMs yield high predictive accuracy in many applications [6,15,16,19,28]. They are supported by strong learning theoretical guarantees [1,9,12,17].

When facing large-scale learning, the applicability of support vector machines (and many other learning machines) is limited by their computational demands. Given n training points, training an SVM with standard dual solvers takes quadratic to cubic time in n [1]. Steinwart [23] established that the number of support vectors is linear in n, and so is the storage complexity of the model as well as the time complexity of each of its predictions. This quickly becomes prohibitive for large n, e.g., when learning from millions of data points.

Due to the prominence of the problem, a large number of solutions was developed. Parallelization can help [29,33], but it does not reduce the complexity of the training problem. One promising route is to solve the SVM problem only locally, usually involving some type of clustering [14,30] or with a hierarchical divide-and-conquer strategy [8,11]. An alternative approach is to leverage the progress in the domain of linear SVM solvers [10,13,32], which scale well to large data sets. To this end, kernel-induced feature representations are approximated

© Springer Nature Switzerland AG 2019
G. Nicosia et al. (Eds.): LOD 2018, LNCS 11331, pp. 329–340, 2019.
https://doi.org/10.1007/978-3-030-13709-0_28

by low-rank approaches [7,20,27,31], either a-priory using random Fourier features, or in a data-dependent way using Nyström sampling.

Budget methods, introducing an a-priori limit $B \ll n$ on the number of support vectors [18,25], go one step further by letting the optimizer adapt the feature space approximation during operation to its needs, which promises a comparatively low approximation error. The usual strategy is to merge support vectors at need, which effectively enables the solver to move support vectors around in input space. Merging decisions greedily minimize the approximation error.

In this paper we propose an effective computational improvement of this scheme. Finding the best merge partners, i.e., support vectors that induce the lowest approximation error when merged, is a rather costly operation. Usually, $\mathcal{O}(B)$ candidate pairs of vectors are considered, and for each pair an optimization problem is solved with an iterative strategy. By modelling the low-dimensional space of (solutions of the) optimization problems explicitly, we can remove the iterative process entirely, and replace it with a simple and fast lookup.

Our results show that merging-based budget maintenance can account for more than half of the total training time. Therefore reducing the merging time is a promising approach to speeding up training. The speed-up can be significant; on our largest data set we reduce the merging time by 65%, which corresponds to a reduction of the total training time by 44%. At the same time, our lookup method is at least as accurate as the original iterative procedure, resulting in nearly identical merging decisions and no loss of prediction accuracy.

The remainder of this paper is organized as follows. In the next section we introduce SVMs and stochastic gradient training on a budget. Then we analyze the computational bottleneck of the solver and develop a lookup smoothed with bilinear interpolation as a remedy. In Sect. 4 we benchmark the new algorithm against "standard" BSGD, and we investigate the influence of the algorithmic simplification on different budget sizes. Our results demonstrate systematic improvements in training time at no cost in terms of solution quality.

2 Support Vector Machine Training

In this section we introduce the necessary background: SVMs for binary classification, and training with stochastic gradient descent (SGD) on a budget, i.e., with a-priori limited number of support vectors.

Support Vector Machines. An SVM classifier is a supervised machine learning algorithm. In its simplest form it linearly separates two classes with a large margin. When applying a kernel function $k : X \times X \to \mathbb{R}$ over the input space X, the separation happens in a reproducing kernel Hilbert space (RKHS). For labeled data $((x_1, y_1), \ldots, (x_n, y_n)) \in (X \times \{-1, +1\})^n$, the prediction on $x \in X$ is computed as

$$\text{sign}\Big(\langle w, \phi(x)\rangle + b\Big) = \text{sign}\left(\sum_{j=1}^{n} \alpha_j k(x_j, x) + b\right)$$

with $w = \sum_{j=1}^{n} \alpha_j \phi(x_j)$, where $\phi(x)$ is an only implicitly defined feature map (due to Mercer's theorem, see also [24]) corresponding to the kernel function fulfilling $k(x, x') = \langle \phi(x), \phi(x')\rangle$. Training points x_j with non-zero coefficients $\alpha_j \neq 0$ are called support vectors; the summation in the predictor can obviously be restricted to this subset. The SVM model is obtained by minimizing the following (primal) objective function:

$$P(w,b) = \frac{\lambda}{2}\|w\|^2 + \frac{1}{n}\sum_{i=1}^{n} L\Big(y_i, \langle w, \phi(x_i)\rangle + b\Big). \tag{1}$$

Here, $\lambda > 0$ is a user-defined regularization parameter and $L(y, \mu) = \max\{0, 1 - y \cdot \mu\}$ denotes the hinge loss, which is a prototypical large margin loss, aiming to separate the classes with a functional margin $y \cdot \mu$ of at least one. The incorporation of other loss functions allows to generalize SVMs to other tasks like multi-class classification, regression, and ranking.

Primal Training. Problem (1) is a convex optimization problem without constraints. It has an equivalent dual representation as a quadratic program (QP), which is solved by several state-of-the-art "exact" solvers like LIBSVM [4] and thunder-SVM [26]. The main challenge is the high dimensionality of the problem, which coincides with the training set size n and can hence easily grow into the millions.

A simple method is to solve problem (1) directly with stochastic gradient descent (SGD), similar to neural network training. When presenting one training point at a time, as done in Pegasos [22], the objective function $P(w,b)$ is approximated by the unbiased estimate

$$P_i(w,b) = \frac{\lambda}{2}\|w\|^2 + L\Big(y_i, \langle w, \phi(x_i)\rangle + b\Big),$$

where the index $i \in \{1, \ldots, n\}$ follows a uniform distribution. The stochastic gradient $\nabla P_i(w,b)$ is an unbiased estimate of the "batch" gradient $\nabla P(w,b)$ but faster to compute by a factor of n, since it involves only a single training point. Starting from $(w,b) = (0,0)$, SGD updates the weights according to

$$(w,b) \leftarrow (w,b) - \eta_t \cdot \nabla P_{i_t}(w,b),$$

where t is the iteration counter. With a learning rate $\eta_t \in \Theta(1/t)$ it is guaranteed to converge to the optimum of the convex training problem [2].

With a sparse representation $w = \sum_{(\alpha, \tilde{x}) \in M} \alpha \cdot \phi(\tilde{x})$ the SGD update decomposes into the following algorithmic steps. We scale down all coefficients α uniformly by the factor $1 - \lambda \cdot \eta_t$. If the margin $y_i(\langle w, \phi(x_i)\rangle + b)$ happens to be less than one, then we add a new point $\tilde{x} = x_i$ with coefficient $\alpha = \eta_t \cdot y_i$ to the model M. With a dense representation holding one coefficient α_i per data point (x_i, y_i) we would add the above value to α_i. The most costly step is the computation of $\langle w, \phi(x_i) \rangle$, which is linear in the number of support vectors (SVs), and hence generally linear in n [23].

SVM Training on a Budget. Budgeted Stochastic Gradient Descent (BSGD) breaks the unlimited growth in model size and update time for large data streams by bounding the number of support vectors during training. The upper bound $B \ll n$ is the budget size. Per SGD step the algorithm can add at most one new support vector; this happens exactly if (x_i, y_i) does not meet the target margin of one (and α_i changes from zero to a non-zero value). After $B + 1$ such steps, the budget constraint is violated and a dedicated budget maintenance algorithm is triggered to reduce the number of support vectors to at most B. The goal of budget maintenance is to fulfill the budget constraint with the smallest possible change of the model, measured by $\|\Delta\|^2 = \|w' - w\|^2$, where w is the weight vector before and w' is the weight vector after budget maintenance. $\Delta = w' - w$ is referred to as the weight degradation.

Budget maintenance strategies are investigated in detail in [25]. It turns out that *merging* of two support vectors into a single new point is superior to alternatives like removal of a point and projection of the solution onto the remaining support vectors. Merging was first proposed in [18] as a way to efficiently reduce the complexity of an already trained SVM. With merging, the complexity of budget maintenance is governed by the search for suitable merge partners, which is $\mathcal{O}(B^2)$ for all pairs, while it is common to apply the $\mathcal{O}(B)$ heuristic resulting from fixing the point with smallest coefficient α_i as a first partner.

When merging two support vectors x_i and x_j, we aim to approximate $\alpha_i \cdot \phi(x_i) + \alpha_j \cdot \phi(x_j)$ with a new term $\alpha_z \cdot \phi(z)$ involving only a single point z. Since the kernel-induced feature map is usually not surjective, the pre-image of $\alpha_i \phi(x_i) + \alpha_j \phi(x_j)$ under ϕ is empty [3,21] and no exact match z exists. Therefore the weight degradation $\Delta = \alpha_i \phi(x_i) + \alpha_j \phi(x_j) - \alpha_z \phi(z)$ is non-zero. For the Gaussian kernel $k(x, x') = \exp(-\gamma \|x - x'\|^2)$, due to its symmetries, the point z minimizing $\|\Delta\|^2$ lies on the line connecting x_i and x_j and is hence of the form $z = hx_i + (1 - h)x_j$. For $y_i = y_j$ we obtain a convex combination $0 < h < 1$, otherwise we have $h < 0$ or $h > 1$. In this paper we merge only vectors of equal label. For each choice of z, the optimal value of α_z is obtained in closed form: $\alpha_z = \alpha_i k(x_i, z) + \alpha_j k(x_j, z)$. This turns minimization of $\|\Delta\|^2 = \alpha_i^2 + \alpha_j^2 - \alpha_z^2 + 2k(x_i, x_j)$ into a one-dimensional non-linear optimization problem, which is solved in [25] with golden section line search. The calculations are further simplified by the relations $k(x_i, z) = k(x_i, x_j)^{(1-h)^2}$ and $k(x_j, z) = k(x_i, x_j)^{h^2}$, which save costly kernel functions evaluations.

Budget maintenance in BSGD usually works in the following sequence of steps, see Algorithm 1: First, x_i is fixed to the support vector with minimal coefficient $|\alpha_i|$. Then the best merge partner x_j is determined by testing B pairs (x_i, x_j), $j \in \{1, \ldots, B + 1\} \setminus \{i\}$. Golden section search is run for each of these steps to determine h to fixed precision $\varepsilon = 0.01$. The weight degradation is computed using the shortcuts mentioned above. Finally, the candidate with minimal weight degradation is selected and the vectors are merged. Hence, although a single golden search search is fast, the need to run it many times per SGD iteration turns it into a rather costly operation.

Algorithm 1. Procedure Budget Maintenance for a sparse model M

---

1 **Input/Output:** model M
2 $(\alpha_{\min}, \tilde{x}_{\min}) \leftarrow \arg\min \{|\alpha| \,|\, (\alpha, \tilde{x}) \in M\}$
3 $WD^* \leftarrow \infty$
4 **for** $(\alpha, \tilde{x}) \in M \setminus \{(\alpha_{\min}, \tilde{x}_{\min})\}$ **do**
5 $\quad m \leftarrow \alpha/(\alpha + \alpha_{\min})$
6 $\quad \kappa \leftarrow k(\tilde{x}, \tilde{x}_{\min})$
7 $\quad h \leftarrow \arg\max \{m\kappa^{(1-h')^2} + (1-m)\kappa^{h'^2} \,|\, h' \in [0,1]\}$
8 $\quad \alpha_z \leftarrow \alpha_{\min} \cdot \kappa^{(1-h)^2} + \alpha \cdot \kappa^{h^2}$
9 $\quad WD \leftarrow \alpha_{\min}^2 + \alpha^2 - \alpha_z^2 + 2 \cdot \alpha_{\min} \cdot \alpha \cdot \kappa$
10 $\quad$ **if** $(WD < WD^*)$ **then**
11 $\quad\quad WD^* \leftarrow WD$
12 $\quad\quad (\alpha^*, \tilde{x}^*, h^*, \kappa^*) \leftarrow (\alpha, \tilde{x}, h, \kappa)$
13 $z \leftarrow h^* \cdot \tilde{x}_{\min} + (1 - h^*) \cdot \tilde{x}^*$
14 $\alpha_z \leftarrow \alpha_{\min} \cdot (\kappa^*)^{(1-h^*)^2} + \alpha^* \cdot (\kappa^*)^{(h^*)^2}$
15 $M \leftarrow M \setminus \{(\alpha_{\min}, \tilde{x}_{\min}), (\alpha^*, \tilde{x}^*)\} \cup \{(\alpha_z, z)\}$

A theoretical analysis of BSGD is provided by [25]. Their Theorem 1 establishes a bound on the error induced by the budget, ensuring that asymptotically the error is governed only by the (unavoidable) weight degradation.

3 Precomputing the Merging Problem

The merging problem for given support vectors x_i and x_j with coefficients α_i and α_j is illustrated in Fig. 1. Our central observation is that the geometry depends only on the (cosine of the) angle between $\alpha_i\phi(x_i)$ and $\alpha_j\phi(x_j)$, and on the relative lengths of the two vectors. These two quantities are captured by the parameters

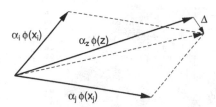

Fig. 1. The merging problem.

- relative length $m = \alpha_i/(\alpha_i + \alpha_j)$
- cosine of the angle $\kappa = k(x_i, x_j)$,

both of which take values in the unit interval. The optimal merging coefficient h is a function of m and κ, and so is the resulting weight degradation $WD = \|\Delta\|^2$. Therefore we can express h and WD as functions of m and κ, denoted as $h(m, \kappa)$

and $WD(m, \kappa)$ in the following. The functions can be evaluated to any given target precision by running the golden section search. Their graphs are plotted in Figs. 2a and b.

If the functions h or WD can be approximated efficiently then there is no need to run a potentially costly iterative procedure like golden section search. This is our core technique for speeding up the BSGD method.

The functions blend between different budget maintenance strategies. While for $\kappa \gg 0$ and for $m \approx 1/2$ it is beneficial to merge the two support vectors, resulting in $h \in (0, 1)$, this is not the case for $\kappa \ll 1$ and $m \approx 0$ or $m \approx 1$, resulting in $h \approx 0$ or $h \approx 1$, which is equivalent to removal of the support vector with smaller coefficient. This means that in order to obtain a close fit that works well in both regimes we may need a quite flexible function class like a kernel method or a neural network, while a simple polynomial function can give poor fits, with large errors close to the boundaries.

A much simpler and computationally very cheap approach is to pre-compute the function on a grid covering the domain $[0, 1] \times [0, 1]$. The values need to be pre-computed only once, and here we can afford to apply golden section search with high precision; we use $\varepsilon = 10^{-10}$. Then, given two merge candidates, we can look up an approximate solution by rounding m and κ to the nearest grid point. The approximation quality can be improved significantly through bilinear interpolation. On modern PC hardware we can easily afford a large grid with millions of points, however, this is not even necessary to obtain excellent results. In our experiments we use a grid of size 400×400.

Bilinear interpolation is fast, and moreover it is easy to implement. When looking up $h(m, \kappa)$ this way, we obtain a plug-in replacement for golden section search in BSGD. However, we can equally well look up $WD(m, \kappa)$ instead to save additional computation steps. Another benefit of WD over h is regularity, see Figs. 2a and b and the following lemma.

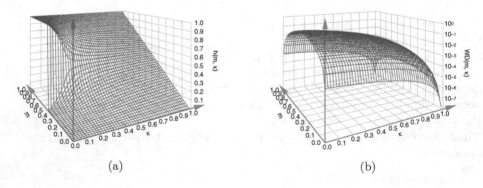

(a)   (b)

Fig. 2. Graphs of the functions $h(m, \kappa)$ (a) and $WD(m, \kappa)$ (b). The latter uses a log scale on the value axis.

Lemma 1. *The functions h and WD are smooth for $\kappa > e^{-2}$. The function h is continuous outside the set $Z = \{1/2\} \times [0, e^{-2}] \subset [0,1]^2$ and discontinuous on Z. The function WD is everywhere continuous.*

Proof. The function $s_{m,\kappa}(h') = m\kappa^{(1-h')^2} + (1-m)\kappa^{h'^2}$ used in line 7 of Algorithm 1 inside the $\arg\max$ expression is a weighted sum of two Gaussian kernels. Depending on the parameters m and κ, it can have one or two modes. It has two modes for parameters in Z, as can be seen from an elementary calculation yielding $s''_{1/2,\kappa}(1/2) > 0 \Leftrightarrow \kappa < e^{-2}$. Due to symmetry, the dominant mode switches at $m = 1/2$. The inverse function theorem applied to branches of $s_{m,\kappa}$ implies that $h(m,\kappa) = \arg\max_{h'}\{s_{m,\kappa}(h')\}$ and $WD(m,\kappa) = (\alpha_i + \alpha_j) \cdot (m^2 + (1-m)^2 - [s_{m,\kappa}(h(m,\kappa))]^2 + 2m(1-m)\kappa)$ vary smoothly with their parameters as long as the same mode is active. The maximum operation is continuous, and so is WD. For each m there is a critical value of $\kappa \leq e^{-2}$ where $s_{m,\kappa}$ switches from one to two modes. We collect these parameter configurations in the set N. On N (in contrast to Z), h is continuous. With the same argument as above, h and WD are smooth outside $N \cup Z$. $\square$

Bilinear interpolation is well justified if the function is continuous, and differentiable within each grid cell. The above lemma ensures this property for $\kappa > e^{-2}$, and it furthermore indicates that for its continuity, interpolating WD is preferable over interpolating h. The regime $\kappa < e^{-2}$ corresponds to merging two points in a distance of more than two "standard deviations" of the Gaussian kernel. This is anyway undesirable, since it can result in a large weight degradation. In fact, if $s_{m,\kappa}$ has two modes, then the optimal merge is close to the removal of one of the points, which is known to give poor results [25].

4 Experimental Evaluation

In this section we evaluate our method empirically, with the aim to investigate its properties more closely, and to demonstrate its practical value. To this end, we'd like to answer the following questions:

1. Which speed-up is achievable?
2. Do we pay for speed-ups with reduced test accuracy?
3. How do results depend on the budget size?
4. How much do merging decision differ from the original method?

To answer these questions we compare our algorithm to "standard" BSGD with merging based on golden section search. We have implemented both algorithms in C++; the implementation is available from the first author's homepage.[1] We train SVM models on the binary classification problems SUSY, SKIN, IJCNN, ADULT, WEB, and PHISHING, covering a range of different sizes. The regularization parameter $C = \frac{1}{n \cdot \lambda}$ and the kernel parameter γ were tuned on a grid of the form $\log_2(C), \log_2(\gamma) \in \mathbb{Z}$ using 10-fold cross-validation. The data sets are summarized in Table 1. SVMs were trained with 20 passes through the data, except for the huge SUSY data, where we used a single pass.

---

[1] https://www.ini.rub.de/the_institute/people/tobias-glasmachers/#software .

Table 1. Data sets used in this study, hyperparameter settings, and test accuracy of the exact SVM model found by LIBSVM.

data set	size	features	C	γ	accuracy	data set	size	features	C	γ	accuracy
SUSY	4,500,000	18	2^5	2^{-7}	79.79%	ADULT	32,561	123	2^5	2^{-7}	84.82%
SKIN	183,793	3	2^5	2^{-7}	99.96%	WEB	17,188	300	2^3	2^{-5}	98.81%
IJCNN	49,990	22	2^5	2^1	98.77%	PHISHING	8,315	68	2^3	2^3	97.55%

To answer the first question, we trained SVM models with BSGD, comparing golden section search (GSS) with our new algorithms looking up $h(m, \kappa)$ (Lookup-h) or $WD(m, \kappa)$ (Lookup-WD). For reference, we also ran golden section search with precision $\varepsilon = 10^{-10}$ (GSS-precise). We used two different budget sizes for each problem.

All methods found SVM models with comparable accuracy as shown in Table 2; in fact, in most cases the systematic differences are below one standard deviation of the variability between different runs.[2] In contrast, the time spent on budget maintenance differs significantly between the methods. In Fig. 3 we provide a detailed breakdown of the merging time, obtained with a profiler.

Lookup-WD and Lookup-h are faster than GSS, which is (unsurprisingly) faster than GSS-precise. The results are very systematic, see Table 3 and Fig. 3. The greatest savings of about 44% of the total training time are observed for the rather large SUSY data set. Although the speed-up can also be insignificant, like for the WEB data, lookup is never slower than GSS. The actual saving depends on the cost of kernel computations and on the fraction of SGD iterations in which merging occurs. The latter quantity, which we refer to as the merging frequency, is provided in Table 3. We observe that the savings shown in Fig. 3 nicely correlate with the merging frequency.

The profiler results provide a more detailed understanding of the differences: replacing GSS with Lookup-h significantly reduces the time for computing $h(m, \kappa)$. Replacing Lookup-h with Lookup-WD removes further steps in the calculation of $WD(m, \kappa)$, but practically speaking the difference is hardly noticeable.

Overall, our method offers a systematic speed-up. The speed-up does not come at any cost in terms of solution precision. This answers the first two questions.

If the budget size is chosen so large that merging is never needed then all tested methods coincide, however, this defeats the purpose of using a budget in the first place. We find that the merging frequency is nearly independent of the budget size as long as the budget is significantly smaller than the number of support vectors of the full kernel SVM model, and hence the fraction of runtime saved is independent of the budget size. The results in Fig. 3 are in line with this expectation, answering the third question.

---

[2] Note that with increasing number of passes (or epochs) the standard deviation does not tend to zero since the training problem is non-convex due to the budget constraint.

Table 2. Test accuracy achieved by the different methods, averaged over 5 runs at different budget sizes.

Data set	Budget size	Test accuracy GSS-precise	Test accuracy GSS-standard	Test accuracy Lookup-h	Test accuracy Lookup-WD
SUSY	100	76.975 ± 1.372	76.628 ± 2.030	76.934 ± 1.426	76.884 ± 1.261
	500	76.989 ± 3.109	75.583 ± 3.0558	75.581 ± 2.558	75.570 ± 3.925
SKIN	100	99.621 ± 0.711	99.629 ± 0.852	99.621 ± 0.201	99.617 ± 0.877
	200	99.868 ± 0.033	99.877 ± 0.053	99.855 ± 0.054	99.754 ± 0.089
IJCNN	100	97.141 ± 0.317	96.807 ± 0.344	97.132 ± 0.371	97.130 ± 0.363
	500	98.138 ± 0.158	98.055 ± 0.334	98.113 ± 0.448	98.070 ± 0.372
ADULT	100	84.234 ± 0,883	84.166 ± 0.701	84.164 ± 0.988	84.200 ± 0.798
	500	84.280 ± 0.800	83.739 ± 1.303	83.836 ± 1.157	83.949 ± 1.001
WEB	100	98.805 ± 0,026	98.793 ± 0.027	98.783 ± 0.045	98.793 ± 0.039
	500	98.809 ± 0,023	98.781 ± 0.047	98.799 ± 0.029	98.807 ± 0.016
PHISHING	100	96.554 ± 0.158	96.254 ± 0.301	96.539 ± 0.242	96.389 ± 0.371
	500	97.555 ± 0.187	97.517 ± 0.292	97.518 ± 0.280	97.525 ± 0.201

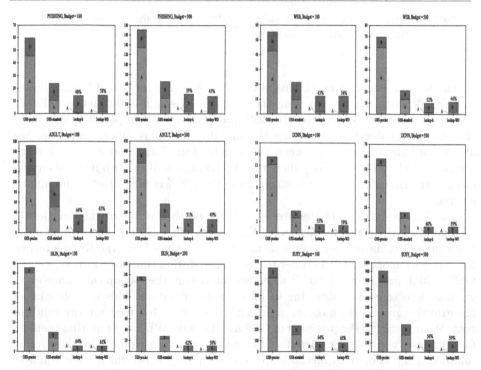

Fig. 3. Breakdown of the merging time in seconds for GSS-precise, GSS, Lookup-h and Lookup-WD. Section A represents the time invested to compute h using either golden section search or lookup. For the Lookup-WD method the same bar represents the look-up of $WD(m, \kappa)$. Section B summarizes all other operations like loop overheads, the computation of α_z, and the construction of the final merge vector z. The numbers on top of the columns for the lookup methods indicate the saving over GSS.

Table 3. Relative improvement of the total training time with respect to golden section search averaged over 5 runs (Lookup-h vs. GSS-standard and lookup-WD vs. GSS-standard), and fraction of merging events for budget size 100 and statistics on the quality of merging decisions (refer to the text for details).

Data set	Budget size	Lookup-h vs. GSS-standard	Lookup-WD vs. GSS-standard	Merging frequency	Equal merging decisions	Factor GSS	Factor lookup-WD
SUSY	100	43.911%	43.396%	43%	93.64%	1.01795	1.00733
	500	39.201%	39.199%				
SKIN	100	20.515%	17.788%	16%	74.31%	1.00047	1.00005
	200	14.173%	14.900%				
IJCNN	100	28.091%	30.372%	17%	91.79%	1.02429	1.00149
	500	30.569%	29.861%				
ADULT	100	21.627%	18.452%	32%	92.54%	1.05064	1.00402
	500	22.334%	22.339%				
WEB	100	3.053%	5.649%	6%	93.77%	1.00255	1.00039
	500	7.483%	0.508%				
PHISHING	100	15.385%	13.946%	21%	96.96%	1.00055	1.00008
	500	7.563%	10.924%				

In the next experiment we have a closer look at the impact of lookup-based merging decisions by investigating the behavior in single iterations, as follows. During a run of BSGD we execute GSS and Lookup-WD in parallel. We count the number of iterations in which the merging decisions differ, and if so, we also record the difference between the weight degradation values. The results are presented in Table 3. They show that the decisions of the two methods agree most of the time, for some problems in more than 99% of all budget maintenance events.

Finally, we investigate the precision with which the weight degradation is estimated by the different methods. While GSS can solve the problem to arbitrary precision, the reference implementation determines $h(m, \kappa)$ only to a rather loose precision of $\varepsilon = 0.01$ in order to save computation time. In contrast, we ran GSS to high precision $\varepsilon = 10^{-10}$ when precomputing the lookup table, however, we may lose some precision due to bilinear interpolation. This loss shrinks as the grid size grows, which comes at added storage cost, but without any runtime cost. We investigate the precision of GSS and Lookup-WD by comparing them to GSS-precise, which is considered a reasonable approximation of the exact minimum of $\|\Delta\|^2$. For both methods we record the factor by which their squared weight degradations exceed the minimum, see Table 3. All factors are very close to one, hence none of the algorithms is wasteful in terms of weight degradation, and indeed Lookup-WD with a grid size of 400×400 is more precise on all 6 data sets. This answers our last question.

5   Conclusion

We have proposed a fast lookup as a plug-in replacement for the iterative golden section search procedure required when merging support vectors in large-scale kernel SVM training. The new method compares favorably to the iterative baseline in terms of training time: it offers a systematic speed-up, resulting in computational savings of up to 65% of the merging time and up to 44% of the total training time, while the training time is never increased. With our method, nearly the full computation time is spent on actual SGD steps, while the fraction of efforts spent on budget maintenance can be reduced significantly. We have demonstrated that our approach results in virtually indistinguishable and even slightly more precise merging decisions. It is for this reason that the speed-up comes at absolutely no cost in terms of predictive accuracy.

Acknowledgments. We acknowledge support by the Deutsche Forschungsgemeinschaft (DFG) through grant GL 839/3-1.

References

1. Bottou, L., Lin, C.J.: Support Vector Machine Solvers, pp. 1–28. MIT Press, Cambridge (2007)
2. Bottou, L.: Large-scale machine learning with stochastic gradient descent. In: Lechevallier, Y., Saporta, G. (eds.) COMPSTAT 2010, pp. 177–186. Physica-Verlag, Heidelberg (2010). https://doi.org/10.1007/978-3-7908-2604-3_16
3. Burges, C.J.: Simplified support vector decision rules, pp. 71–77. Morgan Kaufmann (1996)
4. Chang, C.C., Lin, C.J.: LIBSVM: a library for support vector machines. ACM Trans. Intell. Syst. Technol. **2**, 27 (2011)
5. Cortes, C., Vapnik, V.: Support-vector networks. Mach. Learn. **20**(3), 273–297 (1995)
6. Cui, J., Li, Z., Lv, R., Xu, X., Gao, J.: The application of support vector machine in pattern recognition. IEEE Trans. Control Autom. (2007)
7. Fine, S., Scheinberg, K.: Efficient SVM training using low-rank kernel representations. J. Mach. Learn. Res. **2**(Dec), 243–264 (2001)
8. Graf, H.P., Cosatto, E., Bottou, L., Dourdanovic, I., Vapnik, V.: Parallel support vector machines: the cascade SVM. In: NIPS (2005)
9. Hare, S., et al.: Struck: structured output tracking with kernels. IEEE Trans. Pattern Anal. Mach. Intell. **38**(10), 2096–2109 (2016)
10. Hsieh, C.J., Chang, K.W., Lin, C.J., Keerthi, S.S., Sundararajan, S.: A dual coordinate descent method for large-scale linear SVM. In: ICML (2008)
11. Hsieh, C.J., Si, S., Dhillon, I.: A divide-and-conquer solver for kernel support vector machines. In: International Conference on Machine Learning (ICML), pp. 566–574 (2014)
12. Joachims, T.: Text categorization with support vector machines: learning with many relevant features. In: Nédellec, C., Rouveirol, C. (eds.) ECML 1998. LNCS, vol. 1398, pp. 137–142. Springer, Heidelberg (1998). https://doi.org/10.1007/BFb0026683

13. Joachims, T.: Making large-scale SVM learning practical. In: Advances in Kernel Methods - Support Vector Learning. MIT Press, Cambridge (1999)
14. Ladicky, L., Torr, P.: Locally linear support vector machines. In: International Conference on Machine Learning (ICML), pp. 985–992 (2011)
15. Lewis, D.P., Jebara, T., Noble, W.S.: Support vector machine learning from heterogeneous data: an empirical analysis using protein sequence and structure. Bioinformatics **22**(22), 2753–2760 (2006)
16. Lin, G., Shen, C., Shi, Q., van den Hengel, A., Suter, D.: Fast supervised hashing with decision trees for high-dimensional data. In: The IEEE Conference on Computer Vision and Pattern Recognition (CVPR) (2014)
17. Mohri, M., Rostamizadeh, A., Talwalkar, A.: Foundations of Machine Learning. MIT Press, Cambridge (2012)
18. Nguyen, D., Ho, T.: An efficient method for simplifying support vector machines. In: Proceedings of the 22nd ICML, pp. 617–624 (2005)
19. Noble, W.S.: Support vector machine applications in computational biology. In: Schölkopf, B., Tsuda, K., Vert, J.P. (eds.) Kernel Methods in Computational Biology. MIT Press, Cambridge (2004)
20. Rahimi, A., Recht, B.: Random features for large-scale kernel machines. NIPS **3**(4) (2007)
21. Schölkopf, B., et al.: Input space versus feature space in kernel-based methods. IEEE Trans. Neural Netw. **10**(5), 1000–1017 (1999)
22. Shalev-Shwartz, S., Singer, Y., Srebro, N., Cotter, A.: Pegasos: primal estimated sub-GrAdient SOlver for SVM. Math. Program. **127**(1), 3–30 (2011)
23. Steinwart, I.: Sparseness of support vector machines. J. Mach. Learn. Res. **4**(Nov), 1071–1105 (2003)
24. Vapnik, V.: The Nature of Statistical Learning Theory. Springer, New York (1995). https://doi.org/10.1007/978-1-4757-2440-0
25. Wang, Z., Crammer, K., Vucetic, S.: Breaking the curse of kernelization: budgeted stochastic gradient descent for large-scale SVM training. J. Mach. Learn. Res. **13**, 3103–3131 (2012)
26. Wen, Z., Shi, J., He, B., Li, Q., Chen, J.: Thunder-SVM (2017). https://github.com/zeyiwen/thundersvm
27. Mu, Y., Hua, G., Fan, W., Chang, S.F.: Hash-SVM: scalable kernel machines for large-scale visual classification. In: IEEE Conference on Computer Vision and Pattern Recognition (2014)
28. Yu, J., Xue, A., Redei, E., Bagheri, N.: A support vector machine model provides an accurate transcript-level-based diagnostic for major depressive disorder. Transl. Psychiatry **6**(10), e931 (2016). https://doi.org/10.1038/tp.2016.198
29. Zanni, L., Serafini, T., Zanghirati, G.: Parallel software for training large scale support vector machines on multiprocessor systems. J. Mach. Learn. Res. **7**, 1467–1492 (2006)
30. Zhang, H., Berg, A.C., Maire, M., Malik, J.: SVM-KNN: discriminative nearest neighbor classification for visual category recognition. In: Conference on Computer Vision and Pattern Recognition, vol. 2, pp. 2126–2136. IEEE (2006)
31. Zhang, K., Lan, L., Wang, Z., Moerchen, F.: Scaling up kernel SVM on limited resources: a low-rank linearization approach. In: AISTATS (2012)
32. Zhang, T.: Solving large scale linear prediction problems using stochastic gradient descent. In: International Conference on Machine Learning (2004)
33. Zhu, Z.A., Chen, W., Wang, G., Zhu, C., Chen, Z.: P-packSVM: parallel primal gradient descent kernel SVM. In: IEEE International Conference on Data Mining (2009)

An Unsupervised Learning Classifier with Competitive Error Performance

Daniel N. Nissani (Nissensohn)$^{(\boxtimes)}$

Tel Aviv, Israel
dnissani@post.bgu.ac.il

Abstract. An unsupervised learning classification model is described. It achieves classification error probability competitive with that of popular supervised learning classifiers such as SVM or kNN. The model is based on the incremental execution of small step shift and rotation operations upon selected discriminative hyperplanes at the arrival of input samples. When applied, in conjunction with a selected feature extractor, to a subset of the ImageNet dataset benchmark, it yields 6.2% Top 3 probability of error; this exceeds by merely about 2% the result achieved by (supervised) k-Nearest Neighbor, both using same feature extractor. This result may also be contrasted with popular unsupervised learning schemes such as k-Means which is shown to be practically useless on same dataset.

Keywords: Unsupervised learning · Linear classifiers · Neural models

1 Introduction

Unsupervised learning methods have been explored significantly during the last decades. Part of this work has been oriented towards the solution of the so called Classification function. Most of this research resulted in schemes which are based on either similarity (or alternatively distance) measures (such as the popular k-Means method and its variants, e.g. MacQueen [9]) or on the estimation of density mixtures (e.g. Duda and Hart [1], Bishop [5]); both approaches require the user to supply significant prior information concerning the underlying model (such as number k of distinct classes for both k-Means and parametric models, and the assumed probability distribution function class for the later).

Our work is inspired by the 'cluster assumption': that real world classes are separated by low probability density regions in an appropriate feature space. This assumption has naturally lead us and others to the so called 'valley seeking' methods which have been explored since the 70's (Lewis et al. [11]) and up to the more recent work by Pavlidis et al. [12]. Many of these works, including this last quoted, require the conduction of a density estimation process (parametric or non-parametric); none of them makes *implicit* use of the density mixture as in our present work; and none of them has provided, to the best of our knowledge, satisfactory experimental results on real life challenging problems such as ImageNet dataset classification. It should be noted that these unsupervised learning schemes, relevant to the aforementioned Classification function, should be distinguished from, and should not be confused with,

© Springer Nature Switzerland AG 2019
G. Nicosia et al. (Eds.): LOD 2018, LNCS 11331, pp. 341–356, 2019.
https://doi.org/10.1007/978-3-030-13709-0_29

unsupervised learning schemes relevant to the canonically associated Feature Extraction function, such as Generative Adversarial Networks (GAN, e.g. Donahue et al. [2]), Variational Autoencoders (VAE, e.g. Pu et al. [3]), Restricted Boltzmann Machines (RBM, e.g. Ranzato and Hinton [4]), Transfer Learning (Yosinski et al. [10]), etc.

In the sequel we present initial results on an unsupervised learning classifier exhibiting error performance competitive to that of popular near optimal supervised learning classifiers such as k-Nearest Neighbors (kNN) and Support Vector Machine (SVM). The proposed model is presented in the next Section. In Sect. 3 we present simulation results; in Sect. 4 we summarize our work and make a few concluding remarks.

2   Proposed Model

Our model deals with the unsupervised learning Classifier function of a Pattern Recognition machine. Measurement vectors $v \in R^D$, where D is the dimension of the measurement or observed space, are fed into a Feature Extractor. These vectors may consist of image pixel values, digitized speech samples, etc. These vectors v are mapped by the Feature Extractor into so called feature vectors $x \in R^d$, where d denotes the dimension of the feature space. Such a transformation or mapping is supposedly capable of yielding better inter-class separability properties. The feature vectors x are then fed into a Classifier whose ultimate role is to provide an output code vector y (or equivalently a label, or a class name) providing discrete information regarding the class to which the input measurement vector v belongs. The classifier may be of supervised learning type in which case pre-defined correct labels are fed into it simultaneously with input vectors v during the so called training stage, or of unsupervised learning type, as in our case, where no such labels are presented at any time.

It is a fundamental assumption of our model that the feature space vectors are distributed in accordance with some unknown probability distribution mixture. This mixture is composed of the weighted sum of conditional distribution densities. We assume that the bulks of this mixture occupy a bounded region of feature space. We further assume that said mixture obeys a sufficient condition, informally and loosely herein stated, namely that the intersecting slopes of the weighted conditionals form well defined low density regions, or 'valleys' (the so called 'cluster assumption'). Conditional probability densities should be normalizable by definition; since this forces their density to eventually decay (or vanish) then this sufficient condition is reasonable and commonly met. Note that in spite this assumption our model is not parametric, that is no assumption is made regarding the functional form of the distributions involved, nor about the number of presented classes. The resultant valleys can, in general, take the form of non-linear hyper-surfaces; like any linear classifier (e.g. linear SVM) we finally assume that, by virtue of an appropriate Feature Extraction function, these hyper-surfaces can be reasonably approximated by simple hyperplanes.

Our proposed model is initiated by populating the relevant feature space bounded region, with a certain number, or pool, of hyperplanes. These can be either uniformly randomly, or orderly and uniformly grid-wise distributed within this region. Following this initiation, input feature vectors are presented. Upon arrival of one such input vector, each such hyperplane generates an output signal which indicates whether this input sample resides on one side of the hyperplane or the other. In addition, in response to this input vector, some of the hyperplanes are shifted and/or rotated by small increments, gradually migrating from regions characterized by high probability density to regions of low probability density, and thus converging (in the mean) towards said density mixture 'valleys', by means of a mechanism to be described in the sequel.

It is convenient to commence description of our model by means of a simpler model; refer to Fig. 1 which depicts a 1-dimensional probability density mixture p (x) along with some fundamental entities related to our proposed model. When conditional densities $p(x|C_i)$ of this mixture are sufficiently spaced and have appropriate priors $p(C_i)$ their mixture creates a minimal point (a 'valley') such as point **a** in Fig. 1. The 2-classes mixture density of Fig. 1 can be expressed as

$$p(\mathbf{x}) = p(C_1)p(\mathbf{x}|C_1) + p(C_2)p(\mathbf{x}|C_2) \tag{1}$$

In our simple model of Fig. 1 there exists a single hyperplane (reduced to a single point in our 1-dimensional model) defined by the hyperplanar semi-linear operator (other hyperplanar operators may be considered)

$$y_{t+1} = \max\{0, \mathbf{w}_t \cdot \mathbf{x} - \theta_t\} \tag{2}$$

where the hyperplane weight vector is $\mathbf{w}_t \in R^d$ (d = 1, and $\mathbf{w}_t = 1$ is assumed in this introductory model), the hyperplane threshold variable is $\theta_t \in R$, the t subscript indicates the hyperplane self-time, "." denotes inner product operation, and where individual hyperplane indexes (only one in this model) are omitted for readability.

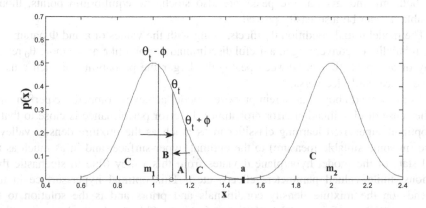

Fig. 1. A 1-dimensional density mixture and proposed model related entities

Hyperplanes hence act in this model as linear discriminator functions. One of these hyperplanes is indicated in Fig. 1 by means of its associated hyperplane threshold variable θ_t. The present state of this variable, in between the distribution mixture two peaks or modes, ensures (with high probability) the model arrival to stable equilibrium, as will momentarily become evident.

We now propose the following simple Shift unsupervised learning rule, applied upon arrival of each input vector **x**:

$$\theta_{t+1} = \begin{cases} \theta_t - \varepsilon, & \theta_t + \Phi \geq \mathbf{w}_t \cdot \mathbf{x} > \theta_t \\ \theta_t + \varepsilon, & \theta_t - \Phi \leq \mathbf{w}_t \cdot \mathbf{x} \leq \theta_t \\ \theta_t, & \text{otherwise} \end{cases} \tag{3}$$

where Φ and ε are two small positive scalar model parameters. In accordance with this rule a new feature vector **x** may either: fall in area **A** in which case θ_t will be Shifted to the left; fall in area **B** which will Shift θ_t to the right; or fall in any of areas **C** which will maintain θ_t unchanged. Since events of class **B** have greater probability than events of class **A** (area of region **B** is greater than area of region **A**) there will be a net 'pressure' to Shift θ_t to the right as indicated by the arrows of different size in Fig. 1. Such Shift operations will tend to carry θ_t from regions of high probability density to regions of low probability density. Once θ_t arrives to the vicinity of the distribution 'valley' (at about $\theta_t = \mathbf{a}$) stochastic stable equilibrium will be achieved: equal pressures will be exerted on both directions.

We note that increasing the Shift step ε may accelerate convergence, but may also increase the probability of jumping over the 'top of the hill' to the left of the leftmost mode, thereafter drifting towards a non-discriminatory and less useful direction. Similarly, increasing Φ may also accelerate convergence (since it reduces probability of inactive C events) but, again, may disrupt convergence direction. Both Φ and ε should be (and were) fine tuned during simulation. Proof of convergence (in the mean) of θ_t to the point **a** is simple, under some idealized assumptions, and is omitted herein. Note that both $\mathbf{m}_1$ and $\mathbf{m}_2$ mixture peaks are also stochastic equilibrium points, though unstable, as can be also easily proved.

The model initial condition θ_0 affects, along with the values of ε and Φ parameters, the probability of converging to a useful discrimination point like **a**: the more θ_0 resides away of any peak (but in between peaks) the larger the probability of such a useful convergence; and vice versa.

As mentioned above our herein proposed model achieves competitive performance in the sense of classification error probability. Its error performance is close to that of an optimal supervised learning classifier in as much as the mixture density valley is close (in some suitable measure) to the optimal hyper-surface, and in as much as the final state of the model hyperplane deviates from this valley (due to stochastic fluctuations, finite valued parameters effects, etc.). This optimal hyper-surface in turn depends on the mixture density conditionals and priors and is the solution to the equation $p(C_1) \, p(\mathbf{x}|C_1) - p(C_2) \, p(\mathbf{x}|C_2) \equiv f(\mathbf{x}) = 0$. If both classes have symmetric identical conditionals (except of course their peaks position) and have equal priors, then the mixture optimal discriminant classifier lies, by symmetry, at $\theta_t = \mathbf{a}$, midway between the two peaks $\mathbf{m}_1$ and $\mathbf{m}_2$. Our proposed method on the other hand, converges in this case (in the mean and neglecting the fore-mentioned deviations) to a minima of $p(\mathbf{x})$, and is thus the zero derivative solution of Eq. (1) (i.e. $p'(\mathbf{x}) = 0$) which results, in said symmetric case, in exactly the same point **a** and thus achieves (in the mean, and neglecting said deviations) optimality. If priors and/or conditionals differ from each other, both solutions (of $f(x) = 0$ and of $p'(x) = 0$) will slowly drift away from each other, and only near-optimality is achieved. The possible distance between the valley and the optimal hyper-surface may be considered the fundamental penalty paid, in lack of labels, for our limited ability to estimate priors and conditionals.

We now turn to revise and modify our model to adapt it to high dimensional feature spaces. The above considerations may be directly extended to these spaces, wherein the valley single point becomes a hyperplane. We will see that all is needed is the addition of a Rotation operation to the fore-mentioned Shift operation, which was described above.

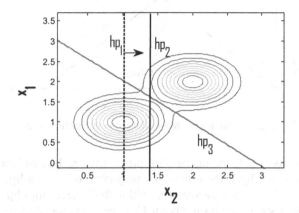

Fig. 2. 2-dimensional feature space – the need for a rotation learning rule

To see why Shift operation alone is not enough refer to Fig. 2 which shows 2 unimodal classes, described by means of their probability iso-density contours in a 2-dimensional space. The discriminating hyperplane initial condition is assumed to be hp_1 and the 'valley' near-optimal hyperplane position is hp_3. It is clear that a sequence of Shift only operations can lead discrimination to an equilibrium point hyperplane like hp_2, but is not capable of converging to the near optimal hp_3. The addition of a Rotation operation is apparently required.

Refer now to Fig. 3 which shows a 2-dimensional space containing 2 distinct classes schematically represented by their means μ_1 and μ_2. These means reside relatively close to and on both sides of a hyperplane hp_1. Figure 3 presents also an assumed near optimally placed hyperplane hp_2, and an input vector x also relatively close to hyperplane hp_1. The hyperplane hp_1 is defined by those vectors x which satisfy the equation $w \cdot x = \theta$ where we have omitted time subscripts for clarity, and where, by convention, the weight vector w is normalized (i.e. $\|w\| = 1$). As is well known, w defines the hyperplane hp_1 orientation relative to the feature space axes and is orthogonal to this hyperplane, while θ defines the hyperplane distance from the axes origin. It is similarly well known that rotation in high dimensional spaces is defined by a 2-plane of rotation, by a rotation point, and by a rotation angle, and that it keeps invariant the $(n - 2)$-subspace orthogonal to this 2-plane (this invariant subspace reduces, when we deal with 3-dimensional spaces, to the familiar 'axis of rotation', a meaningless term in high dimensional spaces).

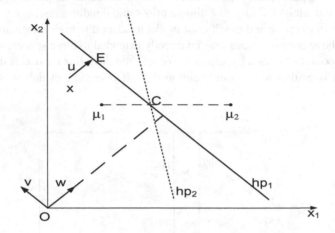

Fig. 3. Entities related to the rotation learning rule

We finally denote by $\mathbf{C}$ the intersection point between $\mathbf{hp_1}$ and the segment connecting $\mathbf{\mu_1}$ and $\mathbf{\mu_2}$, and by $\mathbf{E}$ the orthogonal projection of $\mathbf{x}$ upon $\mathbf{hp_1}$. It is proposed that, upon arrival of some feature vector $\mathbf{x}$, within Φ distance from $\mathbf{hp_1}$, we rotate $\mathbf{hp_1}$ by a small angle $\alpha \ll 1$, around the point $\mathbf{C}$, over a rotation 2-plane defined by the normalized orthogonal vector $\mathbf{u}$ from $\mathbf{x}$ to $\mathbf{hp_1}$ (which equals, up to possible sign inversion, to $\mathbf{w}$) and by the normalized vector $\mathbf{v}$ connecting $\mathbf{E}$ and $\mathbf{C}$; to gain better intuition on this proposal the reader is encouraged to visualize the above defined entities within a 3-dimensional space and corresponding probability density mixture. The model contains, as mentioned above, a pool of hyperplanes. It is convenient to assign to each hyperplane a separate self-timer t (t subscripts omitted above for readability) which advances whenever an incoming feature vector $\mathbf{x}$ falls within Φ distance from said hyperplane. Consolidating, we propose the following unsupervised learning scheme, executed for each and every hyperplane, upon arrival of a feature vector $\mathbf{x}$.

Shift Operation:
Exactly as with our 1-dimensional model, repeated herein for readability:

$$\theta_{t+1} = \begin{cases} \theta_t - \varepsilon, & \theta_t + \Phi \geq \mathbf{w}_t \cdot \mathbf{x} > \theta_t \\ \theta_t + \varepsilon, & \theta_t - \Phi \leq \mathbf{w}_t \cdot \mathbf{x} \leq \theta_t \\ \theta_t, & \text{otherwise} \end{cases} \tag{4}$$

Near Classes Means Estimate:
In real applications the feature space may be populated by many distinct classes. As described above a means estimate of the classes with means close to a hyperplane is

required for the Rotation operation of said hyperplane. Such means estimates can be implemented as weighted averages of input vectors, calculated in separate for both sides of each hyperplane; we allow for the weights to be a function of the distance of $\mathbf{x}$ from the hyperplane, assigning in general a smaller weight to input vectors farther away from it; this allows to disregard distant inputs, possibly belonging to other, faraway, classes. For each and every hyperplane we update c_{t+1}^1 and $\widehat{\mu}_{t+1}^1$ if $\mathbf{w}_t \cdot \mathbf{x} \leq \theta_t$ and update c_{t+1}^2 and $\widehat{\mu}_{t+1}^2$ otherwise:

$$c_{t+1}^j = c_t^j + g(|\mathbf{w}_t \cdot \mathbf{x} - \theta_t|; \beta), \qquad c_0^j = 0, \tag{5}$$

and

$$\widehat{\mu}_{t+1}^j = (c_t^j \widehat{\mu}_t^j + g(|\mathbf{w}_t \cdot \mathbf{x} - \theta_t|; \beta)\mathbf{x})/c_{t+1}^j, \tag{6}$$

where $\widehat{\mu}_t^j$ denote mean estimates, $j = 1$ or 2 denote each of both hyperplane half spaces $\mathbf{w}_t \cdot \mathbf{x} \leq \theta_t$ or $\mathbf{w}_t \cdot \mathbf{x} > \theta_t$ respectively, c_t^j are the cumulative weights associated with this hyperplane, and where $g(|\,.\,|; \beta)$ is a distance dependent weight function with parameter β.

A typical and simple weight function could consist of a uniform positive weight for vectors $\mathbf{x}$ satisfying $\theta_t - \beta \leq \mathbf{w}_t \cdot \mathbf{x} \leq \theta_t + \beta$ with some $\beta > 0$, and zero weight otherwise. We assume the use of this simple weight function in the sequel. To initiate these mean estimates we may simply use the first incoming sample (within appropriate distance from the hyperplane) for one side estimate and a symmetrically reflected virtual point for the other, which results, in the case of the uniform weight function suggested above, in

$$\widehat{\mu}_0^1 = \mathbf{x}, \qquad \widehat{\mu}_0^2 = \mathbf{x} + 2|\mathbf{w}_t \cdot \mathbf{x} - \theta_t|\mathbf{w}_t, \qquad \text{if} \quad \theta_t - \beta \leq \mathbf{w}_t \cdot \mathbf{x} \leq \theta_t \tag{7}$$

$$\widehat{\mu}_0^2 = \mathbf{x}, \qquad \widehat{\mu}_0^1 = \mathbf{x} - 2|\mathbf{w}_t \cdot \mathbf{x} - \theta_t|\mathbf{w}_t, \qquad \text{if} \quad \theta_t < \mathbf{w}_t \cdot \mathbf{x} \leq \theta_t + \beta \tag{8}$$

This initiation scheme allows allocating to $\widehat{\mu}_0^2$ (or alternatively to $\widehat{\mu}_0^1$) some value even though there might exist no class in the neighborhood of the other side of this hyperplane.

C, E, u and v Calculation:
As mentioned above $\mathbf{C}$, $\mathbf{E}$, $\mathbf{u}$ and $\mathbf{v}$ are also required for the implementation of the Rotation operation. Upon arrival of a feature vector $\mathbf{x}$, said calculations are executed only for those hyperplanes for which $\theta_t + \Phi \geq \mathbf{w}_t \cdot \mathbf{x} \geq \theta_t - \Phi$. Inspection of Fig. 3 and simple vector algebra manipulations result in:

$$\mathbf{C} = \widehat{\mu}_t^1 + (\theta_t - \mathbf{w}_t \cdot \widehat{\mu}_t^1) \cdot (\widehat{\mu}_t^2 - \widehat{\mu}_t^1)/(\mathbf{w}_t \cdot (\widehat{\mu}_t^2 - \widehat{\mu}_t^1)) \tag{9}$$

$$\mathbf{E} = \mathbf{x} + (\theta_t - \mathbf{w}_t \cdot \mathbf{x})\mathbf{w}_t/\|\mathbf{w}_t\| \tag{10}$$

$$\mathbf{u} = (\mathbf{E}-\mathbf{x})/\|\mathbf{E}-\mathbf{x}\| = \text{sign}(\mathbf{w}_t \cdot \mathbf{x} - \theta_t) \cdot \mathbf{w}_t/\|\mathbf{w}_t\| \tag{11}$$

where sign(.) denotes the signum function (time subscripts omitted from $\mathbf{C}$, $\mathbf{E}$, $\mathbf{u}$ and $\mathbf{v}$); note the possible sign disagreement between $\mathbf{u}$ and $\mathbf{w}_t$; and

$$\mathbf{v} = (\mathbf{E}-\mathbf{C})/\|\mathbf{E}-\mathbf{C}\| \tag{12}$$

Rotation Operation:
Here too, this operation is executed only for those hyperplanes for which $\theta_t + \Phi \geq \mathbf{w}_t$. $\mathbf{x} \geq \theta_t - \Phi$. For d-dimensional Rotation formulation we follow the neat (and slightly abusive) vector notation of Teoh [6] (the reader is referred there for details):

$$\mathbf{w}_{t+1} = \mathrm{rot}(\mathbf{w}_t)_{P,\alpha,C} = \mathbf{w}_t + [\mathbf{u} \quad \mathbf{v}]\begin{pmatrix} \cos\alpha - 1 & \sin\alpha \\ \sin\alpha & \cos\alpha - 1 \end{pmatrix}\begin{pmatrix} (\mathbf{w}_t \cdot \mathbf{u}) \\ (\mathbf{w}_t \cdot \mathbf{v}) \end{pmatrix} \tag{13}$$

where P is the rotation 2-plane as defined by the vectors $\mathbf{v}$ and $\mathbf{u}$, $\mathbf{C}$ is the rotation point (see Fig. 3), and $\alpha \ll 1$ is a small rotation angle for which we may simplify to $\sin\alpha \approx \alpha$ and $\cos\alpha \approx (1 - \alpha^2/2)$. The sense of rotation (clockwise or counter-clockwise) is set such that the distance between the sample point $\mathbf{x}$ and the rotated hyperplane increases. Similarly to the Shift operation this ensures hyperplane migration (in the mean) toward a lower probability density region as desired.

Rotation also shifts the hyperplane (i.e. changes its distance from O) so it affects its θ_t variable, and this has to be updated too (in addition to the Shift learning rule update, Eq. (4) above); recalling that the point $\mathbf{C}$ is invariant under Rotation and that it belongs to the hyperplane (rotated or not) then

$$\theta_{t+1} = \mathbf{w}_{t+1} \cdot \mathbf{C} \tag{14}$$

Hyperplane Output Code:
The hyperplane output code may take the form (amongst many other options) of Eq. (1) herein repeated for convenience:

$$y_{t+1} = \max\{0, \mathbf{w}_t \cdot \mathbf{x} - \theta_t\} \tag{15}$$

The collection of all scalar outputs y_{t+1} make up together the model output code vector $\mathbf{y}_{t+1} \in R^N$ where N is the size of the hyperplane pool.

Self Timer Update:
As mentioned above a self-timer is conveniently assigned to each hyperplane. This self-timer is incremented whenever an input feature vector $\mathbf{x}$ arrives within distance Φ from said hyperplane, namely (hyperplane index omitted)

$$t \rightarrow \begin{cases} t+1, & \theta_t + \Phi \geq \mathbf{w}_t \cdot \mathbf{x} \geq \theta_t - \Phi \\ t, & \text{otherwise} \end{cases} \tag{16}$$

Just as in our earlier 1-dimensional example, the Shift and Rotation operations will tend (in the mean) to migrate the hyperplanes from regions of high probability density to regions of low probability density. It should be noted that all learning operations (Eqs. (4) to (14) above) executed upon a hyperplane do exclusively depend on variables of this said hyperplane, and of no other; thus, the proposed learning rules are local, and 'Hebbian' in this sense. The Rotate operation related calculations (Eqs. (5) to (14) above) for each hyperplane may be advantageously initiated only after the execution of some pre-defined quantity of executed Shift operations for this hyperplane: this allows each hyperplane to land closer to desired (low density) regions in a first stage, making these later calculations more relevant and the convergence process more efficient.

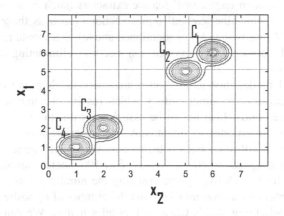

Fig. 4. 2-dimensional example: initial hyperplanes state, uniform orthogonal grid case

Refer now to Fig. 4 which shows a 2-dimensional feature space with 4 classes C_1, C_2, C_3 and C_4 depicted by their iso-density contours. In accordance with our proposed model a pool of hyperplanes is initially distributed across the feature space hypercubic domain of interest $\Omega \subset R^d$ with hypercube edge size o (o $\approx$ 8 in this example) as shown in Fig. 4. The initial state of this pool of hyperplanes may be uniformly random, or ordered in e.g. orthogonal gridlines (as we opted herein). In this later case, an approximate guideline which sets the minimal required density of hyperplanes in feature space is that the distance between neighbor parallel hyperplanes be less than the lower bound for the distance between any two classes' modes peaks. The computational complexity of our proposed model is at most quadratic in the dimension d of the feature space. To see this we note that all described operations for a single hyperplane (Eqs. (4) to (16) above) scale up at most linearly with dimension d, and that the number of hyperplanes itself, needed to create a uniform orthogonal grid such as that of Fig. 4, also scales up linearly with d, overall resulting in quadratic complexity. We note that even though each hyperplane is checked upon arrival of a new feature vector, only a small minority is anticipated to require actual update and full calculations execution, namely only those which are sufficiently close to the incoming vector; thus, great part of the computations are skipped most of the time. In fact our simulations with differing dimension feature spaces indicate a complexity scaling up at approximately $O(d^{1.2})$. As we will see later in the sequel, it is possible to exploit only a small subset of the hyperplane pool in order to perform classification; by so doing, complexity during the classification stage may be further greatly reduced. The memory requirement of our proposed model does not depend on the number of input samples (in contrast with offline unsupervised learning schemes such as parametric models or k-Means) and is quadratic in feature space dimension d: the variables needed to be kept for each hyperplane are all of dimension d (or scalar) and the number of hyperplanes scale up linearly with d, resulting in $O(d^2)$. Since the proposed model operates on a sample by sample basis as mentioned above, then continuous learning and adaptation to slowly changing environments are made possible. Our proposed classifier could be used in conjunction with 'human engineered' feature extractors (such as SIFT in the Machine Vision domain), or 'machine learned' representations (such as those associated with Deep NN, with VAE or with GAN as mentioned above); this would result in an end-to-end unsupervised learning pattern recognition machine, eliminating the need for large sets of labeled data.

The proposed model, just as any other unsupervised learning classifier (e.g. k-Means, etc.), generates an output code **y** which results in some arbitrary 'machine assigned' labels set, and which has to be associated with 'human assigned' labels in order to be usefully interpreted by humans. A brief description of the method we have used for this purpose follows: we allocate, after convergence, a bunch of samples along with their 'human assigned' labels for this task (say 100 labeled samples per Class). For every Class we check each hyperplane counting the number of its consistent outputs (say +1) for samples of this current Class and the number of opposite outputs (say −1) for samples not belonging to this Class (a 1-vs-all scheme). We calculate a weighted sum of these 2 numbers and select the highest score hyperplane as this Class discriminator, associating to it that 'human-assigned' label. We have alternatively built a hierarchy tree scheme (e.g. Genus, Species, etc. levels) and similarly picked up the best hyperplanes at each level; both schemes yielded very similar results.

3   Simulation Results

To demonstrate the ideas and methods presented herein a Matlab based platform was built. Please refer again to Fig. 4. The illustrated 2 pairs (C_1, C_2 and C_3, C_4) of 2-variate normally distributed classes, all have equal $\sigma^2 I$ covariance matrix (where I is the identity matrix), equal priors and shifted means (relative to each other). Inter-means distance and σ^2 were calibrated, so that the optimal classification error probability was approximately 2×10^{-2}. Rough tuning of the parameters was carried out to ensure fast and reliable convergence, ending up in $\varepsilon = 0.0033\ \sigma$, $\Phi = 2\ \sigma$, $\alpha = 0.04$ and $\beta = 8\ \sigma$ values; the linear dependence on σ of ε, Φ and β is useful and intuitively convenient for problem scaling. Hyperplanes pool initial state was set to uniform orthogonal grid with appropriate distances, as already mentioned above and shown in Fig. 4.

Figure 5 shows the final hyperplanes state, after 10000 input samples (~ 2500 vectors for each of said 4 classes) of the 2-dimensional scenario, for which its initial orthogonal grid state was presented in Fig. 4. We notice that many (but not all) hyperplanes have migrated to discriminating states. We see that small groups of hyperplanes, like $\mathbf{hp_3}$ and $\mathbf{hp_4}$, did converge to local low probability density regions, in between classes C_3, C_4 and C_1, C_2 respectively.

The internal world representation $\mathbf{y_t} \in R^N$ of such a model can be interpreted as a hierarchical and multi-resolution distributed code: hyperplanes like $\mathbf{hp_1}$ for example could be viewed as discriminating between carnivores and herbivores (no herbivore classes are shown in Fig. 4), others like $\mathbf{hp_2}$ may distinguish between canines and felines, still others like $\mathbf{hp_3}$ between cats and lions, and so on. Code redundancy is observed: some of the discriminating valleys are occupied by small groups of neighbor, approximately parallel hyperplanes. Similar hyperplanes final state characteristics as noted above may be expected at higher dimensional feature spaces.

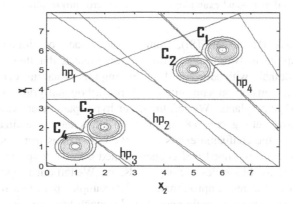

Fig. 5. 2-dimensional example: final hyperplanes state

Figure 6 shows classification error probability for a similar 2 pair categories scenario as described above, except that dimension now is $d = 50$. We note P_{err} convergence (in the mean) from 4×10^{-1} to the optimal 2×10^{-2} after about 6000 input samples (~ 1500 for each of 4 classes). Close to optimal P_{err} was expected in this scenario since we chose a setup of equal priors and identical, mean shifted spherically symmetric (normally distributed) conditionals; this symmetry results in a hyperplanar optimal discrimination surface and a collocated hyperplanar valley. P_{err} in Fig. 6 shows slight but visible fluctuation around a mean value of 2×10^{-2}. In order to mitigate these fluctuations ε and α, rather than staying fixed during model convergence (as was done in the example of Fig. 6) may be made to gradually decrease, either as function of time (sample number) or in some adaptive scheme.

To demonstrate performance on a real life application we chose the ImageNet dataset, a challenging popular benchmark containing over 1000 classes of images, with around 1300 labeled samples per class (Deng et al. [7]). Each sample is a full resolution image with average size of 400×350 pixels. Our proposed model is generic, and operates in conjunction with any reasonable feature extraction method which yields representations fulfilling our above stated assumptions. We used the $(N - 1)$ activation layer of a pre-trained ResNet-50, a variant of which won the ILSVRC 2015

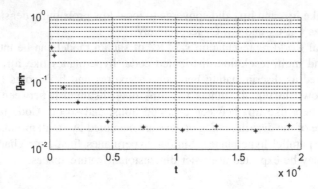

Fig. 6. 50-dimensional example: classification error probability convergence

Classification Task (He et al. [8]), with feature space of dimension d = 2048. Inspection and simple analysis of the feature vectors indicate that they may be enclosed by a hypercubic domain with edge o = 4. Following this visual inspection we placed 4 hyperplanes per dimension on said initial uniform orthogonal grid reaching a total of 2048 × 4 = 8192 hyperplanes. We left the other hyper-parameters values unchanged, as described above, setting σ = 0.8. In order to allow easy visualization and comparative analysis of the confusion errors we picked 50 classes from the ImageNet set, containing a mix of fine-grained and coarse-grained classes: 5 species of Sharks and Rays, Cocks and Hens, 11 species of Song Birds, etc. We left out 100 samples per class for testing and used the rest, approximately 1200 samples per class for training. The training set was run during a single epoch; multi-epoch training did not improve the results further. Model parameters ε and α were kept fixed; their gradual decrease with time did not significantly affect the results. The learning process was relatively fast: about 60000 sample vectors (of d = 2048) in 2.5 h on an i7, 2.7 GHz PC. We tested classification error performance by estimating, using the fore-mentioned test set, Perr (Top-n, n = 1, 3, 5) which indicates the probability that the actual class of a given sample is not contained in the set of n most probable classes.

We have run this same data set, using same feature representations, with k-Nearest Neighbor, a near optimal popular supervised learning benchmark method. The expected performance gap between kNN and its best-in-class alternative (say SVM) is not big for our purpose. The results of this comparative test are brought in Table 1. We note that our proposed method's Perr (Top-3) exceeds by **merely about 2%** that of the kNN near optimal supervised learning classifier. This excess Perr may be called our unsupervised learning penalty or loss. To probe these results a step further we show in Fig. 7 the Confusion Matrices for both our method and kNN; for better readability we show 20 × 20 sub-matrices, but similar behavior is observed in the full matrix. Columns denote 'human assigned' (supposedly ground truth) labels and rows denote our 'machine assigned' labels (numbered 1 to 20 for both).

Table 1. Classification task performance of our proposed method vs. kNN for 50 ImageNet coarse/fine grained classes. The Top-3 and Top-5 kNN figures are approximate calculations

50 classes mix	Top 5 Perr	Top 3 Perr	Top 1 Perr
kNN (k = 3, Euclidean distance)	0.024	**0.043**	0.16
Our method	0.035	**0.062**	0.23

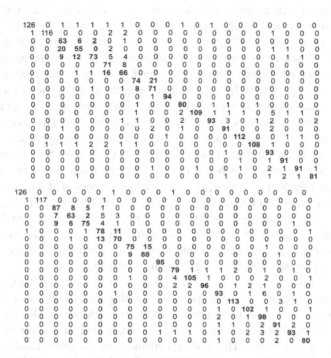

Fig. 7. Our proposed method (top) and kNN (bottom) confusion (Sub-)matrices

Adjacent columns (rows) represent similar classes (close animal species in our case), for example columns 3 to 5 carry Shark species, etc. In an errorless scheme the Confusion Matrix would of course be purely diagonal. We notice that both methods exhibit a strikingly similar behavior; regions of greater confusion, such as Sharks (columns 3 to 5), Rays (6, 7) and Cock/Hen (8, 9) have similar levels of confusion; this confusion is mainly due to separability limitations of the ResNet Deep NN furnished feature space. Regions of good error performance (10 to 20, Song Birds) are also remarkably similar in both methods. The small performance gap between kNN and our proposed method is probably mainly due to the fundamental factors mentioned in Sect. 2 above, namely: non-coincidence between the optimal discriminating hyper-surface and the 'valley' hyper-surface; stochastic fluctuation of our hyperplanes due to finite parameters values, etc.

Given the high dimensionality of the ResNet-50 feature space and the relatively small number of training samples for such a huge space region volume, it is apparent that we have here a case of sparse distribution sampling. One may puzzle how then we could have got such impressive results as presented above; similar good comparative results were also achieved with the MNIST dataset represented by a similarly high dimensionality hand-crafted feature space (benchmarked vs. kNN and SVM, results omitted due to lack of space). A possible explanation is that besides this sparse sampling (sparse in a '1st sense') the feature vectors are also sparse in the sense ('2nd sense') of each containing a relatively large number of zero elements (and few non-zero elements). They are thus confined to low dimensionality manifolds within a much higher dimensional space. When constrained to such manifolds, the space sampling is *effectively* no sparse (in the 1st sense) anymore. We indeed verified that vectors are sparse (in the 2nd sense) by visual inspection of marginal class-conditional densities. An alternative (or additional) reason could be that in spite the insignificant mean number of samples per unit of feature space volume, the share of useful samples which effectively train each hyperplane (approximately 2Φ/o), may typically be (as they were for our above shown values of Φ and o) non-negligible. Evidently more explorative work is required in this area. A similar question regarding learning capability may arise when we notice that the number of model parameters ($\sim$ N d $\sim$ 16e6 in our ImageNet case) is huge relative to the number of training samples (6e4, there); the resolution to this apparent puzzle lies in our view, in the fact that each neuron of our proposed model learns *independently* of each other (as N separate vectors of length d each); this results in d ($\sim$ 2e3) parameters being trained by means of 6e4 samples, a reasonable size; this also stands in contrast with 'conventional' multi-layer neural networks, where all neurons are *concurrently* trained (as one long (N d) parameters vector) so that the efficient training set size should be significantly larger than N d.

It is also of importance to evaluate the ImageNet Classification Task performance of other, potentially competing, unsupervised learning models. We pick for that purpose k-Means, possibly the most popular clustering scheme of all. We choose 10 coarse-grained ImageNet classes; these present a simple challenge to our method which yields Perr (Top-1) = 0.015. The k-Means method on the other hand is practically useless at this trivial task as can be seen in Fig. 8 which presents k-Means Confusion Matrix. We can readily observe that 'human assigned' Class '1' is split amongst 2 k-Means assigned Classes ('1', '7') and single k-Means Class '6' is assigned to 3 'human assigned' Classes ('6', '7', '8'). This is no surprise; in fact, following these k-Means results we have conducted an exhaustive literature search for works reporting unsupervised learning classification results in conjunction with any feature extractor for ImageNet or other challenging, real life applications: we have found none.

50	0	0	0	0	0	0	0	0	0
0	110	0	0	0	0	0	0	0	0
0	0	88	0	1	0	0	0	0	0
0	0	0	86	1	0	0	0	0	0
0	0	2	0	98	1	0	0	2	0
0	0	0	0	0	100	92	14	0	0
51	0	0	0	0	0	1	0	0	3
0	0	0	0	0	0	1	89	0	0
0	4	1	0	5	0	0	4	95	1
0	0	1	0	0	0	0	0	0	99

Fig. 8. k-Means Confusion Matrix for a simple ImageNet Classification Task for which our proposed method achieves Perr (Top-1) = 0.015

It should be probably evident by now to the reader that this proposed model has a natural neural architecture implementation such as that shown in Fig. 9. We note that the resulting neural architecture is feed-forward and 'shallow', consisting merely of a single neural layer.

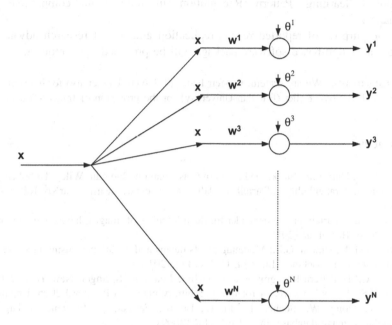

Fig. 9. Proposed model (shallow) neural architecture

4   Concluding Remarks

This work extends the applicability of a linear discriminant surfaces approach to the field of unsupervised learning classifiers. Its main novelty in our view is the *exploitation of the implicit underlying probability density* to train a classifier. We are not aware of any other work which exploits the implicit density for this goal (typically density estimation is carried out). As result of this the proposed method, though stochastic, does not require explicit estimation nor functional form assumption of the probability densities involved. At this initial research stage we have experimentally demonstrated that an unsupervised learning classifier exhibiting low complexity, Hebbian-like local learning rule, online processing, and neural architecture, may achieve competitive classification error performance on a relatively challenging ImageNet Classification Task. Future areas of research may include: test extension to the full ImageNet and other real-life datasets; performance comparison with other supervised learning schemes; feature space sampling sparsity and error performance analysis; convergence analysis; parameters sensitivity analysis; best hyperplane identification analysis; and study of the possibility to extend our model to a near-optimal *supervised* learning variant. Finally, significant progress has been made in recent years in the related fields of unsupervised learning of representations (e.g. VAE [3]) and of supervised learning feature extraction (e.g. GAN [2], DNN Transfer Learning [10]). Integration of either of these 2 types of feature extraction solutions with a classifier model such as ours may bring to life for the first time ever an end-to-end unsupervised learning Pattern Recognition machine with competitive error performance.

For the purpose of reported results replication and model research advance and extension, a Matlab based software package will be provided upon request.

Acknowledgements. We are grateful to Meir Feder (Tel Aviv University) for his support and comments and to Yossi Keller (Bar Ilan University) for the provision of ImageNet data.

References

1. Duda, R.O., Hart, P.E., Stork, D.G.: Pattern Classification, 2nd edn. Wiley, Hoboken (2002)
2. Donahue, J., Krahenbuhl, P., Darrell, T.: Adversarial Feature Learning, arXiv:1605.09782v6 (2017)
3. Pu, Y., et al.: Variational autoencoder for deep learning of images, labels and captions. In: NIPS 2016, Barcelona (2016)
4. Ranzato, M.A., Hinton, G.E.: Modeling pixels means and covariances using factorized third order boltzmann machines. In: IEEE CVPR, June 2010
5. Bishop, C.M.: Pattern Recognition and Machine Learning. Springer, New York (2006)
6. Teoh, H.S.: Formula for vector rotation in arbitrary planes in R^n, April 2005. Unpublished
7. Deng, J., Dong, W., Socher, R., Li, L., Li, K., Fei-Fei, L.: Imagenet: a large-scale hierarchical image database. In: IEEE CVPR (2009)
8. He, K., Zhang, X., Ren, S., Sun, J.: Deep Residual Learning for Image Recognition, arXiv: 1512.03385v1 (2015)
9. MacQueen, J.B.: Some methods for classification and analysis of multivariate observations. In: Proceedings of 5th Berkeley Symposium on Mathematical Statistics and Probability. University of California Press, pp. 281–297 (1967)
10. Yosinski, J., Clune, J., Bengio, Y., Lipson, H.: How transferable are features in deep neural networks? In: Advances in Neural Information Processing Systems 27, NIPS Foundation (2014)
11. Lewis, W., Koontz, G., Fukunaga, K.: A non-parametric valley-seeking technique for cluster analysis. In: Proceedings of the 2nd International Joint Conference on Artificial Intelligence, IJCAI 1971, pp. 411–417 (1971)
12. Pavlidis, N.G., Hofmeyr, D.P., Tasoulis, S.K.: Minimum density hyperplanes. J. Mach. Learn. Res. **17**, 5414–5446 (2016)

A GRASP/VND Heuristic for the Max Cut-Clique Problem

Mathias Bourel, Eduardo Canale, Franco Robledo, Pablo Romero, and Luis Stábile$^{(\boxtimes)}$

Instituto de Matemática y Estadística, IMERL, Facultad de Ingeniería, Universidad de la República, Montevideo, Uruguay
{mbourel,canale,frobledo,promero,lstabile}@fing.edu.uy

Abstract. In Market Basket Analysis, the goal is to understand the human behavior in order to maximize sales. An evident behavior is to buy correlated items. As a consequence, the determination of a set of items with a large correlation with others is a valuable tool for Market Basket Analysis.

In this paper we address a combinatorial optimization problem that formalizes the previous application. Given a simple graph $\mathcal{G} = (V, E)$ (where the nodes are items and links represent correlation), we want to find the clique $\mathcal{C} \subseteq V$ such that the number of links shared between $\mathcal{C}$ and $V - \mathcal{C}$ is maximized. This problem is known in the literature as Max Cut-Clique (MCC).

The contributions of this paper are three-fold. First, the computational complexity of the MCC is established. Second, a full GRASP/VND methodology enriched with a Tabu Search is here developed, where the main ingredients are novel local searches and a Restricted Candidate List that trades greediness for randomization in a multi-start fashion. A Tabu Search is also included in order to avoid locally optimum solutions. Finally, a fair comparison with respect to recent heuristics reveals that our proposal is competitive with state-of-the-art solutions.

Keywords: Market Basket Analysis · Combinatorial optimization · Max Cut-Clique · Metaheuristics

1 Motivation

There is a serious disconnection between the knowledge that academics are producing and the knowledge that practitioners are consuming [7]. A bridge between the science-practice division can be found in Market Basket Analysis (MBA), sometimes known as *affinity analysis* [2]. In synthesis, MBA is a Data Mining technique [1,19] originated in the field of marketing. It has recent applications to other fields, such as bioinformatics [4,5], WWW networks [12], criminal networks [6] and financial networks [13]. The goal of MBA is to identify non-obvious or counterintuitive relationships between groups of products, items, or categories.

© Springer Nature Switzerland AG 2019
G. Nicosia et al. (Eds.): LOD 2018, LNCS 11331, pp. 357–367, 2019.
https://doi.org/10.1007/978-3-030-13709-0_30

The information obtained from MBA can have an important impact in the business strategy and operations. In the specific case of marketing, we can find valuable applications such as product placement, optimal product-line offering, personalized marketing campaigns and product promotions. The analysis is commonly supported by Machine Learning (pattern matching, clustering, feature extraction, statistics), Optimization and Logical rules for association.

This work is focused on a specific combinatorial optimization methodology to assist product placement; however, related applications could be found. The problem under study is called Max Cut-Clique (MCC), and it was introduced by Martins [15]. Given a simple graph $G = (V, E)$ (where the nodes are items and links represent correlation), we want to find the clique $C \subseteq V$ such that the number of links shared between C and $V - C$ is maximized. The MCC has an evident application to product-placement. For instance, the manager of a supermarket must decide how to locate the different items in the different compartments. In a first stage, it is essential to determine the correlation between the different pairs of items, for psychological/attractive reasons. Then, the priceless/basic products (bread, rice, milk and others) could be hidden on the back, in order to give the opportunity for other products in a large corridor (and candies should be at hand by kids as well). Observe that the MCC appears in the first stage, while marketing/psychological aspects play a key role in a second stage for product-placement in a supermarket.

In [15], the author states that the MCC is presumably hard, since related problems such as $MAX - CUT$ and $MAX - CLIQUE$ are both $\mathcal{NP}$-Complete. To the best of our knowledge, there is no formal proof available for the hardness of the MCC in the published scientific literature. Nevertheless, the MCC is systematically addressed by the scientific community with metaheuristics and exact solvers that run in exponential time.

A recent work in the field develops an Iterated Local Search for the MCC [16]. As far as we know, this work belongs to the state-of-the-art techniques for the MCC. The authors find optimal solutions for most instances under study, and suggest a rich number of applications.

The contributions of this paper can be summarized in the following items:

1. The $\mathcal{NP}$-Completeness of the MCC is established (Sect. 2).
2. A hybrid GRASP/VND heuristic enriched with Tabu Search is developed to address the MCC (Sect. 3).
3. A fair comparison with a state-of-the-art heuristic is presented using DIMACS benchmark (Sect. 4).

2   Computational Complexity

The cornerstone in computational complexity is Cook's Theorem [8] and Karp reducibility among combinatorial problems [14].

Stephen Cook formally proved that the joint satisfiability of an input set of clauses in disjunctive form is the first $\mathcal{NP}$-Complete decision problem [8]. Furthermore, he provided a systematic procedure to prove that a certain problem is

$\mathcal{NP}$-Complete. Specifically, it suffices to prove that the decision problem belongs to set $\mathcal{NP}$, and that it is at least as hard as an $\mathcal{NP}$-Complete problem. Richard Karp followed this hint, and presented the first 21 combinatorial problems that belong to this class [14]. In particular, $MAX-CLIQUE$ belongs to this list. The reader is invited to consult an authoritative book in Complexity Theory, which has a larger list of $\mathcal{NP}$-Complete problems and a rich number of bibliographic references [10].

Here, we formally prove that the MCC is at least as hard as $MAX - CLIQUE$. Let us denote $|\mathcal{C}|$ the cardinality of a clique $\mathcal{C}$, and $\delta(\mathcal{C})$ denotes the corresponding cutset induced by the clique (or the set) $\mathcal{C}$.

Definition 1 (MAX-CLIQUE).

GIVEN: a simple graph $G = (V, E)$ and a real number K.
QUESTION: is there a clique $\mathcal{C} \subseteq V$ such that $|\mathcal{C}| \geq K$?

For convenience, we describe MCC as a decision problem:

Definition 2 (MCC).

GIVEN: a simple graph $G = (V, E)$ and a real number K.
QUESTION: is there a clique $\mathcal{C} \subseteq G$ such that $|\delta(\mathcal{C})| \geq K$?

Theorem 1. *The MCC belongs to the class of $\mathcal{NP}$-Complete problems.*

Proof. We prove that the MCC is at least as hard as $MAX - CLIQUE$. Consider a simple graph $G = (V, E)$ with order $n = |V|$ and size $m = |E|$. Let us connect a large number of M hanging nodes, to every single node $v \in V$. The resulting graph is called H (see Fig. 1 for an example). If we find a polynomial-time algorithm for MCC, then we can produce the max cut-clique in H. But observe that the Max Cut-Clique $\mathcal{C}$ in H cannot include hanging nodes, thus it must belong entirely to G. If a clique $\mathcal{C}$ has cardinality c, then the clique-cut has precisely $c \times M$ hanging nodes. By construction, the cut-clique must maximize the number of hanging nodes, if we choose $M \geq m$. As a consequence, c must be the $MAX - CLIQUE$. We proved that the MCC is at least as hard as $MAX - CLIQUE$, as desired. Since MCC belongs to the set of $\mathcal{NP}$ Decision problems, it belongs to the $\mathcal{NP}$-Complete class. ∎

Theorem 1 promotes the development of heuristics in order to address the MCC.

3   Methodology

GRASP and Tabu Search are well known metaheuristics that have been successfully used to solve many hard combinatorial optimization problems. GRASP is an iterative multi-start process which operates in two phases [17]. In the Construction Phase a feasible solution is built whose neighborhood is then explored in the Local Search Phase. Tabu Search [3,11] is a strategy to prevent local

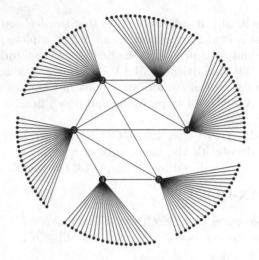

Fig. 1. Construction of H with $M = 21$ hanging nodes.

search algorithms getting trapped in locally optimal solutions. A penalization mechanism called Tabu List is considered to avoid returning to previously visited solutions. For a complete description of these methods the reader is referred to the works of Glover and Laguna [11] and Resende and Ribeiro [17]. The reader is invited to consult the comprehensive Handbook of Metaheuristic for further information [18].

Here, we develop a GRASP/VND methodology enriched with Tabu Search in order to avoid getting trapped in previous visited solutions. In the following, the pseudocode of our Hybrid Metaheuristic (HM) for the Max Cut-Clique is presented. It follows the traditional two-phase GRASP template enriched with a Variable Neighborhood Descent (Lines 4–5). A Tabu Search strategy is included in order to enhance feasible solutions. The tabu list $\mathcal{T}$ stores tabu nodes (Line 2), discarding previous solutions. Essentially, the most frequent nodes involved in all solutions after the second phase of Variable Neighborhood Descent (VND) are not considered for further solutions during θ iterations, whenever we reach θ^{max} consecutive iterations without improvement. The most frequent nodes are selected if they appear more than ϕ times since the last tabu list refresh. The real numbers ϕ and θ are uniformly chosen at random in the interval $[1, \theta^{max}]$, being θ^{max} a parameter of the algorithm. The specific GRASP phases for the MCC are described in detail in the following subsections.

3.1 Construction Phase - *Clique*

The construction phase of the proposed algorithm is depicted in Algorithm 2. Let us denote by $\mathcal{C}$ the clique under construction, $\delta(U)$ and $\Delta(U)$ the minimum and maximum degree of the node-set U. The clique $\mathcal{C}$ is initially empty (Line 1), and a multi-start process is considered (Line 2). A Restricted Candidate

---

Algorithm 1. HM PSEUDOCODE

    Input: α, θ^{max}, maxIter, $\mathcal{G}$
    Output: $\mathcal{C}^*$
1: $\mathcal{C}^* \leftarrow \emptyset$
2: $\mathcal{T} \leftarrow \emptyset$
3: **for** iter = 1 **to** maxIter **do**
4:    $\mathcal{C} \leftarrow$ CLIQUE$(\alpha, \mathcal{T}, \mathcal{G})$
5:    $\mathcal{C} \leftarrow$ VND$(\mathcal{C}, \mathcal{T}, \mathcal{G})$
6:    $\mathcal{T} \leftarrow$ UPDATE$(\mathcal{T}, \theta^{max}, \mathcal{C})$ ▷ Tabu List
7:    **if** $|E'(\mathcal{C})| > |E'(\mathcal{C}^*)|$ **then**
8:       $\mathcal{C}^* \leftarrow \mathcal{C}$
9: **return** $\mathcal{C}^*$

List, RCL, is defined in Line 3. Observe that the RCL includes nodes with the highest degree, and α trades greediness for randomization. During the *While* loop of Lines 4–11, a singleton $\{i\}$ is uniformly picked from the RCL (Line 5), and the maximum clique $\mathcal{C}'$ is built using all the nodes from the set $\mathcal{C} \cup \{i\}$ (see Line 6). The best solution is updated if necessary (Lines 7–8). Observe that the process is finished only if we meet $MAX\_ATTEMPTS$ without improvement (Lines 9–11). The reader can appreciate that the output $\mathcal{C}$ is the best feasible clique during the whole process (Line 12).

Algorithm 2. CLIQUE

    Input: α, $\mathcal{T}$, $\mathcal{G}$
    Output: $\mathcal{C}$
1: $\mathcal{C} \leftarrow \emptyset$
2: *improving* = MAX_ATTEMPTS
3: $RCL \leftarrow \{v \in V - \mathcal{C} : |E'(v)| \geq \Delta(V - \mathcal{C}) - \alpha(\Delta(V - \mathcal{C}) - \delta(V - \mathcal{C}))\}$
4: **while** *improving* > 0 **do**
5:    $i \leftarrow selectRandom(RCL)$
6:    $\mathcal{C}' \leftarrow [\mathcal{C} \cap N(i)] \cup \{i\}$
7:    **if** $|E'(\mathcal{C}')| > |E'(\mathcal{C})|$ **then**
8:       $\mathcal{C} \leftarrow \mathcal{C}'$
9:       *improving* $\leftarrow$ MAX_ATTEMPTS
10:   **else**
11:      *improving* $\leftarrow$ *improving* $- 1$
12: **return** $\mathcal{C}$

3.2 Local Search Phase - *VND*

The goal is to combine a rich diversity of neighborhoods in order to obtain an output that is locally optimum solution for every feasible neighborhood. Five neighborhood structures are considered to build a VND [9].

- **Remove:** a singleton $\{i\}$ is removed from a clique $\mathcal{C}$.
- **Add:** a singleton $\{i\}$ is added from a clique $\mathcal{C}$.
- **Swap:** if we find $j \notin \mathcal{C}$ such that $\mathcal{C} - \{i\} \subseteq N(j)$, we can include j in the clique and delete i (swap i and j).
- **Cone:** generalization of Swap for multiple nodes. The clique $\mathcal{C}$ is replaced by $\mathcal{C} \cup \{i\} - \mathcal{A}$, being $\mathcal{A}$ the nodes from $\mathcal{C}$ that are non-adjacent to i.
- **Aspiration:** this movement offers the opportunity of nodes belonging to the Tabu List to be added.

The previous neighborhoods take effect whenever the resulting cut-clique is increased. It is worth to remark that **Add**, **Swap**, and **Aspiration** are taken from a previous ILS [16]. However, our VND is enriched with 2 additional neighborhood structures, named **Remove** and **Cone**. Observe that the Tabu list works during the potential additions during **Add**, **Swap** and **Cone**. On the other hand, **Aspiration** provides diversification with an *opportunistic unchoking* process: it picks nodes from the Tabu List instead.

For the remaining four local searches, there is an efficient way to determine whether there is an improvement with respect to some neighbor-set. Specifically, the Test Lemmas 1 to 4 are useful to determine the improvements for **Remove**, **Add**, **Swap** and **Cone** movements, respectively. We call Aspiration Test to Lemma 2 but applied in a different domain (specifically, the candidate nodes must belong to the Tabu List).

Lemma 1 (Remove). $|\delta(\mathcal{C} - \{i\})| > |\delta(\mathcal{C})|$ *iff* $|\delta(i)| < 2(|\mathcal{C}| - 1)$.

Proof.

$$
\begin{aligned}
|\delta(\mathcal{C} - \{i\})| &= |\delta(\mathcal{C})| + |\mathcal{C}| - 1 - (|\delta(i)| - (|\mathcal{C}| - 1)) \\
&= |\delta(\mathcal{C})| + |\mathcal{C}| - 1 - |\delta(i)| + |\mathcal{C}| - 1 \\
&= |\delta(\mathcal{C})| + 2(|\mathcal{C}| - 1) - |\delta(i)| \\
&> |\delta(\mathcal{C})|,
\end{aligned}
$$

where the last inequality holds iff $2(|\mathcal{C}| - 1) - |\delta(i)| > 0$.   ■

Lemma 2 (Add). $|\delta(\mathcal{C} \cup \{i\})| > |\delta(\mathcal{C})|$ *iff* $|\delta(i)| > 2|\mathcal{C}|$.

Proof.

$$
\begin{aligned}
|\delta(\mathcal{C} \cup \{i\})| &= |\delta(\mathcal{C})| - |\mathcal{C}| + |\delta(i)| - |\mathcal{C}| \\
&= |\delta(\mathcal{C})| + |\delta(i)| - 2|\mathcal{C}| \\
&> |\delta(\mathcal{C})|,
\end{aligned}
$$

where the last inequality holds iff $|\delta(i)| > 2|\mathcal{C}|$.   ■

Lemma 3 (Swap). $|\delta(\mathcal{C} - \{j\} \cup \{i\})| > |\delta(\mathcal{C})|$ *iff* $|\delta(i)| > |\delta(j)|$.

Proof.

$$|\delta(\mathcal{C} - \{j\} \cup \{i\})| = |\delta(\mathcal{C})| - |\delta(j)| + 2(|\mathcal{C}| - 1) + |\delta(i)| - 2(|\mathcal{C}| - 1)$$
$$= |\delta(\mathcal{C})| - |\delta(j)| + |\delta(i)|$$
$$> |\delta(\mathcal{C})|,$$

where the last inequality holds iff $|\delta(i)| > |\delta(j)|$.   ∎

Lemma 4. (Cone). $|\delta(\mathcal{C} - \mathcal{A} \cup \{i\})| > |\delta(\mathcal{C})|$ *iff* $|\delta(i)| > |\delta(\mathcal{A})| - 2|\mathcal{C} - \mathcal{A}|(|\mathcal{A}| - 1)$.

Proof.

$$|\delta(\mathcal{C} - \mathcal{A} \cup \{i\})| = |\delta(\mathcal{C})| + |\mathcal{A}||\mathcal{C} - \mathcal{A}| - (|\delta(\mathcal{A})| - |\mathcal{A}||\mathcal{C} - \mathcal{A}|) - 2|\mathcal{C} - \mathcal{A}| + |\delta(i)|$$
$$= |\delta(\mathcal{C})| + 2|\mathcal{A}||\mathcal{C} - \mathcal{A}| - |\delta(\mathcal{A})| - 2|\mathcal{C} - \mathcal{A}| + |\delta(i)|$$
$$= |\delta(\mathcal{C})| + 2|\mathcal{C} - \mathcal{A}|(|\mathcal{A}| - 1) - |\delta(\mathcal{A})| + |\delta(i)||\delta(\mathcal{C} - \mathcal{A} \cup \{i\})|$$
$$> |\delta(\mathcal{C})|$$

where the last inequality holds iff $|\delta(i)| > |\delta(\mathcal{A})| - 2|\mathcal{C} - \mathcal{A}|(|\mathcal{A}| - 1)$.   ∎

The Flow Diagram of our VND is presented in Fig. 2. The ordered sequence of local searches are **Remove, Add, Swap, Cone** and **Aspiration** moves. Once an improvement is obtained, the process restarts from the beginning. Observe that, in the output, a locally optimum solution under all neighborhood structures is met.

4   Computational Results

In order to test the performance of the algorithm, a fair comparison with respect to an Iterated Local Search solution [16] is carried out using DIMACS benchmark. The test was executed on an Intel Core i7, 2.4 GHz, 8 GB RAM.

Table 1 reports the performance of our HM algorithm for each instance[1]. All instances were tested using 100 runs with $\alpha = \frac{1}{2}$, $MAX\_ATTEMPTS = \lfloor \frac{|V|}{10} \rfloor$, $\theta^{max} = 10$. The values remarked using bold letters from column $|E'(\mathcal{C})|$ indicate that the best solution known was reached according to [16].

Following the terminology, max_iter represents the number of iterations considered in the algorithm, $|E'(\mathcal{C})|$, $|\mathcal{C}|$ and $Time$ represent maximum cut-clique size found, best solution and the CPU time for the Best solution found. The same columns are reported for an averaging over 100 runs.

The reader can appreciate that our HM algorithm meets the best solution known so far in all cases. On the one hand, HM is a more powerful strategy than ILS, since the local search from the latter are completely included in the former. On the other, the computational effort is increased using HM. Even though a

---

[1] All the scripts are available at the following URL: https://www.fing.edu.uy/~lstabile/mcc-octave-source.zip.

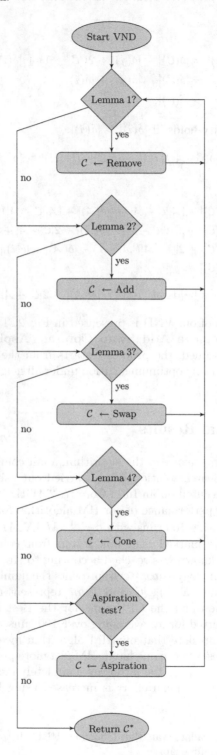

Fig. 2. Flow diagram for the local search phase - VND.

Table 1. Results of the algorithm for the MCC problem

Instances			Parameters	Best			Average	
Name	n	Density	max_iter	$|E'(\mathcal{C})|$	$|\mathcal{C}|$	Time (s)	$|E'(\mathcal{C})|$	Time (s)
c-fat200-1	200	0.071	10	**81**	9	0.1	81	0.3
c-fat200-2	200	0.163	10	**306**	17	0.5	306	0.8
c-fat200-5	200	0.426	10	**1892**	43	3	1892	4.9
c-fat500-1	500	0.036	10	**110**	10	0.5	110	2.4
c-fat500-2	500	0.073	10	**380**	19	3	380	5.8
c-fat500-5	500	0.186	10	**2304**	48	10	2304	10.8
c-fat500-10	500	0.374	10	**8930**	94	38	8930	65
p_hat300-1	300	0.244	100	**789**	8	129	787	905
p_hat300-2	300	0.489	100	**4637**	25	8	4636	3659
p_hat300-3	300	0.744	1000	**7740**	36	469	7556	3992
p_hat500-1	500	0.253	100	**1621**	9	13	1621	694
p_hat500-2	500	0.505	100	**11539**	36	16	11401	723
p_hat500-3	500	0.752	1000	**18859**	50	679	18855	723
p_hat700-1	700	0.249	100	**2606**	11	305	2602	439
p_hat700-2	700	0.498	1000	**20425**	44	79	20425	839
p_hat700-3	700	0.748	1000	**33480**	62	945	33468	1807
p_hat1000-1	1000	0.245	1000	**3556**	10	216	3556	355
p_hat1000-2	1000	0.490	10000	**31174**	46	2124	31174	2538
p_hat1000-3	1000	0.744	10000	**51259**	65	2687	53256	3584
p_hat1500-1	1500	0.253	1000	**6018**	11	399	6018	904
p_hat1500-2	1500	0.506	10000	**67486**	65	2482	67486	2942
p_hat1500-3	1500	0.754	10000	**112873**	94	1174	112872	23162
keller4	171	0.649	100	**1140**	11	9	1140	11
keller5	776	0.752	10000	**15184**	27	1956	15183	1167
keller6	3361	0.818	100000	**159608**	59	26362	158423	321731
c125_9	125	0.899	1000	**2766**	34	102	2766	253
c250_9	250	0.899	1000	**8123**	44	426	8123	831
c500_9	500	0.901	10000	**22691**	57	2354	22652	4469
c1000_9	1000	0.901	10000	**57149**	68	3924	56038	4125
c2000_5	2000	0.500	10000	**16106**	16	23472	16082	23472
c2000_9	2000	0.900	50000	**136769**	79	37472	135001	45472
c4000_5	4000	0.500	50000	**36174**	18	31196	35891	38119
MANN_a9	45	0.927	1000	**412**	16	4	412	145
MANN_a27	378	0.990	10000	**31284**	126	309	31244	548
MANN_a45	1035	0.996	50000	**236406**	344	46881	235072	52112
MANN_a81	3321	0.999	50000	**2436894**	1098	73213	2433624	96743

globally optimum is not formally proved for some instances, the null gap between ILS and our solution suggests an evidence of optimality.

The results described in this section reflect that our GRASP/VND methodology is competitive with state-of-the-art solutions for the MCC. We underscore the simplicity of implementation conducted by simple building blocks (solution construction procedures and local search methods).

5   Conclusions and Trends for Future Work

Several business models can be represented by Market Basket Analysis (MBA). A relevant marketing approach is to find a subset of items that are strongly correlated with the others. This intuition is formalized by means of a combinatorial optimization problem, called Max Cut-Clique (MCC). In this paper the $\mathcal{NP}$-Completeness of MCC is established. Then, a GRASP/VND methodology enriched with Tabu Search is developed to address the MCC. A fair comparison confirms that our approach is competitive with state-of-the art solutions.

As future work, we want to implement our solution into a real-life product-placement scenario. In a first stage, we need historical information to determine the links between pairs of items. Finally, the physical location of the items must be determined using a complementary geometrical problem with constraints. The solution could consider multi-constrained clustering in order to include categories for the items, or other Machine Learning techniques to determine profiles for the customers, according to the product under study. After the real implementation, the feedback of sales in a period is a valuable metric of success.

Acknowledgements. This work is partially supported by Project 395 CSIC I+D *Sistemas Binarios Estocásticos Dinámicos*. We would like to thank the reviewers for their insightful comments that simplified the readability of this paper.

References

1. Agrawal, R., Imieliński, T., Swami, A.: Mining association rules between sets of items in large databases. SIGMOD Rec. **22**(2), 207–216 (1993)
2. Aguinis, H., Forcum, L.E., Joo, H.: Using market basket analysis in management research. J. Manag. **39**(7), 1799–1824 (2013)
3. Amuthan, A., Thilak, K.D.: Survey on Tabu search meta-heuristic optimization. In: 2016 International Conference on Signal Processing, Communication, Power and Embedded System (SCOPES), pp. 1539–1543, October 2016
4. Bader, G.D., Hogue, C.W.V.: An automated method for finding molecular complexes in large protein interaction networks. BMC Bioinform. **4**, 2 (2003)
5. Brohée, S., van Helden, J.: Evaluation of clustering algorithms for protein-protein interaction networks. BMC Bioinform. **7**(1), 488 (2006)
6. Bruinsma, G., Bernasco, W.: Criminal groups and transnational illegal markets. Crime Law Soc. Change **41**(1), 79–94 (2004)
7. Cascio, W.F., Aguinis, H.: Research in industrial and organizational psychology from 1963 to 2007: changes, choices, and trends. J. Appl. Psychol. **93**(5), 1062–1081 (2008)

8. Cook, S.A.: The complexity of theorem-proving procedures. In: Proceedings of the Third Annual ACM Symposium on Theory of Computing, STOC 2071, pp. 151–158. ACM, New York (1971)
9. Duarte, A., Mladenović, N., Sánchez-Oro, J., Todosijević, R.: Variable neighborhood descent. In: Martí, R., Panos, P., Resende, M. (eds.) Handbook of Heuristics, pp. 1–27. Springer, Cham (2016). https://doi.org/10.1007/978-3-319-07153-4_9-1
10. Garey, M.R., Johnson, D.S.: Computers and Intractability: A Guide to the Theory of NP-Completeness. W. H. Freeman and Company, New York (1979)
11. Glover, F., Laguna, M.: Tabu Search. Kluwer Academic Publishers, Norwell (1997)
12. Henzinger, M., Lawrence, S.: Extracting knowledge from the world wide web. Proc. Natl. Acad. Sci. **101**(Suppl. 1), 5186–5191 (2004)
13. Hüffner, F., Komusiewicz, C., Moser, H., Niedermeier, R.: Enumerating isolated cliques in synthetic and financial networks. In: Yang, B., Du, D.-Z., Wang, C.A. (eds.) COCOA 2008. LNCS, vol. 5165, pp. 405–416. Springer, Heidelberg (2008). https://doi.org/10.1007/978-3-540-85097-7_38
14. Karp, R.M.: Reducibility among combinatorial problems. In: Miller, R.E., Thatcher, J.W. (eds.) Complexity of Computer Computations, pp. 85–103. Plenum Press (1972)
15. Martins, P.: Cliques with maximum/minimum edge neighborhood and neighborhood density. Comput. Oper. Res. **39**, 594–608 (2012)
16. Martins, P., Ladrón, A., Ramalhinho, H.: Maximum cut-clique problem: ILS heuristics and a data analysis application. Int. Trans. Oper. Res. **22**(5), 775–809 (2014)
17. Resende, M.G.C., Ribeiro, C.C.: Optimization by GRASP - Greedy Randomized Adaptive Search Procedures. Computational Science and Engineering. Springer, New York (2016). https://doi.org/10.1007/978-1-4939-6530-4
18. Salhi, S.: Handbook of metaheuristics. J. Oper. Res. Soc. **65**(2), 320–320 (2014)
19. Tan, P.-N., Steinbach, M., Kumar, V.: Introduction to Data Mining. Addison-Wesley Longman Publishing Co. Inc., Boston (2005)

A Modelica-Based Simulation Method for Black-Box Optimal Control Problems with Level-Set Dynamic Programming

Ping Qiao[✉], Yizhong Wu, and Qi Zhang

School of Mechanical Science and Engineering,
Huazhong University of Science and Technology, Wuhan 430074, China
qiaopingde@hust.edu.cn

Abstract. A system model in practice may be a black-box function. However, most of the current research on optimal control problems is conducted under the condition that the specific expression of the model is known, and there is a lack of research on the optimal control problem of black-box models. Based on Modelica language and corresponding simulation platform, this paper gets the simulation data from a Modelica model with the serialization of parallel simulation and uses level-set dynamic programming (DP) algorithm to calculate the cost-to-go function recursively. In order to retrieve the sequence of optimal control variables and corresponding optimal state trajectory, two methods are proposed, namely the method based on continuous simulation and the method that approximates state transfer equations locally with a sequence of Radial Basis Functions (RBFs). As an example, an academic case is analyzed. The result proves the effectiveness of the proposed method in solving the optimal control problem of the black-box models.

Keywords: Black-box · Non-causal · Modelica-based · Level-set DP · RBF

1 Introduction

At present, numerical methods for solving optimal control problems can be divided into three approaches: indirect method, direct method, and dynamic programming [1]. The indirect methods use the Pontryagin's Maximum Principle to convert the optimal control problem to the boundary value problem [2] and then solve it using corresponding approaches. The direct methods are based on discretization of the state variables or control variables and convert the optimal control problem to a nonlinear programming problem [3], which can be solved by some nonlinear programming solvers. Based on the principle of optimality, dynamic programming leads to the Hamilton-Jacobi-Bellman equation that can be approximately computed [4, 5].

Based on the Modelica language [6–8], the Jmodelica.org platform [9] uses pseudo spectral collocation methods and local collocation methods to solve the optimal control problem and can solve trajectory optimization and parameter optimization. All the above methods and the platform that we discussed are only suitable for solving the optimal control problem with specific expressions of its differential-algebraic equations (DAEs).

© Springer Nature Switzerland AG 2019
G. Nicosia et al. (Eds.): LOD 2018, LNCS 11331, pp. 368–380, 2019.
https://doi.org/10.1007/978-3-030-13709-0_31

However, there are many practical situations where a mathematical model is not available. Another case refers to computer programs where commonly only input/output information is available by simulation. The Modelica model established by non-causal modeling on the Modelica software platform or FMU model derived from other platforms is such an example. The advantage of non-causal models is that there is no limit to the direction of solving the equations when declaring them. Thus the equation has more flexibility and functionality than the assignment statement. Due to this feature, it is convenient to reuse the model, which is conducive to constructing more complex models and models closer to the real world. However, the models of non-causal modeling and the FMU models derived from other platforms should be considered as black box because it is difficult to get the clear data flow like DAEs. Unfortunately, the representation and solution of optimal control problems are mostly based on the DAEs. The optimal control problem of these models is called the optimal control problem of black-box models. Moreover, the research on the optimal control problem is rarely involved in black-box models.

This paper studies the optimal control problem of black-box models in Modelica language and corresponding simulation platform. The contribution of this paper is to compute the cost-to-go function using the simulation data and level-set dynamic programming (DP) algorithm. One of the highlights of the paper is parallel simulation serialization that greatly reduces the simulation time. In the calculation of the optimal state trajectory, two methods are proposed. One is to calculate the next state point based on the current state point by the sequence of model simulation. The other is to approximate state transfer equations locally by establishing a series of Radial Basis Functions (RBFs) [10]. Compared with the exact solution of the same white-box model in a given example, a good effect is obtained on the optimal control problem of the black-box models without the mathematical expression. The example also shows that the two methods to calculate the optimal state trajectory are effective and the second method is more efficient.

This brief is organized in the following way. Section 2 defines the optimal control problem and Sect. 3 reviews dynamic programming algorithm and level-set DP. The main contribution of this article is given in Sect. 4, where the modeling and simulation of the optimal control problem and the detailed implementation using level-set DP are explained. The two methods used in the calculation of optimal state trajectory are also introduced here. Finally, a case study demonstrates that the methods proposed in the article can be used to solve the optimal control problem of black-box models.

2   Optimal Control Problem

A special class of optimal control problems with a fixed final time and a partially constrained final state may be summarized as follows: Find an admissible control sequence $u_k, k = 0, 1, \ldots, N$ that solve the following problem [11].

$$Minimize \, J = \phi(x_N) + \sum_0^{N-1} L_k[x_k, u_k]$$

$$\begin{aligned} Subject\,to \quad & x_{k+1} = f_k[x_k, u_k] \\ & x_k \in X_k \subseteq \mathbb{R}^n \\ & x_0 = x_{ic} \\ & x_N \in T \\ & u_k \in U_k \subseteq \mathbb{R}^m \\ & for\,all\,k = 0, 1, \ldots, N. \end{aligned} \qquad (1)$$

We consider the dynamic programming algorithm to solve the optimal control problem. The continuous-time model needs to be discretized since dynamic programming algorithm is a numerical algorithm. And time is divided to N stages and the function $\phi(x_N)$ is the final cost. Applying the control signal u_k to the practical system at discrete time k causes the cost $L_k[x_k, u_k]$ named the stage cost. The dynamic system has n state variables and m control variables. We divide the time to N instances now, and it is assumed that the control variables are time-invariant during every time instance k. However, we must notice that L_k and f_k can be time-variant at different time instance. f_k is called as state transfer equations, which is the evolution process from one state to another. x_k is the current state point and x_{k+1} is the next state point. The dynamic system given by (1) is based on this assumption and its solution is obtained through the simulation in Modelica language and its software environment. The last point we need to discuss is the constraints about the dynamic system. X_k is the time-variant admissible state variables set and U_k is the time-variant admissible control variables set. x_{IC} gives the initial condition, and the final state value is constrained to the target set T.

In terms of the discrete state variables set X_k, it can be represented by

$$X_k = \{x_1^k, x_2^k, \ldots x_j^k, \ldots, x_q^k\}, \quad q = \prod_i^{n-1} q_i. \, n$$ is the dimension of state variables, q_i is the

number of points used to discretize state variable i. and q is the total number of state points. The subscript j in x_j^k denotes the state point with time k and state index j. The discrete representation of control space is analogous to the representation of state space.

It can be denoted as the set $U_k = \{u_1^k, u_2^k, \ldots u_j^k, \ldots, u_p^k\}, \quad p = \prod_i^{m-1} p_i.$

3   Basic DP and Level-Set DP

3.1   Basic DP

On the basis of (2) derived from the Bellman equation, the optimal cost-to-go function $J_k(x^i)$ at point x^i at time k, which represents the optimal cumulative cost from the current state to the final state, can be evaluated by proceeding backwards in time.

$$J_k(x^i) = \min_{u_k \in U_k} \{ L_k(x_k^i, u_k) + J_{k+1}(f_k(x_k^i, u_k)) \} \tag{2}$$

However, J_{k+1} in (2) only records the value of the optimal cost-to-go function on discrete grid points and the state point $f_k(x_k^i, u_k)$ is continuous. In order to evaluate $J_{k+1}(f_k(x_k^i, u_k))$, approximant treatments are necessary to be used. Classical approximation strategies are nearest neighbors and linear interpolation.

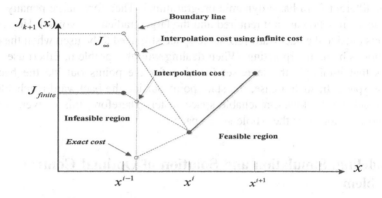

Fig. 1. The interpolation error at the boundary in 1D

When initializing the cost-to-go function, the value of the state point that violates the terminal constraint is set to infinity. Sundström [12] used the concept of the backward-reachable space and linear interpolation to evaluate $J_{k+1}(f_k(x_k^i, u_k))$ and describes the situation where the state point $f_k(x_k^i, u_k)$ is in the vicinity of the boundaries between the backward-reachable space and non-backward-reachable space. In such a case, the grid points used for interpolation include feasible infeasible points. Since the value of the infeasible point is infinite, the result of interpolation is also infinite, which brings great error.

Taking into account the errors caused by the above practices, the cost value of the state point which violates terminal constrains is changed from infinity to limited. This algorithm is called basic dynamic programming. However, this method results in a steep gradient in the cost-to-go function and distorts the solution as Fig. 1 shows.

3.2 Level-Set DP

In order to reduce the numerical errors mentioned above, a method of level-set DP is proposed in [13]. The level-set DP introduces a new function I, its definition is as follows.

$$I : X \subseteq R^n \to R \tag{3}$$

The positive and negative functions represent the external and internal parts of the backward-reachable space respectively.

$$S = \{x \in X | I(x) \leq 0\} \tag{4}$$

When solving the optimal control problem, the level-set DP algorithm calculates the level-set and the cost-to-go functions parallel in each time instance. This detailed algorithm estimates the level-set function can be seen in the literature [13].

The algorithm can tell whether the state point is in the backward-reachable space, which is different from basic dynamic programming. Therefore, large penalty costs in the cost-to-go function are not required and the large gradient is avoided. However, the grid points outside the backward-reachable space still need to be used when the cost-to-go function is being interpolating. When dealing with this problem, Elbert uses control variables that minimize the level-set function for the points outside the backward-reachable space. In such a case, the state point outside the backward-reachable space can approach the backward-reachable space faster. Therefore, this delivers a smooth cost-to-go function over the whole state space.

4   Modeling, Simulation and Solution of Optimal Control Problems

As mentioned earlier, Modelica language supports non-causal modeling. Compared with declarative modeling, this feature of Modelica language makes modeling more convenient. Simultaneously, this work is based on Mworks, which is a platform based on Modelica language to support multi-domain modeling, simulation, analysis and optimization [14]. The method proposed in this paper is for the optimal control problem of the black-box models. However, taking into account the verification of the correctness of the solution results, the description of the proposed method is performed using an example of known mathematical expression. The modeling of the optimal control problem is needed to be introduced here. In order to solve the optimal control problem of the black-box models through the level-set DP algorithm, the simulation of the Modelica model describing the optimal control problem in every time instance is needed. The finer the time division, the more simulations are needed. For this reason, time consumption will be an unacceptable cost when the model is complex. Hence, the situation that the dynamic constraint equations don't contain time in an explicit way is only considered.

4.1   Modeling of Optimal Control Problems

The description of the optimal control problem is shown in Sect. 2. Here we mainly introduce the modeling of the optimal control problem with differential algebraic equations as an example. The Modelica model framework is established in this paper as shown in Fig. 2. Two state variables and two control variables are in the example.

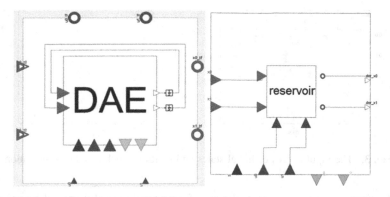

Fig. 2. The Modelica model framework and the nested DAE model (Color figure online)

The two red triangles on the left side of the DAE rectangle in the figure represent the state variables. The small white triangles on the right represent the derivatives of the state variables, which are equal to the corresponding state variables after the 1/s integration blocks. The initial value block and terminal block of these variables are also given. The nested DAE block is shown in Fig. 2. The black-box DAEs model is established by drag and drop. In addition to the state variables and control variables, there are some related parameters for use.

4.2 Implementation of Parallel Simulation Serialization

As shown in (1), the differential algebraic equations become the ordinary differential equations when the control variables are fixed in the time instance and the initial state values are given. Thus, a simulation can be carried out. The level-set DP algorithm is based on grid sampling. For state variables at grid points, all discrete control variables are traversed to obtain the values of state variables at the end of the time instance. The current problem is the simulation of the same DAEs model under different simulation conditions (initial state variables and control variables are different). Obviously, if the DAEs model needs to be simulated every time the simulation conditions are changed, it will be very cumbersome and time-consuming. A parallel simulation serialization method is proposed to realize the series connection of different simulation conditions so that simulation results under all simulation conditions can be obtained only through one simulation.

The main basis for implementing this assumption is the reinit function in Modelica language, which can reset the state variables value at the time of the event. Based on this function, the state variable points and the control variable points obtained from the grid sampling are strung into the input curves of these variables. These curves are piecewise constant functions and the time interval is equal to the time instance used by the level-set DP algorithm. Each dimension corresponds to an input curve as the Fig. 3 shows.

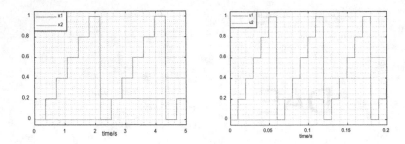

Fig. 3. The input curves of initial states and control variables after serialization

In the time instance $t_k \sim t_k + 1$, the transformed ordinary differential equations are simulated by the Mworks platform. The values of state variables obtained by the simulation are recorded in the time instance $t_k + 1$. After the state variables of the DAEs are reset to the value in the input curves of state variables in the next time instance, the output values of state variables under the corresponding simulation condition are obtained through the next simulation.

By serializing the parallel simulation of different simulation conditions of the same model, a large amount of time for the simulation solver to be set up and turned on is saved, which lays the foundation for solving the optimal control problem of black-box models.

4.3   The Calculation of the Optimal State Trajectory

After the level-set DP is completed, the optimal cost-to-go function corresponding to all grid points x at the discrete time point is obtained. In order to get the sequence of optimal control variables and the corresponding optimal state trajectory that allows the system to achieve the lowest cost or maximum benefit from the initial states, a forward simulation from the initial states needs to be conducted. However, the trajectory cannot be retrieved directly because the DAEs are black-box and the specific expressions of state transfer equations are unknown. Two methods are proposed in this article. One is to calculate the next state point based on the current state point by the sequence of simulations. The other is to approximate state transfer equations locally by establishing a series of local RBFs based on the simulation data of Modelica model.

Based on the Sequence of Simulations
This method is similar to the simulation in step Sect. 4.2. The difference is that the state variables are known as the initial states of the simulation for each time instance here. In each time instance, the simulation values of state variables corresponding to all discrete control variables are obtained through one simulation. The specific steps are as follows:

- Initialization of state variables

$$x_o = \begin{cases} x_{ic} & if \ k = 0 \\ x_k^{op} & else \end{cases} \tag{5}$$

- Traverse all discrete control variables to obtain the corresponding simulation value of state variables.

$$x_{k+1} = S(x_o, u_k) \qquad for\ all\ u_k \in U_k \tag{6}$$

S represents the mapping between and the initial value of state variables and control variables to the simulation result, i.e.:

$$S:\ R^n \times R^m \to R^n \tag{7}$$

- Find the optimal control variables in every time instances

$$u_k^{op} = \begin{cases} \arg\min\{L_k(x_o, u_k) + J_{k+1}(x_{k+1})\}, & if\ I_{k+1}(x_{k+1}) \leq 0 \\ \arg\min\{I_{k+1}(x_{k+1})\} & else \end{cases} \tag{8}$$

- Calculate the value of state variables using the optimal control variables

$$x_{k+1}^{op} = S(x_o, u_k^{op}) \tag{9}$$

Repeat these steps to obtain the sequence of optimal control variables and optimal state trajectory. The results obtained by this method are relatively accurate. However, the next simulation can be only started after the current simulation is over. There, the time cost is huge when the time division of the algorithm is very dense. To solve this problem, the following local approximation method of transfer equations is proposed.

Local Approximation of Transfer Equations Based on the Sequence of RBFs

This method is essentially an approximation method. However, the establishment of a global RBF requires a large number of sampling points to meet the model accuracy, which results in a huge time consumption when doing a large number of valuations of the approximate model. In general, the number of sampling points when building a local approximation model is much smaller than the number of sampling points for establishing a suitable global model. And in the local area, the local RBF is more accurate than the global RBF. Simultaneously, the estimation speed is faster.

There is no need to re-simulate here and the data are all from the simulation of the original model before level-set DP. Compared with the continuous simulation method introduced before, the efficiency has been greatly improved, which can be reflected in the following example. The specific steps are as follows:

- Initialization of state variables

$$x_o = \begin{cases} x_{ic} & if\ k = 0 \\ x_k^{op} & else \end{cases} \tag{10}$$

- Determine the response surface in current time instance

Find the adjacent points of state variables on the discrete grid. The total number of these points is 2^n. Then check whether the current state point is within the active area of the last RBF. If the current state point is not in this area, sample the discrete points of the control variables corresponding to these state grid points. Then take these state grid points and the corresponding sampled control grid points as inputs to obtain a locally active response surface model (RSM) in the neighborhood of x_o. The graphic display of the image is shown in Fig. 4. The range of the response surface model is defined as a hyper-rectangle by these 2^n points adjacent to x_o.

$$RSM_k = \begin{cases} RSM_{k-1} & if\ k \geq 1\ and\ x_o \in \varepsilon(x_{k-1}) \\ RSM_{new}(D_{\varepsilon(x_o)}) & else \end{cases} \tag{11}$$

$\varepsilon(\bullet)$ indicates the neighborhood where the state point is located and $D_{\varepsilon(x_o)}$ represents the training data set sampled in the current local area.

- Traverse all discrete control variables to obtain the corresponding simulation value of state variables

$$x_{k+1} = R(x_o, u_k) \qquad for\ all\ u_k \in U_k \tag{12}$$

R represents the transfer equation of the current time instance using the RSM.

- Find the optimal control variables in every time instance

$$u_k^{op} = \begin{cases} \arg\min\{L_k(x_o, u_k) + J_{k+1}(x_{k+1})\}, & if\ I_{k+1}(x_{k+1}) \leq 0 \\ \arg\min\{I_{k+1}(x_{k+1})\} & else \end{cases} \tag{13}$$

- Calculate the value of state variables using the optimal control variables

$$x_{k+1}^{op} = S(x_o, u_k^{op}) \tag{14}$$

Repeat steps 1–5 to obtain the sequence of optimal control variables and optimal state trajectory.

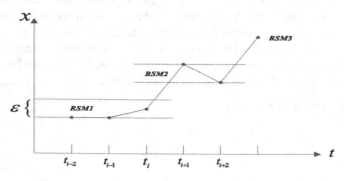

Fig. 4. Possible sequence of the RBF response surfaces

5 Case Study: Simple Dynamic System

To illustrate the effectiveness of the proposed method, a simple example from [13] is used here. The simple system can be introduced as in Fig. 5. The state variables represent the amount of water in the two reservoirs. The first control variable decides the total amount of water flowing into the two reservoirs and the second control variable determines the water distribution of the two reservoirs. When the terminal time arrives, the specified terminal water level needs to be reached in the two reservoirs, using a minimum amount of water. There is a problem that needs to be concerned in the example, i.e. the two reservoirs have a leak, where the leak rate is proportional to the amount of water in the reservoir.

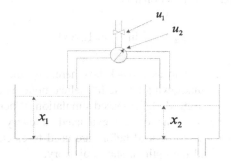

Fig. 5. The simple dynamic system

The system with two state variables and two control variables is described by the following dynamic equations:

$$\dot{x}_1 = -\frac{1}{2}x_1(t) + u_1(t) \cdot u_2(t)$$
$$\dot{x}_2 = -\frac{1}{2}x_2(t) + u_1(t) \cdot (1 - u_2(t))$$

$$(15)$$

The path constrains are:

$$
\begin{aligned}
x(t) &\in [0,1] \times [0,1] & t &\in [0,2] \\
u(t) &\in [0,1] \times [0,1] & t &\in [0,2]
\end{aligned}
$$

$$(16)$$

and the initial and final conditions are:

$$
\begin{aligned}
x_1(0) &= x_2(0) = 0 \\
x_1(2) &= x_2(2) \geq 0.5
\end{aligned}
$$

$$(17)$$

The terminal time in this example is $2s$ and the cost function that needs to be minimized is:

$$J = \int_0^2 u_1(t) + 0.1 \cdot |u_2(t) - 0.5| dt \qquad (18)$$

The analytic solution of this problem is given as follows:

$$u_1(t) = \begin{cases} 0, & if \quad t < 0.6137 \\ 1, & if \quad t > 0.6137 \end{cases}$$

$$u_2(t) = 0.5 \qquad t \in [0, 2] \qquad (19)$$

The value of the optimal cost functional is:

$$J_a = -2\ln(\frac{1}{2}) \approx 1.3863 \qquad (20)$$

We treat this dynamic system as a black box here. In other words, the concrete expression of the system is unknown after the Modelica model is established. The data used for level-set DP come from Modelica-based simulation. When the optimal cost-to-go function and the level-set function are evaluated at every node at every time instance, the two methods as introduced before are used to calculate the sequence of optimal control variables and the optimal state trajectory.

In terms of the level-set DP used, the discretization of these variables directly affects the accuracy of the solution. In this example, the time discretization is $\Delta t = 0.01s$. The state discretization is chosen to be $N_x = 51 \times 51$ and the control discretization is $N_u = 21 \times 21$.

Figures 6 and 7 shows the sequence of optimal control variables and optimal trajectory respectively, which can be obtained by both methods in Sect. 4.3. The solution with the discretization introduced before is close to the analytic solution. Although the two methods can obtain similar solutions, the efficiency of the latter is much higher than that of the former, which can be seen from Table 1.

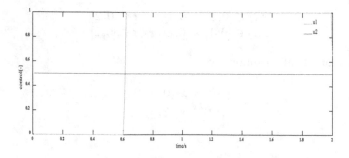

Fig. 6. The sequence of optimal control variables

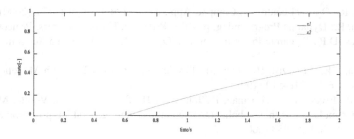

Fig. 7. The corresponding optimal state trajectory

Table 1. Time used with the two methods at similar accuracy for this example

	Simulation-based	RBF-based
The numbers of simulations or RBF	200	24
Time[s]	140.703	2.384

6 Conclusions

This paper simulates the black-box model under different simulation conditions on the Modelica platform by parallel simulation serialization. After obtaining the simulation data, the level-set function and the optimal cost-to-go function are calculated according to the level-set dynamic programming algorithm. Then using the two methods proposed in the paper to calculate the optimal state trajectory and the sequence of control variables, the latter is much more efficient than the former.

After establishing the black-box model, the proposed method does not require the specific mathematical expression of the model in the algorithm process. Simultaneously, the level-set DP algorithm has better accuracy in solving the optimal control problem with terminal constraints. The algorithm does not need global optimality conditions and does not need to calculate the related gradient. Compared with other optimal control algorithms, it is more suitable for black- box situations. However, due to the inherent curse of dimensionality of dynamic programming algorithm, the method proposed in the paper has its limitations. Both the model simulation and the level-set DP consume a lot of time in high-dimensional problems.

Acknowledgments. Financial support from the National Natural Science Foundation of China under Grant No. 51575205 is gratefully acknowledged.

References

1. Nikoobin, A., Moradi, M.: Indirect solution of optimal control problems with state variable inequality constraints: finite difference approximation. Robotica **35**(1), 50–72 (2017)
2. Nikoobin, A., Moradi, M.: Optimal balancing of robot manipulators in point-to-point motion. Robotica **29**(2), 233–244 (2011)

3. Sargent, R.: Optimal control. J. Comput. Appl. Math. **124**(1–2), 361–371 (2000)
4. Bellman, R.: Dynamic Programming, p. 1957. Princeton University Press, Princeton (1957)
5. Bertsekas, D.P.: Dynamic Programming and Optimal Control. Athena Scientific, Belmont (1995)
6. Mattsson, S.E., Elmqvist, H., Otter, M.: Physical system modeling with Modelica. Control Eng. Pract. **6**(4), 501–510 (1998)
7. Sahlin, P., Eriksson, L., Grozman, P., Johnsson, H., Shapovalov, A., Vuolle, M.: Whole-building simulation with symbolic DAE equations and general purpose solvers. Build. Environ. **39**(8), 949–958 (2004)
8. Wetter, M.: Modelica-based modelling and simulation to support research and development in building energy and control systems. J. Build. Perform. Simul. **2**(2), 143–161 (2009)
9. Åkesson, J., Årzén, K.-E., Gäfvert, M., Bergdahl, T., Tummescheit, H.: Modeling and optimization with Optimica and JModelica. org—languages and tools for solving large-scale dynamic optimization problems. Comput. Chem. Eng. **34**(11), 1737–1749 (2010)
10. Buhmann, M.D.: Radial Basis Functions: Theory and Implementations. Cambridge University Press, Cambridge (2003)
11. van Berkel, K., de Jager, B., Hofman, T., Steinbuch, M.: Implementation of dynamic programming for optimal control problems with continuous states. IEEE Trans. Control Syst. Technol. **23**(3), 1172–1179 (2015)
12. Sundström, O., Ambühl, D., Guzzella, L.: On implementation of dynamic programming for optimal control problems with final state constraints. Oil Gas Sci. Technol.-Revue de l'Institut Français du Pétrole **65**(1), 91–102 (2010)
13. Elbert, P., Ebbesen, S., Guzzella, L.: Implementation of dynamic programming for n dimensional optimal control problems with final state constraints. IEEE Trans. Control Syst. Technol. **21**(3), 924–931 (2013)
14. Chen, X., Wei, Z. (eds.): A new modeling and simulation platform-MWorks for electrical machine based on Modelica. In: International Conference on Electrical Machines and Systems, ICEMS 2008. IEEE (2008)

A Clonal Selection Algorithm for Multiobjective Energy Reduction Multi-Depot Vehicle Routing Problem

Emmanouela Rapanaki[1], Iraklis-Dimitrios Psychas[1], Magdalene Marinaki[1], Yannis Marinakis[1], and Athanasios Migdalas[2(✉)]

[1] School of Production Engineering and Management, Technical University of Crete, Chania, Greece
emmarap@hotmail.com, ipsychas102@gmail.com, magda@dssl.tuc.gr, marinakis@ergasya.tuc.gr
[2] Industrial Logistics, Luleå Technical University, 97187 Luleå, Sweden
athmig@ltu.se

Abstract. Clonal Selection Algorithm is a very powerful Nature Inspired Algorithm that has been applied in a number of different kind of optimization problems since the time it was first published. Also, in recent years a growing number of optimization models have been proposed that are trying to reduce the energy consumption in vehicle routing. In this paper, a new variant of Clonal Selection Algorithm, the Parallel Multi-Start Multiobjective Clonal Selection Algorithm (PMS-MOCSA) is proposed for the solution of a Vehicle Routing Problem variant, the Multiobjective Energy Reduction Multi-Depot Vehicle Routing Problem (MERMDVRP). In the formulation four different scenarios are proposed where the distances between the customers and the depots are either symmetric or asymmetric and the customers have either demand or pickup. The algorithm is compared with two other multiobjective algorithms, the Parallel Multi-Start Non-dominated Sorting Differential Evolution (PMS-NSDE) and the Parallel Multi-Start Non-dominated Sorting Genetic Algorithm II (PMS-NSGA II) for a number of benchmark instances.

Keywords: Vehicle Routing Problem · Clonal Selection Algorithm · NSGA II · NSDE · VNS

1 Introduction

The **Vehicle Routing Problem** (VRP) is one of the most famous optimization problems and its main goal is the design of the best routes for a selected number of vehicles in order to serve a set of customers in the best possible way according to some selected criteria which vary with the situation in case. As the interest of researchers and decision makers for the solution of different variants

© Springer Nature Switzerland AG 2019
G. Nicosia et al. (Eds.): LOD 2018, LNCS 11331, pp. 381–393, 2019.
https://doi.org/10.1007/978-3-030-13709-0_32

of VRP continuously increases, a number of more realistic and thus more complicated versions and formulations of it have been proposed and a number of more sophisticated algorithms are used for their solutions [19].

The combination of more than one objective functions in the formulation of a Vehicle Routing Problem variant could produce a more realistic problem. Thus, in the recent years there is an increasing number of research works that propose formulations with more than one criteria. The resulting VRPs are Multiobjective Vehicle Routing Problems. In the present paper, we propose a formulation of the problem that combines the existence of several depots, the possibility of pickups and deliveries, and the simultaneous reduction of the fuel consumption. Both the symmetric and, essentially the more realistic asymmetric cases are considered. Also in recent years, a number of evolutionary, swarm intelligence and other nature inspired algorithms have been proposed for the solution of the VRP both for the single-objective and the multi-objective case. In the present paper, we propose a Multiobjective variant of the Clonal selection algorithm, the Parallel Multi-Start Multiobjective Clonal Selection Algorithm (PMS-MOCSA), is proposed for the solution of the Multiobjective Energy Multi-Depot Vehicle Routing Problem (MEMDVRP). The algorithm is compared with two other evolutionary algorithms, the Parallel Multi-Start Non-dominated Sorting Differential Evolution (PMS-NSDE) [15] and the Parallel Multi-Start Non-dominated Sorting Genetic Algorithm II (PMS-NSGA II) [16].

In recent years there is a growing number of papers devoted to the solution of multi-depot vehicle routing problems [13] or energy consumption vehicle routing problems [10,12,17]. In the present paper, three objective functions are used in all four **Multiobjective Route-based Fuel Consumption Vehicle Routing Problems** presented. The first objective function is the same one that was presented in Psychas et al. [16]. This function is used for the minimization of the total travel and service **time** needed and is given by the following equation:

$$\min OF1 = \sum_{i=I_1}^{n} \sum_{j=1}^{n} \sum_{\kappa=1}^{m} (t_{ij}^{\kappa} + s_j^{\kappa}) x_{ij}^{\kappa}, \tag{1}$$

where t_{ij}^{κ} is the time needed to visit customer j immediately after customer i using vehicle κ, s_j^{κ} is the service time of customer j using vehicle κ, n is the number of nodes, m is the number of homogeneous vehicles and the depots are a subset $\Pi = \{I_1, I_2, \ldots I_\pi\}$ of the set of the n nodes where denoted by $i = j = I_1, I_2, \ldots I_\pi$ (π is the number of homogeneous depots). The set of nodes is then $\{I_1, I_2, \ldots I_\pi, 2, 3, \ldots, n\}$.

The second objective function is used for the minimization of the **Route based Fuel Consumption** (RFC) for the case in which the vehicle performs only deliveries. It takes into account real life route parameters such as weather conditions or uphills and downhills or driver's behavior. This objective function is given by the equation below:

$$\min OF2 = \sum_{h=I_1}^{I_\pi} \sum_{j=2}^{n} \sum_{\kappa=1}^{m} c_{hj} x_{hj}^{\kappa} (1 + \frac{y_{hj}^{\kappa}}{Q}) r_{hj} + \sum_{i=2}^{n} \sum_{j=I_1}^{n} \sum_{\kappa=1}^{m} c_{ij} x_{ij}^{\kappa} (1 + \frac{y_{i-1,i}^{\kappa} - D_i}{Q}) r_{ij}, \tag{2}$$

with the maximum capacity of the vehicle denoted by Q, the i customer has demand equal to D_i and $D_{I_1} = D_{I_2} = ... = D_{I_\pi} = 0$, x_{ij}^κ denotes that the vehicle κ visits customer j immediately after customer i with load y_{ij}^κ and $y_{I_1 j}^\kappa = \sum_{i=I_1}^{n} D_i$ for all vehicles as the vehicle begins with load equal to the summation of the demands of all customers assigned in its route and c_{ij} is the distance from node i to node j. The parameter r_{ij} corresponds to the route parameters from the node i to the node j and it is always a positive number. Due to the fact that it may be $r_{ij} \neq r_{ji}$ the product $c_{ij} r_{ij}$ leads to an asymmetric formulation of the whole problem. A value of r_{ij} less than 1 corresponds to the case in which the route from i to j is a downhill or the wind is back-wind or the driver drives with smooth shifting. A value of r_{ij} larger than 1 corresponds to the case in which the route from i to j is an uphill or the wind is a head-wind or the driver drives with aggressive shifting. If the $r_{ij} = 1 \forall (i, j)$ belonging to the route, then the problem is a symmetric problem.

The third objective function is used for the minimization of the **Route based Fuel Consumption** (RFC) for the case in which the vehicle performs only pick-ups along its route (see Psychas et al. [15]) and is given by the following equation:

$$\min OF3 = \sum_{h=I_1}^{I_\pi} \sum_{j=2}^{n} \sum_{\kappa=1}^{m} c_{hj} x_{hj}^\kappa r_{hj} + \sum_{i=2}^{n} \sum_{j=I_1}^{n} \sum_{\kappa=1}^{m} c_{ij} x_{ij}^\kappa (1 + \frac{y_{i-1,i}^\kappa + D_i}{Q}) r_{ij}, (3)$$

where r_{ij} is the route parameters as in the OF2, and $y_{I_1 j}^\kappa = 0$ for all vehicles as the vehicle begins with empty load. In this case the D_i denotes the pick-up amount of the customer i. The only difference from the previous functions is that we now have more than one depots, defined by a subset $\Pi = \{I_1, I_2, \ldots I_\pi\}$ of the set of the n nodes, and denoted by $i = j = I_1, I_2, \ldots I_\pi$ (π is the number of the homogeneous depots). COnsequently, the set of nodes is then $\{I_1, I_2, \ldots I_\pi, 2, 3, \ldots, n\}$. It is assumed that each vehicle returns always to the depot from where it starts and it does not visits any other depot during along its route. Thus, there are no transitions between the depots (for example from I_1 to I_3). The constraints of the problems are [15]:

$$\sum_{j=I_1}^{n} \sum_{\kappa=1}^{m} x_{ij}^\kappa = 1, i = I_1, \cdots, n \quad (4)$$

$$\sum_{i=I_1}^{n} \sum_{\kappa=1}^{m} x_{ij}^\kappa = 1, j = I_1, \cdots, n \quad (5)$$

$$\sum_{j=I_1}^{n} x_{ij}^\kappa - \sum_{j=I_1}^{n} x_{ji}^\kappa = 0, i = I_1, \cdots, n, \kappa = 1, \cdots, m \quad (6)$$

$$\sum_{j=I_1, j\neq i}^{n} y_{ji}^\kappa - \sum_{j=I_1, j\neq i}^{n} y_{ij}^\kappa = D_i, i = I_1, \cdots, n, \kappa = 1, \cdots, m, \, for \, deliveries \quad (7)$$

$$\sum_{j=I_1, j \neq i}^{n} y_{ij}^{\kappa} - \sum_{j=I_1, j \neq i}^{n} y_{ji}^{\kappa} = D_i, \ i = I_1, \cdots, n, \ \kappa = 1, \cdots, m, \ for \ pick - ups \quad (8)$$

$$Qx_{ij}^{\kappa} \geq y_{ij}^{\kappa}, i, j = I_1, \cdots, n, \kappa = 1, \cdots, m \quad (9)$$

$$x_{ij}^{\kappa} = \begin{cases} 1, & \text{if } (i, j) \text{ belongs to the route} \\ 0, & \text{otherwise} \end{cases} \quad (10)$$

Constraints (4) and (5) require that each customer must be visited only by exactly one vehicle; constraints (6) ensure that each vehicle that arrives at a node must also leave from that node. Constraints (7) and (8) indicate that the reduced (if it concerns deliveries) or increased (if it concerns pick-ups) load (cargo) of the vehicle after it visits a node is equal to the demand of that node. Constraints (9) are used to limit the maximum load carried by the vehicle and to force y_{ij}^{κ} to become zero when $x_{ij}^{\kappa} = 0$, while constraints (10) ensure that only one vehicle visits each customer. It should be noted that the problems solved in this paper are symmetric (where $r_{ij} = 1 \forall (i, j)$ holds) or asymmetric (where $r_{ij} \neq r_{ji} \forall (i, j)$ holds). This paper is organized into three sections. In the next section, the proposed algorithm is described in detail, while in Sect. 3 the computation results are presented. The last section provides concluding remarks and future research.

2   Parallel Multi-Start Multiobjective Clonal Selection Algorithm

Artificial Immune Systems (AIS) [5,7] are inspired by the natural immune system. The AIS algorithms are classified into three categories [1]: (a) the Positive/Negative Selection algorithm [11], (b) the Clonal Expansion and Selection algorithm [8,9] and (c) the Network Algorithms [18]. For analytical informations, surveys and applications about artificial immune systems please see [1–7,14].

In the Parallel Multi-Start Multiobjective Clonal Selection Algorithm (PMS-MOCSA), initially a population (X) of solutions (antibodies) are selected as follows: the solutions of initial population are placed in the Antibody Best table. Subsequently, the Pareto Front of the initial population is created from this table. In every iteration, in order to find the best antibodies in the current population of antibodies (Fb), the non-dominated antibodies (Pareto antibodies) are found. Then, we sort the antibodies as follows: Initially, the average of the cost values for each objective function is calculated. Then, the Euclidean distance between the average and each of the non-dominated solutions of antibodies is calculated. Finally, the solutions are classified according to their distance from the average. After all the above, the Fb non-dominated solutions of antibodies are classified again as follows: The closest to the average antibody is placed in the first antibody position of the list of the Fb antibodies. The most distant from the average is placed in the second antibody position of the list. The second closest antibody to the average is placed in the third position of the list and the fourth position gets occupied by the second most distant antibody from the average, etc. Due to this ordering a sufficient number of clones will be produced from both extreme non-dominated solutions of Pareto front and from those which are in the center of Pareto front.

Then, a number of clones from Fb antibodies are produced by using the following function:

$$\sum_{i=1}^{Fb} \text{round}\left(\frac{b_1 W}{i}\right) \tag{11}$$

where b_1 is equal to 1. We also add ten extra randomly generated antibodies to the total of clones in order to increase exploration abilities.

After the creation of the clones, a random number in the interval $(0, 1)$ is produced for each clone. If this number is less or equal to the $Mr = 0.5$, then a hypermutation phase is applied to the corresponding clone. In this process, a number of customers (nodes) in the route represented by the clone is chosen randomly and their positions are permuted. If the random number generated is greater than Mr, the receptor editing process is applied to the clone solution. This process is realized using the 2-opt method.

When the above processes have been completed for all clones, then the cost for every objective function for each clone is calculated and the non-dominated clone solutions (Pareto clone solutions) are selected. If the number of the non-dominated clone solutions is less than or equal to W, then randomly selected antibodies from the current population of antibodies are replaced by non-dominated clone solutions. In every other case, all the antibodies are replaced by the non-dominated clone solutions. For the improvement of the new antibodies, a Variable Neighborhood Search algorithm (VNS) is applied to every antibody. The resulting antibodies form the population in the next iteration.

As mentioned previously, the proposed algorithm works with a number of different (X) populations. In each iteration and for each population all the processes that described earlier are applied in parallel (with the term "parallel" we mean that the populations are processed independently from each other). Thus, each population has its own Pareto front. However, in the final iteration all solutions of these Pareto fronts are combined into one population and a new final Pareto Front is produced. The proposed algorithm is compared with two other evolutionary multiobjective optimization algorithms, the Parallel Multi-Start Non-dominated Sorting Differential Evolution (PMS-NSDE1) algorithm [15] and the Parallel Multi-Start Non-dominated Sorting Genetic Algorithm II (PMS-NSGA II) algorithm [16]. For all the necessary information about the algorithms PMS-NSDE1 and PMS-NSGA II, please see Psychas et al. [15,16].

3   Computational Results

The algorithms were implemented in Visual C++ and were tested on the same set of instances. The data were created according to Psychas et al. [15]. The only difference is that in this research we use three depots with evenly distributed customers, which means that in an instance with 100 customers, the first customer is the first depot, the second depot has coordinates equal to $x_2 = y_1$ and

$y_2 = x_1$, where (x_1, y_1) are the coordinates of the first depot, the third depot has coordinates between $(100, 100)$ and $(500, 500)$ depending on the instance and, finally, 33 customers are allocated in each depot. The parameters we used to Parallel Multi-Start Multiobjective CSA are selected after testing different values. We selected those that gave the best computational results. The selected parameters for the other two algorithms are taken from Psychas et al. [15]. The parameters of Parallel Multi-Start Multiobjective CSA are: (1) number of individuals for each initial population equal to 100, (2) number of generations equal to 500, (3) number of initial populations equal to 10, (4) Mr = 0.5 and (5) $\beta = 1$.

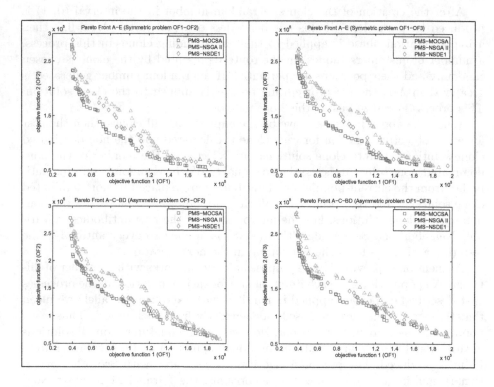

Fig. 1. Pareto fronts of the four algorithms for the instances "A-E" and "A-C-BD".

Then, the algorithms were tested for ten combinations of the two objective functions (OF1-OF2 or OF1-OF3) each, five times. For the comparison of the three algorithms, we use four different evaluation measures: M_k which is the range that the front extends, L that measures the number of solutions of the Pareto front, Δ measure which includes information about both spread and distribution of the solutions and the coverage measure (C measure). For further details about the measures, please refer to [16]. In Table 1, the results of the proposed algorithm for the ten instances, for each problem, for five executions

Table 1. Results of the first three measures in ten instances for five executions using the proposed algorithm

Multiobjective asymmetric delivery route based fuel consumption VRP

	Execution 1			Execution 2			Execution 3			Execution 4			Execution 5			Average			Best run	
	L	M_k	Δ	L	M_k	Δ	L	M_k	Δ	L	M_k	Δ	L	M_k	Δ	L	M_k	Δ	L	M_k	Δ
A-B-CD	62	608.70	0.66	60	610.08	0.73	63	577.84	0.55	64	580.71	0.53	58	579.86	0.54	61.40	591.44	0.60	**64**	580.71	0.53
A-C-BD	51	593.14	0.76	58	602.44	0.62	61	606.48	0.76	50	579.29	0.54	68	597.10	0.62	57.60	595.69	0.66	68	597.10	**0.62**
A-D-BE	50	607.29	0.66	64	604.52	0.60	59	613.32	0.63	59	597.16	0.56	46	615.36	0.69	55.60	607.53	0.63	**64**	**604.52**	0.60
A-E-BD	64	596.31	0.52	59	589.31	0.57	57	589.62	0.59	61	604.27	0.65	65	592.97	0.62	61.20	594.50	0.59	**64**	596.31	**0.52**
B-C-AD	58	609.17	0.64	47	593.33	0.71	47	600.97	0.70	53	593.42	0.72	54	598.42	0.72	51.80	599.06	0.70	58	**609.17**	0.64
B-D-AC	60	603.77	0.71	57	521.72	0.55	56	571.21	0.53	61	607.76	0.56	49	531.38	0.53	56.60	567.17	0.58	61	607.76	0.56
B-E-AD	50	548.29	0.56	61	606.11	0.56	52	593.80	0.70	54	591.32	0.71	58	594.46	0.54	55.00	586.79	0.61	61	606.11	**0.56**
C-D-AE	50	592.44	0.53	71	599.24	0.68	53	596.45	0.62	65	589.67	0.65	54	578.96	0.52	58.60	591.35	0.60	71	599.24	0.68
C-E-AB	46	579.20	0.58	59	594.15	0.55	58	590.30	0.62	56	546.01	0.57	62	576.29	0.62	56.20	577.19	0.59	**59**	594.15	**0.55**
D-E-BC	58	598.44	0.53	55	593.18	0.74	46	599.47	0.62	64	597.70	0.63	45	535.02	0.65	53.60	584.76	0.64	**58**	**598.44**	0.53

Multiobjective symmetric delivery route based fuel consumption VRP

	L	M_k	Δ	L	M_k	Δ	L	M_k	Δ	L	M_k	Δ	L	M_k	Δ	L	M_k	Δ	L	M_k	Δ
A-B	73	608.59	0.66	62	611.94	0.72	60	602.00	0.66	60	589.45	0.60	71	611.70	0.59	65.20	604.74	0.65	**71**	**611.70**	0.59
A-C	70	609.97	0.66	71	597.86	0.58	72	600.56	0.61	66	615.67	0.63	67	614.84	0.69	69.20	607.78	0.63	**72**	600.56	0.61
A-D	78	589.98	0.72	81	576.68	0.66	67	580.20	0.56	79	584.42	0.71	71	574.39	0.59	75.20	581.14	0.65	**78**	589.98	0.72
A-E	65	596.01	0.59	68	593.60	0.67	61	581.82	0.65	66	592.68	0.67	65	601.41	0.64	65.00	593.11	0.65	**65**	601.41	0.64
B-C	58	536.06	0.66	73	621.22	0.70	69	602.41	0.47	63	594.65	0.54	53	523.23	0.65	63.20	575.51	0.60	**73**	**621.22**	0.70
B-D	62	608.56	0.72	72	593.97	0.61	62	578.88	0.66	63	611.85	0.69	71	595.56	0.64	66.20	597.77	0.66	**72**	593.97	0.61
B-E	60	594.91	0.68	57	608.20	0.56	51	540.50	0.61	64	616.37	0.60	59	608.97	0.61	58.20	593.79	0.61	**64**	**616.37**	0.60
C-D	61	595.05	0.54	73	609.02	0.69	66	580.98	0.67	71	589.66	0.59	66	594.34	0.56	67.40	593.81	0.61	**73**	**609.02**	0.69
C-E	78	600.56	0.50	57	604.89	0.75	64	551.48	0.71	64	605.58	0.68	64	608.88	0.67	65.40	594.28	0.66	**78**	600.56	**0.50**
D-E	75	619.27	0.64	62	599.92	0.57	66	601.14	0.60	73	597.19	0.60	82	612.39	0.58	71.60	605.98	0.60	**82**	612.39	0.58

Multiobjective asymmetric pick-up route based fuel consumption VRP

	L	M_k	Δ	L	M_k	Δ	L	M_k	Δ	L	M_k	Δ	L	M_k	Δ	L	M_k	Δ	L	M_k	Δ
A-B-CD	60	608.30	0.61	48	614.71	0.55	62	574.93	0.67	58	593.83	0.70	62	575.73	0.55	58.00	593.50	0.61	48	**614.71**	**0.55**
A-C-BD	65	593.37	0.64	56	595.90	0.64	64	608.93	0.60	67	589.86	0.61	72	614.65	0.64	64.80	600.54	0.63	**72**	**614.65**	0.64
A-D-BE	54	610.68	0.54	52	632.40	0.68	50	588.74	0.55	57	599.81	0.63	59	604.09	0.58	54.40	607.14	0.60	54	610.68	**0.54**
A-E-BD	57	579.46	0.56	60	588.89	0.48	60	575.82	0.64	62	611.69	0.57	50	590.89	0.66	61.20	589.35	0.58	**62**	611.69	0.57
B-C-AD	52	533.47	0.63	55	580.36	0.69	66	600.06	0.62	52	533.02	0.50	62	599.76	0.60	57.40	569.33	0.61	**66**	600.06	0.62
B-D-AC	44	534.00	0.61	72	569.85	0.61	38	606.86	0.67	51	535.29	0.65	55	528.83	0.57	52.00	554.96	0.62	**72**	569.85	0.61
B-E-AD	48	594.05	0.72	73	584.76	0.63	56	611.58	0.64	50	611.35	0.62	65	587.64	0.57	58.40	597.88	0.64	**65**	587.64	**0.57**
C-D-AE	62	585.05	0.61	60	594.94	0.55	57	579.30	0.63	58	595.58	0.69	63	582.20	0.55	60.00	587.42	0.61	**60**	594.94	**0.55**
C-E-AB	45	596.44	0.57	62	592.53	0.63	59	592.09	0.61	60	581.63	0.51	62	604.50	0.54	57.60	593.44	0.57	**62**	604.50	**0.54**
D-E-BC	59	579.13	0.62	58	512.10	0.67	66	572.10	0.57	56	582.51	0.65	75	573.38	0.56	62.80	563.84	0.61	**75**	573.38	**0.56**

Multiobjective symmetric pick-up route based fuel consumption VRP

	L	M_k	Δ	L	M_k	Δ	L	M_k	Δ	L	M_k	Δ	L	M_k	Δ	L	M_k	Δ	L	M_k	Δ
A-B	59	616.83	0.70	62	600.53	0.55	75	620.04	0.65	58	586.58	0.64	56	603.97	0.70	62.00	605.59	0.65	75	620.04	0.65
A-C	65	612.67	0.59	67	597.46	0.62	69	606.80	0.63	60	606.44	0.54	59	604.22	0.68	64.00	605.52	0.61	**65**	612.67	**0.59**
A-D	64	578.51	0.57	74	600.68	0.69	63	597.29	0.57	72	572.20	0.62	71	595.77	0.55	68.80	588.89	0.60	**74**	**600.68**	0.69
A-E	73	611.32	0.65	72	603.71	0.64	58	596.27	0.64	63	581.63	0.63	76	582.02	0.62	68.40	594.99	0.63	**76**	582.02	0.62
B-C	55	589.65	0.77	65	592.57	0.64	68	579.04	0.63	55	531.97	0.52	64	602.47	0.56	61.40	579.14	0.63	**64**	602.47	**0.56**
B-D	89	593.98	0.64	66	598.44	0.76	60	602.14	0.65	76	602.77	0.64	57	602.99	0.67	69.60	600.06	0.67	**89**	593.98	**0.64**
B-E	63	614.74	0.63	59	592.94	0.75	57	539.22	0.62	60	598.21	0.64	57	614.36	0.64	59.20	591.89	0.66	**63**	614.74	0.63
C-D	71	591.01	0.65	70	605.05	0.67	71	579.98	0.69	64	582.92	0.76	67	587.76	0.63	68.60	589.35	0.68	**71**	**591.01**	0.65
C-E	69	598.29	0.67	68	602.78	0.56	69	598.09	0.62	61	611.81	0.58	57	592.74	0.63	64.80	600.74	0.61	**68**	602.78	**0.56**
D-E	67	608.18	0.60	70	597.40	0.52	76	604.25	0.66	62	619.25	0.54	82	615.17	0.62	71.40	608.85	0.59	**70**	597.40	**0.52**

for each one of the instances, and for the three first measures are presented. In Tables 2 and 3, the average Results and Best Runs of the proposed algorithms and of the other two algorithms used in the comparisons are presented.

Table 2. Average Results and Best Runs of all algorithms used in the comparisons.

	Algorithms	Asymmetric delivery FCVRP			Asymmetric pick-up FCVRP		
		L	M_k	Δ	L	M_k	Δ
A-B-CD	PMS-MOCSA	61.40(**64**)	591.44(580.71)	0.60(0.53)	58.00(48)	593.50(**614.71**)	0.61(**0.55**)
	PMS-NSGA II	56.40(62)	592.33(598.84)	0.61(0.54)	59.80(**63**)	598.92(608.39)	0.61(0.65)
	PMS-NSDE	50.00(59)	598.17(**604.03**)	0.61(**0.53**)	46.00(49)	598.81(605.87)	0.61(0.62)
A-C-BD	PMS-MOCSA	57.60(68)	595.69(597.10)	0.66(**0.62**)	64.80(**72**)	600.54(**614.65**)	0.63(0.64)
	PMS-NSGA II	61.80(**72**)	594.92(602.67)	0.61(0.68)	58.80(56)	604.25(613.08)	0.59(0.64)
	PMS-NSDE	49.40(56)	594.19(**603.86**)	0.63(0.64)	44.40(50)	597.25(612.67)	0.63(**0.62**)
A-D-BE	PMS-MOCSA	55.60(**64**)	607.53(**604.52**)	0.63(0.60)	54.40(54)	607.14(610.68)	0.60(**0.54**)
	PMS-NSGA II	54.20(54)	601.74(**597.52**)	0.61(**0.55**)	54.60(**63**)	602.82(**610.88**)	0.61(0.60)
	PMS-NSDE	46.80(51)	591.33(585.05)	0.68(0.58)	46.20(53)	594.36(608.66)	0.64(0.58)
A-E-BD	PMS-MOCSA	61.20(**64**)	594.50(596.31)	0.59(**0.52**)	61.20(**62**)	589.35(611.69)	0.58(0.57)
	PMS-NSGA II	57.20(56)	595.30(604.81)	0.58(0.56)	56.40(60)	591.73(**615.11**)	0.58(0.53)
	PMS-NSDE	53.80(58)	589.49(595.89)	0.66(0.66)	45.40(49)	598.88(578.64)	0.64(**0.52**)
B-C-AD	PMS-MOCSA	51.80(58)	599.06(**609.17**)	0.70(0.64)	57.40(**66**)	569.33(600.06)	0.61(0.62)
	PMS-NSGA II	51.20(**64**)	596.11(602.55)	0.61(0.60)	53.60(60)	602.63(**609.71**)	0.63(0.65)
	PMS-NSDE	42.20(47)	593.19(586.51)	0.63(0.56)	44.00(47)	596.09(594.45)	0.69(**0.61**)
B-D-AC	PMS-MOCSA	56.60(61)	567.17(607.76)	0.58(0.56)	52.00(**72**)	554.96(569.85)	0.62(0.61)
	PMS-NSGA II	54.60(**63**)	593.43(**618.00**)	0.61(**0.53**)	51.40(46)	587.73(**606.69**)	0.60(**0.51**)
	PMS-NSDE	43.20(48)	591.59(611.63)	0.71(0.72)	44.80(50)	592.20(586.71)	0.66(0.58)
B-E-AD	PMS-MOCSA	55.00(**61**)	586.79(606.11)	0.61(**0.56**)	58.40(**65**)	597.88(587.64)	0.64(**0.57**)
	PMS-NSGA II	50.80(57)	597.52(599.73)	0.56(0.56)	50.20(55)	594.89(600.52)	0.60(0.58)
	PMS-NSDE	41.80(47)	596.38(**611.30**)	0.66(0.61)	41.00(46)	598.64(**622.57**)	0.66(0.66)
C-D-AE	PMS-MOCSA	58.60(**71**)	591.35(599.24)	0.60(0.68)	60.00(**60**)	587.42(594.94)	0.61(**0.55**)
	PMS-NSGA II	53.60(49)	597.24(589.81)	0.58(**0.53**)	51.80(59)	597.78(**610.64**)	0.63(0.59)
	PMS-NSDE	42.40(42)	594.55(**610.84**)	0.63(0.60)	44.40(51)	594.87(591.47)	0.70(0.71)
C-E-AB	PMS-MOCSA	56.20(**59**)	577.19(594.15)	0.59(**0.55**)	57.60(62)	593.44(604.50)	0.57(**0.54**)
	PMS-NSGA II	55.00(47)	592.95(**602.86**)	0.61(0.57)	51.00(**65**)	595.58(608.99)	0.65(0.74)
	PMS-NSDE	39.40(39)	592.59(597.07)	0.67(0.65)	47.20(47)	593.76(**612.40**)	0.63(0.57)
D-E-BC	PMS-MOCSA	53.60(**58**)	584.76(**598.44**)	0.64(0.53)	62.80(**75**)	563.84(573.38)	0.61(**0.56**)
	PMS-NSGA II	43.60(48)	526.74(503.05)	0.62(**0.53**)	50.80(56)	581.08(**602.31**)	0.66(0.69)
	PMS-NSDE	42.00(44)	582.50(589.22)	0.68(0.64)	43.60(52)	581.38(588.35)	0.60(0.64)

In all tables, in column Average, the average of all five executions in each measure is presented and in the column Best Run, the best computational results of all executions in each measure are presented. In addition, the Best Runs, in Tables 2 and 3, are in parentheses. Also, the best results of each of the three measures in each of the ten combinations are signed with bold letters. Finally, Fig. 1 presents the Pareto Fronts of the symmetric delivery problem using objective functions OF1-OF2 ((a) part of the figure) and of the symmetric pick-up problem using objective functions OF1-OF3, in the (b) part of the figure, for the instance "A-E" for all algorithms are presented, while in the (c) part of the figure, the Pareto Fronts of the asymmetric delivery problem using objective functions OF1-OF2, and of the asymmetric pick-up problem using objective functions OF1-OF3, ((d) part of the figure), for the instance "A-C-BD" for all algorithms. In Tables 4 and 5 the results of the C measure in all cases are presented (bold letters means best results).

Table 3. Average Results and Best Runs of all algorithms used in the comparisons.

		Symmetric delivery FCVRP			Symmetric pick-up FCVRP		
A-B	PMS-MOCSA	65.20(**71**)	604.74(**611.70**)	0.65(0.59)	62.00(75)	605.59(620.04)	0.65(0.65)
	PMS-NSGA II	56.40(61)	602.89(603.41)	0.66(0.62)	58.60(**79**)	596.88(605.27)	0.66(**0.60**)
	PMS-NSDE	44.60(45)	605.73(596.04)	0.67(**0.55**)	44.60(46)	602.96(**622.10**)	0.68(0.62)
A-C	PMS-MOCSA	69.20(**72**)	607.78(600.56)	0.63(0.61)	64.00(**65**)	605.52(612.67)	0.61(**0.59**)
	PMS-NSGA II	62.60(66)	609.36(**611.80**)	0.63(**0.54**)	56.60(63)	606.55(**615.56**)	0.64(0.61)
	PMS-NSDE	51.40(57)	596.73(602.61)	0.65(0.69)	49.20(44)	603.51(607.51)	0.65(0.61)
A-D	PMS-MOCSA	75.20(**78**)	581.14(589.98)	0.65(0.72)	68.80(**74**)	588.89(**600.68**)	0.60(0.69)
	PMS-NSGA II	54.40(57)	592.85(587.50)	0.67(**0.63**)	58.60(66)	580.64(582.86)	0.66(0.60)
	PMS-NSDE	46.20(40)	581.70(**591.46**)	0.66(0.67)	47.00(47)	579.02(577.96)	0.66(**0.58**)
A-E	PMS-MOCSA	65.00(**65**)	593.11(601.41)	0.65(0.64)	68.40(**76**)	594.99(582.02)	0.63(0.62)
	PMS-NSGA II	51.20(51)	598.85(**608.34**)	0.62(**0.60**)	61.00(74)	598.56(598.90)	0.65(0.59)
	PMS-NSDE	44.60(49)	604.13(608.11)	0.68(0.63)	43.80(43)	602.41(**605.73**)	0.62(**0.58**)
B-C	PMS-MOCSA	63.20(**73**)	575.51(**621.22**)	0.60(0.70)	61.40(**64**)	579.14(602.47)	0.63(**0.56**)
	PMS-NSGA II	58.80(55)	591.49(602.55)	0.65(**0.55**)	56.20(61)	597.28(590.74)	0.65(0.61)
	PMS-NSDE	49.80(60)	589.78(610.81)	0.66(0.56)	42.00(49)	589.82(**613.07**)	0.72(0.64)
B-D	PMS-MOCSA	66.00(**72**)	597.77(593.97)	0.66(0.61)	69.60(**89**)	600.06(593.98)	0.67(**0.64**)
	PMS-NSGA II	58.60(56)	595.91(**609.20**)	0.64(0.63)	55.80(54)	592.55(**608.67**)	0.65(0.66)
	PMS-NSDE	43.20(45)	595.65(595.43)	0.69(**0.60**)	41.20(54)	582.58(600.87)	0.70(0.68)
B-E	PMS-MOCSA	58.20(**64**)	593.79(**616.37**)	0.61(0.60)	59.20(**63**)	591.89(614.74)	0.66(0.63)
	PMS-NSGA II	57.60(59)	603.48(607.87)	0.61(**0.60**)	58.60(**63**)	603.78(**618.11**)	0.63(**0.55**)
	PMS-NSDE	47.80(45)	603.11(609.40)	0.67(0.60)	44.40(46)	580.23(573.63)	0.66(0.58)
C-D	PMS-MOCSA	67.40(**73**)	593.81(**609.02**)	0.61(0.69)	68.60(**71**)	589.35(**591.01**)	0.68(0.65)
	PMS-NSGA II	56.60(59)	587.97(604.68)	0.63(0.64)	51.20(57)	586.43(587.41)	0.59(0.60)
	PMS-NSDE	42.80(51)	577.40(594.58)	0.65(**0.61**)	46.80(51)	586.32(573.07)	0.66(**0.57**)
C-E	PMS-MOCSA	65.40(**78**)	594.28(600.56)	0.66(**0.50**)	64.80(**68**)	600.74(602.78)	0.61(**0.56**)
	PMS-NSGA II	60.40(63)	599.00(592.77)	0.64(0.58)	60.40(61)	607.48(613.53)	0.65(0.60)
	PMS-NSDE	49.40(53)	594.95(**611.88**)	0.73(0.75)	52.20(60)	601.83(**615.53**)	0.67(0.72)
D-E	PMS-MOCSA	71.60(**82**)	605.98(612.39)	0.60(0.58)	71.40(**70**)	608.85(597.40)	0.59(**0.52**)
	PMS-NSGA II	60.20(52)	601.63(610.06)	0.67(0.66)	59.00(53)	606.62(617.50)	0.63(0.57)
	PMS-NSDE	49.80(46)	604.82(**619.96**)	0.66(**0.55**)	49.00(54)	615.41(**622.74**)	0.67(0.64)

In general, based on all Tables (Tables 1, 2, 3, 4 and 5), from the comparison of the three algorithms we conclude that considering the L measure the PMS-MOCSA algorithm performs better than the other two algorithms as it performs better in 82.5% of the instances while the PMS-NSGA II performs better than the other algorithms in 17.5% of the instances. Considering the M_k measure, PMS-NSGA II algorithm performs better than the other algorithms in 37.5% of the instances, while PMS-NSDE1 and PMS-MOCSA perform better than the other algorithms in 35% and 27.5% of the instances, respectively. Also, considering the Δ measure, the PMS-MOCSA algorithm performs better than the other algorithms in 40% of the instances, while algorithms PMS-NSDE1 and PMS-NSGA II perform better than the other algorithms in 30% of the instances, each one of them. Finally, considering the C measure the PMS-MOCSA algorithm performs slightly better than the other two algorithms as it performs better in 100% of the instances. According to the average numbers of the results, the PMS-MOCSA algorithm produce Pareto front with more solutions and better distribution than the other algorithms. The PMS-NSGA II algorithm produce more extend Pareto fronts and the Pareto fronts produced from PMS-MOCSA algorithm dominates the Pareto fronts produced from the other two algorithms.

Table 4. Results of the C measure for the three algorithms in ten instances in all symmetric problems

OF1-OF2	Multiobjective asymmetric delivery route based fuel consumption VRP					
A-B-CD	CSA	NSDE1	NSGA II	**B-D-AC**	CSA	NSDE1	NSGA II
CSA	-	0.81	0.90	CSA	-	0.94	0.87
NSDE1	0.05	-	0.90	NSDE1	0	-	0.63
NSGA II	0.03	0.08	-	NSGA II	0.05	0.17	-
A-C-BD	CSA	NSDE1	NSGA II	**B-E-AD**	CSA	NSDE1	NSGA II
CSA	-	0.95	0.90	CSA	-	0.74	0.95
NSDE1	0	-	0.83	NSDE1	0.13	-	0.93
NSGA II	0	0.04	-	NSGA II	0.03	0	-
A-D-BE	CSA	NSDE1	NSGA II	**C-D-AE**	CSA	NSDE1	NSGA II
CSA	-	0.82	1	CSA	-	0.74	0.94
NSDE1	0.14	-	0.93	NSDE1	0.20	-	0.82
NSGA II	0	0.02	-	NSGA II	0.04	0.07	-
A-E-BD	CSA	NSDE1	NSGA II	**C-E-AB**	CSA	NSDE1	NSGA II
CSA	-	0.93	0.93	CSA	-	0.59	0.87
NSDE1	0.08	-	0.91	NSDE1	0.25	-	0.83
NSGA II	0.02	0	-	NSGA II	0.14	0.13	-
B-C-AD	CSA	NSDE1	NSGA II	**D-E-BC**	CSA	NSDE1	NSGA II
CSA	-	0.79	0.89	CSA	-	0.68	0.60
NSDE1	0.19	-	0.94	NSDE1	0.09	-	0.40
NSGA II	0.14	0	-	NSGA II	0.36	0.64	-
OF1-OF2	Multiobjective symmetric delivery route based fuel consumption VRP					
A-B	CSA	NSDE1	NSGA II	**B-D**	CSA	NSDE1	NSGA II
CSA	-	0.67	0.97	CSA	-	0.93	0.96
NSDE1	0.20	-	0.97	NSDE1	0.01	-	0.84
NSGA II	0.03	0	-	NSGA II	0	0.07	-
A-C	CSA	NSDE1	NSGA II	**B-E**	CSA	NSDE1	NSGA II
CSA	-	0.86	1	CSA	-	0.76	0.83
NSDE1	0.03	-	0.98	NSDE1	0.11	-	0.90
NSGA II	0	0	-	NSGA II	0.06	0.07	-
A-D	CSA	NSDE1	NSGA II	**C-D**	CSA	NSDE1	NSGA II
CSA	-	0.65	0.91	CSA	-	0.90	0.98
NSDE1	0.33	-	0.82	NSDE1	0.01	-	0.90
NSGA II	0.04	0.10	-	NSGA II	0	0.06	-
A-E	CSA	NSDE1	NSGA II	**C-E**	CSA	NSDE1	NSGA II
CSA	-	1	1	CSA	-	0.75	0.94
NSDE1	0	-	0.86	NSDE1	0.18	-	0.86
NSGA II	0	0.06	-	NSGA II	0.03	0.15	-
B-C	CSA	NSDE1	NSGA II	**D-E**	CSA	NSDE1	NSGA II
CSA	-	0.77	0.91	CSA	-	0.65	0.96
NSDE1	0.18	-	0.91	NSDE1	0.26	-	0.98
NSGA II	0.05	0.10	-	NSGA II	0.07	0	-

Table 5. Results of the C measure for the three algorithms in ten instances in all asymmetric problems

OF1-OF3	Multiobjective asymmetric pick-up route based fuel consumption VRP						
A-B-CD	CSA	NSDE1	NSGA II	**B-D-AC**	CSA	NSDE1	NSGA II
CSA	-	0.78	0.76	CSA	-	0.80	0.98
NSDE1	0.15	-	0.76	NSDE1	0.10	-	0.98
NSGA II	0.13	0.14	-	NSGA II	0	0	-
A-C-BD	CSA	NSDE1	NSGA II	**B-E-AD**	CSA	NSDE1	NSGA II
CSA	-	0.86	0.93	CSA	-	0.89	0.85
NSDE1	0.13	-	0.82	NSDE1	0.17	-	0.84
NSGA II	0	0.06	-	NSGA II	0.05	0.11	-
A-D-BE	CSA	NSDE1	NSGA II	**C-D-AE**	CSA	NSDE1	NSGA II
CSA	-	0.83	0.90	CSA	-	0.94	0.90
NSDE1	0.15	-	0.87	NSDE1	0.02	-	0.83
NSGA II	0	0.02	-	NSGA II	0.03	0.02	-
A-E-BD	CSA	NSDE1	NSGA II	**C-E-AB**	CSA	NSDE1	NSGA II
CSA	-	0.59	0.83	CSA	-	0.89	1
NSDE1	0.34	-	0.82	NSDE1	0.15	-	0.86
NSGA II	0.03	0.10	-	NSGA II	0	0.04	-
B-C-AD	CSA	NSDE1	NSGA II	**D-E-BC**	CSA	NSDE1	NSGA II
CSA	-	0.91	0.90	CSA	-	0.62	0.89
NSDE1	0.02	-	0.68	NSDE1	0.39	-	0.73
NSGA II	0.06	0.17	-	NSGA II	0.01	0.17	-
OF1-OF3	Multiobjective symmetric pick-up route based fuel consumption VRP						
A-B	CSA	NSDE1	NSGA II	**B-D**	CSA	NSDE1	NSGA II
CSA	-	0.76	0.96	CSA	-	1	0.98
NSDE1	0.12	-	1	NSDE1	0	-	0.89
NSGA II	0	0	-	NSGA II	0	0.02	-
A-C	CSA	NSDE1	NSGA II	**B-E**	CSA	NSDE1	NSGA II
CSA	-	0.75	0.97	CSA	-	0.91	0.81
NSDE1	0.18	-	0.90	NSDE1	0.11	-	0.71
NSGA II	0.05	0	-	NSGA II	0.02	0.11	-
A-D	CSA	NSDE1	NSGA II	**C-D**	CSA	NSDE1	NSGA II
CSA	-	0.62	0.94	CSA	-	0.94	0.96
NSDE1	0.24	-	0.94	NSDE1	0.03	-	0.81
NSGA II	0.08	0.04	-	NSGA II	0.01	0.14	-
A-E	CSA	NSDE1	NSGA II	**C-E**	CSA	NSDE1	NSGA II
CSA	-	0.98	0.97	CSA	-	0.87	0.90
NSDE1	0.01	-	0.85	NSDE1	0.10	-	0.92
NSGA II	0	0.12	-	NSGA II	0.04	0.03	-
B-C	CSA	NSDE1	NSGA II	**D-E**	CSA	NSDE1	NSGA II
CSA	-	0.88	1	CSA	-	0.89	0.98
NSDE1	0.03	-	1	NSDE1	0.04	-	0.94
NSGA II	0	0	-	NSGA II	0	0	-

4   Conclusions and Future Research

In this paper, we proposed an algorithm (PMS-MOCSA) for solving four newly formulated multiobjective fuel consumption multi-depot vehicle routing problems (symmetric and asymmetric pick-up and symmetric and asymmetric delivery cases). The proposed algorithm was compared with other two algorithms the PMS-NSDE1 and PMS-NSGA II. In general, for the four different problems, the PMS-MOCSA algorithm performs slightly better than the other two algorithms in the most measures, as we analyzed in the Computational Results section. As expected, the behavior of the algorithms was slightly different when a symmetric and an asymmetric problem was solved. Our future research will be, mainly, focused on PMS-MOCSA algorithm in other multiobjective combinatorial optimization problems.

References

1. Brabazon, A., O'Neill, M.: Biologically Inspired Algorithms for Financial Modeling. Natural Computing Series. Springer, Berlin (2006). https://doi.org/10.1007/3-540-31307-9
2. Cutello, V., Nicosia, G.: An immunological approach to combinatorial optimization problems. In: Garijo, F.J., Riquelme, J.C., Toro, M. (eds.) IBERAMIA 2002. LNCS (LNAI), vol. 2527, pp. 361–370. Springer, Heidelberg (2002). https://doi.org/10.1007/3-540-36131-6_37
3. Cutello, V., Nicosia, G.: Multiple learning using immune algorithms. In: Proceedings of 4th International Conference on Recent Advances in Soft Computing, RASC, pp. 102–107 (2002)
4. Cutello, V., Nicosia, G., Pavia, E.: A parallel immune algorithm for global optimization. In: Kłopotek, M.A., Wierzchoń, S.T., Trojanowski, K. (eds.) Intelligent Information Processing and Web Mining. AINSC, vol. 35, pp. 467–475. Springer, Berlin (2006). https://doi.org/10.1007/3-540-33521-8_51
5. Dasgupta, D. (ed.): Artificial Immune Systems and Their Application. Springer, Heidelberg (1998). https://doi.org/10.1007/978-3-642-59901-9
6. Dasgupta, D., Niño, L.F.: Immunological Computation: Theory and Applications. CRC Press, Taylor and Francis Group, Boca Raton (2009)
7. De Castro, L.N., Timmis, J.: Artificial Immune Systems: A New Computational Intelligence Approach. Springer, Heidelberg (2002)
8. De Castro, L.N., Von Zuben, F.J.: The clonal selection algorithm with engineering applications. In: Workshop on Artificial Immune Systems and Their Applications (GECCO 2000), Las Vegas, NV, pp. 36–37 (2000)
9. De Castro, L.N., Von Zuben, F.J.: Learning and optimization using the clonal selection principle. IEEE Trans. Evol. Comput. 6(3), 239–251 (2002)
10. Demir, E., Bektaş, T., Laporte, G.: A review of recent research on green road freight transportation. Eur. J. Oper. Res. 237(3), 775–793 (2014)
11. Forrest, S., Perelson, A., Allen, L., Cherukuri, R.: Self-nonself discrimination in a computer. In: Proceedings of the 1994 IEEE Symposium on Research in Security and Privacy, pp. 202–212. IEEE Computer Society Press, Los Alamitos (1994)
12. Lin, C., Choy, K.L., Ho, G.T.S., Chung, S.H., Lam, H.Y.: Survey of green vehicle routing problem: past and future trends. Expert Syst. Appl. 41(4), 1118–1138 (2014)

13. Montoya-Torres, J.R., Franco, J.L., Isaza, S.N., Jimenez, H.F., Herazo-Padilla, N.: A literature review on the vehicle routing problem with multiple depots. Comput. Ind. Eng. **79**, 115–129 (2015)
14. Pavone, M., Narzisi, G., Nicosia, G.: Clonal selection - an immunological algorithm for global optimization over continuous spaces. J. Global Optim. **53**(4), 769–808 (2012)
15. Psychas, I.D., Marinaki, M., Marinakis, Y., Migdalas, A.: Non-dominated sorting differential evolution algorithm for the minimization of route based fuel consumption multiobjective vehicle routing problems. Energy Syst. **8**, 785–814 (2016)
16. Psychas, I.D., Marinaki, M., Marinakis, Y., Migdalas, A.: Minimizing the fuel consumption of a multiobjective vehicle routing problem using the parallel multi-start NSGA II algorithm. In: Kalyagin, V., Koldanov, P., Pardalos, P. (eds.) NET 2014. PROMS, vol. 156, pp. 69–88. Springer, Cham (2016). https://doi.org/10.1007/978-3-319-29608-1_5
17. Srivastava, S.K.: Green supply-chain management: a state-of the-art literature review. Int. J. Manag. Rev. **9**(1), 53–80 (2007)
18. Timmis, J., Neal, M.: A resource limited artificial immune system for data analysis. In: Bramer, M., Preece, A., Coenen, F. (eds.) Research and Development in Intelligent Systems XVII, vol. 14, pp. 19–32. Springer, London (2000). https://doi.org/10.1007/978-1-4471-0269-4_2
19. Toth, P., Vigo, D.: Vehicle Routing: Problems, Methods and Applications. MOS-Siam Series on Optimization, 2nd edn. SIAM, Philadelphia (2014)

Big Data Privacy by Design Computation Platform

Rui Claro[1,2] (ID), José Portêlo[2](✉) (ID), Miguel L. Pardal[1,3] (ID), and Raquel Pinho[2] (ID)

[1] Instituto Superior Técnico, Universidade de Lisboa, Lisbon, Portugal
[2] Altran Portugal, Lisbon, Portugal
jose.portelo@altran.com
[3] INESC-ID Lisboa, Lisbon, Portugal

Abstract. We live in the age of Big Data, and personal user data, in particular, is necessary for the operation and improvement of healthcare services. Many times, the capture and use of personal data are not made explicit to the users, but they are central to the business model of companies. However, each person's right to privacy needs to be respected.

With the goal of reconciling these two conflicting needs, we designed and implemented a proof-of-concept platform for performing privacy-preserving computations. In particular, we implemented privacy-preserving versions of Machine Learning algorithms, namely Decision Trees, k-Means, Logistic Regression, and Support Vector Machines, using Secure Multi-party Computations with Homomorphic Encryption and Garbled Circuits. For each combination of Machine Learning algorithms with Secure Multi-party Computation techniques, we present the reasoning behind our choices and their potential consequences in terms of performance.

The ultimate goal is to provide *Privacy-Preserving Computation as a Service*. With this platform, we wish to contribute to the faster integration of solutions developed by the scientific community in enterprise systems, thus reducing the time required for innovation to reach products used by many people where privacy improvements are urgently needed.

Keywords: Privacy-preserving computations · Machine Learning · Big Data · Secure-multi-party computations · Privacy-preserving platform

1 Introduction

The term *Big Data* means that there are vast amounts of data being analysed and processed by companies every day [7]. Through this data processing, meaningful information can be obtained to improve existing systems or to discover new approaches in business models. Machine Learning (ML) algorithms in the context of Big Data processing can produce significant results, so that it is possible to do knowledge learning from datasets in order to predict future labels (i.e. classes of data) or clusters for new data. An example of this can be seen in the field of

G. Nicosia et al. (Eds.): LOD 2018, LNCS 11331, pp. 394–405, 2019.
https://doi.org/10.1007/978-3-030-13709-0_33

Healthcare, where it can be beneficial to analyse patient records from different hospitals in order to identify inefficiencies and develop best practices [8]. For example, Google Deepmind is developing ML algorithms for faster patient triage and admission processes in hospitals[1], and IBM Watson is supporting medical personnel consider treatment options for their patients[2].

There are restrictions to the processing of personal data, such as patient data. *Privacy* can be defined as the ability or right of an individual to protect his/her personal information, and extends the ability or right to prevent invasions on the personal space of said individual [2]. If patient data can be processed with privacy then they can enable novel applications and scientific breakthroughs in Healthcare. By combining ML algorithms and privacy-preserving techniques, it is possible to create Data Mining processes that, not only allow for knowledge learning on large datasets, but also help maintain a level of privacy desired by individuals and compliant with existing legislation [5].

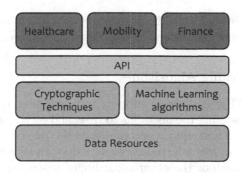

Fig. 1. Conceptual view of the platform.

In Fig. 1 we present the conceptual view of our platform. The *data resources* represent the datasets that are used in the classification process. The data processing itself is done using the combination of ML algorithms and cryptographic techniques for performing privacy-preserving computations. The Application Programming Interface (API) layer abstracts details and provides the operations of the platform itself, which allow a simplified building of applications and data visualizations. The use-cases describe the various subjects that can be addressed using this platform, and allow us to place it in real-world scenarios that have high impact and demand in Big Data operations. More use-cases are possible beyond Healthcare, Mobility and Finance, as the platform is designed for general use.

In this work, we present a proof-of-concept platform for privacy-preserving distributed ML computations without resorting to trusted third parties. With it, we aim to give users a platform that provides simplified access to privacy-preserving techniques that can be used to meet privacy requirements in data

---

[1] https://deepmind.com/applied/deepmind-health/.

[2] https://www.mskcc.org/about/innovative-collaborations/watson-oncology.

processing. We provide a detailed comparison of four ML algorithms, namely: Decision Trees (DT), k-Means, Logistic Regression (LR) and Support Vector Machines (SVM); combined with two Secure Multi-party Computations (SMPC): Garbled Circuits (GC) and Homomorphic Encryption (HE); presenting details on how this can be performed. We performed an experimental evaluation of the adapted ML algorithms using publicly available datasets, and compare the results with a baseline.

The paper is further organized as follows. Section 2 describes the ML algorithms and the SMPC techniques used in the platform. In Sect. 3, we present the related work. Section 4 describes the design of the platform, detailing the adjustments done to ML algorithms. In Sect. 5, we present the experimental results. Finally, in Sect. 6 we present the conclusions and propose future work.

2   Background

In this section we present the ML algorithms for which we wish to develop privacy-preserving implementations and the used SMPC techniques.

2.1   Machine Learning Algorithms

Decision Trees (DT): A decision support tool composed of nodes and leaves, with each node representing the decisions to take, and each leaf representing class labels. Classification of a sample is accomplished by traversing the tree from the top, comparing the features selected on each node with its respective threshold, and choosing one branch or the other accordingly, repeating the process until a leaf is reached. At each tree node, a decision is computed using:

$$f_{\mathrm{DT}}(x_i) = x_i \overset{?}{\geq} \theta_j \tag{1}$$

where x_i is the feature value of interest of the testing sample and θ_j is the decision threshold of node j. If the output is 0, the left hand child is selected; if it is 1, the right hand child is selected.

Support Vector Machines (SVM): An SVM model represents the samples as points in space, mapped so that the margin between the two classes is as wide as possible. The vectors that define this margin are called Support Vectors (SVs). The classification of new samples in SVM is done using the scoring function in Eq. 2, where each testing sample x is attributed to a prediction label.

$$f_{\mathrm{SVM}}(x) = \sum_{i=1}^{m} \alpha_i K(x_{SV}^{(i)}, x) + b \tag{2}$$

where α_i is the coefficient associated with the support vector $x_{SV}^{(i)}$, K is the kernel function chosen, and b is a scalar number.

k-Means: An iterative algorithm, with two distinct steps. (1) Each instance is assigned to a cluster, by calculating the Euclidean distance, d_E, between that instance and each centroid. Then, the lowest distance indicates which cluster the instance is assigned to. (2) Each centroid is updated to be the mean of all the instances assigned to it. The algorithm stops when the centroids no longer change position. The classification of a new sample is done by computing the d_E of the new sample with each centroid, discovering which is closer. The predicted label of the sample is computed as described in Eq. 3:

$$f_{\text{k-M}}(x) = \underset{C}{\arg\min}\, d_E(x, C_j) \tag{3}$$

where C are the centroids of each cluster and x is the testing sample.

Logistic Regression (LR): A statistical model that analyses a dataset in order to determine an outcome. This binary LR model is used to estimate the probability of a binary response based on one or more variables. The classification of samples is done using the following equation:

$$f_{\text{LR}}(x) = \beta_0 + \sum_{i=1}^{m} \beta_i x_i \tag{4}$$

where β_0 is the intercept from the linear regression, β_i are each regression coefficient that is multiplied by each feature of the sample, and x is the testing sample.

2.2 Privacy-Preserving Techniques

Garbled Circuits (GC). [14] allow two mutually mistrusting parties to evaluate a function over their private inputs without resorting to a trusted third party. GC allows two parties holding inputs x and y to evaluate an arbitrary function $f(x, y)$ without leaking any information about their inputs beyond what is inferred from the function output. The idea behind GC is that one party prepares an encrypted version of a circuit that computes $f(x, y)$ and the second party then computes the output of the circuit without learning any intermediate values.

Homomorphic Encryption (HE). [11] is a cryptographic technique that allows computations to be carried with the ciphertext, so that, when decrypted, the resulting plaintext reflects the computation made. In other words, HE allows making some computation over the ciphertext, for example, addition, without decrypting it, and the result is the same as making that computation on the plaintext. This is of great importance because it allows chaining multiple services that make computations on a ciphertext, without the need to expose the data to those services. Homomorphic cryptosystems can be classified into two distinct groups: Partially Homomorphic Cryptosystems (PHE), where there is only one operation that is allowed by the homomorphic property (ex: addition, multiplication, XOR); and Fully Homomorphic Cryptosystems (FHE), where it is possible to perform both addition and multiplication.

3 Related Work

Although well known Big Data platforms such as Apache Hadoop[3] and MongoDB[4] have been around for a while, most (if not all) of them were designed without Data Privacy concerns in mind. We envision a Big Data platform following a Privacy by Design approach, where data privacy is taken into consideration on every development step.

In many cases there are privacy-preserving versions of ML algorithms, for example, k-Means [12], LR [4] or SVM [13], but they are not made available in a platform.

There are also works on designing platform architectures focused on the protection of privacy in location-based services, and describing privacy-preserving algorithms for them [1], but these works do not actually perform any implementation or testing of such solutions.

4 Platform Design

We structured the platform design in two major parts: a non-privacy-preserving *baseline* and the *privacy-preserving implementation*. The former allows comparing the effects of the privacy-preserving techniques in terms of performance.

4.1 Non-privacy-preserving Components

While designing the first part of our platform, the focus was to build a baseline so that meaningful observations could be achieved, while also paving the way to build the privacy-preserving approach. The models that we implemented allowed us to later adapt the prediction step of the ML algorithms for GC and HE, while also giving insight on which technique to use for each algorithm.

4.2 Privacy-Preserving Components

The privacy-preserving part consisted of adjustments to the evaluation processes of the ML algorithms in order to be compatible with two privacy-preserving techniques: GC and HE. These two techniques offer different means to obtain privacy-preserving computations. GC builds ciphered boolean circuits where most operations are possible to implement. However, arithmetic operations require a large number of logic gates, creating an overhead that makes GC very slow for those operations. So, for some of the ML algorithms, we used an HE system, since it offers arithmetic operations as core operations. The following sections describe the chosen combinations.

---

[3] http://hadoop.apache.org/.

[4] https://www.mongodb.com/.

Garbled Circuits and Decision Trees. The process of evaluating a DT in a privacy-preserving context is similar to evaluating it in the usual manner, as described in Eq. 1. The main differences are: basic operations such as comparisons are replaced with logic gates; and the evaluation of the DT involves evaluating every single node in it, to disclose the least possible information caused by observation of the computations. Figure 2 shows the computations done inside each node of the DT.

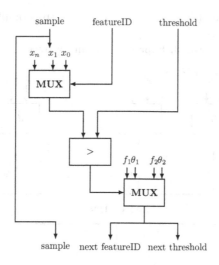

Fig. 2. Boolean circuit of each node in a DT.

Another aspect to mention is that the trees are always complete, i.e., the number of nodes n is always the maximum possible, and can be defined as $n = 2^{h+1} - 1$, where h is the *height* of the tree. Even though in most cases this will lead to an exponential increase of the number of nodes with increasing tree depth, we feel this is necessary to prevent information leaks due to an attacker being able to know the different path depths. Figure 3 shows the implications of this expansion.

Garbled Circuits and k-Means. The process of evaluating the k-Means algorithm in a privacy-preserving manner is similar to evaluating in the usual manner. The operations in the prediction step of the algorithm were transformed into boolean circuits, with logic gates representing operations. In Fig. 4 we show the circuit we have designed to represent the k-Means prediction, where d_E represents the Euclidean distance between testing sample x and each centroid C_i.

Homomorphic Encryption and Logistic Regression. In order to use a FHE system, the prediction function for LR described in Eq. 4 must be converted to:

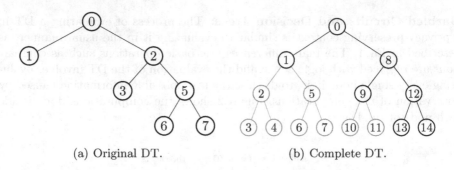

(a) Original DT.   (b) Complete DT.

Fig. 3. Expansion of binary trees.

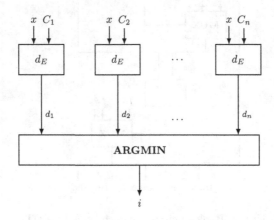

Fig. 4. Boolean circuit of the k-Means prediction.

$$f_{\text{LR,FHE}}(x) = D_k\left(E_k(\beta_0) + \sum_{i=0}^{m} E_k(\beta_i) \cdot E_k(x_i) \right) \tag{5}$$

where E_k represents the encryption operation and D_k represents the decryption operation using the key k.

Converting Eq. 5 to be computed using a PHE system is straightforward, but this can only be done under two assumptions: (1) the data to be evaluated (x) and the model parameters $(\beta_0, \beta_1, \ldots, \beta_m)$ must come from two different parties, and (2) the owner of the model parameters must be the one processing the data. Under these assumptions, the linear prediction function for a additive PHE system becomes:

$$f_{\text{LR,PHE}}(x) = D_k\left(E_k(\beta_0) \cdot \prod_{i=1}^{m} E_k(x_i)^{\beta_i} \right) \tag{6}$$

Homomorphic Encryption and Support Vector Machines. For the SVM algorithm, we only considered the linear kernel, as it simplifies the scoring function. The Eq. 2 is then simplified to the following:

$$f_{\text{SVM}}(x) = \sum_{i=1}^{m} \alpha_i x_{SV}^{(i)} x + b = \sum_{i=1}^{m} \alpha_i \sum_{j=1}^{n} x_j x_{SV}^{(i,j)} + b \qquad (7)$$

To compute this function using a FHE system, we must convert it to:

$$f_{\text{SVM,FHE}}(x) = D_k \left(\sum_{i=1}^{m} E_k(\alpha_i) \cdot \sum_{j=i}^{n} E_k(x_j) \cdot E_k(x_{SV}^{(i,j)}) + E_k(b) \right) \qquad (8)$$

where E_k represents the encryption operation and D_k represents the decryption operation using the key k.

Like before, converting it to be computed using a PHE system is equally straightforward, and under the same two assumptions, the scoring function for a additive PHE system becomes:

$$f_{\text{SVM,PHE}}(x) = D_k \left(\prod_{i=1}^{m} \left(\prod_{j=1}^{n} E_k(x_i)^{x_{SV}^{(i,j)}} \right)^{\alpha_i} \cdot E_k(b) \right) \qquad (9)$$

4.3   Architecture

The combination of the components above helped us create a data processing architecture for a privacy-preserving ML platform, presented in Fig. 5.

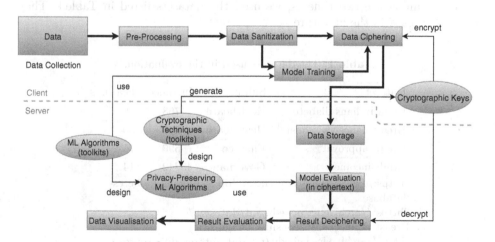

Fig. 5. Data processing architecture for the platform.

We assume that only two parties exist: the client and the server. The client represents a user or an individual who owns data and wishes store them and to perform some processing over them, but does not have the capabilities to do so (e.g.: real-time processing, fault tolerance systems, scalable environment). Despite this, the client wishes to keep these data private. The server represents a cloud or service provider who has the capabilities to perform such processing. To achieve the privacy goals of both parties, the client pre-processes, sanitizes and encrypts the data and models before sending them to the server. The model training is performed in the usual manner. The model evaluation process is the main focus of our work, and it is where the privacy-preserving techniques are deployed. At the end of the flow, the platform produces the prediction results.

With this architecture, we aim at providing companies a way to integrate their Big Data systems processes with privacy-preserving ML algorithms, allowing them to provide additional data privacy guarantees to their clients.

5   Experimental Results

This section presents the evaluation results. The objective of the experimental evaluation is to answer two important questions: (1) How accurate is the prediction *versus* the baseline system? (2) How easily can the platform be adapted to different size and context of the datasets?

The datasets were split into three sets: training (70%), validation (15%) and testing (15%) sets. The training step of the baseline ML algorithms was performed using the scikit-learn toolkit for Python[5]. The GC results were obtained using the VIPP toolkit [10]. The results using FHE were obtained using the HElib toolkit [6]. The results using PHE were obtained using our own implementation of the Paillier cryptosystem [9].

For running the experiments, we used the datasets listed in Table 1. They are widely used in the literature.

Table 1. The datasets used in the evaluation.

Dataset	Subject	Instances	Features
Pima Indians diabetes[a]	Healthcare	768	8
Breast cancer wisconsin[b]	Healthcare	569	30
Credit approval[c]	Finance	690	15
Adult income[d]	Governance	48842	14

[a]https://www.kaggle.com/uciml/pima-indians-diabetes-database
[b]https://archive.ics.uci.edu/ml/datasets/Breast+Cancer+Wisconsin+(Diagnostic)
[c]http://archive.ics.uci.edu/ml/datasets/credit+approval
[d]https://archive.ics.uci.edu/ml/datasets/adult

---

[5] http://scikit-learn.org/.

5.1   Accuracy

After analysing the results obtained with the privacy-preserving ML algorithm implementations using GC, we verified that changing the number of bits for the actual numeric precision of the data and model parameters affects the accuracy of the results. The absolute error percentage values for the experiments on DT and k-Means are presented in Tables 2 and 3, respectively. It is to be noted that this error is computed versus the baseline prediction results, not the prediction labels from the dataset.

Table 2. GC + DT. Average absolute label prediction error vs. the baseline.

Bits	Pima Indians	Breast cancer	Credit approval	Adult income
8	1.88%	0.55%	8.70%	0.00%
12	0.00%	0.13%	1.11%	0.00%
16	0.00%	0.13%	0.31%	0.00%
20	0.00%	0.13%	0.31%	0.00%
24	0.00%	0.13%	0.31%	0.00%

Table 3. GC + k-Means. Average absolute label prediction error vs. the baseline.

Bits	Pima Indians	Breast cancer	Credit approval	Adult income
8	2.03%	3.07%	0.05%	0.02%
12	0.39%	0.85%	0.00%	0.00%
16	0.29%	0.72%	0.00%	0.00%
20	0.29%	0.72%	0.00%	0.00%
24	0.00%	0.00%	0.00%	0.00%

Analysing the obtained results, we can conclude that the loss of prediction performance caused by using the privacy-preserving versions of the ML algorithms is not relevant, as long as at least 16 bits are used to represent the data. Since both DT and k-Means only output an integer representing the label, and not a real number, the visible effect of changing the number of bits is minimal.

After analysing the results obtained using the PHE and FHE systems, we verified that all predicted labels and almost all function evaluation outputs match the baseline. The few examples when an exact match does not happen come mostly from the SVM scoring evaluation function implemented in HElib, and are most likely caused by the accumulation of the intrinsic noise generated every time an operation is performed between two ciphertexts. Therefore, we can conclude that our privacy-preserving versions of the ML algorithms using PHE and FHE have no relevant loss of prediction performance.

5.2   Discussion

Although we did not compare the performance of GC and HE directly, for instance by choosing a ML algorithm and implementing it using both privacy-preserving techniques, it is clear that the HE approach is adequate for ML algorithms that rely on arithmetic operations, and the GC approach is adequate for ML algorithms that rely on non-arithmetic operations.

We did not perform a detailed computational times analysis because we feel that an evaluation of execution times is less relevant, as it is extremely dependent on the hardware and toolkits used (which evolve through time), while the evaluations we performed in terms of accuracy are not. We are, however, aware of very efficient GC implementations for evaluating DT, but many of them do not scale adequately with the DT size [3]. We did not consider such implementation because we experimented with fully expanded DT of considerable depth.

An important remark on our experiments with GC is related to our choice to only analyse fully expanded DT instead of the original ones, in order to prevent any information leakage regarding the shape of the original tree. However, in most cases this causes an exponential growth of the number of nodes with increasing tree depths, leading to proportional increases in both the execution times and the communication costs.

Another important conclusion made possible by our experiments with HE is *when* each of the techniques should be used. We verified that PHE is, in fact, usable in practice but under some restrictions (e.g.: if there is no need for complex composition of operations and if data is separated between client and server), while FHE is more flexible but still too computationally expensive.

With our implementation, we were able to understand that, despite the fact that GC and HE are very different techniques, they can be used in almost the same manner. The main difference is that the ML algorithms must be adapted differently for each one. The tweaks done to the algorithms presented in Sect. 4.2 allowed us to implement privacy-preserving versions of them and running them in the same manner as the non-privacy-preserving approach.

We were also able to produce results with datasets from varied contexts, such as Healthcare or Finance, and of very different sizes, without the need to specifically adapt the algorithms for them. With this, we have shown that the platform can be used for different application domains.

6   Conclusions and Future Work

This paper presented a platform to perform privacy-preserving ML computations to be applied in Big Data applications. We discussed the existing techniques that provide the level of privacy compliance with the laws in force and matched those techniques with the most commonly used ML algorithms. We evaluated the solution by comparing two SMPC techniques: GC and HE. Overall, we produced a proof-of-concept platform that provides a unified and simplified API for privacy-compliant ML. This shortens the distance between the scientific community that develops the techniques and the companies that employ them in products that impact many people. With our approach, the most recent scientific advances in privacy-preserving technologies can be applied faster in enterprise applications.

For future work, we propose the following points to enhance the functionalities of the platform and its performance: extend the platform to work with more ML algorithms (ex: Neural Networks or Naive Bayes), so that the platform can be used for more purposes (ex: Deep Learning); optimize the SMPC techniques used, to improve the performance of the platform; implement and test the SMPC techniques using other toolkits, also to improve the performance of the platform.

Acknowledgements. Work supported by Portuguese national funds through Fundação para a Ciência e a Tecnologia (FCT) with reference UID/CEC/50021/2013 (INESC-ID).

References

1. Abbas, F., Hussain, R., Son, J., Oh, H.: Privacy preserving cloud-based computing platform (PPCCP) for using location based services. In: Proceedings of the 2013 IEEE/ACM 6th International Conference on Utility and Cloud Computing, pp. 60–66. IEEE Computer Society (2013)
2. Anderson, R.: Security Engineering. Wiley, Hoboken (2008)
3. Bost, R., Popa, R.A., Tu, S., Goldwasser, S.: Machine learning classification over encrypted data. In: NDSS, vol. 4324, p. 4325 (2015)
4. Chaudhuri, K., Monteleoni, C.: Privacy-preserving logistic regression. In: Advances in Neural Information Processing Systems, pp. 289–296 (2009)
5. D'Acquisto, G., Domingo-Ferrer, J., Kikiras, P., Torra, V., de Montjoye, Y.A., Bourka, A.: Privacy by design in big data: an overview of privacy enhancing technologies in the era of big data analytics. European Union Agency for Network and Information Security (2015)
6. Halevi, S., Shoup, V.: HElib-an implementation of homomorphic encryption. Cryptology ePrint Archive, Report 2014/039 (2014)
7. Lee, I.: Big data: dimensions, evolution, impacts, and challenges. Bus. Horiz. **60**(3), 293–303 (2017)
8. Lu, R., Zhu, H., Liu, X., Liu, J., Shao, J.: Toward efficient and privacy-preserving computing in big data era. IEEE Network **28**(4), 46–50 (2014)
9. Paillier, P.: Public-key cryptosystems based on composite degree residuosity classes. In: Stern, J. (ed.) EUROCRYPT 1999. LNCS, vol. 1592, pp. 223–238. Springer, Heidelberg (1999). https://doi.org/10.1007/3-540-48910-X_16
10. Pignata, T.: Garbled circuit designer and executer from the visual information processing and protection (VIPP) research group (2012). http://clem.dii.unisi.it/~vipp/index.php/software/135-garbledcircuit
11. Rivest, R.L., Adleman, L., Dertouzos, M.L.: On data banks and privacy homomorphisms. Found. Secure Comput. **4**(11), 169–180 (1978)
12. Upmanyu, M., Namboodiri, A.M., Srinathan, K., Jawahar, C.V.: Efficient privacy preserving K-means clustering. In: Chen, H., Chau, M., Li, S., Urs, S., Srinivasa, S., Wang, G.A. (eds.) PAISI 2010. LNCS, vol. 6122, pp. 154–166. Springer, Heidelberg (2010). https://doi.org/10.1007/978-3-642-13601-6_17
13. Vaidya, J., Yu, H., Jiang, X.: Privacy-preserving SVM classification. Knowl. Inf. Syst. **14**(2), 161–178 (2008)
14. Yao, A.C.C.: How to generate and exchange secrets. In: 1986 27th Annual Symposium on Foundations of Computer Science, pp. 162–167. IEEE (1986)

Assessing Accuracy of Ensemble Learning for Facial Expression Recognition with CNNs

Alessandro Renda[1,2](✉) ⓘ, Marco Barsacchi[1,2] ⓘ, Alessio Bechini[1] ⓘ,
and Francesco Marcelloni[1] ⓘ

[1] Department of Information Engineering, University of Pisa,
Via G. Caruso, 56122 Pisa, Italy
{alessio.bechini,francesco.marcelloni}@unipi.it
[2] University of Florence, Florence, Italy
{alessandro.renda,marco.barsacchi}@unifi.it

Abstract. Automatic facial expression recognition has recently attracted the interest of researchers in the field of computer vision and deep learning. Convolutional Neural Networks (CNNs) have proved to be an effective solution for feature extraction and classification of emotions from facial images. Further, ensembles of CNNs are typically adopted to boost classification performance.

In this paper, we investigate two straightforward strategies adopted to generate error-independent base classifiers in an ensemble: the first strategy varies the seed of the pseudo-random number generator for determining the random components of the networks; the second one combines the seed variation with different transformations of the input images. The comparison between the strategies is performed under two different scenarios, namely, training from scratch an ad-hoc architecture and fine-tuning a state-of-the-art model. As expected, the second strategy, which adopts a higher level of variability, yields to a more effective ensemble for both the scenarios. Furthermore, training from scratch an ad-hoc architecture allows achieving on average a higher classification accuracy than fine-tuning a very deep pretrained model. Finally, we observe that, in our experimental setup, the increase of the ensemble size does not guarantee an accuracy gain.

Keywords: Facial expression recognition ·
Convolutional Neural Network · Ensemble learning

1 Introduction

One of the most powerful communication tools is represented by human expressions: out of all the information exchanged in an oral communication, facial expressions account for 55%, whereas the plain language only for 7% [26]. Moreover, in 1971, Ekman et al. [4] showed that members of both preliterate and

© Springer Nature Switzerland AG 2019
G. Nicosia et al. (Eds.): LOD 2018, LNCS 11331, pp. 406–417, 2019.
https://doi.org/10.1007/978-3-030-13709-0_34

literate cultures use the same facial expression to convey any specific emotion. Human facial expressions of emotion are universal, related to biological and evolutionary factors rather than cultural or environmental ones. A set of distinctive patterns of the facial muscles characterizes each one of the so-called basic emotions: happiness, sadness, anger, fear, surprise, disgust. Thus, the Facial Expression Recognition (FER) problem has attracted the attention of the Computer Vision and Machine Learning communities: the ability to automatically perform FER over human facial images opens up the possibility to develop several applications in different fields, from Human Computer Interaction to Data Analytics, emotional health and sentiment analysis [17].

The core of an automatic FER system is represented by the *feature extraction* functionality, aimed at extracting a representative and discriminating set of features from the original facial images. Real-world applications ask for robust feature extractors, able to cope with image variations typical of an "in-the-wild" setting [3], such as occlusions, different head poses and illumination conditions. Hand-crafted feature extractors turned out to be inadequate for this challenging scenario, lacking the ability to generalize on incoming images: thus, the need has arisen for new, more flexible methods. As Deep Learning methods obtained excellent results in a wide variety of similar problems [15,16], their application in the context of FER has been explored as well. Convolutional Neural Networks (CNNs) can be regarded as one of the most popular models used for this purpose; they autonomously learn a hierarchical representation of the features of the original images [11]. The success of recent classification systems relies on the use of large collections of labeled data for training: 2012 ImageNet [2], for example, is a dataset of 1.4 million images with 1000 classes. On the other hand, annotating a large dataset of facial expression images is a difficult and time consuming task: FER2013 is one of the largest datasets of this kind built so far, and contains 35,887 images of different subjects.

A general, effective solution for boosting classification performance is represented by *ensemble techniques*, which combine multiple, *diverse* base learners (networks in our case). Several strategies have been proposed for the production of error-independent networks and for merging their classification outputs [5,12,24] but, to the best of our knowledge, in the FER context, their relative effectiveness has not been adequately investigated.

The present work is aimed at shedding light on the effectiveness of two simple techniques to generate diversity among the base classifiers of an ensemble: *Seed Strategy*, i.e. varying only the seed of the random number generator in the learning procedure of each network, and *Preprocessing Strategy*, combining the seed variation with different transformations of the input images. It is important to underline that different scenarios can be considered, and we perform this analysis in two of them: (i) training from scratch an ad-hoc architecture, *CNN10-S* (S stands for scratch), and (ii) fine-tuning a pre-trained state of the art model *VGG16-FT* (FT stands for fine-tuning). Both architecture were chosen for their recognized importance in the literature and availability to the research community [18,19]. It is worth pinpointing that the paper focus is on experimentally comparing different ensemble strategies, instead of achieving the best absolute accuracy on the FER-2013 dataset.

The remainder of the paper is organized as follows: in Sect. 2 we describe the typical approaches for training CNNs. In Sect. 3 we provide a detailed description of our experimental framework, from the used datasets to the proposed ensemble strategies, along with the scenarios for the comparison. In Sect. 4 we discuss the results of the experimentation, and finally Sect. 5 concludes the paper.

2   Brief Introduction to CNN Training Approaches

CNN [16] is a class of feed-forward neural networks: it is a convenient choice for input data with known topology, such as 2D or 3D pixel matrices that represent grayscale or RGB images, respectively.

As a Machine Learning model, the supervised learning procedure for CNNs aims to minimize the training error by experimenting a labeled dataset. However, the real objective is to perform well on new, unseen examples. To evaluate this generalization capability, a validation set is used during the training: several techniques are typically adopted to reduce the discrepancy between training and validation errors, such as dropout [21], data augmentation [23], and weight regularization [7]. Besides these, gathering and annotating more data is one of the best practices to reduce the risk of overfitting, but this is often difficult and time-consuming for many applications.

In the present work, we refer to a well-known, medium-size dataset (FER2013, described in Sect. 3.1). Against this background, two scenarios are taken into account: training an ad-hoc model from scratch, and using a pre-trained model. We tackled the FER problem following both the approaches.

2.1   Training a Model from Scratch

All the weights in the model are randomly initialized: they characterize the behaviour of every action unit. Along the training, an optimization algorithm, typically based on stochastic gradient descent (SGD) [7], iteratively updates the weights in order to minimize a cost function. In this scenario, the capacity of the model is carefully tuned, considering the limited size of the dataset.

2.2   Using a Pre-trained Architecture

Training from scratch a novel architecture on datasets of limited size has recently become unpopular [1]. Instead, a highly effective approach can be based on exploiting the pre-training of a large network, with higher capacity, over a big dataset, and then re-purposing such a network for the application of interest. Indeed, modern CNNs for Computer Vision show a common behaviour [25]: the features extracted in the first layers are quite standard and do not depend on the specific image dataset, while the high level features are strongly related to the considered task. Weights in the first layer typically learn filters that resemble fixed patterns, such as edge detectors, color blobs detectors, Gabor filters, etc. In the last few years, this approach has gained popularity mainly for two reasons: the availability of big labeled datasets for classification tasks, e.g. ImageNet with 1.4M images, and the availability of pretrained state of the art models such as VGG [20], Inception [22], and ResNet [10]. In our work, we use an already pretrained VGG16 model.

3   Experimental Setup

In this section, we describe the dataset used in the present work and the experimental approach. We recall that we want to compare two strategies for generating variability among base classifiers in an ensemble. In order to evaluate the general validity of the results, we perform the comparison in two typical scenarios: training from scratch an ad-hoc architecture, and fine-tuning a state of the art model. Experiments have been carried out over a server equipped with Nvidia GTX 1080 Ti with 11 GB Memory.

3.1   FER-2013 Facial Expression Dataset

The Facial Expression Recognition 2013 (FER-2013) dataset [8] has been chosen for our experiments because it is the most commonly adopted for this task, as reviewed in [19]: it is one of the largest collections of *in-the-wild* facial images consisting of 35.887 images from 7 classes: Neutral (6197), Anger (4945), Disgust (547), Fear (5121), Happiness (8988), Sadness (6076), and Surprise (4001). The official split of FER-2013 has been used after the removal of 11 black images, and it consists of a training set with 28699 images, a validation set with 3588 images, and 3589 images as test set.

The classification accuracy on FER2013 represents the performance measure of the models used in the present work. To the best of our knowledge, the best model achieves a 75.2% accuracy on FER2013 [19], while the average human accuracy on FER2013 is 65%.

3.2   Ensemble Design Strategies

There are two possible approaches for the design of an ensemble of neural networks [5,24]: the *implicit* (or *direct*) method aims to generate an ensemble of error-independent base classifiers by introducing one or more sources of variability. The *explicit* (or *overproduce and choose*) method involves a further optimization step: a subset of networks is selected from an initial large set by optimizing an error diversity measure out of selected base classifiers. In order to keep our model as simple as possible, we consider two direct ensemble design strategies: Seed Strategy and Preprocessing Strategy. We combine the outputs computed by the base classifiers by using the most common aggregation schemes: average and majority voting. For each strategy, a fixed-size ensemble of nine networks is used.

Seed Strategy (SE). The training procedure makes an extensive use of random choices, namely in the following operations: (i) initial distribution of weights; (ii) shuffle of the dataset; (iii) data augmentation; (iv) dropout.

This strategy thus exploits the simplest way to piece together an ensemble of diverse single CNNs: it sets a different seed value for the random number generator used in building up each base classifier, thus ensuring diversity across the members of the ensemble.

Preprocessing Strategy (PS). As proposed in [13], it makes use of another source of variability across the networks in the ensemble: a *preprocessing* layer is added before the CNN input stage. Nine networks are obtained by combining seed variability and preprocessing variability. Three different seeds are used in order to generate three networks for each of the following groups (Fig. 1):

- networks fed with the original, unchanged images (*default*);
- networks fed with images that underwent histogram equalization (histEq), which show an enhanced contrast with respect to the original ones [6];
- networks fed with images that underwent illumination normalization (iNor): it results in a smoothed version of the illumination-induced variations of the original images [9].

Fig. 1. The three versions of a sample image from FER2013 dataset adopted in the Preprocessing Strategy. *Left:* default, original image. *Center:* image modified by histogram equalization. *Right:* image modified by illumination normalization.

3.3   Two Scenarios of Interest: Adopted Models and Parametrization

The proposed strategies are evaluated on two typical scenarios: training from scratch an ad-hoc architecture, and fine-tuning a pre-trained model. Hereafter we describe the relative model, the preprocessing stage, the learning procedure, and the specific data augmentation step used to obtain a wider training set.

CNN10-S: Training from Scratch an Ad-hoc Architecture

Model. We trained from scratch a classical feed-forward CNN (Fig. 2): it is a 10-layers network resulting in 1,769,447 trainable parameters. It mimics the VGG-B architecture [20] by the Visual Geometry Group of the University of Oxford, modified with batch normalization layers and dropout layers according to the specification proposed by [19].

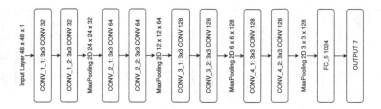

Fig. 2. Scheme of the architecture used in CNN10-S (10 is the depth, S stands for "scratch"). The network is entirely trained from scratch.

Preprocessing Stage. After applying one of the transformations described in Sect. 3.2, a global mean value μ, and a global standard deviation value σ were evaluated over the training set. The normalization step was performed by subtracting μ and dividing by σ. The transformation was then applied to every training, validation, and test image.

Learning Procedure. Following the approach proposed in [19], we used a stochastic gradient descent procedure (momentum = 0.9) to minimize the loss function; it is composed by a cross-entropy term and a L2 regularization term ($\lambda = 0.0001$). The batch size is 200 and the minimum number of epochs is 300. Since then, validation accuracy is monitored by stopping the training procedure after awaiting 20 epochs since the last improvement. The learning rate is a piecewise constant function of the training step (boundaries: [12000, 18000, 24000, 30000, 36000], values: [0.1, 0.05, 0.025, 0.0125, 0.00625, 0.003125]).

Data Augmentation. To artificially increase the training set size, every input image undergoes the following transformations: zero-padding from 48×48 to 54×54, and selection of a random crop of size 48×48; random horizontal flip with probability 0.5.

VGG16-FT: Fine-Tuning a Pretrained Model

Model. The reference pretrained model used in our framework is described in [18] and has been released as Caffe model by the Visual Geometry Group:

- the architecture is the VGG16 [20]: it is a 16-layers network resulting in 134,289,223 parameters; a dropout layer is added after FC_7 layer to reduce overfitting.
- the available weights of the model have been obtained by pretraining on a dataset for face recognition: the authors in [18] proposed a method for collecting and annotating 2.6M images from 2.6K different identities.

Since transfer learning is more successful when the source task and the target task are more similar, this pretrained model perfectly fits on our case-study, i.e. the classification of emotion from facial images.

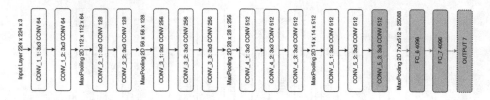

Fig. 3. The VGG16-FT architecture. During the first training step, only the output layer (dotted box on the right) is updated. During the second training step, the layers represented with filled boxes are fine-tuned.

Preprocessing Stage. The pretrained architecture, with a $224 \times 224 \times 3$ input layer, requires that grayscale images of FER2013 are upscaled to 224×224 and replicated on three channels. Images are then zero-centered by subtracting the global mean value μ.

Learning Procedure. The following steps are performed:

– the original output layer is removed (it was designed for another classification task). We add our custom output layer consisting of 7 units with softmax activation. Dropout is added before the output layer to reduce overfitting.
– *Step 1: Training the output layer.* The whole network, except the newly added output layer, is kept frozen, i.e. weights are not updated during training. Since the output weights are randomly initialized, the loss function is high in the first steps: including the convolutional layers in the learning procedure would damage the representations previously learned by such layers, because of a large error signal back-propagating through the network. The classifier is trained for 5 epochs using the Adam optimizer [14] with a learning rate of 0.00005 and a categorical cross-entropy loss function. In this step, the number of trainable parameters is 28,679. The batch size is 64.
– *Step 2: Fine-Tuning.* As shown in Fig. 3 all the hidden layers after *conv5_3* are unfrozen. Learning rate is halved and a new training procedure jointly fine-tunes these layers and the output added layer. In this step, the number of trainable parameters raises to 121,934,343.

Data Augmentation. To enable a fair comparison among the two scenarios, we perform the same data augmentation adopted for CNN10-S: images are zero-padded to size 256×256. A random crop of size 224×224 is extracted from either the padded image or its horizontally flipped version.

4   Experimental Results

In this section we show the experimental results: the performance of the proposed models is evaluated in terms of accuracy on the FER2013 test set. For both the ensemble strategies and the model architectures, three groups of nine networks were trained and evaluated in order to assess the stability of the measures.

Being able to rely on batches of 27 networks, we thus further investigated the accuracy obtained increasing the ensemble size. Results are reported in Table 1 and summarized in Fig. 4.

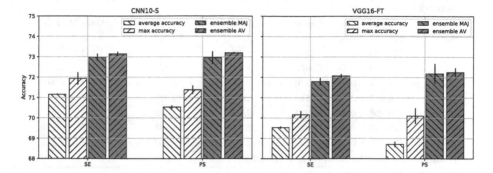

Fig. 4. Base classifiers (white bars) and ensemble (filled bars) accuracy for both strategies. In each group, the former white bar represents average of base classifiers accuracy, the latter represents the best base learners accuracy; the former filled bar represents ensemble accuracy using average voting, the latter represents ensemble accuracy using majority voting. *Left:* trained from scratch CNN10-S. *Right:* fine-tuned VGG16-FT.

Comparison Between CNN10-S and VGG16-FT. A first result is that the architecture CNN10-S (Table 1A), trained from scratch, achieves better performance than the architecture VGG16-FT (Table 1B), used as a pretrained model with fine-tuning (Fig. 4): the discrepancy between the accuracy values (between 1% and 2%) is confirmed both in terms of base and ensemble classifiers. Nevertheless, the two scenarios share common trends: they are analyzed in the following paragraphs.

Accuracy of Base Classifiers. For each strategy we can rely on three groups of nine networks. The low standard deviation of the average accuracy inter-groups suggests that results are fairly stable, independently of the strategy and the architecture. Both using CNN10-S and VGG16-FT, we observed that networks of SE strategy achieve better performance than networks of PS strategy (Fig. 4, white bars). Furthermore, the intra-group analysis suggests that networks of the PS strategy have a higher standard deviation, especially for VGG16-FT. Indeed, we noticed that the introduction of histogram equalization and illumination normalization leads to a slight performance drop compared to the use of default images.

Ensemble Accuracy. We define the ensemble gain as the difference between ensemble accuracy and average base classifier accuracy: combining preprocessing

Table 1. Accuracy of base classifiers and ensembles (average and majority voting), and the respective ensemble gains, are reported for each repeated measure. The mean and the standard deviation values are reported for each strategy.

A. Results obtained by training CNN10-S from scratch					
Network architecture CNN10-S					
	Base classifiers	Ensemble AV.	Gain AV.	Ensemble MAJ.	Gain MAJ.
SE	71.165 ± 0.520	73.084	1.919	72.862	1.697
	71.180 ± 0.495	73.279	2.099	72.862	1.681
	71.109 ± 0.275	73.057	1.947	73.224	2.114
Mean ± Std	71.152 ± 0.031	73.140 ± 0.099	1.989 ± 0.079	72.982 ± 0.171	1.831 ± 0.201
PS	70.465 ± 0.585	73.140	2.675	72.973	2.508
	70.496 ± 0.472	73.224	2.727	72.583	2.087
	70.645 ± 0.649	73.224	2.579	73.335	2.690
Mean ± Std	70.535 ± 0.078	73.196 ± 0.039	2.660 ± 0.062	72.964 ± 0.307	2.428 ± 0.253
B. Results obtained by fine-tuning VGG16-FT					
Network architecture VGG16-FT					
	Base classifiers	Ensemble AV.	Gain AV.	Ensemble MAJ.	Gain MAJ.
SE	69.546 ± 0.320	71.942	2.396	71.580	2.034
	69.462 ± 0.333	72.137	2.675	71.775	2.313
	69.583 ± 0.487	72.137	2.554	72.026	2.443
Mean ± Std	69.530 ± 0.050	72.072 ± 0.092	2.542 ± 0.114	71.793 ± 0.182	2.263 ± 0.170
PS	68.636 ± 1.187	72.388	3.752	72.527	3.892
	68.592 ± 0.941	71.942	3.350	71.496	2.904
	68.886 ± 0.403	72.416	3.529	72.527	3.641
Mean ± Std	68.705 ± 0.130	72.249 ± 0.217	3.544 ± 0.165	72.184 ± 0.486	3.479 ± 0.419

and seed variability ensures a higher gain value than just varying the seed. Nevertheless, in both the scenarios, PS strategy and SE strategy lead to very close ensemble performances (Fig. 4, filled bars). Despite being based on a deeper model, ensemble learning in VGG16-FT proves to be more effective than in CNN10-S, since it shows higher ensemble gain.

Even if the adopted aggregation schemes (average and majority voting) lead to comparable results, average voting shows slightly higher performance: in our framework, with low intra-group accuracy variability, average voting represents the proper choice. Indeed, majority voting is typical less sensitive to the output of a single base classifier since it considers only the predicted labels.

It is worth noting that each ensemble achieves better performance than the best base classifier composing it.

Increasing the Number of Base Classifiers. Let A, B, C be the three groups of networks produced for each strategy. We could rely on 3 ensembles of 9 networks (A, B, C), 3 ensembles of 18 networks (AB, AC, BC), and 1 ensemble of 27 networks (ABC). Figure 5 shows the results.

PS strategy shows a slight boost in performance with CNN10-S (+0.110%), but a drop using VGG16-FT (−0.028%). On the other hand, SE strategy shows a promising trend with VGG16-FT (+0.316%) while decreases with CNN10-S (−0.139%). Values in brackets are obtained by subtracting the accuracy value of the 27-nets ensemble and the 9-nets ensemble, considering average voting.

From the above considerations and by analyzing the trend shown in the plot, it is not possible to state that the increase of the number of base classifiers considerably improves, in general, the performance of our ensembles. Further, using the proposed model and the adopted parametrization, the training procedure is extremely time-consuming.

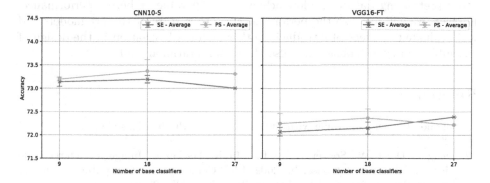

Fig. 5. Ensemble accuracy values versus number of base classifiers in the ensemble. Mean and standard deviation of three values are available for the ensembles with 9 and 18 networks, while a single value is available for the ensemble with 27 networks. For each strategy, we considered only average voting. *Left:* trained from scratch CNN10-S. *Right:* fine-tuned VGG16-FT.

5   Conclusion

In this paper we evaluated the performance of two design strategies for generating ensembles of CNNs used to tackle the FER problem, namely the Seed Strategy and the Preprocessing Strategy. The former generates diversity among base classifiers by simply varying the seed; the latter combines different values of the pseudorandom number generator with the introduction of different transformations of the input images.

Using a well known medium-sized dataset (FER2013), we carried out our comparison following two approaches: training an ad-hoc model from scratch (CNN10-S) and fine-tuning a pretrained model (VGG16-FT).

Results have shown that the ad-hoc architecture is an appropriate choice for the considered task, since it performs better than the fine-tuned model, both considering base classifiers and ensemble accuracy. Nevertheless, using a pretrained model requires less effort.

In the presented experimental setup, Seed Strategy and Preprocessing Strategy achieve comparable results using both the approaches (CNN10-S and

VGG16-FT). However, the variability induced by the Preprocessing strategy allows obtaining significantly higher ensemble gain than using the solely seed variation.

To the best of our knowledge, this is the first work that analyze the effectiveness of simple ensemble strategies using Deep Learning approaches for the FER task. Since we did not make specific assumptions based on the facial images, it could represent a starting point for further investigation also in other Computer Vision classification tasks.

In future work, we will investigate if other models, which use the same or other pretraining datasets, allow achieving comparable or better performance. Further, we will analyze the performance of other state of the art models and will evaluate the effect of introducing other factors of variation in the design of ensemble strategies, considering also their computational load.

References

1. Chollet, F.: Deep Learning with Python. Manning Publications Co., Shelter Island (2017)
2. Deng, J., Dong, W., Socher, R., Li, L.J., Li, K., Fei-Fei, L.: ImageNet: a large-scale hierarchical image database. In: 2009 IEEE Conference on Computer Vision and Pattern Recognition. CVPR 2009, pp. 248–255. IEEE (2009)
3. Dhall, A., Goecke, R., Joshi, J., Sikka, K., Gedeon, T.: Emotion recognition in the wild challenge 2014: baseline, data and protocol. In: Proceedings of the 16th International Conference on Multimodal Interaction, pp. 461–466. ICMI 2014. ACM (2014). https://doi.org/10.1145/2663204.2666275
4. Ekman, P., Friesen, W.V.: Constants across cultures in the face and emotion. J. Pers. Soc. Psychol. **17**(2), 124 (1971)
5. Giacinto, G., Roli, F.: Design of effective neural network ensembles for image classification purposes. Image Vis. Comput. **19**(9), 699–707 (2001)
6. Gonzalez, R.C., Woods, R.E.: Digital Image Processing, 3rd edn. Prentice-Hall Inc., Upper Saddle River (2006)
7. Goodfellow, I., Bengio, Y., Courville, A.: Deep Learning. MIT Press, Cambridge (2016). http://www.deeplearningbook.org
8. Goodfellow, I.J., et al.: Challenges in representation learning: a report on three machine learning contests. In: Lee, M., Hirose, A., Hou, Z.-G., Kil, R.M. (eds.) ICONIP 2013. LNCS, vol. 8228, pp. 117–124. Springer, Heidelberg (2013). https://doi.org/10.1007/978-3-642-42051-1_16
9. Gross, R., Brajovic, V.: An image preprocessing algorithm for illumination invariant face recognition. In: Kittler, J., Nixon, M.S. (eds.) AVBPA 2003. LNCS, vol. 2688, pp. 10–18. Springer, Heidelberg (2003). https://doi.org/10.1007/3-540-44887-X_2
10. He, K., Zhang, X., Ren, S., Sun, J.: Deep residual learning for image recognition. In: Proceedings of the IEEE Conference on Computer Vision and Pattern Recognition, pp. 770–778 (2016)
11. Hertel, L., Barth, E., Käster, T., Martinetz, T.: Deep convolutional neural networks as generic feature extractors. In: 2015 International Joint Conference on Neural Networks (IJCNN), pp. 1–4, July 2015. https://doi.org/10.1109/IJCNN.2015.7280683

12. Ju, C., Bibaut, A., van der Laan, M.J.: The relative performance of ensemble methods with deep convolutional neural networks for image classification. arXiv preprint arXiv:1704.01664 (2017)
13. Kim, B.K., Dong, S.Y., Roh, J., Kim, G., Lee, S.Y.: Fusing aligned and non-aligned face information for automatic affect recognition in the wild: a deep learning approach. In: Proceedings of the IEEE Conference on Computer Vision and Pattern Recognition Workshops, pp. 48–57 (2016)
14. Kingma, D.P., Ba, J.: Adam: a method for stochastic optimization. In: ICLR (2015). https://arxiv.org/abs/1412.6980
15. Krizhevsky, A., Sutskever, I., Hinton, G.E.: ImageNet classification with deep convolutional neural networks. In: Advances in Neural Information Processing Systems, pp. 1097–1105 (2012)
16. LeCun, Y., Kavukcuoglu, K., Farabet, C.: Convolutional networks and applications in vision. In: Proceedings of 2010 IEEE International Symposium on Circuits and Systems (ISCAS), pp. 253–256. IEEE (2010)
17. Martinez, B., Valstar, M.F.: Advances, challenges, and opportunities in automatic facial expression recognition. In: Kawulok, M., Celebi, M.E., Smolka, B. (eds.) Advances in Face Detection and Facial Image Analysis, pp. 63–100. Springer, Cham (2016). https://doi.org/10.1007/978-3-319-25958-1_4
18. Parkhi, O.M., Vedaldi, A., Zisserman, A., et al.: Deep face recognition. In: BMVC, vol. 1, p. 6 (2015)
19. Pramerdorfer, C., Kampel, M.: Facial expression recognition using convolutional neural networks: state of the art. arXiv preprint arXiv:1612.02903 (2016)
20. Simonyan, K., Zisserman, A.: Very deep convolutional networks for large-scale image recognition. In: ICLR (2015). https://arxiv.org/abs/1409.1556
21. Srivastava, N., Hinton, G.E., Krizhevsky, A., Sutskever, I., Salakhutdinov, R.: Dropout: a simple way to prevent neural networks from overfitting. J. Mach. Learn. Res. 15(1), 1929–1958 (2014)
22. Szegedy, C., Vanhoucke, V., Ioffe, S., Shlens, J., Wojna, Z.: Rethinking the inception architecture for computer vision. In: Proceedings of the IEEE Conference on Computer Vision and Pattern Recognition, pp. 2818–2826 (2016)
23. Wang, J., Perez, L.: The effectiveness of data augmentation in image classification using deep learning. Technical report (2017)
24. Wen, G., Hou, Z., Li, H., Li, D., Jiang, L., Xun, E.: Ensemble of deep neural networks with probability-based fusion for facial expression recognition. Cogn. Comput. 9(5), 597–610 (2017). https://doi.org/10.1007/s12559-017-9472-6
25. Yosinski, J., Clune, J., Bengio, Y., Lipson, H.: How transferable are features in deep neural networks? In: Advances in Neural Information Processing Systems, pp. 3320–3328 (2014)
26. Zhang, T.: Facial expression recognition based on deep learning: a survey. In: Xhafa, F., Patnaik, S., Zomaya, A.Y. (eds.) IISA 2017. AISC, vol. 686, pp. 345–352. Springer, Cham (2018). https://doi.org/10.1007/978-3-319-69096-4_38

Processing Online SAT Instances
with Waiting Time Constraints
and Completion Weights

Robinson Duque[1](✉), Alejandro Arbelaez[2], and Juan Francisco Díaz[1]

[1] Universidad del Valle, Cali, Colombia
{robinson.duque,juanfco.diaz}@correounivalle.edu.co
[2] Cork Institute of Technology, Cork, Ireland
alejandro.arbelaez@cit.ie

Abstract. In online scheduling, jobs arrive over time and information about future jobs is typically unknown. In this paper, we consider online scheduling problems where an unknown and independent set of Satisfiability (SAT) problem instances are released at different points in time for processing. We assume an existing problem where instances can remain unsolved and must start execution before a waiting time constraint is met. We also extend the problem by including instance weights and used an existing approach that combines the use of machine learning, interruption heuristics, and an extension of a Mixed Integer Programming (MIP) model to maximize the total weighted number of solved instances that satisfy the waiting time constraints. Experimental results over an extensive set of SAT instances show an improvement of up to 22.3× with respect to generic ordering policies.

1 Introduction

Typical deterministic scheduling models are usually based on the assumption that all problem data are known in advance. However, in real world problems, this assumption is not always adequate. For instance, job processing times may be unknown and subject to fluctuations; job arrivals (i.e, job releases) are often random events that can not be known in advance; jobs might have different requirements or constraints for processing (e.g., weights, due dates, etc.) [10].

In online scheduling, jobs arrive over time and depending on the assumptions, the decision-maker might become aware of some of the job data on its arrival. For instance, the processing time of a job might be presented on every job arrival or remain unknown until its completion [10]. However, we recall that many computational problems, including the Satisfiability (SAT) problem, display a high runtime variability. SAT is one of the fundamental problems in computer science and has received a lot of attention, specially for solving large-scale computational problems and numerous solver algorithms have been proposed including local search techniques, linear programming, methodologies based on statistical physics [2], immune algorithms [3], connected components of a graph [9], etc.

© Springer Nature Switzerland AG 2019
G. Nicosia et al. (Eds.): LOD 2018, LNCS 11331, pp. 418–430, 2019.
https://doi.org/10.1007/978-3-030-13709-0_35

In this paper, we use supervised machine learning to estimate SAT instance processing times and take this information into account to make decisions. We extend an existing scheduling approach proposed in [5] to study an online scenario where released instances have weights, and must start execution before a given maximum waiting time. In Sect. 2, we introduce a notation to represent our online over time scheduling problem. In Sect. 3 we present the computational approach for online processing of weighted combinatorial problems under single and multiple machine configurations. In Sect. 4 we empirically evaluate our models and present results. Finally, Sect. 5 presents our conclusions.

2   Online Scheduling Problem

Machine scheduling refers to an allocation of jobs within a set of machines given a series of constraints in order to optimize a specific criterion. Typical machine scheduling approaches assume that every job j is completed at some point and usually minimize some completion time criterion. For instance, the most common is to minimize the makespan (i.e., the time between the start of the first activity and the end of the last one) [10]. In online scheduling problems, the number of jobs to be processed is unknown and no information is given about future jobs. We recall that *online clairvoyant approaches* present all the relevant data of a job once it is released. However, *online non-clairvoyant* approaches might reveal some relevant data when jobs are released or remain unknown until job completion [10].

In contrast to traditional approaches that assume job completions, in [5] the authors assume that some jobs can be interrupted from running and remain unsolved at the end of the schedule. The authors study a problem where combinatorial problem instances arrive over time and must start execution before a maximum waiting time is met. Thus, the objective consisted in maximizing the total number of solved instances within certain limited time (e.g., using cloud rented resources). Cloud computing provides on-demand resources and services over a network that are often offered with a pricing model that lets you pay for the services that you use [7]. This kind of computation offers an interesting opportunity to solve combinatorial problems. However, attempting to solve a single job might consume all the rented computational time.

In general, in [5] the authors proposed a three phase approach that combines the use of machine learning, interruption heuristics, and an algorithm to run a MIP model over a bounded queue. The authors reported important improvements that go over 12.2× more solved instances than generic ordering policies. In this paper, we study an extension to the problem and evaluated the impact of adding weights to the instances. In general, instances arrive over time for processing, each with a random completion weight and must start execution before a maximum waiting time is met. Therefore, the objective consists in maximizing the total weighted number of solved instances (unsolved instances add no weight to the objective function).

Machine scheduling problems are usually described using the standard $\alpha|\beta|\gamma$ classification scheme proposed in [6] by Graham et al. (resp, α describes the machine environment; β provides details of the processing constraints; and γ describes the objective function). As in [5], we use an extension of the Graham notation to represent the problems. Furthermore, we also train regression and classification models in order to use them in our scheduling approach to estimate processing times or classify instances according to their empirical hardness:

$$1|online, \overrightarrow{wt_j}, \overset{\approx}{\tilde{s}_j}, \overset{\approx}{\tilde{p}_j}, interrupt| \sum \overrightarrow{w_j}S_j$$
$$Pm|online, \overrightarrow{wt_j}, \overset{\approx}{\tilde{s}_j}, \overset{\approx}{\tilde{p}_j}, interrupt| \sum \overrightarrow{w_j}S_j \tag{1}$$

The notation from the scheduling problems in (1) is then described as follows. The top arrow denotes that certain information is presented on the arrival of a job (e.g., $\overrightarrow{p_j}$ if the processing time is presented at the arrival as in a clairvoyant approach). Additionally, if a value is somehow estimated, it is denoted with an approximation symbol (e.g., $\overset{\approx}{\tilde{p}_j}$ for a semi-clairvoyant approach).

1 and Pm denote single and parallel identical machine environments. *online* denotes that jobs arrive over time and we have no prior knowledge about the released jobs; we assume that *release dates* $(\overrightarrow{r_j})$ are not known in advance; $(\overrightarrow{wt_j})$ denotes the *maximum waiting time* that an instance can remain in the system waiting to be attended; $(\overrightarrow{w_j})$ represents the weighted cost of solving a given instance i; $(\overset{\approx}{\tilde{p}_j})$ is the prediction of the processing time of an algorithm A on instance j; $(\overset{\approx}{\tilde{s}_j})$ denotes the prediction of the machine learning model indicating whether a given job j is solvable or not by algorithm A within time t. (See Sect. 3.1 for further details). Finally, *interrupt* denotes that a running job j may be interrupted, losing all the work done on it and becoming available to be rescheduled.

Our objective function consists in maximizing the total weighted number of solved jobs $(\overrightarrow{w_j}S_j)$. A job j is considered to be *attended*, if its waiting time in the system is less than $(\overrightarrow{wt_j})$ and it is processed for some time greater than 0 (i.e., instances that do not satisfy the waiting time constraint are discarded). Additionally, we assume that a given job j is solved if the system finds a solution for j within the assigned processing time. In Eq. (2), (ST_j) represents the time when job j starts being processed (resp. (ET_j) represents the end time). (WT_j) denotes the time that a job waits to start being processed from the time it is released (i.e., $WT_j = ST_j - \overrightarrow{r_j}$):

$$S_j = \begin{cases} 1, & \text{if } WT_j \leq \overrightarrow{wt_j} \wedge solve(j, ST_j, ET_j) \\ 0, & otherwise \end{cases} \tag{2}$$

3 Online Computational Approach

In this paper, we use the same approach as proposed in [5] and extend the existing MIP model to optimize the new objective function. We use a training/testing phase to create regression and classification models; scheduling policies to tackle the online problems presented in Sect. 2; instance interruption heuristics to mitigate inaccurate scheduling decisions.

3.1   Training/Testing Phase

Supervised machine learning can be used to learn a performance criteria from problems to algorithms using features extracted from the instances. In [8], the authors describe a problem instance j with a list of m features $d_j = [z_1, ..., z_m]$. To compute the training dataset, they run a set of n instances for a run time limit t and compute the vector of features d_j and the processing time p_j. Interestingly, the authors divide the features into four categories according to the complexity to collect the descriptors: trivial, cheap, moderate, and expensive.

In particular, we used the vector of features and runtime data for SAT problems introduced in [8] and we limited our approach to the usage of a single algorithm (i.e., MiniSAT). We used their datasets to train random forest models $(R : \{d_j\} \rightarrow \{\widetilde{p}_j\})$ that estimate the runtime of algorithm A on an instance j. We also used the data to create classification models $(C : \{d_j\} \rightarrow \{\widetilde{s}_j\})$ to estimate whether an algorithm A is able to solve an instance j within time t. For the classification models, we replaced processing times with Boolean values $s_j = (p_i < t)$, thus timeouts can be represented with 0's and solved instances with 1's (i.e., false and true respectively). Finally, in order to avoid situations in which calculating the features adds a big overhead to the schedule, we take advantage of the *trivial and cheap* features to Train-Test our models as studied in [5].

3.2   Scheduling Approach

Online scheduling is a challenging combinatorial problem and solving the scheduling problem might end up adding a considerably big overhead. Additionally, due to the uncertainty of job arrivals and processing times, the approach must be able to deal with continuous changes as instances arrive or get solved. Interestingly, in [11] the authors study the integration of long-term queuing policies with short-term scheduling, in the context of dynamic scheduling problems. They showed that combining long-term guidance from queuing theory with short-term combinatorial decision making, outperforms individual queuing and scheduling approaches for a dynamic flow-shop problem.

In [4], the authors proposed a MIP model to tackle an online problem that attempts to maximize the number of solved SAT instances. In particular, the authors used MIP every time the system needed to select a job for execution by creating an execution schedule. Extensive experimental results indicate that there is an important improvement w.r.t. a set of popular approaches. However, in practice, when the number of released instances in queue becomes large, using MIP has a negative impact on the system due to the added overhead of solving the scheduling problem. Later in [5], the authors implemented a bounded queue that keeps the size of the scheduling problem relatively small.

In this section we present the approach proposed in [5] that combines long-term queuing policies with short-term scheduling. We also extend their MIP model and study other ordering policies to tackle problems with weighted

Algorithm 1. SJF-MIP$(j, f, Q^{rel}, Q^{int}, Q^{sch}$, *MIP-model, k)*

1: Discard (i.e., Delete) all jobs that break the waiting time constraint $\overrightarrow{wt_j}$ from Q^{rel}, Q^{sch}, and Q^{int}.
2: If either queue (Q^{rel} or Q^{sch}) have jobs, then go to Step 4. Otherwise, continue to Step 3.
3: If Q^{int} is not empty, return the interrupted job with the shortest estimated processing time $\widetilde{p}_j$. Otherwise, wait for Q^{rel} to get an online (released) instance from a user and go to Step 4.
4: If the schedule queue Q^{sch} is empty, move at most the first k instances from Q^{rel} to Q^{sch} and go to Step 6. Otherwise, move to Step 5.
5: If the first job in Q^{rel} has a smaller processing time than the last job in Q^{sch}, move all the jobs from Q^{sch} to Q^{rel} and go to Step 4. Otherwise, go to Step 7.
6: If Q^{sch} has two or more jobs, run the MIP model with the jobs in the queue to compute the execution ordering of the jobs in Q^{sch}. Then go to Step 7.
7: Return the job with the highest priority in Q^{sch}.

---

instances as described in Sect. 2. Below we present four long-term queuing policies and then introduce a hybrid scheduling policy that will be evaluated in our experiments section:

1. **First Come First Serve (FCFS):** jobs are sequenced and processed in the same order as they are released (i.e., $\overrightarrow{r_j}$).
2. **Shortest Waiting Time First (SWTF):** jobs are sequenced and processed in ascending order by using the maximum waiting time constraint (i.e., $\overrightarrow{wt_j}$).
3. **Shortest Job First (SJF):** jobs are sequenced and processed in ascending order by using regression models to estimate runtimes (i.e., $\widetilde{p}_j$).
4. **Shortest Weighted Processing Time (SWPT):** it is a commonly used policy in offline problems where jobs are scheduled by the lowest ratio (p_j/w_j). In our approach we use runtime estimations and the weight of each instance (i.e., $\widetilde{p}_j/\overrightarrow{w_j}$).

SJF-MIP Hybrid Scheduling Policy: Using MIP over all the released instances in queue can be computationally expensive as shown in [4], therefore, this approach implements a bounded queue to reduce the time that MIP takes to schedule instances for execution and can be used together with regression and classification models for the problems presented in Sect. 2. Namely, this approach implements 3 priority queues and Algorithm 1 describes how to select an instance for execution when a machine becomes available:

- Q^{rel} - *(released instances queue)* priority queue that stores online released instances using a SJF policy based on runtime estimations.
- Q^{int} - *(interrupted instances queue)* priority queue that stores instances that are interrupted from running and might be scheduled for execution again. Instances are stored using a SJF ordering policy.

– Q^{sch} - (*scheduled instances queue*) queue that has a fixed capacity k to limit the number of instances that our MIP model will schedule for execution.

In general, *Algorithm SJF-MIP* runs MIP on a bounded queue Q^{sch} of size k only when released instances in Q^{rel} have smaller estimated processing times than the last instance scheduled by MIP in Q^{sch}. Next, we propose a modification of the former MIP model in [5] for multiple machines in order to schedule instances for the problems presented in Sect. 2.

MIP-Model for Weighted Combinatorial Problems in Multiple Machines: This MIP model maximizes the total weighted number of solved instances that satisfy the waiting constraints. It can be used with either all instances or only with instances classified as solvable with a classification model:

Indices and sets:

– J: set of instances (also jobs) in the queue;
– i, j: instances $(i, j \in J)$
– M: set of machines to process the instances;
– m: machine index $(m \in M)$

Parameters:

– $\vec{r_j}$: release time of instance j;
– $\tilde{\tilde{p}}_j$: runtime of instance j (estimation using a regression model);
– $\vec{wt_j}$: maximum expected waiting time of instance j.
– $\vec{w_j}$: weight of instance j.
– ct: cost of solving an instance;
– $time$: time when the system becomes available;
– $nextET_m$: time when machine m is expected to become available;

Decision variables:

– ST_j: start time of instance j;
– ET_j: end time of instance j;
– WT_j: waiting time in the system of instance j;
– AT_j: boolean variable used to determine whether an instance j is attended before the maximum waiting time or not;
– X_j^m: boolean variable used to determine whether an instance j is assigned to machine m or not;

Maximize:

$$\sum_{j \in J} (ct * \vec{w_j} - \tilde{\tilde{p}}_j) * AT_j \tag{3}$$

Subject to:

$$ST_j \geq \vec{r_j} \wedge ST_j \geq time + 1 \quad \forall j \in J \tag{4}$$

$$WT_j = ST_j - \vec{r_j} \quad \forall j \in J \tag{5}$$

$$AT_j = \begin{cases} 1, & \text{if } WT_j \leq \vec{wt_j} \\ 0, & otherwise \end{cases} \quad \forall j \in J \tag{6}$$

$$ET_j = ST_j + (\tilde{\tilde{p}}_j * AT_j) \quad \forall j \in J \tag{7}$$

$$\sum_{m \in M} X_j^m = 1 \quad \forall j \in J \tag{8}$$

$$ST_j * X_j^m \geq (nextET_m + 1) * X_j^m \quad \forall j \in J, \forall m \in M \tag{9}$$

$$ET_j \leq (ST_i - 1) \vee ST_j \geq (ET_i + 1)$$
$$\forall i, j \in J, \forall m \in M \quad | \quad i \neq j \wedge X_i^m = X_j^m = 1 \tag{10}$$

$$ST_j, ET_j, WT_j \geq 0 \quad \forall j \in J \tag{11}$$

$$AT_j, X_j^m \in \{0, 1\} \quad \forall j \in J \tag{12}$$

The objective function of the *MIP-model* is influenced by Constraint (6) which enforces that only instances with a valid waiting time can be marked as attended. Thus, the objective of the model is to compute a schedule that maximizes the total weighted number of solved instances. We added a cost value ct to the objective function, in order to prioritize instances with higher weights $\vec{w_j}$ and smaller processing time estimations $\tilde{\tilde{p}}_j$.

Constraint (4) implies that every instance has to start after its arrival time and after the system becomes available. Constraint (5) calculates the waiting time of each instance. It is also used to determine if an instance is attended or not in constraint (6). Constraint (7) calculates the end time of an instance j. Such end time, depends mainly on its start time and on the runtime estimation to solve such instance. However, it can also assign an end time equals to the start time (i.e., 0 time for processing) when an instance is not marked as attended (i.e., $ET_j = ST_j$, if $AT_j = 0$).

Constraint (8) guarantees that every instance is assigned to a single machine m. Additionally, when an instance is assigned to machine m, Constraint (9) makes sure that the start time of such instance is greater than the estimated time for the machine to become available. The estimation of when a machine m becomes available can be calculated by keeping track of when each machine started processing an instance j plus the runtime estimation. Finally, the disjunctive constraints (10) ensure that every pair of instances i and j, both assigned to machine m (i.e., $X_i^m = X_j^m = 1$) do not overlap.

3.3   Instance Interruption Heuristics

As in [5], we study two heuristics to interrupt instances from running in order to overcome the impact of inaccurate predictions:

N - **The Naive Interruption:** represents an execution where the selected instance is executed to completion or until a cap time limit t is met. We will denote this heuristic with N in the experiments.

H_n - **The Heuristic Based Execution:** let Q denote all un-executed and alive jobs in (Q^{rel} and Q^{sch}). Also, let P denote all the currently running jobs that could be interrupted. Then, H_n denotes that a running instance $j \in P$ can be interrupted if there are n or more instances in Q that would not be attended if j continues its regular execution. Equation (13) depicts how we make such decision. We estimate the end time of a running instance j by adding the time when the instance started to be processed and the estimated runtime ($ET_j = ST_j + \tilde{\tilde{p}}_j$). Additionally, a waiting instance $i \in Q$, will not be attended if ($ET_j > \overrightarrow{r_i} + \overrightarrow{wt_i}$). Since processing times might be overestimated, to avoid early interruptions, the execution of instance j is extended by adding $\tilde{\tilde{p}}_i$ to the formulation:

$$NotExecuted_{ij} = \begin{cases} 1, & \text{if } ET_j > (\overrightarrow{r_i} + \overrightarrow{wt_i} + \tilde{\tilde{p}}_i) \\ 0, & otherwise \end{cases} \quad \forall i \in Q, \forall j \in P$$

$$\sum_{i \in Q} NotExecuted_{ij} \leq n \quad \forall j \in P \tag{13}$$

In general, Constraint (13) must be satisfied during execution of all jobs $j \in P$. For instance, the usage of heuristic H_4 means that any running instance j can be interrupted if 4 or more instances in Q are detected not to be attended if j continues to run. Notice that the heuristic requires to estimate the end time ET_j of a running instance j. Since j is already running, then its start time ST_j is already known but the processing time $\tilde{\tilde{p}}_j$ is an estimation.

4   Experiments

In this paper, we use the LaScILab cluster from the Universidad del Valle in Cali-Colombia (http://lascilab.univalle.edu.co/). This cluster features 320 Cores and 768 Gb of RAM memory accessed through HTCondor, a distributed batch computing system that supports High Throughput Computing (HTC) [12].

To evaluate the online approach proposed in Sect. 3, we use the same set of instances and algorithm runtime data as those used and reported in [8]. We recall that our study is limited to the usage of a single algorithm/solver per dataset type. The SAT instance dataset comprises data about MiniSAT with a time limit t of one hour. The dataset was collected from the international SAT competitions and races from 2002 to 2010. It contains 7012 instances that includes industrial (INDU), hand crafted (HAND), and random problems (RAND). We also used the combination of all SAT instances into a single dataset (INDU-HAND-RAND). Complete details of the instances and runtimes are available at www.cs.ubc.ca/labs/beta/Projects/EPMs/.

Table 1. Configuration values for test simulations

M	Interr.	r_j	wt_j	Dataset/solver	Scheduling policies
1	N	1 s–10 s	0 s–30 s	INDU/Minisat	FCFS(R) & FCFS(RC)
2	H_2	1 s–30 s	0 s–90 s	HAND/Minisat	SWTF(R) & SWTF(RC)
4	H_4	1 s–60 s	0 s–180 s	RAND/Minisa	SJF(R) & SJF(RC)
	H_8	1 s–120 s	0 s–240 s	INDU HAND RAND/Minisat	SWPT(R) & SWPT(RC)
			240 s–480 s		SJF-MIP(R) & SJF-MIP(RC)

To train our regression models we performed a log transformation of the runtimes. To train the classification models we managed to label timeouts as *not solvable* and valid runtimes as *solvable* (i.e., 0 or 1 for instances running for a cap time limit t). Moreover, we randomly split the instance sets into 30% for training and 70% for testing. We also use the authors categories (i.e., trivial and cheap features) to present our results and use a random forest implementation from Weka (version 3.8) with its default hyperparameters to train-test our models.

We assigned random weights from 1 to 100 to each SAT instance in the Test set. Table 1 shows the configuration of our experimental evaluations. In particular, we tested our approach with 1, 2, and 4 machines and experimented by combining different interruption heuristics, release dates, waiting times, datasets, and scheduling policies. We also experimented with four interruption heuristics. Namely, naive (N) and three configurations for (H_n) as proposed in Sect. 3.3. We configured our evaluations by releasing instances (in the Test set) one at a time with random inter-arrival interval times ranging 1 s to 120 s; the waiting time for each instance is randomly distributed within five intervals ranging from 0 s to 480 s; and tested five scheduling policies (i.e, FCFS, SWTF, SJF, SWPT, SJF-MIP) using both Regression (R) and Regression-Classification (RC) models. We recall that we use regression to estimate the runtime of a given instance and classification to only use instances classified as solvable within time t. Additionally, we ran each simulation (i.e., each configuration combination) 5 times using different instance orderings and calculated the average total weighted number of solved instances for each simulation.

4.1 Interruption Heuristics Analysis

In Fig. 1 we reported the total average weighted number of solved instances of the INDU-HAND-RAND dataset under a single machine using four interruption heuristic configurations. The results were obtained across all the simulations using *Real values*, *Cheap*, and *Trivial* features to Train the regression/classification models. "*Real values*" refers to models that are able to do

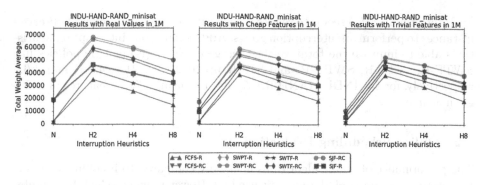

Fig. 1. Graphic with average total weighted number of solved instances using four interruption heuristics under a single machine configuration.

perfect classifications and runtime estimations, in order to establish a comparison reference.

It can be observed that the interruption heuristic with the highest total weight average is (H_2) independently from the features that we use to train our Random Forest models. As expected, a model with perfect predictions (i.e., actual runtimes) lead to higher weight averages. Moreover, such results tend to decrease as we move from models trained with *Cheap* features to *Trivial* . We attribute this behaviour to the fact that models trained with *Cheap* features reported a higher correlation coefficient (CC) and a lower root mean square error (RMSE) than those trained using *Trivial* features. CC is a value between -1 and 1 where 1 is a perfect correlation, 0 is no correlation, and -1 is an inverse correlation. On the other hand, RMSE is used to measure the discrepancies between true values and the estimated ones (i.e., lower RMSE are better). For instance, for the industrial dataset (INDU), the correlation of the regression model is 0.90 (*Cheap*) vs. 0.78 (*Trivial*) (resp. 0.80 vs 1.14 for the RMSE). We also observed the same behaviour across all the datasets studied here.

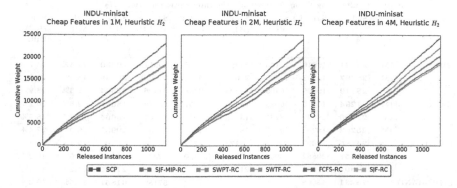

Fig. 2. Policy comparison against the SCP using cumulative weights as the number of released instances grows.

Interestingly, (H_n) seems to reduce its effectiveness as the number of observed instances to perform an interruption grows. Additionally, our interruption heuristic is able to improve the total weight average of Naive (N) generic policies like SWTF-R, FCFS-R, SWPT-R, and SJF-R up to 22.3×, 21×, 4.7×, and 4.6× respectively, for the INDU-HAND-RAND dataset using *Cheap* features in a single machine.

4.2   Online Scheduling Evaluation

The performance of online algorithms is usually compared to its offline counterpart version assuming all data in advanced [1]. However, when scheduling combinatorial problems, this comparison against an all-knowledgeable adversary seems unrealistic, specially because there are unknown processing times in the testing data. Preliminary experiments showed that our hybrid policy SJF-MIP(CR) using classification and regression behaves better than any other policy if working with an accurate predictive model. Thus, every policy can be compared with the **Semi Clairvoyant Policy (SCP)** that is a version of our SJF-MIP(RC) that is non-anticipative and features perfect predictions.

We now move our attention to Fig. 2, where we reported the average weighted number of solved instances as the number of released instances increases for the INDU dataset. The results are presented across three machine configurations with the best known interruption heuristic (i.e., H_2) using regression and classification models trained with *Cheap* features. Interestingly, it can be observed that our hybrid approach (i.e., SJF-MIP(RC)) is typically closer to the SCP than other policies. It also seems that the rate of solved instances tend to grow slower as the number of machines is incremented. We also observed that SWPT-RC and SJF-RC reported nearly the same cumulative weights and are typically better than SWTF-RC and FCFS-RC which present similar results.

Table 2. Experiments summary reporting the average weights of the best three policies

30%–70% partition results using *Cheap* features and the interruption heuristic H_2

Dataset	M	SCP	SJF (R)	SJF (RC)	SWPT (R)	SWPT (RC)	SJF-MIP (R)	SJF-MIP (RC)	SJF-MIP (RC) Diff.	SJF (RC) Diff.
INDU	1M	**23787**	16910	18143	16987	18398	19911	**20350**	14.4%	23.7%
	2M	**25397**	19403	19708	20051	19865	**21659**	21450	15.5%	22.4%
	4M	**28719**	19696	20409	20298	20468	21406	**22304**	22.3%	28.9%
HAND	1M	**21294**	14515	16985	14478	16757	16426	**18616**	12.6%	20.2%
	2M	**22426**	18042	18231	18065	18045	19601	**19664**	12.3%	18.7%
	4M	**25202**	18220	19550	18471	19620	19855	**20953**	16.9%	22.4%
RAND	1M	**24885**	15003	**22055**	15173	21762	18298	21795	12.4%	11.4%
	2M	**26657**	20882	23947	20755	23490	22331	**24918**	6.5%	10.2%
	4M	**30604**	21843	26590	21543	26234	22033	**27165**	11.2%	13.1%
INDU-HAND	1M	**73231**	45558	59116	46540	57702	57189	**61861**	15.5%	19.3%
	2M	**77430**	60879	64227	60897	63461	66629	**69743**	9.9%	17.1%
	4M	**87802**	62829	68778	63077	67823	66483	**72513**	17.4%	21.7%

We recall that our MIP approach reported a few milliseconds every time we use it since it runs over a bounded queue. Thus, the total overhead added by MIP is not significant (e.g., less than 10s in total for the INDU-HAND-RAND experiments that contain the greatest number of instances released overtime).

Finally, Table 2 summarizes the results of our experiments using *Cheap* features. For this table, we reported the total average weights per test and machine configuration of the best three scheduling policies. We included two columns to compare the total weight differences of SJF(RC) and SJF-MIP(RC) with respect to the SCP (i.e., lower percentages are desirable). It can be observed that our hybrid approach SJF-MIP(RC) is the overall winner followed by our SJF-MIP(R) approach and reported smaller differences with respect to the SCP. For instance, for INDU in 1M, our hybrid approach SJF-MIP(RC) reported weights of 14.4% below the SCP whilst SJF(RC) reported 23.7%.

5   Conclusions

In this paper, we have presented an online approach to maximize the total weighted number of SAT combinatorial problem instances subject to waiting time constraints. We extended an existing approach for online scheduling of combinatorial problems that consisted of three parts. Namely, training/testing models for processing time estimations; implementation of a hybrid scheduling policy using SJF and a MIP model to maximize weighted combinatorial problems; usage of instance interruption heuristics to mitigate inaccurate predictions. We tested our approach with a well-known SAT dataset and reported considerably big improvements of up to 22.3× when using interruption heuristics with long term policies. Additionally, our hybrid approach observed results that are closer to a semi clairvoyant policy (SCP) featuring perfect estimations.

References

1. Anderson, E.J., Potts, C.N.: Online scheduling of a single machine to minimize total weighted completion time. Math. Oper. Res. **29**(3), 686–697 (2004)
2. Angione, C., Occhipinti, A., Nicosia, G.: Satisfiability by Maxwell-Boltzmann and Bose-Einstein statistical distributions. ACM J. Exp. Algorithmics (JEA) **19**, 1–4 (2014)
3. Cutello, V., Nicosia, G.: A clonal selection algorithm for coloring, hitting set and satisfiability problems. In: Apolloni, B., Marinaro, M., Nicosia, G., Tagliaferri, R. (eds.) NAIS/WIRN -2005. LNCS, vol. 3931, pp. 324–337. Springer, Heidelberg (2006). https://doi.org/10.1007/11731177_39
4. Duque, R., Arbelaez, A., Díaz, J.F.: Off-line and on-line scheduling of SAT instances with time processing constraints. In: Solano, A., Ordoñez, H. (eds.) CCC 2017. CCIS, vol. 735, pp. 524–539. Springer, Cham (2017). https://doi.org/10.1007/978-3-319-66562-7_38
5. Duque, R., Arbelaez, A., Díaz, J.F.: Online over time processing of combinatorial problems. Constraints **23**(3), 1–25 (2018)

6. Graham, R.L., Lawler, E.L., Lenstra, J.K., Kan, A.R.: Optimization and approximation in deterministic sequencing and scheduling: a survey. Ann. Discrete Math. **5**, 287–326 (1979)
7. Grossman, R.L.: The case for cloud computing. IT Prof. **11**(2), 23–27 (2009)
8. Hutter, F., Xu, L., Hoos, H.H., Leyton-Brown, K.: Algorithm runtime prediction: Methods & evaluation. Artif. Intell. **206**, 79–111 (2014)
9. Nicosia, G., Conca, P.: Characterization of the $\#k$–SAT problem in terms of connected components. In: Pardalos, P., Pavone, M., Farinella, G.M., Cutello, V. (eds.) MOD 2015. LNCS, vol. 9432, pp. 257–268. Springer, Cham (2015). https://doi.org/10.1007/978-3-319-27926-8_23
10. Pinedo, M.L.: Scheduling: Theory, Algorithms, and Systems, 5th edn. Springer International Publishing, New York City (2016)
11. Terekhov, D., Tran, T.T., Down, D.G., Beck, J.C.: Integrating queueing theory and scheduling for dynamic scheduling problems. J. Artif. Intell. Res. **50**, 535–572 (2014)
12. Thain, D., Tannenbaum, T., Livny, M.: Distributed computing in practice: the condor experience. Concur.- Pract. Exp. **17**(2–4), 323–356 (2005)

Variable Selection and Outlier Detection in Regularized Survival Models: Application to Melanoma Gene Expression Data

Eunice Carrasquinha[1,2]([✉]) [iD], André Veríssimo[1,2] [iD], Marta B. Lopes[1,2] [iD], and Susana Vinga[1,2] [iD]

[1] IDMEC, Instituto Superior Técnico, Universidade de Lisboa,
Avenida Rovisco Pais, 1, 1049-001 Lisbon, Portugal
eunice.trigueirao@tecnico.ulisboa.pt
[2] INESC-ID, Instituto Superior Técnico, Universidade de Lisboa,
Rua Alves Redol, 9, 1000-029 Lisbon, Portugal

Abstract. The importance of gene expression data analysis for onco-logical diagnosis and treatment has become widely accepted in recent years. One of the main associated challenges is the development of mathematical and statistical methods for data analysis to improve prognosis and guide treatment decisions. One of the difficulties that researchers face when dealing with gene expression datasets concerns their high-dimensionality. In this context, the goal of this work is to reduce the dimensionality of gene expression data using regularization techniques such as Lasso and Elastic net, complemented with DegreeCox, a network-based regularization method for survival analysis recently proposed. Also identification of long or short-term survivors (outliers) may lead to the detection of new prognostic factors, and the Rank Product test is used to identify those observations. An example based on the The Cancer Genome Atlas (TCGA) Melanoma dataset is presented, where the covariates are patients' gene expression. The application of data reduction techniques to the Melanoma dataset enabled the selection of relevant genes over a range of parameters evaluated, with 5 in common between elastic net regularization and DegreeCox for one of the two models further evaluated. Moreover, a long term survivor was detected as outlier by the Rank Product test, being systematically highly ranked for the martingale residuals of the models evaluated.

Keywords: High-dimensional data · Regularization techniques ·
Gene expression dataset · Outlier detection

The authors thank the European Union Horizon 2020 under grant agreement No. 633974 (SOUND project) and the Portuguese Foundation for Science & Technology (FCT) under projects UID/CEC/50021/2013, UID/EMS/50022/2013, PTDC/EMS-SIS/0642/2014, IF/00653/2012, SFRH/BD/97415/2013.

G. Nicosia et al. (Eds.): LOD 2018, LNCS 11331, pp. 431–440, 2019.
https://doi.org/10.1007/978-3-030-13709-0_36

1   Introduction

One of the challenges that scientists face nowadays is to deal with high-dimensional datasets, specially when the number of covariates (p) greatly exceeds the number of observations (n), $p \gg n$. This type of data is present in many fields of science and is in constant outgrowth due to the technology development. A particular case in the medical field is gene expression data of oncological diseases. In the last years, scientists attempt to deal with high-dimensional gene expression data, in order to bring more information into the diagnosis of oncological patients.

In this context, traditional statistical techniques for the estimation of the parameters cannot be applied, due to the inherent ill-posed inverse problem. When dealing with thousands of variables, dimensionality reduction is a crucial initial step, leading to distinct models depending on the variable selection method used [5]. Identifying the relevant variables or biomarkers precisely have become a challenge for the further advancement of the medical field. A solution to cope with this dimensionality problem is the use of additional constraints in the cost function optimization. Regularized optimization techniques ([13–15]) are widely used in most regression models, particularly in survival analysis by constraining the Cox's proportional hazards model. Lasso, elastic net and other sparsity methods have been successfully applied with such idea. Although leading to more interpretable models, these methods still do not fully profit from the relationships between the features, specially when these can be represented through graphs. Following these ideas, the DegreeCox [14], a method that applies network-based regularizers to infer Cox proportional hazard models when the features are genes and the outcome is patient survival, has been proposed.

Another important aspect of gene expression datasets, in the context of survival analysis, is the identification of patients that are long or short-term survivors, i.e., outliers. The detection of outliers can lead to the discovery of new prognostic factors. However, different models are obtained depending on the technique used to reduce the dimensionality of the data, and therefore different outliers can be obtained [5]. To overcome this issue, the Rank Product test is used, as a consensual method for the different models obtained.

The main goal of this work is first, to evaluate the predictive performance of survival analysis in a gene expression dataset using different dimensionality reduction techniques, namely the Lasso, elastic net and DegreeCox; second, to detect of outlier patients using the Rank Product test based on multiple models obtained by resampling the features. The gene expression dataset studied is the Melanoma cancer dataset obtained from The Cancer Genome Atlas (TCGA).

2   High-Dimensional Survival Data

In this section the techniques used to reduce the dimensionality of gene expression for cancer data, elastic net [15] and DegreeCox [14], are described. Since the gene expression data that is considered is approached from the survival point of view, a brief introduction to this subject is also given.

2.1   Survival Analysis and the Cox Regression Model

Survival analysis, which studies the time until an event of interest occurs, is used in many fields of science, in particular in the medical area. The event may be death, the relapse of a tumour, or the development of a disease. The response variable is the time until that event, called survival or event time, which can be censored, i.e. not observed on all individuals present in the study.

There are different ways of modelling this type of data, one of most widely used due to its flexibility is the Cox regression model [6], which is based on a semi-parametric likelihood, able to deal with censored data and assumes that the hazard function $h(t)$ at time t is:

$$h(t; \mathbf{x}) = h_0(t) \exp(\mathbf{x}^T \boldsymbol{\beta}), \tag{1}$$

where $\boldsymbol{\beta} = (\beta_1, ..., \beta_p)$ are the unknown regression coefficients, which represent the covariate effect in the survival, $\mathbf{x} = (x_1, ..., x_p)$ is the covariate vector associated to an individual and $h_0(t)$ represents the baseline hazard.

The Cox regression model is called a semi-parametric regression model, because the baseline hazard function, $h_0(t)$, is not specified. This contributes for the flexibility of the model. The semi-parametric likelihood function is given by

$$L(\boldsymbol{\beta}) - \prod_{i=1}^{n} \left[\frac{\exp(\mathbf{x}_i^T \boldsymbol{\beta})}{\sum_{j \geq i} \exp(\mathbf{x}_j^T \boldsymbol{\beta})} \right]^{\delta_i}, \tag{2}$$

where δ_i is the censored indicator.

The unknown regression coefficients, $\boldsymbol{\beta}$, are obtained by maximizing

$$l(\boldsymbol{\beta}) = \sum_{i=1}^{n} \delta_i \left\{ \mathbf{x}_i^T \boldsymbol{\beta} - \log \left[\sum_{j \geq i} \exp(\mathbf{x}_j^T \boldsymbol{\beta}) \right] \right\}, \tag{3}$$

the partial log-likelihood function.

To obtain the unknown regression coefficients, $\boldsymbol{\beta}$, the baseline hazard, $h_0(t_i)$ has to be estimated. [3], proposed

$$\hat{h}_0(t_i) = \frac{1}{\sum_{j \geq i} \exp(\mathbf{x}_i^T \boldsymbol{\beta})}, \tag{4}$$

to obtain the estimators for the baseline hazard. Therefore the total log-likelihood function in Eq. (3) is the following

$$l(\boldsymbol{\beta}, h_0) = \sum_{i=1}^{n} - \exp(\mathbf{x}_i^T \boldsymbol{\beta}) H_0(t_i) + \delta_i \left[\log(h_0(t_i)) + \mathbf{x}_i^T \boldsymbol{\beta} \right] \tag{5}$$

where $H_0(t_i)$ is the cumulative baseline hazard function.

When we have high-dimensional datasets ($p \gg n$) the estimation procedure, for the Cox regression model exhibits identifiability problems, leading to multiple possible solutions with a large number of non-zero parameters. In the literature there are a vast number of techniques to overcome this problem, providing a sparse estimate of β. The two variable selection techniques chosen to reduce the dimensionality namely Elastic-net [15] and DegreeCox [14], are presented.

2.2   Elastic Net

The elastic net, is a regularization technique proposed by [15] in order to restrict the solution space by imposing sparsity and small coefficients to the parameters, combining the L_1 and L_2 norms. It can be used in different types of regression models, particularly in the Cox's regression model for survival data by penalizing the log-likelihood function.

The penalized partial log-likelihood function (3) with a weighted sum of the L_1 and L_2 norms is given by

$$l(\beta) = \sum_{i=1}^{n} \left\{ \delta_i x_i^T \cdot \beta - \delta_i \left[\log \sum_{j \geq i} (x_i^T \cdot \beta) \right] \right\} + \lambda \Psi(\beta) \tag{6}$$

where,

$$\lambda \Psi(\beta) = \lambda \left(\alpha ||\beta||_1 + \frac{1}{2}(1 - \alpha)||\beta||_2^2 \right) \tag{7}$$

where the parameter that controls the penalization of the weights is given by λ, and the balance between L_1 and L_2 norms is given by α, with $0 \leq \alpha \leq 1$. The Ridge and Lasso, regularization techniques, are a particular case. For $\alpha = 0$, Eq. (6) leads to the Ridge regression, for $\alpha = 1$ leads to the Lasso regression.

2.3   DegreeCox

The DegreeCox, proposed by [14], is a network-based regularization technique for the Cox's regression model. This method combines the partial log-likelihood function of the Cox's regression model (5) with degree regularization, which conveys a vertex centrality information of the network. Each vertex in the network represents a gene and to obtain the corresponding vertex centrality information, two networks are constructed from the data using Pearson's correlation and covariance. The DegreeCox adds a network degree-based constraint to the Cox's regression model as a weight of η in the regularization function defined in θ with $\eta_i = \beta_i \theta_i$. This is an extension of DegreeCox that allows to use both the L_1 and L_2 norm in the regularization function, where

$$\lambda \Psi(\eta) = \lambda \left(\alpha ||\eta||_1 + (1 - \alpha)||\eta||_2^2 \right) \tag{8}$$

is the extended cost function, with vector θ given by

$$\theta_i = \max(\mathbf{Adj}) - \sum_{j=1}^{P} Adj_{ij}, \tag{9}$$

representing the vertex weighted degree. The resulting vector was scaled between 0 and 1 and then transformed using a double exponential heuristic to scale exponentially the coefficients of the vector that is defined by

$$\theta_i' = \gamma + 10^{\exp(\theta_i)-1} - 1 \tag{10}$$

where γ is the minimum value attributed to all vertices that avoids vertices with 0 weights in the regularization function.

The aim of this network-based regularization technique is to identify a set of genes that correlate with survival and also have a relevant role as a hub in the underlying network.

3   Outlier Detection

In survival analysis, outliers are defined as individuals with a survival time too high or too short. The identification of those individuals has gain great importance in the medical field due to the fact that they potentially allow the discovery of new prognostic factors can be found [9].

There are in the literature many attempts to detect outliers in survival data ([9] and [12]). However, the most common are based on the residuals. In this work, the residuals chosen for outlier detection were the martingale residuals.

In the context of this work, with high-dimensional datasets (gene expression data), dimensionality reduction is a first step in the analysis, which may lead to different models depending on the variable selection method used [5]. Depending on the methodology used to reduce the dimensionality of the data, different models are obtained and, consequently, distinct outliers are identified. To overcome this problem [5] proposed the Rank Product (RP) test to identify the outliers that are consistently highly ranked in each of the different models obtained.

3.1   Martingale Residuals

The Martingale residuals [12] are very useful to identify outlying observations in survival analysis, and are asymmetric distributed. The martingale residual for the ith individual is given by

$$r_{\hat{M}_i} = \delta_i - \hat{H}_0(t_i) \exp(\hat{\beta}^T \mathbf{x}_i), \tag{11}$$

with values between $-\infty$ and 1. These residuals reveal the individuals that are not well fitted to the model. i.e., individuals who lived too long or died too soon, when compared to other individuals with the same covariate pattern.

3.2   Rank Product Test

The Rank Product (RP) test is a non-parametric statistical technique and has been used in many types of experiments in order to combine the experiments

([2] and [4]). The use of the RP test to detect outliers was recently proposed by the authors in the context of survival analysis [5] and logistic regression [8] applied to gene expression data. Given different sub-models based on the original dataset lead to different sets of observations, the rational behind the application of the RP test in this context is to identify the observations that are consistently classified as outliers by each sub-model.

Formally, let the number of individuals, n, and the number of different sub-models where the outlier detection method was performed, k. Let Z_{ij} be a measure of the outlyingness of the ith individual in the jth sub-model, with $1 \leq i \leq n$ and $1 \leq j \leq k$.

For each Z_{ij}, the deviance rank is defined by

$$R_{ij} = \text{rank}(Z_{ij}), \qquad 1 \leq R_{ij} \leq n. \tag{12}$$

The ranks for each sub-model are determined and the RP is obtained,

$$RP_i = \prod_{j=1}^{k} R_{ij}. \tag{13}$$

The lowest rank suggest that the individual is more outlier that the others. After obtain the p-values [7], the False Discovery Rate (FDR) [11] was performed in order to avoid type-I errors, due to the multiple testing.

4   Data Analysis

To illustrate the problem of handle high-dimensional data and consequently identify outlying observations, the Melanoma cancer gene expression dataset from The Cancer Genome Atlas (TCGA) (http://cancergenome.nih.gov/) was used. Melanoma cancer is the most aggressive skin cancer, and is also one of the fastest increasing cancers in terms of incidence worldwide [1]. Due to the very low survival rate of this type of cancer, the study of new prognostic factors aiming at improving the survival of a patient has gained great importance. The Melanoma cancer dataset used is based on gene expression data, i.e., RNA-Seq data, of oncological patients and is constituted by 84 observations measured over 52, 746 covariates. The Melanoma dataset comprises four types of tissues: primary solid tumor, metastatic, additional metastatic and solid tissue normal. For the purpose of this analysis, primary solid tumor was considered.

The clinical data was cleaned, only cases for which the number of days of follow-up and days to death matched were included in the analysis. The same process was performed for days to death and vital status, where some cases had as status deceased, but a missing days to death.

The Melanoma dataset was analyzed in two steps. First, different regularization techniques were performed in order to reduce the dimensionality of the data and identify a sub-set of relevant genes. Second, outlier detection was performed based on the different models obtained.

Before applying the dimensionality reduction techniques, an univariate survival analysis for each of the 52, 746 gene expression was done. The log-rank test [10] was performed, and for a 20% level of significance 52, 669 were significant, to ensure that key genes will not be missed. Notice that to perform the log-rank test the individuals were divided in two groups: high and low-risk, based on their risk median value. Based on those gene expressions, two techniques, elastic net and the DegreeCox, were used to reduce the dimensionality of the data.

By varying the regularization parameters, α and λ, and the cut-off of the DegreeCox, several models were obtained. Due to the fact that for the Melanoma dataset the event (death) only occurred for 10 individuals, the cross-validation, usually used in regularization techniques, to optimize the λ, did not produce reliable results. To overcome this problem, a sequence of λ values from 0.01 to 0.1 was considered. Regarding the α parameter, the values considered were a sequence from 0.1 to 1 (Lasso). For the DegreeCox network-based regularization technique, the covariance matrix was used to obtain the network, and sequence of the cut-off from 0 to 0.08 were considered. Based on the combination of the parameters, 900 models were obtained.

In a first exploratory phase and based on preliminary results regarding the complexity of the solutions in terms of the number of selected genes, we present the results for two models.

The parameters of the models chosen are:

- Model 1: $\alpha = 1$, $\lambda = 0.02$ and degree cut-off $= 0.02$;
- Model 2: $\alpha = 0.5$, $\lambda = 0.04$ and degree cut-off $= 0.01$;

For each model chosen, a more detailed analysis of the genes selected was done. Also, identification of outlying observations, based on the martingale residuals, will be presented based on the RP test.

All the analysis were performed in R and the Melanoma dataset is available at http://web.ist.utl.pt/~susanavinga/TCGA-melanoma.

4.1   Variable Selection Results

From a general point of view, the higher the α and λ parameters, the less variables are selected. For model 1, the parameters chosen for the regularization technique (Lasso) were $\alpha = 1$ and $\lambda = 0.02$. The cut-off for the DegreeCox was 0.02, since as we increased the value, very few variables were selected. In fact, whenever Lasso was used, for all λ and cut-off of the degree considered, no intersection of genes was detected. The genes selected by model 1, 12 for the DegreeCox and 13 for the Lasso, with no intersection, are the following:

- Lasso: *WASHC4, FAM47DP, RGS7BP, MICB, RNU6-80P, ZNF726, RNU4-61P, ERVFRD-1, PABPC4L, AC129507.4, ZNF415P1, AL024498.2* and *AL139351.1*;
- DegreeCox: *TOP2B, SLC25A36, RLF, GCC2, SENP7, ETAA1, SOCS5, RAB33B, CEP83, ZDHHC17, ZNF527* and *RNU6-1048P*.

Figure 1 shows the survival curves for each model. In both cases a difference for high and low risk individuals was statistically significant, as assessed by the log-rank test.

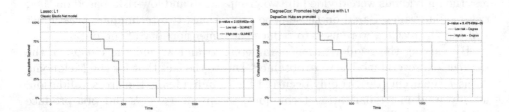

Fig. 1. Kaplan Meier curves for model 1, using Lasso and DegreeCox.

For model 2, the parameters chosen for the regularization technique were $\alpha = 0.5$ and $\lambda = 0.04$. The cut-off for the Degree was 0.01. In this case, for the parameters considered, 246 genes were selected for the regularization technique (elastic net), and 35 for the network-based DegreeCox. From the gene selected, five were common to both methodologies: *ANKRD12*, *ZNF569*, *ZNF28*, *ZNF891* and *NPM1P26*.

Figure 2 shows the survival curves for each technique. In both cases a difference for high and low risk individuals was also statistically significant.

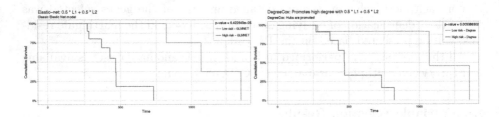

Fig. 2. Kaplan Meier curves for model 2, using Elastic-net and DegreeCox.

Following the variable selection performed, the identification of outlying observations was performed. The results are described next.

4.2   Outlier Detection Results

To overcome the fact that different models are obtained depending on the parameter value chosen, a sampling approach was applied to identify outliers. The resampling algorithm used ([5] and [8]) randomly chooses 1000 genes (without replacement) from the melanoma cancer dataset. A reduced set of selected genes was obtained, based on the elastic net regularization. Different α parameters were considered. The Cox regression model is then performed on the reduced gene set, and the corresponding martingale residuals determined. The observations were sorted according to their outlyingness. The process was repeated 10 times, yielding 10 models to be considered in the RP test.

Table 1. Top 10 of the outliers obtained for the resampling technique for 10 models, selecting 1000 genes sorted by q-value, with $\alpha = 0.5$.

ID	Model 1	Model 2	Model 3	Model 4	Model 5	Model 6	Model 7	Model 8	Model 9	Model 10	RP	p-values	q-values
26	12	2	1	2	7	1	6	6	3	1	3.63E+04	2.10E-05	0.0005
71	1	8	7	6	4	8	9	2	5	5	4.84E+06	5.56E-03	0.0639
40	3	5	2	3	17	6	3	3	17	6	8.43E+06	9.18E-03	0.0704
34	13	7	5	5	2	7	1	7	8	10	1.78E+07	1.74E-02	0.0999
62	8	12	9	9	3	2	4	11	7	4	5.75E+07	4.27E-02	0.1964
17	11	3	6	8	5	13	5	4	6	8	9.88E+07	6.23E-02	0.2046
43	6	1	17	18	8	19	2	13	1	13	9.43E+07	6.03E-02	0.2046
72	2	19	3	4	1	23	20	21	19	7	5.86E+08	1.80E-01	0.5184
11	10	6	10	10	6	9	11	8	9	12	3.08E+09	3.81E-01	0.9310
13	21	21	21	7	23	4	21	1	2	19	4.76E+09	4.45E-01	0.9310

Table 2. Top 10 of the outliers obtained for the resampling technique for 10 models, selecting 1000 genes sorted by q-value, with $\alpha = 1$.

ID	Model 1	Model 2	Model 3	Model 4	Model 5	Model 6	Model 7	Model 8	Model 9	Model 10	RP	p-values	q-values
26	1	4	4	7	2	1	8	4	2	4	5.73e+04	3.87e-05	0.0009
17	5	6	11	1	5	5	10	10	12	3	2.97e+07	2.61e-02	0.1978
40	2	19	20	2	1	16	3	2	6	22	1.93e+07	1.85e-02	0.1978
43	16	7	5	15	16	3	4	1	18	2	5.81e+07	4.30e-02	0.1978
77	3	5	1	5	3	21	1	23	20	23	5.00e+07	3.86e-02	0.1978
13	18	1	2	4	18	17	7	20	23	1	1.42e+08	7.90e-02	0.2596
62	4	11	10	11	4	6	5	5	5	8	1.16e+08	6.93e-02	0.2596
34	6	9	7	9	7	7	9	9	9	7	8.51e+08	2.18e-01	0.6259
71	10	10	12	16	13	4	6	6	7	6	1.51e+09	2.84e-01	0.7262
11	7	8	9	8	6	9	12	11	10	10	2.87e+09	3.71e-01	0.8525

The results displayed in Tables 1 and 2 show that the observation 26 (a long term survivor) is considered an outlier for the 10 different models obtained. Notice that, for other values of α considered, this observation had always a high rank, a strong evidence of being an outlier. From the RP test, we were able identify a consensual list of putative outliers in the dataset.

5   Conclusion

The aim of this work was to address the challenges high-dimensionality datasets bring. In particular, we explored the application of different frameworks to model and feature selection of RNA-seq Melanoma oncological survival data. Model and gene selection was achieved through elastic net and also DegreeCox, a network-based regularization method. Although the results are dependent on the specific parameters considered, it was nevertheless possible to identify systematically a group of outlier observations through the rank product ensemble test. One of the major difficulties in this study was the optimization of the regularization parameters, since approximately 88% of the data was censored, which hampered the cross-validation procedure. Future analysis will include the clinical evaluation of the selected genes with respect to oncobiology knowledge and the improvement of ensemble methods for model selection for highly censored data.

References

1. Braun-Falco, O., Plewig, G., Wolff, H.H., Burgdorf, W.H.C.: Melanocytic lesions. Dermatology. Springer, Berlin (2000). https://doi.org/10.1007/978-3-642-97931-6
2. Breitling, R., Armengaud, P., Herzykr, P.: Rank products: a simple, yet powerful, new method to detect differentially regulated genes in replicated microarray experiments. FEBS Lett. **573**, 83–92 (2004)
3. Breslow, N.: Discussion on professor Cox's paper. J. Roy. Stat. Soc.: Ser. B **34**, 216–217 (1972)
4. Caldas, J., Vinga, S.: Global meta-analysis of transcriptomics studies. PLoS One **9**(2) (2014). https://doi.org/10.1371/journal.pone.0089318
5. Carrasquinha, E., Veríssimo, A., Lopes, M., Vinga, S.: Identification of influential observations in high-dimensional cancer survival data through the rank product test. BioData Min. **11**(1) (2018). https://doi.org/10.1186/s13040-018-0162-z
6. Cox, D.R.: Regression models and life-tables. J. Roy. Stat. Soc.: Ser. B (Methodol.) **34**(2), 187–220 (1972). http://www.jstor.org/stable/2985181
7. Heskes, T., Eisinga, R., Breitling, R.: A fast algorithm for determining bounds and accurate approximate p-values of the rank product statistic for replicate experiments. BMC Bioinformatics **15**, 367 (2014). https://doi.org/10.1186/s12859-014-0367-1
8. Lopes, M., Veríssimo, A., Carrasquinha, E., Casimiro, S., Beerenwinkel, N., Vinga, S.: Ensemble outlier detection and gene selection in triple-negative breast cancer data. BMC Bioinformatics (2018). https://doi.org/10.1186/s12859-018-2149-7
9. Nardi, A., Schemper, M.: New residuals for Cox regression and their application to outlier screening. Biometrics **55**(2), 523–529 (1999). http://www.jstor.org/stable/2533801
10. Peto, R., Peto, J.: Asymptotically efficient rank invariant test procedures. J. Roy. Stat. Soc.: Ser. A (Gen.) **135**(2), 185–207 (1972). http://www.jstor.org/stable/2344317
11. Storey, J.D.: A direct approach to false discovery rates. J. Roy. Stat. Soc. B **13**(2), 216–225 (2002)
12. Therneau, T., Grambsch, P.M., Fleming, T.R.: Martingale-based residuals for survival models. Biometrika **77**(1), 147–160 (1990). http://www.jstor.org/stable/2336057
13. Tibshirani, R.: Regression shrinkage and selection via the Lasso. J. Roy. Stat. Soc.: Ser. B **58**(1), 267–288 (1996)
14. Veríssimo, A., Oliveira, A.L., Sagot, M.F., Vinga, S.: DegreeCox - a network-based regularization method for survival analysis. BMC Bioinformatics **17**(16), 449 (2016). https://doi.org/10.1186/s12859-016-1310-4
15. Zou, H., Hastie, T.: Regularization and variable selection via the elastic net. J. Roy. Stat. Soc.: Ser. B **67**(2), 301–320 (2005)

Methods of Machine Learning
for Censored Demand Prediction

Evgeniy M. Ozhegov and Daria Teterina[✉]

National Research University Higher School of Economics, Perm 614070, Russia
tos600@gmail.com, dvteterina@gmail.com

Abstract. In this paper, we analyze a new approach for demand prediction in retail. One of the significant gaps in demand prediction by machine learning methods is the unaccounted sales data censorship. Econometric approaches to modeling censored demand are used to obtain consistent and unbiased estimates of parameters. These approaches can also be transferred to different classes of machine learning models to reduce the prediction error of sales volume. In this study we build two ensemble models to predict demand with and without demand censorship, aggregating predictions for machine learning methods such as Linear regression, Ridge regression, LASSO and Random forest. Having estimated the predictive properties of both models, we test the best predictive power of the models with accounting for the censored nature of demand.

Keywords: Demand · Censorship · Machine learning · Prediction

1 Introduction

The grocery retail market has been under the close scrutiny of economists over the past few decades. A surge of interest to this field occurred in the late 90's when the Nilson and IRI Marketing Research companies began to collect individual data on purchases of retail chains visitors [6]. Advances in individual data availability drew the researchers' attention to the methods of machine learning. Analysis of «big consumer data» revealed the huge potential of machine learning methods for working with massive data sets, both in terms of the number of observations and predictors [6]. A number of scientists, including [1,2,8] have shown greater predictive power of machine learning methods compared to traditional econometric approach. Therefore, today, when solving the problem of demand predicting, analysts' preference is often given to machine learning.

The publication was prepared within the framework of the Academic Fund Program at the National Research University Higher School of Economics (HSE) in 2018–2019 (grant No 18-01-0025 and by the Russian Academic Excellence Project "5-100").

G. Nicosia et al. (Eds.): LOD 2018, LNCS 11331, pp. 441–446, 2019.
https://doi.org/10.1007/978-3-030-13709-0_37

However, despite the significant breakthrough made by scientists in the demand prediction due to the methods of machine learning, there are still a lot of gaps, filling of which can improve the predictive quality of models. One of such white spots is demand censorship. Economists talk about censored demand or a corner solution in demand system when the number of product purchases desired by consumers on a certain price is negative, leading to significant share of zero observed sales [5]. To date, there are a number of works devoted to censored demand prediction using traditional econometric approaches (for example, [3,7]), as well as several studies on demand forecasting (without censorship) using machine learning methods (e.g. [9]), at the same time, there are no works that combine censorship and machine learning methods. Therefore, in our work, we fill this gap by constructing an ensemble model for censored demand prediction using machine learning methods and empirically check its better predictive properties on the data of the retail food chain.

In this paper we analyze the demand for one product category (pasta) on the purchases data provided by the Russian regional retail food chain. The sample size is 800000 daily observations for various brands of pasta. Since more than 60% of pasta daily sales are equal to zero, one needs to account for demand censorship.

We propose an estimator for demand prediction that allows to use the potential capacity of machine learning methods as well as to account for the data censorship. The estimator is based on the idea of combining several simple predictors into constrained linear ensemble models. Censoring accounting is carried out due to algorithm of separate prediction of censored and uncensored parts of demand applied for each of simple estimators. It should be noted that all censored models separately (Linear regression, Ridge regression, LASSO regression and Random Forest) have better predictive properties than the same models without censorship consideration; and the models combination via weighted linear regression, in turn, allows to improve the prediction accuracy even more. Thus, the prediction error for an ensemble model with censoring turned out to be equal to 1.11, while for the ensemble without censorship – 1.16. The developed algorithm can be applied for demand prediction in retail as well as in others spheres where the optimal inventory management and accurate prediction of sales volume are required.

2 Data

The study is conducted on the data, provided by the Russian regional grocery chain. Pasta product category is selected for analysis. The choice of pasta is justified by the high frequency of purchases of this product and the breadth of the product range. The initial data from the grocery chain sales represents the full information on the pasta purchases from 2009 to 2014. The size of the analyzed sample, formed on the basis of the initial data, is 800000 observations. An observation reflects a stock keeping unit (SKU) that was available in a certain store on a specific date. It is known how many units of a each SKU were purchased each day and at what price it were sold. More than 60% of sales were zero (See Fig. 1). This leads to the necessity of censorship accounting. In order to obtain better predictive quality of a demand model, we use the product catalog to recover product characteristics for each SKU. Thus, for each purchase we collect the colour and shape of pasta, the flour type, the volume and type of packaging, the origin country, the brand name. In addition to all of the above, for each observation we trace the format of the store where the purchase was made and promotion indicator.

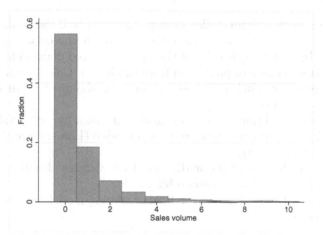

Fig. 1. Fraction histogram of pasta sales

3   Methodology

The general regression task is to predict sales volume of some product. In a linear regression form a model is as follows:

$$y_{jmt} = X_{jmt}\beta + \epsilon_{jmt} \tag{1}$$

where y_{jmt} is a volume of the j-th product sales in the store m on the day t, X_{jmt} is a matrix of attributes including log of the price, product characteristics, promotional indicators and time attributes (dummies for a month, a year, an intra-week seasonality and holidays), ϵ_{jmt} is an idiosyncratic shock to each product, market and time.

According to the literature ([2, 8, 9]), machine learning methods are better able to cope with demand prediction due to the better out-of-sample fits without loss of in-sample fit quality [2]. Therefore, to achieve the most accurate prediction, four machine learning methods are used in the research. In this study, we assume to partially follow the algorithm described in the [2] research, generalizing it by adding the stages of estimating censored models similar to [4]. The main steps of the empirical part of the study are as follow:

1. Split the data randomly into three groups for the subsequent double cross-validation, where 25% of the data falls into the test sample, 15% in the validation, and 60% in the training set.
2. Construct indicator variable $d_{jmt} = I\{y_{jmt} > 0\}$ for sales censorship.
3. Train a classification model for censorship dummy d using explanatory variables X.
4. Classify observation in a training set by probability threshold α into «censored» and «uncensored» ones.
5. Train a model for continuous («uncensored») part of train set splitted by a threshold α;
6. Combine predictions from models of steps (3) and (5). If the predicted dummy for censorship by classification model is 0 or prediction on a continuous part of demand by model (5) is below 0 then the predicted demand is 0, otherwise the prediction is equal to prediction from model (5). Calculate RMSE on test set for a given threshold α. Choose optimal threshold α to split by based on validation set RMSE;
7. Take prediction obtained from optimal threshold α on a validation set obtained from various classes of prediction models (Linear regression, LASSO, Ridge, Random Forest);
8. Train an ensemble model on predictions from various classes of models and obtain their weights for final ensemble model;
9. Calculate RMSE on a test set for final ensemble model and particular predictive models.

In the next part we compare models with and without censorship accounting. To train models without censorship accounting we treat all observations as uncensored, skip estimation steps (2–4) and treat optimal α as 0.

4 Results

Since more than 60% of sales are zero, we should check the parameter estimates for the need of use the censored regression model, testing for a bias in a simple linear regression framework (1) versus the censored regression model. The parameter estimates for this two specifications are presented in Table 1.

Due to the reported results, the effect of price in the model with censorship accounting is greater in absolute value. This supports the underestimation of the parameters estimates in the uncensored model. Moreover, censored linear model has better predictive properties in terms of out-of-sample RMSE. After evaluating the parameters of the basic linear model, the sales volume variable is fitted in the training set by four models (Linear regression, Ridge regression, Lasso regression and Random Forest). Then, for every model the measure of the prediction quality is calculated (Table 2).

Finally, models included in the ensemble with positive linear weights estimated by constrained linear model. The results of constrained linear regressions estimation for both ensembles, with and without censorship accounting, are presented in Table 2 as models weights. According to the estimation results, both

Table 1. Results for linear regression with and without censorship accounting.

Variable	Linear regression	Censored linear regression
Log. of price	−2.624***	−3.430***
	(0.014)	(0.020)
N	800000	800000
k	95	95
Test sample RMSE	1.256	1.220

Notes: Parameters estimates are presented in table cells, standard errors in parenthesis. Significance level is $p^{***} < 0.01$, N is the number of observations, k is the number of parameters. Brands, forms of pasta, country of origin, package type, colour of pasta, type of flour, time attributes (year, month, day of the week, holiday), store type are included in the model as control variables.

ensemble models with and without censorship accounting has better performance than any of the evaluated models individually. Moreover, the ensemble model accounting censorship of the data, has a better predictive power, which is indicated by the comparatively smaller RMSE.

Table 2. RMSE for models with and without censorship accounting.

Model	RMSE		Weight in ensemble	
	Without censorship accounting	With censorship accounting	Without censorship accounting	With censorship accounting
Linear regression	1.256	1.220	1%	3%
Ridge regression	1.255	1.218	13%	11%
Lasso regression	1.244	1.203	42%	39%
Random forest	1.198	1.164	44%	47%
Ensemble model	1.163	1.114		
t-stat = 3.22	p-value = 0.01			

Note: t-statistics and its p-value corresponds to the significance of difference between
RMSE in ensemble models with and without censorship accounting. Standard error
is calculated from bootstrap distribution of RMSE difference on 1000 replications.

5   Conclusion

The demand estimation in retail is quite developed in academic literature; nevertheless, there are still some gaps and contentious issues which generate debates among researchers. In particular, the potential of machine learning methods for censored demand prediction has not been researched so far. This paper fills this void by introducing new prediction algorithm dealing with censored demand. Having based on previous demand studies reporting that machine learning methods have more predictive power [2,8], and allowing for censorship of data leads to more consistent and less biased estimates [4,7], we propose an estimator for demand prediction that allows us to use the potential capacity of machine learning methods as well as to consider the data censorship. The research is based on the idea of comparing the prediction accuracy of machine learning models with and without censorship accounting and combining various estimators into constrained linear ensemble models.

According to the results obtain, two vital conclusions can be drawn: firstly, we showed the better quality of machine learning methods combination for solving the prediction problem in retail demand. Secondly, we proved better predictive properties of models that take into account censored nature of the retail data.

Since the research is conducted on the basis of real FMCG retail chain data, we can assert that the result obtained has practical significance for retailers. Thus, the results of the study can be used by the seller to establish the optimal price for goods with different characteristics and at various time periods, as well as for optimal inventory management.

References

1. Agrawal, D., Schorling, C.: Market share forecasting: an empirical comparison of artificial neural networks and multinomial logit model. J. Retail. **72**(4), 383–408 (1996)
2. Bajari, B., Nekipelov, D., Ryan, S., Yang, M.: Machine learning methods for demand estimation. Am. Econ. Rev. **105**(5), 481–485 (2015)
3. Chernozhukov, V., Fernandez-Val, I., Kowalski, A.E.: Quantile regression with censoring and endogeneity. J. Econom. **186**(1), 201–221 (2015)
4. Chernozhukov, V., Hong, H.: Three-step censored quantille regression and extramarital affairs. J. Am. Stat. Assoc. **97**(459), 872–882 (2002)
5. Ozhegov, E.M., Ozhegova, A.: Bagging prediction for censored data: application for theatre demand. In: van der Aalst, W.M.P., et al. (eds.) AIST 2017. LNCS, vol. 10716, pp. 197–209. Springer, Cham (2018). https://doi.org/10.1007/978-3-319-73013-4_18
6. Richards, T., Bonnet, C.: Models of consumer demand for differentiated products. TSE Working Papers, no. 16–741, Toulouse School of Economics (2016)
7. Tobin, J.: Estimation of relationships for limited dependent variables. Econometrica **26**(1), 24–36 (1958)
8. Varian, H.: Big data: new tricks for econometrics. J. Econ. Perspect. **28**(2), 3–27 (2014)
9. Witten, I., Frank, E., Hall, M., Pal, C.: Data Mining: Practical Machine Learning Tools and Techniques, 4th edn. Morgan Kaufmann Publishers Inc., Burlington (2016)

N-Gram Representation for Web Service Description Classification

Christian Sánchez-Sánchez[1(✉)] and Leonid B. Sheremetov[2]

[1] Universidad Autónoma Metropolitana, Unidad Cuajimalpa,
Mexico City, Mexico
csanchez@correo.cua.uam.mx
[2] Instituto Mexicano del Petróleo, Mexico City, Mexico
sher@imp.mx

Abstract. Despite increasing availability of Web Services (WS), their auto-matic processing (classification, grouping or composition) slows down because of the difficulty to read the WSDL service descriptions without related technical knowledge. Categorizing services for automatic service discovery and compo-sition has become a challenging problem. The paper argues that n-gram repre-sentation of the data extracted from the different sections of the WSDL description (types, messages and operations) along with the weighing scheme can benefit the classification of services. Experiments are carried out with three different classifiers over available collections of WS descriptions. It is shown that such representations as word bigrams or letter trigrams extracted from WSDL Operations and Types service description features with TF-IDF as n-gram weighting scheme, can improve automatic WS classification.

Keywords: N-gram representation · Web service classification ·
Term-weighting

1 Introduction

Nowadays, the use of Web Services (WS) becomes common due to their interoper-ability and reusability, allowing considerable cost reduction during software develop-ment phase. Some years ago, Ratnatsingam [1] warned that WS had become an evolutionary step in designing distributed applications. Despite the time already elapsed, many of the research problems mentioned in that paper, are still present. WS contain encapsulated descriptions of their functionality (i.e., operations, messages, data types and binding usually known as description features) defined as an abstract interface by means of using standard Web Services Description Language (WSDL). In spite of the fact that the WSDL description is a structured document, for a common Internet user it is hard to understand its content.

In order to locate a service, users appeal for using WS registries and repositories such as Universal Description Discovery and Integration (UDDI), however their search functionality still is relatively simple and fails to account for relationship between WS and users' real needs [2]. With the rapidly increasing number of services, the development of

© Springer Nature Switzerland AG 2019
G. Nicosia et al. (Eds.): LOD 2018, LNCS 11331, pp. 447–459, 2019.
https://doi.org/10.1007/978-3-030-13709-0_38

methods and tools helping in the process of WS categorization and search for their further automatic discovery and composition has become a challenge [2, 3].

In this paper, the effect of the n-gram representation of the data extracted from different sections of the WSDL service description (types, messages and operations) and term weighing scheme of the WS classification is explored. Three different representations are compared (letter n-grams, word n-grams and Bag of Words) using three different weighing schemes: Boolean, Term Frequency (TF) and Term Frequency-Inverse Document Frequency (TF-IDF) with the purpose of analyzing whether the representation influences classification results. For the WS classification three different classification algorithms: Naïve Bayes (NB), Support Vector Machines (SVM) and Decision Trees (DT) are tested. WSDL standard collections, namely OWLS-v3 (TC3), v4 (TC4) and ASSAM, are used for experiments.

The rest of the paper is organized as follows. Section 2 presents a related work concerning the WS classification. Section 3 describes how a WSDL document is formed and how it can be represented. Section 4 outlines the service categorization methodology. Section 5 describes used datasets, experimental settings and obtained results. Finally, Sect. 6 depicts conclusions and some future work directions.

2   Related Work

The common way to classify WS is selecting and assigning a category from the standard taxonomy called the United Nations Standard Products and Service Code (UNSPSC). In this taxonomy, it is only possible to assign a business category to a WS while many users also want to find information related to its functionality. To relate a service to a specific context, Wang et al. proposed to extract a subtree from the UNSPSC taxonomy, where the categories assigned to the WS are contained, to be treated as domain concepts [4]. In addition to the UNSPSC taxonomy, WordNet was used in that work to provide semantic similarity of concepts to weight the terms in the vector space model. A set of vectors was generated using all the concepts (representing the WS) for the training phase of the SVM algorithm. Once a model is obtained, it is used to classify new services.

Other researchers like Liang et al. [5], have used the UNSPSC to assign a class to a WS. For doing that, the terms contained in the metadata of the WSDL documents were used to generate a tree structure representation of the WS. After that, the underlying semantic relations among metadata structures, such as, terms co-occurrences of words taken from the input, output and function descriptions of the WSDL document and the taxonomy, were considered.

Another approach proposed by Yang and Zhou [6], is based on words extraction from the WSDL. In that paper, the OWLS-TC4 dataset was used along with an external resource to identify abbreviations, nouns and verbs. The pre-processing step included splitting, eliminating stop words and tags (web, service, input, output), and stemming. Four different classification algorithms such as Support Vector Machines (SVM), Naive Bayes (NB), Decision Tree (DT) C4.5 and Neural Networks were used. Only words from names of services, operations, inputs and outputs were extracted under a

TF-IDF weighting scheme. The best results for classification were obtained using output name words and applying the C4.5 DT algorithm.

In the paper by Nisa and Qamar [7], the authors extracted service name, service documentation, messages, ports and schema from WSDL documents in order to classify WS using text mining. Maximum entropy was used for WS classification, and a comparison of the accuracy through different categories was done. The best results were obtained with the WSDL Schemas information included. Classification results were improved when lemmatization and word splitter were applied to the extracted data at the preprocessing step. Unfortunately, the used dataset is not available for comparison.

Another approach was proposed by Saha et al. [8], were a WS representation based on a Tensor Space Model (TSM) was developed, in order to capture the internal structure of WSDL documents. The method consisted of selecting a set of relevant tags from a WSDL document. For each tag, a tensor was built using all words under that particular tag. Then a classification algorithm was applied for each tensor. Finally, all the information was combined using rough sets.

External resources to categorize WS are also frequently used to classify WS. Sharma et al. [9], proposed a classification approach based on the OWL-S semantic description as well as syntactic information presented within the service description by combining machine learning techniques (SVM and K-nearest neighbors), data mining, logical reasoning, statistical methods and measures of semantic relations. The authors reported a 97% accuracy using the extracted semantic information and applying the SVM classifier. Nevertheless, that approach cannot be widely used at the moment since there are few WS that are annotated semantically.

Qamar et al. [10] proposed an approach to categorize WS by employing the ensemble of three classifiers: NB, DT and SVM. First, the descriptions' pre-processing (word splitting and lemmatization) was used, followed by feature selection (stop and function Word removal), and majority vote based ensemble. An average accuracy of 92% was obtained over 3738 WS distributed over five context fields.

It should be noticed that the majority of the described approaches used the Bag of Words (BOW) representation and some of them use additional tools to add extra terms for constructing a more elaborated representation. On the contrary, the proposed approach is based on the WSDL descriptions analysis. It does not depend on any external resources during the classification phase (only WSDL extracted data were used) and extends the use of WSDL document representations.

3 Web Service Description Language (WSDL) and Representation

The WSDL provides a model and an XML format for describing WS [11]. Though this description is formed by different kinds of elements, the majority of them are included in four main components: Types, Messages, Operations and Binding (Fig. 1).

The main WSDL elements are [11]:

- Definition: The root element of all WSDL documents, which defines the name of the WS.
- Data types: The types to be used in the messages.
- Message: It is an abstract definition of the data presented either as an entire document or as arguments to be mapped to a method invocation.
- Operation: The abstract definition of the operation for a message that will accept and process the message.
- Port type: It is an abstract set of operations mapped to one or more end-points. End-points can be mapped to multiple transports through various bindings.
- Binding: The concrete protocol, data formats and messages defined for a particular port type.
- Port: It is a combination of a binding and a network address.
- Service: It is a collection of related end-points that map mainly the binding to the port.

```
▶ <wsdl:types>...</wsdl:types>
▶ <wsdl:message name="get_RECOMMENDEDPRICE_COLORResponse">...</wsdl:message>
▶ <wsdl:message name="get_RECOMMENDEDPRICE_COLORRequest">...</wsdl:message>
▶ <wsdl:portType name="AutoRecommendedpricecolorSoap">...</wsdl:portType>
▶ <wsdl:binding name="AutoRecommendedpricecolorSoapBinding" type="tns:AutoRecomm
▶ <wsdl:service name="AutoRecommendedpricecolorService">...</wsdl:service>
</wsdl:definitions>
```

Fig. 1. WSDL main components.

The content inside the elements and their attributes are commonly represented as strings. For example, the content inside the attribute name of the Operation element is shown in Fig. 2.

```
▼ <wsdl:portType name="AutoPriceSoap">
  ▼ <wsdl:operation name="get_PRICE">
      <wsdl:input message="tns:get_PRICERequest"></wsdl:input>
      <wsdl:output message="tns:get_PRICEResponse"></wsdl:output>
  </wsdl:operation>
```

Fig. 2. An excerpt of the content of an operation element

As we can see, the Operation contains the string "get_PRICE". To process the content, e.g. for automatic classification or grouping of services, this string can be extracted and presented in different ways. Three different ways of representing the information contained in the elements of the WSDL, are proposed:

1. Words (Bag of Words). This representation requires dividing the string into words, taking advantage of the fact that it is common for programmers to form names of

methods or parameters by concatenating words with uppercase letters, underscore or dot representing where each word starts. The words contained in the string can be separated by string pre-processing with the help of Wordnet. The string "get_-PRICE" gets the words "get" and "price". The state of the art approaches mostly use this representation.

2. Word n-grams. This representation is based on dividing the string into groups of n contiguous words. For example, the new string "get price" is obtained from the string "get_PRICE".

3. Letter n-grams. This representation consists of dividing the string into groups of n contiguous characters. For example, from the string "get_PRICE", the following trigrams are obtained: "get", "et_", "t_p", "_pr", "pri", "ric", "ice".

The first and the second representations are language dependent and require additional tools to identify the language and word recognition, which implies the use of more resources for pre-processing. The advantage of the third representation is that it is language independent because neither for extracting nor for comparing these kind of n-grams it is necessary to identify words.

4 WS Categorization Methodology

WS classification based on WSDL descriptions uses text categorization methodology, composed of several weighting schemes, representation methods and classification algorithms described below.

4.1 Documents' Classification and Weighting Schemes

Text categorization is the task of assigning a value to each pair $<d_j, c_i> \in D \times C$, where D is a domain of documents and $C = \{c_1, \ldots, c_{|C|}\}$ is a set of predefined categories. A classification algorithm is able to extract patterns from a training set, those patterns are useful to create a model (rule or hypothesis) to classify new documents [12]. Each document is represented usually as a vector d_j of weighted terms, where each vector component (dimension) represents a data (term) t_k extracted from the documents. The vector's dimension T is defined by all the data extracted from all the documents.

For each document represented as a vector $d_j < w(t_{1j}), \ldots, w(t_{|T|j}) >$, a weight w (t_{ij}) is assigned to each vector component. There are several proposals for computing the weight $w(t_{ij})$ of each term (i.e., the importance of each term/word). Among the most successful weighting strategies are: the Boolean weight, term frequency (TF) and relative term frequency (TF-IDF) [13]:

1. Boolean weighing:

$$w(t_{kj}) = \begin{cases} 1 \ if \ tf_{kj} > 0 \\ 0 \ otherwise \end{cases}, \tag{1}$$

where tf_{kj} is the number of times the term t_k appears within the document d_j.

2. Term Frequency Weighting:

$$w(t_{kj}) = \log(1 + tf_{kj}) \qquad (2)$$

3. Relative Term Frequency Weighting:

$$w(t_{kj}) = TF * \log\left(\frac{|D|}{\{d_j \in D : t_{kj} \in d_j\}}\right), \qquad (3)$$

where $|D|$ is the number of documents within the dataset.

For the experiments, the data extracted from WSDL documents, were configured based on the following representations: Words (BOW), Word n-grams and Letter n-grams. It is important to explain the reason why these representations were used. BOW has been a traditional form for representing documents and it was the most used representation in reviewed related works [12]. Word n-grams allow to store words that occur together in a sequence. In the approach proposed by Flores [14], trigrams (letter n-grams) worked well to find source code plagiarism. WSDL element content (operations, messages and types) is written similarly to the conventions used to name methods and parameters within the analyzed source code, so the same n-gram configuration was tested.

4.2   Classification Algorithms

In order to analyze the effect of the n-gram representation and weighing scheme in the classification, three different classification algorithms were tested:

- Naïve Bayes is a classification algorithm based on the Bayes theorem, where the independence of the predictors is assumed.
- Sequential Minimal Optimization for SVM is a simple algorithm that solves the problem of quadratic programming. Support Vector Machines (SVM) require the quadratic programming solution that can be solved by means of this technique, which decomposes the problem into smaller problems that are solved analytically.
- Decision Trees is an algorithm that creates a model in the form of a tree. It works by breaking the data set into smaller subsets, by which it evaluates the attributes (predictors) according to the information gain to form a tree.

The following metrics are commonly used in Information Retrieval to evaluate the classification;

- Accuracy is the ratio between the number of relevant documents and the number of retrieved documents.
- Recall is the ratio between the number of relevant documents obtained compared to all relevant documents.

- F-measure (score) is the measure used to obtain a weighted single value of the Accuracy and the Recall:

$$F - measure = 2\left(\frac{Accuracy * Recall}{Accuracy + Recall}\right) \tag{4}$$

5 Experiment Settings and Results

This paper explores different representations of the information extracted from the WSDL elements or features (Types, Messages and PortType-Operations) individually and in combination for the purpose of their suitability for the classification of WS. The following sub-sections describe the datasets, experiments setup and obtained results using the classifiers described above.

5.1 Datasets Description

The experiments were carried out on 3 different collections of WS:

1. The first collection (ASSAM), which contains WS descriptions extracted from Xmethods and Salcentral, was taken from the work reported by Hess et al. [15]. The collection includes 26 main classes, making a total of 814 WS descriptions. To provide the experiments, some modifications were made to the original collection. For example, the "unlabelled" class that contains uncategorized descriptions was deleted, the classes were flattened, that is, the elements of the subclasses were passed to the main class and classes that had very few elements were eliminated. The sub-collection used for experiments, included 9 classes, which contain 386 documents, ranging from 23 to 64 documents in each.
2. The other collection was that used by Klusch et al. [16]. It is important to mention that the second dataset has different available versions, in this paper version 3 (TC3) and version (TC4) were used. The TC3 version is comprised of 7 classes and a total of 1006 WSDL documents, ranging from 34 to 355 documents in each. While the TC4 version is conformed of 9 classes and a total of 1082 WSDL documents, from 16 to 355 in each class.

5.2 Experimental Setup

At the first phase of experiments, the information of the following features was extracted from each WSDL:

- Operations (Ops), which involved extracting data from the XML-elements: port-Type, operation and service of its XML-attribute called Name.
- The messages (Msgs) that involved extracting data contained in: the message XML-elements of its XML-attributes and the XML-element Part of its Name and Type XML-attributes.

- Data types (Types), which involved extracting data from the XML-elements contained in the Complex and Simple Types, where the strings contained in its XML-attributes (Name and Type) were extracted for each element.

The data extracted from the features were the following: letter n-grams, BOW and word n-grams. Each of the extractions served to form the different data sets and respective configurations described below. It is worth to mention that the following combinations of the attributes were also tested: Operations and Messages (OpsMsgs), Operations and Types (OpsTypes) and Operations, Messages and Types (OpsMsgsTypes).

At the second phase, a pre-processing for each type of extracted data was carried out as described below.

- Words (BOW) preprocessing: the splitting of strings into words was done by separation by first capital letter or by punctuation marks. Numbers and special characters (%, #, @ and so on) were eliminated, while capital letters were transformed into lowercase. Once the strings were separated, the WordNet tool [17] was used to identify words.
- Word n-grams preprocessing: a preprocessing was performed similar to that of the words but instead of storing the different words, bigrams of words were taken.
- Letter n-grams preprocessing: the same preprocessing of the strings was done to later split them into trigrams. The reason for selecting trigrams was a configuration that had given good results to detect plagiarism [14], which used names of methods and parameters specified in a similar way to those of the WS.

The obtained sets of data are summarized in Table 1.

Table 1. Attribute configurations for different representations

	Ops	Msgs	Types	OpsMsgs	OpsTypes	OpsTypesMsgs
Words	W1ij	W2ij	W3ij	W4ij	W5ij	W6ij
N-grams of letters	NL1ij	NL2ij	NL3ij	NL4ij	NL5ij	NL6ij
N-grams of words	NW1ij	NW2ij	NW3ij	NW4ij	NW5ij	NW6ij

In Table 1, the index i represents a version for each WSDL collection, which the information was obtained from (ASSAM, TC3 and TC4), while the index j refers to a weighing scheme (Boolean, TF or TF-IDF). For example, if we consider the W1TC3Boolean collection (W_{1ij} configuration), we can say that it is the dataset that has, for each WSDL from the TC3 collection, a vector whose attributes are the words and the values of the each vector elements are: 1 if the word exists in the operations, content within the WSDL, and 0 otherwise.

In total 162 different samples (6 configurations * 3 representations * 3 weighting schemes * 3 datasets) were classified.

WEKA implementation of Naïve Bayes, SMO y Decision Tree (J48) classification algorithms for data mining tasks with default settings were used for experiments [18].

Ten folds cross validation was used to evaluate classification algorithms over 162 dataset samples. For each experiment consisting of a classification algorithm (Sequential Minimal Optimization, Decision trees, and Naïve Bayes) and particular representation (letter n-grams, word n-grams and BOW), accuracy, recall and F-measure (reported in the following sub-sections) were obtained. It is important to mention that BOW, as the state-of-the-art most used representation, is used in the following subsections as a reference to compare the competitiveness of the results obtained using n-grams representations.

5.3   WS Classification Results Using the SMO Classifier

Tables 2 and 3 resume the experimental results (F-score) of the SMO classifier obtained for n-grams of letters and words respectively for experimental datasets.

As it can be seen from Table 2, for ASSAM dataset the highest F-measure (0.0565) was obtained by extracting n-grams of letters from Operations and Types (OperationsTypes) contents with the TF-IDF weighing scheme. It is important to say that in the sub-collection used, there are classes that contain elements that overlap.

Table 2. F-measure for the SMO classifier for n-grams of letters (NL) representation of experimental datasets

	ASSAM-NL		TC3-NL		TC4-NL	
Attributes	TF	TF-IDF	TF	TF-IDF	TF	TF-IDF
Msgs	0.444	0.468	0.566	0.559	0.538	0.537
Operations	0.468	0.499	0.809	0.813	0.799	0.795
Types	0.467	0.48	0.959	0.959	0.955	0.957
OperationsMsgs	0.474	0.493	0.812	0.814	0.796	0.807
OperationsTypes	0.530	**0.565**	0.96	**0.963**	0.961	0.962
OperationsMsgsTypes	0.526	0.558	0.961	0.958	0.962	**0.965**

Table 3. F-measure for the SMO classifier for n-grams of words (NW) representation of experimental datasets

	TC3-NW		TC4-NW	
Attributes	TF	TF-IDF	TF	TF-IDF
Msgs	0.567	0.565	0.541	0.539
Operations	0.805	0.796	0.788	0.774
Types	0.956	0.959	0.954	0.955
OperationsMsgs	0.809	0.809	0.791	0.791
OperationsTypes	0.961	0.963	0.964	**0.966**
OperationsMsgsTypes	0.964	**0.965**	0.963	0.963

For TC3 dataset, word n-grams extracted from Operations, Messages and Types (OperationsMsgsTypes) with a TF-IDF weighing scheme obtained the highest F-measure of 0.965 (Table 3). However, after applying a corrected T-Test, no significance was found with respect to extracting only the word n-grams from OperationsTypes under the same weighing scheme. Similarly, no statistical significance was found with the results obtained from extracting the word n-grams from OperationsMsgsTypes and OperationsTypes under a TF weighing scheme. Comparing these results with the obtained using letter n-grams for the OperationsTypes (under TF or TF-IDF), no statistical significance was found, and apparently word bigrams and letter trigrams were equally useful. Something similar to the previous case, regarding word bigrams and letter trigrams, happened for the TC4 dataset.

In other words, for the three datasets acceptable results were obtained by extracting letter n-grams from OperationsTypes under the TF-IDF weighing scheme. One of the advantages of this type of n-grams is that they are language independent. The drawback is that more trigrams, than words, are obtained. For TC3 and TC4 datasets, the word n-grams also worked under the TF-IDF weighing scheme where the information from Operations and Types is combined. It is important to consider that this scheme requires more resources in pre-processing, including a words identification tool. During the classification using BOW representation with the data extracted from OperationsMsgsTypes, OperationTypes and Types, similar results were obtained (F-measure close to 0.95). Apparently, using BOW representations, the weighting scheme didn't influenced the results, probably because the difference of word frequencies was not significant. For this experiment, it can be said that using n-gram representation the obtained F value was higher (see Table 3).

5.4 WS Classification Results Using the Decision Trees Classifier

The following results were obtained with the DT classifier (Table 4). For the ASSAM data collection, the highest F-measure was 0.466 through the OperationsMsgsTypes representation and the TF-IDF weighing scheme, however no statistical significance was found with the result of only selecting OperationsTypes.

Table 4. F-measure for the DT classifier for n-grams of letters (NL) and words (NW) representations of experimental datasets

	ASSAM-NL		TC3-NW		TC4-NW	
Attributes	TF	TF-IDF	TF	TF-IDF	TF	TF-IDF
Msgs	0.376	0.367	0.557	0.548	0.523	0.517
Operations	0.410	0.416	0.762	0.637	0.742	0.623
Types	0.392	0.429	0.922	0.925	0.916	0.925
OperationsMsgs	0.410	0.397	0.77	0.773	0.753	0.752
OperationsTypes	0.435	0.461	0.919	**0.932**	0.915	**0.93**
OperationsMsgsTypes	0.420	**0.466**	**0.932**	0.925	0.928	0.929

For the TC3 collection the highest results were obtained extracting word n-grams from OperationTypes and OperationMsgsTypes under TF-IDF and TF weighted schemes respectively. In this case, representation of the letter n-gram closest to the highest value of F was OperationMsgsTypes with a TF weighing scheme with an F-measure of 0.926. Whilst, it was obtained the same value of F equal to 0.926 using BOW OperationTypes under TF weighing.

For the TC4 collection, the highest values of F (0.93) were obtained under the words n-gram representation with the content of OperationTypes under the TF-IDF weighing scheme (See Table 4). For the letter n-gram representation, the highest value was obtained with the contents of Types under the TF weighing scheme with an F-measure of 0.914. Also a similar F-measure of 0.929 was obtained for BOW and the word n-gram representations of OperationMsgsTypes.

5.5   WS Classification Results Using the Naïve Bayes Classifier

For the case of the Naïve Bayes classifier, the highest value of F (0.539) for ASSAM collection was obtained under the letter n-grams representation with the contents of Operations and Types under a Boolean weighing scheme (see Table 5).

For the TC3 collection, the highest F value was obtained using word n-gram of OperationMsgsTypes under a TF weighing scheme. Testing letter n-gram representation (not shown in Table 5), the highest value of F (0.889) was obtained using OperationsMsgsTypes under a TF-IDF weighing scheme. However, with the BOW representation using the Types content with a Boolean weighing scheme, F-measure of 0.912 was obtained.

Table 5. F-measure for the Naïve Bayes classifier for n-grams of letters (NL) and words (NW) representations of experimental datasets

	ASSAM-NL		TC3-NW		TC4-NL	
Attributes	TF	Boolean	TF	TF-IDF	TF	TF-IDF
Msgs	0.397	0.418	0.457	0.535	0.442	0.502
Operations	0.485	0.525	0.61	0.629	0.613	0.634
Types	0.449	0.498	0.891	0.87	0.901	0.881
OperationsMsgs	0.452	0.501	0.601	0.601	0.6	0.601
OperationsTypes	0.501	**0.539**	0.915	0.894	0.911	0.89
OperationsMsgsTypes	0.512	0.537	**0.938**	0.922	**0.939**	**0.922**

For the TC4 collection, the highest F-measure of 0.939 was obtained using OperationMsgsTypes with a TF weighted scheme. Using BOW representation (under Boolean weighting scheme), the F-measures of 0.912 for the contents of Types and OperationsMsgsTypes and of 0.911 for OperationsTypes combination were obtained.

6   Conclusions and Future Work

In this paper, n-gram representations through the use of WSDL features, their combinations and weighing schemes were compared in terms of their usefulness for classification of WS descriptions. The main contribution is the analysis of how the word and letter n-gram representations, which are not commonly used, influence classification performance without using any information external to WSDL descriptions. The studied representations include: (a) feature selection, based on WSDL operations and types data obtained from all WSDL data, (b) data extraction from WSDL, as word bigrams or letter trigrams format and using (c) TF-IDF as n-gram weighting scheme. This configuration (in most of the cases) obtained competitive results through all the experiments.

Apparently, the Types attribute turned out to be very important to classify services; just taking the content of the Msgs attribute is not so effective. The best results were obtained using the SMO classification algorithm for SVM. For example, using Word Bigrams extracted from Operations and Types (over TC3 dataset) with a TF-IDF weighting scheme, F-measure of 0.963 was achieved that is a competitive performance against the results reported in [4] and the results obtained using BOW representation.

On the other hand, the language independent representation that obtained acceptable results in most cases, was a letter trigram also using a combination of Operations and Types features. The F-measure of 0.926 was obtained classifying TC3 dataset, it also outperforms the results reported in [4].

Weighing scheme with the highest value of F varied among collections, that is why, a selection of terms among the attributes is proposed for the future work to evaluate, which weighing scheme improves the results. Other types of representations, e.g. Distributional Representations like term co-occurrence representation (TCOR), allowing working with low frequencies of terms and ambiguity, along with different configurations of the classification algorithms should also be studied.

References

1. Ratnasingam, P.: The importance of technology trust in web services security. Inf. Manag. Comput. Secur. **10**(5), 255–260 (2002)
2. Batra, S., Bawa, S.: Web service categorization using normalized similarity score. Int. J. Comput. Theory Eng. **2**(1), 139–141 (2010)
3. Balasubramanian, D.L., Murugaiyan, S.R., Sambasivam, G., Vengattaraman, T., Dhavachelvan, P.: Semantic web service clustering using concept lattice: multi agent based approach. Int. J. Eng. Technol. **5**(5), 3699–3714 (2013)
4. Wang, H., Shi, Y., Zhou, X., Zhou, Q., Shao, S., Bouguettaya, A.: Web service classification using support vector machine. In: 22nd IEEE International Conference on Tools with Artificial Intelligence (ICTAI), vol. 1, pp. 3–6. IEEE Computer Society, Arras (2010)
5. Liang, Q., Li, P., Hung, P., Wu, X.: Clustering web services for automatic categorization. In: 2009 IEEE International Conference on Services Computing, SCC 2009, pp. 380–387. IEEE Computer Society, Bangalore (2009)
6. Yang, J., Zhou, X.: Semi-automatic web service classification using machine learning. Int. J. u-and e-Serv. Sci. Technol. **8**(4), 339–348 (2015)

7. Nisa, R., Qamar, U.: A text mining based approach for web service classification. Inf. Syst. e-Bus. Manag. **13**(4), 751–768 (2015)
8. Saha, S., Murthy, C.A., Pal, S.K.: Classification of web services using tensor space model and rough ensemble classifier. In: An, A., Matwin, S., Ras, Z.W., Slezak, D. (eds.) ISMIS. LNCS, vol. 4994, pp. 508–513. Springer, Heidelberg (2008). https://doi.org/10.1007/978-3-540-68123-6_55
9. Sharma, S., Lather, J.S., Dave, M.: Semantic approach for Web service classification using machine learning and measures of semantic relatedness. Serv. Oriented Comput. Appl. **10** (3), 221–231 (2016)
10. Qamar, U., Niza, R., Bashir, S., Khan, F.H.: A majority vote based classifier ensemble for web service classification. Bus. Inf. Syst. Eng. **58**(4), 249–259 (2016)
11. WSDL. https://www.w3.org/TR/2007/REC-wsdl20-20070626/. Accessed 30 Mar 2018
12. Sebastiani, F.: Machine learning in automated text categorization. ACM Comput. Surv. (CSUR) **34**(1), 1–47 (2002)
13. Salton, G., Buckley, C.: Term-weighting approaches in automatic text retrieval. Inf. Process. Manag. **24**(5), 513–523 (1988)
14. Flores, E.: Reutilización de código fuente entre lenguajes de programación. Master's thesis, Universidad Politécnica de Valencia, Valencia, España, February 2012
15. Hess, A., Johnston, E., Kushmerick, N.: Machine learning techniques for annotating semantic web services. Citeseer (2005)
16. Klusch, M., Fries, B., Sycara, K.: OWLS-MX: a hybrid semantic web service matchmaker for OWL-S services. Int. J. Web Semant. **7**(2), 121–133 (2009)
17. Miller, G.A., Beckwith, R., Fellbaum, C.D., Gross, D., Miller, K.: WordNet: an online lexical database. Int. J. Lexicogr. **3**(4), 235–244 (1990)
18. Frank, E., Hall, M.A., Witten, I.H.: Data Mining: Practical Machine Learning Tools and Techniques, 4th edn. Morgan Kaufmann, Los Altos (2016)

Lookahead Policy and Genetic Algorithm for Solving Nurse Rostering Problems

Peng Shi[✉] and Dario Landa-Silva

School of Computer Science, ASAP Research Group, University of Nottingham,
Nottingham, UK
{peng.shi,dario.landasilva}@nottingham.ac.uk

Abstract. Previous research has shown that value function approximation in dynamic programming does not perform too well when tackling difficult combinatorial optimisation problems such as multi-stage nurse rostering. This is because the large action space that needs to be explored. This paper proposes to replace the value function approximation with a genetic algorithm in order to generate solutions for the dynamic programming stages. Then, the paper proposes a hybrid approach that generates sets of weekly rosters with a genetic algorithm for consideration by the lookahead procedure that assembles a solution for the whole planning horizon of several weeks. Results indicate that this hybrid between a genetic algorithm and the lookahead policy mechanism from dynamic programming exhibits a more competitive performance than the value function approximation dynamic programming investigated before. Results also show that the proposed algorithm ranks well in respect of several other algorithms applied to the same set of problem instances. The intended contribution of this paper is towards a better understanding of how to successfully apply dynamic programming mechanisms to tackle difficult combinatorial optimisation problems.

Keywords: Hybrid algorithm · Genetic algorithm ·
Lookahead policy evaluation · Dynamic programming ·
Nurse rostering problem

1 Introduction

Dynamic programming (DP) is a divide-and-conquer optimisation approach in which a problem is solved by splitting it into a set of sub-problems. The solution to each sub-problem is recorded in case the same sub-problem is faced later in the search. However, as the size of the input problem increases, the split can result in a large number of sub-problems. This means that implementations of dynamic programming require large memory to store information about solved sub-problems and long computation time to evaluate solutions. This is usually called the *curse of dimensionality* in the dynamic programming algorithms. To make the search more efficient, Approximate Dynamic Programming (ADP)

© Springer Nature Switzerland AG 2019
G. Nicosia et al. (Eds.): LOD 2018, LNCS 11331, pp. 460–471, 2019.
https://doi.org/10.1007/978-3-030-13709-0_39

considers only a small part of the search space based on the use of approximation functions [1]. Solutions obtained by ADP are expected to be close to optimality while using shorter computational time than DP.

Nurse rostering is a difficult combinatorial optimisation problem for which many solution techniques have been proposed in the literature [2,3]. In our previous research, the suitability of ADP to solve the Nurse Rostering Problem (NRP) was investigated by approaching NRP as a Markov Decision Process [4]. The approximation function focused on selecting actions that satisfy the principle of optimality [5] but not all were covered. That approach was evaluated using a subset of problem instances from the Nurse Scheduling Problem Library (NSPLib) [6]. Experimental results indicated that the performance of the implemented ADP was competitive with various heuristic algorithms from the literature. However, the performance of that ADP algorithm was not very good when tackling the multi-stage NRP proposed as part of the Second International Nurse Rostering Competition (INRC-II). In the *single-stage NRP*, all information about the weekly staffing requirements is known in advance, and then a schedule for the full planning horizon (several weeks) is produced. In the *multi-stage NRP*, the staffing requirements for future weeks are unknown when solving each week, and then a schedule is produced for one week at a time, hence the schedule for each week has an effect on the scheduling of future weeks. An ADP approach that incorporates a combined policy function for solving the multi-stage NRP was proposed later [7]. Experimental results showed an improved performance on tackling problem instances with 4 or 8 weeks planning horizon. However the computational time for solving each instance is longer than the other approaches (all of them heuristics) from the competition.

It has been observed that more than 60% of the computational time spent by our latest ADP implementation is used to produce the solutions in each stage. This has been the motivation for developing an improved way to generate good solutions but in considerably shorter time. Then, in the present paper a population-based optimisation technique, namely a Genetic Algorithm (GA), is implemented to replace the value function approximation used in our previous work. A GA is a heuristic approach that evolves a population of solutions using crossover and mutation operators [8]. The resulting technique is a hybrid method that uses the GA to produce a pool of solutions in each stage (week roster) and the lookahead policy selects the most promising candidate solution for each stage in order to construct a schedule for the whole planning period.

The combination of dynamic programming and GAs has been investigated before in the literature. Early works such as [9] proposed dynamic programming to produce new solutions after the crossover operation in a GA. The rationale for that methodology is the assumption that good solutions tend to have a lot of common in their structure. Then, the common genes between two offspring solutions after crossover were identified and dynamic programming was then applied to produce a new solution based on this common structure. The solution produced in this way was then passed to the next generation in the GA. In a more recent work, dynamic programming was used to evaluate the fitness value of chromosomes when solving a bi-objective cell formation problem [10].

Most hybrid algorithms combining dynamic programming and GAs in the literature follow the design of a GA as the driving technique and then dynamic programming is used to evaluate part of the procedure. The design proposed in this paper is different because the whole methodology is driven by the dynamic programming paradigm and the GA is used to tackle the sub-problems. That is, the GA generates solutions for the weekly problem and the lookahead policy evaluates the effect of those solutions on the future stages of the problem. The best schedule generated by the GA for a given week is usually not the best to guarantee the best overall solution. The power of the proposed approach is precisely in the GA producing a set of solutions from which the lookahead policy can choose the most suitable to construct a full schedule of the best quality. Details of the proposed hybrid algorithm are given in Sect. 2. Section 3 describes the experimental settings and results. Section 4 concludes the paper and outlines future work.

2   Overview of the Hybrid Algorithm

2.1   Proposed Hybrid Algorithm

Function (1) represents the general procedure of dynamic programming for solving a multi-stage optimisation problem M. In this function, T represents the number of stages to solve M. The requirements of problem stage M_t, and the pre-condition information ν_t, are the input for $F(.)$ to obtain stage solutions where ν_t is a representation of all solutions explored before stage t, $\overline{V}$ is a fitness function and s_t is an individual stage solution. Once stage problem M_t is solved, s_t will be transferred into ν_{t+1} as a new pre-condition information for the next stage. A solution of M is a combination of one s_t at each stage and the objective is to obtain the one with minimum overall cost.

$$V(M) = \min \sum_{t=1}^{T} \overline{V}(s_t | s_t \in F(M_t, \nu_t)) \tag{1}$$

A similar procedure to the one described above can be implemented to tackle the multi-stage nurse rostering problem in this paper. T is the number of weeks or stages in the rostering problem. Stage problem M_t can be seen as a single-stage nurse rostering problem and the aim is to produce a schedule that satisfies weekly constraints. ν_t is a schedule comprising the individual solutions for all previous stages (weeks). s_t is a weekly schedule and the fitness value $\overline{V}(s_t)$ gives the quality (constraint violations) of s_t.

However, since nurse rostering problem is an NP-hard combinatorial optimisation problem, applying dynamic programming to solve it demands huge computational effort. In order to address this issue, an approximation function can be applied to obtain a solution s_t that is not only a good solution to the current stage problem but it is also good considering the following stages. This is the basis for the proposed hybrid algorithm. Solution s_t is obtained by the genetic algorithm and the future effect of this solution on the following stages of the problem is evaluated through a lookahead procedure. The overall framework of this hybrid algorithm is exhibited in Algorithm 1. The following subsections explain the algorithm in detail.

Algorithm 1. Hybrid of Lookahead Policy and Genetic Algorithm

1: Initialise population C
2: $\forall c \in C$, calculate $CV(c)$
3: **while** stopping criteria not reached **do**
4:      **for** every selected parents (c_1, c_2) **do**
5:          $(ch_1, ch_2) = c_1 \oplus c_2$
6:          $ch_1' = Mutation(ch_1)$
7:          $ch_2' = Mutation(ch_2)$
8:          Calculate $CV(ch_1')$ and $CV(ch_2')$
9:          $Replace(C, c_1, c_2, ch_1', ch_2')$
10: Initialise $LK(C) = 0$
11: **for** each $c \in C$ **do**
12:     $\{Sol_1, \cdots, Sol_{pe}\} = Simulate(c)$
13:     $LK(c) = \sum CV(Sol_1) + \cdots + CV(Sol_{pe})$
14:     $V(c) = CV(c) + LK(c)$
15: Return $argmin_{c \in C} V(c)$

2.2   Genetic Algorithm Component

The GA is in steps 1–9 of Algorithm 1 and its output is a population of solutions C. A chromosome $c \in C$ represents a weekly schedule using an indirect encoding. The length of c is the number of nurses and each gene is an index indicating the valid shift pattern assigned to the corresponding nurse. A valid shift pattern (*vsp*) is a pre-constructed feasible (satisfies hard constraints) nurse's weekly roster. Nurses may have different individual requirements hence the number of *vsp* could be different for different nurses. As part of our approach, we build a set of *vsp* for each nurse (this procedure is from our previous work [7]). A full weekly schedule is decoded from the chromosome based on this set of *vsp*. An example of this encoding and decoding scheme is shown in Fig. 1 with 3 nurses and 2 shifts. In this example, E and L is an abbreviation for early and late shift respectively, and empty blocks indicate a day-off.

At the start of the GA in Algorithm 1, the initial population C is constructed randomly and the constraints violations value $CV(c)$ for each individual in the population is calculated. Note that later in step 14 of the algorithm, the fitness value $V(C)$ for each solution is given by the sum of the corresponding constraint violations $CV(c)$ from the GA phase and the future estimation $LK(c)$ from the lookahead phase.

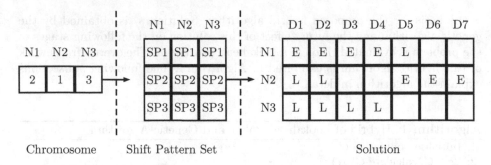

Fig. 1. Example of the indirect chromosome encoding used in the GA component.

In steps 4–9 of Algorithm 1, a number of generations are executed where the population is evolved towards better solutions. The GA uses the three typical operators to generate new solutions or offspring: *Selection, Crossover* and *Mutation*. The *Selection* operator implemented here chooses parents through an elitist-tournament selection procedure that works as follows. All chromosomes are sorted in a non-increasing order of their fitness value. The best chromosome is saved for the next generation (this is the elitist mechanism). Then, a double-elimination tournament as illustrated in Fig. 2 is used to select two parents. Tournament selection is widely used in the implementation of GAs because it applies selection pressure to keep the best individuals while also promoting diversity in the chromosomes for the next generation. With this selection approach half of the current population is selected for the following operations in the GA.

Once the two parents are selected as described above, two offspring ch_1 and ch_2 are produced by applying the crossover operator $\oplus$ which combines genes from the two parents. The widely used uniform crossover operator is implemented here [8]. In this operator each gene for the offspring is chosen at random from the two corresponding genes in the parents.

The mutation operator is then applied with some probability (mutation rate) to the generated offspring. The aim of the mutation operator is to maintain the diversity in the population. The mutation operator works on a chromosome gene by gene. A commonly used mutation operator is a *swap* that exchanges the content between two genes in the chromosome. The mutation operator implemented here is a neighbourhood-swap. The values of two consecutive genes b_i and b_{i+1} are exchanged. For the last gene in the chromosome, the swap is made with the first gene in the chromosome. For example, an offspring $ch_1 = \{3, 7, 2, 11, 5, 1\}$ will result in offspring $ch_1' = \{7, 3, 11, 5, 2, 1\}$ after 3 *swap* operations. Each gene in a chromosome is an integer value representing a valid shift in the nurse's *vsp*. Since nurses could have different *vsp* size, it is possible that the mutated offspring is infeasible. In the example above, the value in the third gene of ch_1 changed from 2 to 11 after mutation. This would be infeasible if the third nurse has only 8 valid shift patterns for example. Hence, a simple repair is implemented where

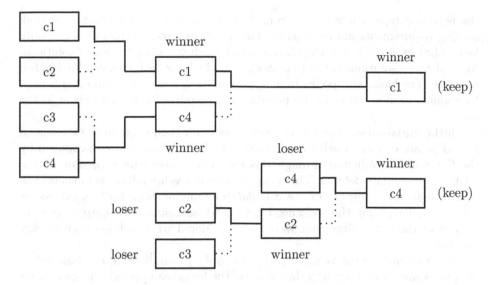

Fig. 2. Selection by double-elimination tournament where each C represents an individual chromosome.

a gene is assigned a random valid value (no larger than vsp) if the mutation resulted in an infeasible shift assignment. The full mutation procedure is shown in Algorithm 2.

Algorithm 2. Neighbourhood-Swap Operator

---
for every b_i in chromosome c **do**
   if probabilty of mutation is met **then**
      Swap(b_i, b_{i+1}), $i < length(c)$ or Swap(b_i, b_1), otherwise.
      if $b_i < vsp^i_{max}$ **then**
         $b_i = Random(vsp^i_{max})$

After the mutation process in complete, the constraint violations value CV is calculated for each offspring. Then, a *Replace* procedure takes place where the new offspring is added to the population replacing the parents.

Two stopping criteria are used here and the GA terminates once any of them is satisfied. One stopping criterion is the maximum number of generations and the other one is that the best chromosome so far has not changed after a number of generations.

2.3   Lookahead Policy Evaluation

This component is in steps 10–14 of Algorithm 1. In the multi-stage NRP, the best solution Sol_{best} produced by the GA in a stage is not guaranteed to be the best weekly schedule for the complete overall roster, once the future week staffing requirements are considered. This is because some constraints can only be checked until the last stage. Since the GA produces a population of solutions, some of those solutions other than Sol_{best} might be a better choice for the full schedule. The lookahead policy from approximate dynamic programming is used to evaluate each solution in the population through a lookahead period in the future.

In the initialisation, the future requirements for each stage in the lookahead period pe are defined. Each chromosome c in the final population produced by the GA will be evaluated through this lookahead procedure according to the future requirements defined. The future estimation value $LK(c)$ is initialised to 0 in step 10. The purpose of the $Simulate(c)$ function is to build a full roster $\{Sol_1, \cdots, Sol_{pe}\}$ for the lookahead period pe assessing each chromosome in respect of the constraints that were not considered when solving each weekly problem.

A full simulation solution set $\{Sol_1, \cdots, Sol_{pe}\}$ of individual c is built when the procedure terminates in the last stage of the lookahead period. The constraint violations of each single solution in this set will be calculated and updated in $LK(c)$. The fitness value $V(c)$ is then the sum of $LK(c)$ and $CV(c)$, note that $V(c)$ is to be minimised. The chromosome c with the lowest $V(c)$ is the final output of the whole algorithm and the decoded solution is recorded for the next solving stage.

3   Experiments and Results Analysis

In this section we present experiments to assess the performance of the proposed hybrid approach. The selected problem instances are described in Subsect. 3.1. Experimental settings for generating results are given in Subsect. 3.2. Subsection 3.3 compares the performance of the proposed approach to our previous method. Full experimental results are discussed in Subsect. 3.4. The proposed hybrid algorithm described in Sect. 2 was implemented in Java (JDK 1.7) and all computations were performed on an Intel (R) Core (TM) i7 CPU with 3.2 GHz and 6 GB of RAM.

3.1   Problem Instances

The problem instances used were selected from the Second International Nurse Rostering Competition [11]. Three types of instances are available defined by a set of files, scenario file, week data files and initial history files. The scenario file provides scenario information and requirements for the whole planning horizon. There are 10 week data files that define the specific requirement of each week. There are 4 initial history files that define the constraints for the rostering of the first week. With these files, a variety of problem instances with different planning horizons and conditions can be produced. For the aforementioned competition, a set of instances was provided to compare the various proposed approaches.

Even after the competition, that set of problem instances continues to be used by researchers as a benchmark to test algorithms for the NRP. In the set of instances used here, the planning horizon is either 4 or 8 weeks with the number of nurses ranging from 5 up to 100. Details about these set of problem instances are available at [12].

3.2 Experimental Settings

The parameter settings used for the genetic algorithm (GA) are listed in Table 1. These values were obtained through preliminary experimentation and no sophisticated mechanism to set parameter values was explored given that the aim of the GA is not to generate the best possible solution for a given stage of the problem, but instead to generated a population of good quality solutions for the lookahead policy evaluation. Experimental results in the rest of this section use the same set of parameter values.

Table 1. Genetic algorithm parameter settings

Population size	250
Crossover rate	55%
Mutation rate	10%
Maximum number of generations	50000
Maximum number of idle generations with no change in best chromosome	5000
Number of runs per instance	50

As described above, solutions produced by the GA to the weekly problems are evaluated through the lookahead procedure. In respect of the length of the lookahead period (pe), there is a trade-off between the quality of solutions and the computation required for the lookahead policy evaluation. Following our previous work in [7] the length is set to $pe = 3$ for 4-week scenarios and $pe = 7$ for 8-week scenarios.

3.3 Performance Comparison on Solving the Stage Problem

First, we compare the performance of the GA against the Value Function Approximation (VFA) from our previous paper [7] on solving the stage (weekly) problem. Table 2 shows summarised results from solving each stage problem instance 50 times. Column *Min.* presents the minimum (best) objective values obtained by each algorithm. Average values and standard deviation values are summarised in columns *Avg.* and *Std. Dev.* respectively. Column *time* presents the average computational time in minutes.

As can be seen from column *Min.*, the best objective values obtained by the two approaches are relatively close to each other. The genetic algorithm obtained

Table 2. Summary of results produced by the genetic algorithm described here and the value function approximation from [7] when solving weekly test instances from the Second International Nurse Rostering Competition. Time is reported in minutes.

	Genetic algorithm				Value function approximation			
Instance	Min.	Avg.	Std. Dev	Time	Min.	Avg.	Std. Dev	Time
n005w4_1	455	458.8	19.712	0.210	450	455.5	23.573	7.46
n005w4_2	435	439.3	25.137	0.213	435	439.6	31.578	7.19
n005w4_3	530	536.6	30.861	0.220	530	537.8	33.584	7.85
n012w8_1	1230	1241.8	117.862	1.456	1235	1251.2	185.683	18.545
n012w8_2	1540	1553.6	185.407	1.471	1540	1555.0	254.673	19.643
n012w8_3	1515	1525.1	127.593	1.509	1515	1528.3	186.460	18.730
n021w4_1	1725	1739.4	187.683	0.916	1815	1833.6	235.256	11.235
n021w4_2	2150	2162.8	168.974	0.976	2150	2166.2	205.574	12.085
n021w4_3	1940	1955.0	265.053	0.954	2035	2052.7	385.678	11.586

slightly better results than the value function approximation on the instances with larger number of nurses.

The average value *Avg.* obtained from multiple independent runs helps to estimate the overall performance of both algorithms in solving the weekly problem. As can be seen from the Table, the average value for the genetic algorithm is slight smaller than the one for the value function approximation. The standard deviation *Std. Dev* indicates the spread in the range of solution quality values obtained by each algorithm in the 50 runs. These values are much smaller for the genetic algorithm than for the value function approximation. This gives an indication of an overall more stable performance by the GA. Hence, replacing the value function approximation with the genetic algorithm for solving the weekly problems should results in an improvement in the performance of the hybrid approach.

Moreover, as it can be seen from the Table, the computational times for the genetic algorithm are much shorter than those for the value function approximation. Hence, the genetic algorithm achieved as good as or better solutions than the value function approximation but in considerably shorter time. In summary, the implemented genetic algorithm is a better approach to tackle the stage problem as part of the proposed hybrid solution method for solving the multi-stage nurse rostering problem.

3.4   Performance Comparison on Solving the Full Problem

Table 3 presents the results of the proposed hybrid approach on tackling the full multi-stage problem instances of the competition. The values in column *Gap* correspond to the difference in objective value between the given approach (GA-Lookahead or ADP-CP) and the Best result from the competition. A mark '+'

Table 3. Quality of solutions produced by the proposed hybrid approach combining a Genetic Algorithm with a Lookahead Policy (GA-Lookahead) and the previous method Combined Policy Adaptive Dynamic Programming (ADP-CP). The best and worst values from the competition results (achieved by a variety of algorithms) are also reported for comparison. The best values produced by our approaches are indicated in bold. The Gap value is reported as the difference in the objective value to the best from the competition results. The Rank value indicates the position of the approach with respect to all the results (from different algorithms) submitted for the competition.

Instance	GA-Lookahead	Gap	Rank	ADP-CP	Gap	Rank	Best	Worst
n030w4_1	2000	+255	5	**1780**	+235	4	1745	9850
n030w4_2	2130	+195	5	**1610**	+175	4	1935	10605
n030w8_1	**2940**	+645	5	4830	+2535	14	2295	21185
n030w8_2	**2380**	+480	5	4855	+2955	14	1900	21145
n040w4_1	**2075**	+350	7	3270	+1545	14	1765	14680
n040w4_2	**2235**	+325	6	3735	+1825	14	1910	14460
n040w8_1	**3755**	+650	4	9305	+6200	15	3105	35010
n040w8_2	**3735**	+760	6	8975	+6000	15	2975	33000
n050w4_1	**1890**	+365	6	3535	+2010	14	1525	17745
n050w4_2	**1955**	+475	6	3030	+1550	12	1480	15380
n050w8_1	**6630**	+1070	5	8965	+3405	12	5560	43040
n050w8_2	**6630**	+1155	5	8420	+2945	11	5475	42765
n060w4_1	**3455**	+625	9	12282	+9452	15	2830	19230
n060w4_2	**3540**	+590	6	15019	+12004	16	2950	20400
n060w8_1	**4010**	+1170	6	9720	+6880	15	2840	44130
n060w8_2	**4505**	+1305	6	10160	+6960	15	3200	44430
n080w4_1	**4130**	+655	6	18350	+14875	15	3474	26935
n080w4_2	**4130**	+595	6	16885	+13350	15	3535	27210
n080w8_1	**6735**	+1890	6	35975	+31130	15	4845	64915
n080w8_2	**6765**	+1660	6	38800	+33695	16	5105	66515
n100w4_1	**2350**	+905	6	16045	+14600	16	1445	33740
n100w4_2	**2915**	+845	6	17885	+15815	16	2070	33465
n100w8_1	**5115**	+2020	8	35690	+32595	16	3095	85260
n100w8_2	**5505**	+2370	7	35440	+32305	16	3135	87445
n120w4_1	**3385**	+915	7	22960	+20490	16	2470	36235
n120w4_2	**3435**	+905	6	22065	+19535	15	2530	36320
n120w8_1	**6145**	+2590	7	39170	+35615	15	3555	83590
n120w8_2	**6315**	+2880	7	41350	+37915	15	3435	82145

next to a Gap value indicates that the obtained solution cost value is greater than the best known. The values in column *Rank* indicate the ranking achieved by the proposed algorithm when compared to all the algorithms participating in the competition. Comparing the hybrid **GA-Lookahead** method proposed in this paper to our previous approach **ADP-CP**, it is clear that the proposed approach performs significantly better except in the first two problem instances.

There is no paper reporting fully on the results achieved by all the approaches in the INRC-II competition. So it is difficult to have an accurate comparison between our approaches and the several algorithms in the competitions. This is because results in the competition website as verified by the competition committee seem to be different from the results reported by the competition participants. Here we compare against results reported by the competition participants.

The two right-most columns of Table 3 show the *Best* and *Worst* values for each problem instance from the various algorithms in the competition. The gap achieved by the GA-Lookahead has decreased significantly with respect to the gap achieved by the previous approach ADP-CP. Even though these values of the gap to the best known solutions are still not negligible, the ranking of the proposed hybrid algorithm when compared to the combined performance of all the algorithms in the competition is better for about 10 positions. It is important to emphasise that the collection of best results for the set of competition instances has been obtained by several algorithms. Hence, the hybrid GA-Lookahead algorithm achieving a good overall ranking across all instances is a significant accomplishment.

4   Conclusion

In this paper we proposed a hybrid algorithm by combining a genetic algorithm with lookahead policy from dynamic programming to tackle the multi-stage nurse rostering problem. In this problem, a stage is defined as a week and the roster of each week is constructed while assuming that the staff requirements for the future weeks are not known. Also, when constructing the roster for a week, the historical information from the previous weeks needs to be considered. Previous research investigated approximate dynamic programming with a combined policy function to solve this problem. In the hybrid algorithm proposed here, a genetic algorithm is applied to tackle the weekly problem. The genetic algorithm produces a set of rosters for the week while not considering the global constraints. The lookahead policy then evaluates each of the rosters in respect of the future demand. That is, the lookahead procedure tries to select the roster that performs the best considering the future weeks and the history from the previous weeks among population. The lookahead policy then assembles a roster for the whole planning horizon. The algorithm is tested on solving a set of problem instances from the Second International Nurse Rostering Competition. Results produced by the proposed approach are compared to a previous method based on approximate dynamic programming with combined policy function and to all the results submitted to the competition. The improvement achieved with the proposed GA-Lookahead algorithm is considerable when compared to the previous approximate dynamic programming method. The intended contribution of this paper is to progress the understanding of how dynamic programming mechanisms can be successfully used to tackle difficult combinatorial optimisation problems.

References

1. Powell, W.B.: Approximate Dynamic Programming: Solving the Curses of Dimensionality, vol. 703. Wiley, New York (2007)
2. Van den Bergh, J., Beliën, J., De Bruecker, P., Demeulemeester, E., De Boeck, L.: Personnel scheduling: a literature review. Eur. J. Oper. Res. **226**(3), 367–385 (2013)
3. Burke, E.K., De Causmaecker, P., Berghe, G.V., Van Landeghem, H.: The state of the art of nurse rostering. J. Sched. **7**(6), 441–499 (2004)
4. Shi, P., Landa-Silva, D.: Dynamic programming with approximation function for nurse scheduling. In: Pardalos, P.M., Conca, P., Giuffrida, G., Nicosia, G. (eds.) MOD 2016. LNCS, vol. 10122, pp. 269–280. Springer, Cham (2016). https://doi.org/10.1007/978-3-319-51469-7_23
5. Davis, S.G., Reutzel, E.T.: A dynamic programming approach to work force scheduling with time-dependent performance measures. J. Oper. Manag. **1**(3), 165–171 (1981)
6. Maenhout, B., Vanhoucke, M.: NSPLib - a nurse scheduling problem library: a tool to evaluate (meta-)heuristic procedures. In: O.R. in Health, pp. 151–165. Elsevier (2005)
7. Shi, P., Landa-Silva, D.: Approximate dynamic programming with combined policy functions for solving multi-stage nurse rostering problem. In: Nicosia, G., Pardalos, P., Giuffrida, G., Umeton, R. (eds.) MOD 2017. LNCS, vol. 10710, pp. 349–361. Springer, Cham (2018). https://doi.org/10.1007/978-3-319-72926-8_29
8. Back, T.: Evolutionary Algorithms in Theory and Practice. Oxford University Press, Oxford (1996)
9. Yagiura, M., Ibaraki, T.: The use of dynamic programming in genetic algorithms for permutation problems. Eur. J. Oper. Res. **92**(2), 387–401 (1996)
10. Mohammadi, M., Forghani, K.: A hybrid method based on genetic algorithm and dynamic programming for solving a bi-objective cell formation problem considering alternative process routings and machine duplication. Appl. Soft Comput. **53**, 97 110 (2017)
11. Ceschia, S., Dang, N.T.T., De Causmaecker, P., Haspeslagh, S., Schaerf, A.: Second international nurse rostering competition (INRC-II)–problem description and rules–. arXiv preprint arXiv:1501.04177 (2015)
12. INRC-II the second nurse rostering competition. http://mobiz.vives.be/inrc2/. Accessed 23 May 2016

Image-Based Fashion Product Recommendation with Deep Learning

Hessel Tuinhof[1]([✉]), Clemens Pirker[2], and Markus Haltmeier[3]

[1] InvoiceFinance, 's-Hertogenbosch, Netherlands
hessel@invoicefinance.nl
[2] Department of Strategic Management, Marketing and Tourism,
University of Innsbruck, Innsbruck, Austria
clemens.pirker@uibk.ac.at
[3] Department of Mathematics, University of Innsbruck, Innsbruck, Austria
markus.haltmeier@uibk.ac.at

Abstract. We develop a two-stage deep learning framework that recommends fashion images based on other input images of similar style. For that purpose, a neural network classifier is used as a data-driven, visually-aware feature extractor. The latter then serves as input for similarity-based recommendations using a ranking algorithm. Our approach is tested on the publicly available Fashion dataset. Initialization strategies using transfer learning from larger product databases are presented. Combined with more traditional content-based recommendation systems, our framework can help to increase robustness and performance, for example, by better matching a particular customer style.

Keywords: Recommendation · Deep learning ·
Convolutional neural networks · Similarity recommendation

1 Introduction

Identifying products a specific customer likes most can significantly increase the earnings of a company [15]. Clearly, recommending suitable products in E-commerce increases the probability of a customer's purchase. Additionally, offering too many products can reduce the probability that a potential customer performs a purchase at all. Finally, knowing and subsequently targeting customer preferences increases the medium- and long-term commitment of the customer to the company, which is a key factor to profitability [3,17]. Prior studies demonstrate that recommendation engines help consumers to make better decisions, reduce search efforts and find the most suitable prices [5].

One possibility to infer knowledge about customer preferences is via specific questioning in customer surveys. However, this is not always possible and customer responses may not be correct or sufficient for accurately describing preferences. In this work, we follow a different, data-driven approach, where customer preferences are automatically extracted from available information on the

© Springer Nature Switzerland AG 2019
G. Nicosia et al. (Eds.): LOD 2018, LNCS 11331, pp. 472–481, 2019.
https://doi.org/10.1007/978-3-030-13709-0_40

customer. More specifically, we focus on fashion products and develop a method that only requires a single input image to return a ranked list of similar-style recommendations.

1.1   Proposed Recommendation System

The proposed recommendation system operates in a two-stage mode. In the first step, we train a convolutional neural network (CNN) to solve specific image classification tasks. The trained CNN is then used as a problem-specific feature extractor, where the features serve as inputs for the ranking system. While in this paper we work with fashion products, similar recommendation systems can be employed for other product categories as well.

Image data provides a wealth of information on a visually-aware feature level, e.g. edges and color blobs. Plenty of image processing techniques exist to extract such low-level features [14]. Deep learning provides a technique to extract hidden higher-level features by composing several convolutional layers. Therefore CNNs are a natural choice to provide fashion product recommendations based solely on image data. Compared to classical content-based recommendation, which is mainly based upon descriptive metadata like manually annotated product tags or user reviews, our approach relies on visual information.

1.2   Relation to Previous Work

There are at least two main approaches for product recommendations: collaborative filtering and content-based filtering. Whereas the former relies on historical user-item interactions, the latter tries to relate user profiles and item descriptors. A recent deep learning approach is the neural collaborative filtering framework proposed in [7], which generalizes the matrix factorization technique used extensively in collaborative filtering methods. Others like [6] employ a hybrid approach, where a matrix factorization based predictor is combined with a deep learning model that extracts visual features as well as latent non-visual user features. A recent thorough overview on deep learning-based recommender systems can be found in [20].

The success of CNNs for computer vision tasks like object classification, detection and segmentation [4] gives reason to decouple classical product recommendation solutions from its extensive user-item interaction data usage requirement. Therefore our method uses product image data, which, for example in E-commerce, is readily available. This also allows to mitigate the cold start problem of collaborative filtering and classical content-based recommender systems. Closely related to our approach are the works [1,16]. Due to the high degree of subjectivity related to fashion articles, general recommender systems usually perform poorly in fashion recommendation tasks. We show that recommendation systems purely relying on visual features are reasonable as they are able to provide highly visually appealing recommendations of similar style. This can also be helpful in the case of new customers, where no historical user data is yet available. It can also be integrated in existing content-based systems, for example, to account for a particular or desired style of a customer, or to address the cold-start problem.

1.3   Outline

The remainder of the paper is structured as follows. Section 2 presents the proposed product recommendation method. In particular, we give details on the used network architectures, the used ranking algorithm and describe the Fashion dataset. In Sect. 3 we present some numerical results. The paper concludes with a short discussion in Sect. 4.

2   Methods

2.1   Fashion Dataset

Throughout this paper, we work with a subset of the publicly available Fashion[1] dataset [12]. In order to obtain high-quality ground-truth labels for category type and texture attributes, we design a labeling questionnaire on the crowdsourcing platform CrowdFlower[2]. Every image is labeled by a total maximum of five human operators. To be a valid label at least three human operators have to agree. Each labeling task consists of five images to be labelled, one of which is a simple test image. If a human operator fails a test more than twice, she is no longer allowed to continue. Separate datasets for category and texture classification have been created.

The used class labels for category types are blouse, dress, pants, pullover, shirt, shorts, skirt, top, T-shirt. For the texture attributes we use the labels graphic, plaid, plain, spotted, striped. Figure 1 shows the frequency distributions for the two datasets. The category type dataset contains 11 851 and the texture attributes dataset 7342 images. Further characteristics can be found in Table 1.

Table 1. Summary of the datasets created from the Fashion dataset. The third column indicates the total amount of class labels for the respective dataset.

Dataset	Classification		Samples		
	Type	No.	Total	Train	Val
Fashion category	Multinomial	9	11,851	9,480	2,371
Fashion texture	Multinomial	5	7,342	5,873	1,469

---

[1] http://imagelab.ing.unimore.it/fashion_dataset.asp.
[2] www.crowdflower.com.

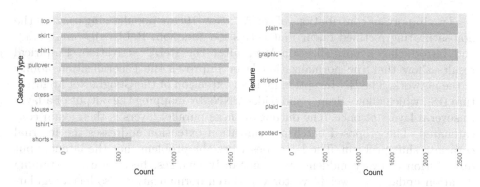

Fig. 1. Frequency distributions for the category type and texture attributes datasets created from the Fashion dataset.

2.2 Proposed Framework

Our method composes of a trained CNN classifier used as image feature extractor and a modification of the k-nearest neighbors (k-NN) algorithm used for ranking in feature space.

■ **Classification via CNNs:** In the first step, we train separate CNNs to predict the category and texture type. Each of the CNNs can be written as

$$\mathcal{N}_i(\mathbb{W}_i, \cdot) \triangleq \mathcal{S}_i(\mathbb{V}_i, \mathcal{F}_i(\mathbb{U}_i, \cdot)) \quad \text{for } i = 1, 2. \tag{1}$$

Here $\mathbb{W}_i = (\mathbb{U}_i, \mathbb{V}_i)$ are weight vectors, $\mathcal{S}_i(\mathbb{V}_i, \cdot)$ are fully connected softmax output layers that actually perform classification and $\mathcal{F}_i(\mathbb{U}_i, \cdot)$ are the CNNs without the last layer. The latter are used as feature extractor.

■ **Ranking in feature space:** After training and evaluating the performance of these classifiers, we remove the softmax output layer $\mathcal{S}_i$ of each model. The remaining CNNs are then concatenated and $\mathcal{F} = [\mathcal{F}_1, \mathcal{F}_2]$ is used to extract the feature vector $\mathcal{F}(\mathbf{X})$ of any input image $\mathbf{X} \in \mathbb{R}^{N \times N}$. We then use the k-NN algorithm to search for the closest items to $\mathcal{F}(\mathbf{X})$ in feature space.

Details on the employed CNNs and the k-NN algorithm for ranking are presented below.

2.3 Network Architectures

A wealth of CNN architectures are available today. In this section we briefly discuss the two architectures that we use in our work: AlexNet and batch-normalized Inception (BN-Inception). The AlexNet and BN-Inception are both standard architectures and well established. AlexNet has been chosen as a benchmark to compare against deeper, more complex networks like the BN-Inception. AlexNet consists of 8 layers and BN-Inception of 34. Both use an image of size 224×224 as input.

Two important contributions of the AlexNet [10] are popularizing usage of the non-saturating rectified linear unit activation function, $\text{ReLU}(x) \triangleq \max(0, x)$, and introducing a normalization layer after the ReLU activation. Empirical results show that the normalization layer improves the generalization ability of the network. The BN-Inception [8] is an extension of the GoogLeNet architecture [18], which allows deeper and wider CNNs by mapping the output of a layer to several layers at once. The output of these parallel layers is then again concatenated. The proposed batch normalization extension addresses the internal covariate shift problem. The latter describes the problem that the latent input distribution of every hidden layer constantly changes, because every training iteration updates the weight vector $\mathbb{W}_i$. Batch normalization also has a regularization effect.

2.4 Network Training

In order to adjust $\mathcal{N}_i(\mathbb{W}_i, \cdot)$ to the particular classification task, the weight vector $\mathbb{W}_i$ is selected depending on a set of training data $\mathcal{T}_i \triangleq \{(\mathbf{X}_n, \mathbf{Y}_n)\}_{n=1}^{N_i}$. For this purpose, the weights are adjusted in such a way, that the overall error of $\mathcal{N}_i(\mathbb{W}_i, \cdot)$ made on the training set is small. This is achieved by minimizing the error function

$$E(\mathbb{W}_i) \triangleq \sum_{n=1}^{N_i} d(\mathcal{N}_i(\mathbb{W}_i, \mathbf{X}_n), \mathbf{Y}_n) + \lambda \|\mathbb{W}_i\|^2 , \tag{2}$$

where d is a distance measure that quantifies the error made by the network function $\mathcal{N}_i(\mathbb{W}_i, \cdot)$ for classifying the n-th training sample.

To stabilize the weight computation in (2), we add a L^2-regularization term $\lambda \|\mathbb{W}_i\|^2$ with regularization parameter $\lambda \geq 0$. As is common for classification with neural networks, we use the cross entropy for the loss function d. The actual minimization of (2) is performed by stochastic gradient descent.

2.5 Ranking by k-NN

The k-NN algorithm can be used as simple ranking algorithm. For that purpose, consider the feature space $\mathbb{R}^p$ and denote with $d_2(\mathbf{f}, \mathbf{g}) = \|\mathbf{f} - \mathbf{g}\|_2$ the Euclidean distance of two feature vectors. Let $\{\mathbf{f}_1, \ldots, \mathbf{f}_m\}$ be a training set of feature vectors. A k-NN algorithm then solves some regression or classification task at $\mathbf{f} \in \mathbb{R}^p$ using the k closest training features. This can be implemented by first computing an enumeration $\pi(\mathbf{f}): \{1, 2, \ldots, m\} \to \{1, 2, \ldots, m\}$ satisfying $d_2(\mathbf{f}, \mathbf{f}_{\pi(\mathbf{f})(i)}) \leq d_2(\mathbf{f}, \mathbf{f}_{\pi(\mathbf{f})(i+1)})$. We use the permutation $\pi(\mathbf{f})$ as ranking output for the input feature $\mathbf{f}$. To reduce memory requirements of the k-NN ranking, we use an implementation that employs a balltree search [13].

3 Results

In this section we present results for the image classification and similarity recommendation with the proposed framework.

Table 2. Summary of the datasets created from the DeepFashion Attribute Prediction dataset used for pretraining. The third column indicates the total amount of class labels for the respective dataset.

Dataset	Classification		Samples		
	Type	No.	Total	Train	Val
DeepFashion category	Multinomial	46	289,222	231,377	57,845
DeepFashion texture	Multinomial	156	111,405	89,124	22,281

3.1 Pretraining

To overcome difficulties arising from the relative small size of the Fashion dataset, we use the concept of transfer learning [4,19]. For that purpose, we pretrain the classification models on a larger dataset (namely, the DeepFashion Attribute Prediction[3] dataset, [11]) containing 289 222 garment images. A full summary of the dataset can be found in Table 2.

For pretraining we use AlexNet and BN-Inception architectures. For the AlexNet we minimize (2) with stochastic gradient descent using batch size of 64, regularization parameter $\lambda = 0.0005$, learning rate 0.01 and momentum 0.9. For training the BN-Inception we use the ADAM [9] algorithm with batch size of 32, $\lambda = 0$, and learning rate 0.001. Following [4], we use early stopping as an efficient regularization technique to prevent overfitting. We therefore stop training AlexNet/BN-Inception after 9/8 and 17/13 epochs for the category and texture classification, respectively.

Additional to the cross entropy loss, we use the evaluation metrics accuracy,

$$\text{accuracy}(\mathbf{y}, \hat{\mathbf{y}}) \triangleq \frac{1}{N} \sum_{n=1}^{N} \mathbf{1}_{y_n}(\hat{y}_n), \tag{3}$$

and top-K accuracy, which is defined as in Eq. (3) with a slightly modified indicator function such that top-K predicted classes are incorporated. Table 3 shows accuracy, top-K-accuracy and loss evaluated on the test set for both AlexNet and BN-Inception. The BN-Inception achieves higher accuracy and better generalization ability. Therefore, we only use the BN-Inception architecture for classification on the Fashion dataset.

3.2 Classification

For the final classification models we train the BN-Inception by minimizing (2) on the Fashion dataset with ADAM, where the weights are initialized using the ones from the pretraining stage. Due to the small size of the Fashion dataset, we add L^2-regularization with $\lambda = 0.0001$ to the loss function and also reduce the batch size to 16.

---

[3] http://mmlab.ie.cuhk.edu.hk/projects/DeepFashion/AttributePrediction.html.

Table 3. Pretraining results: the left table depicts results for the category classification and the right table for the texture classification.

Category	AlexNet	BN-Inception
Accuracy	0.57	0.63
Top-5	0.79	0.84
Loss	1.48	1.27

Texture	AlexNet	BN-Inception
Accuracy	0.28	0.32
Top-3	0.62	0.66
Loss	3.00	2.82

Several image augmentation techniques are applied in order to effectively increase dataset size. These include random rotations with a maximum rotation angle of ± 3 for the category type model and ± 8 for the texture attributes model, random changes of HSL color channels within a range of $[-6, 6]$, a shear transformation with random shear factor within $[-0.25, 0.25]$, random aspect ratio changes within a range of $[0.875, 1.125]$ and random vertical flips. The random augmentations are applied to the training set every epoch anew. This allows to train longer without overfitting too fast. Following early stopping regularization, we stop training the category type and texture attributes classification models after 15 and 4 epochs respectively. Table 4 summarizes the final training results. The top-K accuracy metric is however excluded due the smaller number of class labels in the Fashion datasets.

Table 4. Final BN-Inception classification results on the Fashion datasets for category and texture.

	Category	Texture
Accuracy	0.87	0.80
Loss	0.42	0.61

3.3   Similarity Recommendation

The CNN classifiers are used as feature extractors and return feature vectors $\mathcal{F}_i(\mathbf{X})$ of size $d = 1024$ for any input image. The feature extractors are applied to a set of $n = 19\,422$ test images. These corresponding feature vectors are concatenated and stacked to obtain a $n \times 2d$ feature matrix. The k-NN ranking algorithm is applied to the feature matrix. For the recommendation task, it is now sufficient to extract the features from an input image, submit them to the k-NN ranking algorithm and return the top-k matching style recommendations. In Fig. 2 we present several query images and corresponding top-5 recommendations. Subjectively, the top-5 recommendations indeed look quite similar to the query images. In the top row a query image from the dataset itself is used. This corresponding top-5 recommendations demonstrate that if the image appears in the dataset it is actually most similar to itself. Similar results have been

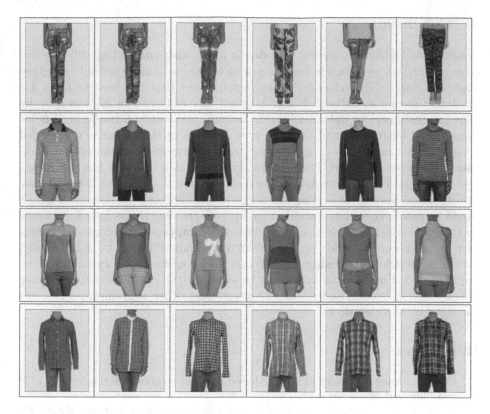

Fig. 2. k-NN recommendation ranking. First column displays the query images and columns 2–6 display the predicted five nearest neighbors, where column 2 is the most similar.

obtained in other performed tests. Other than that, an implicit objective metric for recommendation quality can be found by means of the classification accuracies reported in Table 4. The definition of a precise objective evaluation criterion, however, remains difficult due to the inherent subjectivity of recommendation quality. This also makes comparison with other methods quite challenging. The computationally most time-consuming part in the application of the proposed recommendation system is the evaluation of the CNN classifiers.

In our implementation, we have implemented the CNNs in MXNet [2] using its Python API. Running on a desktop PC with an Intel i7-6850K CPU and a NVIDIA 1080Ti GPU, the whole image processing pipeline applied to a given input image only takes fractions of a second. Note that the potentially time-consuming network training is done before a new input image is provided to the recommendation system, which therefore allows fast online product recommendation.

4   Conclusion

We presented a visually-aware, data-driven and rather simple but still effective recommendation system for fashion product images. The proposed two-stage approach uses a CNN classifier to extract features that are used as input for similarity recommendations. It can be used, for example, in E-commerce by allowing customers to upload a specific fashion image and then offering similar items based on texture and category type features of the customer's uploaded image. Additional feature extractors, e.g. trained on gender or color classification tasks, can be easily added. Furthermore, generalization to other domains makes sense, e.g. music recommendation based on raw music data, but needs further investigation. Several interesting extensions of our approach are possible. First, it would be promising to integrate the two separate training stages into a single one and provide end-to-end deep learning-based fashion product recommendations. In particular, consideration should be given to Siamese networks. Additionally, hybrid approaches combining image-based and content-based systems will be implemented. Finally, it is important to evaluate the customer impact of our image-based approach and its extensions against other recommender systems through customer surveys.

References

1. Chen, L., Yang, F., Yang, H.: Image-based product recommendation system with convolutional neural networks (2017)
2. Chen, T., et al.: MXNET: a flexible and efficient machine learning library for heterogeneous distributed systems. arXiv preprint arXiv:1512.01274 (2015)
3. Dick, A.S., Basu, K.: Customer loyalty: toward an integrated conceptual framework. J. Acad. Market. Sci. **22**(2), 99–113 (1994)
4. Goodfellow, I., Bengio, Y., Courville, A.: Deep Learning. MIT, Cambridge (2016)
5. Häubl, G., Murray, K.B.: Double agents: assessing the role of electronic product recommendation systems. MIT Sloan Manag. Rev. **47**(3), 8–12 (2006)
6. He, R., McAuley, J.: VBPR: visual Bayesian personalized ranking from implicit feedback. In: AAAI, pp. 144–150 (2016)
7. He, X., Liao, L., Zhang, H., et al.: Neural collaborative filtering. In: WWW 2017, pp. 173–182 (2017)
8. Ioffe, S., Szegedy, C.: Batch normalization: accelerating deep network training by reducing internal covariate shift. In: ICML, pp. 448–456 (2015)
9. Kingma, D.P., Ba, J.: Adam: a method for stochastic optimization. arXiv:1412.6980 (2014)
10. Krizhevsky, A., Sutskever, I., Hinton, G.E.: Imagenet classification with deep convolutional neural networks. In: NIPS, pp. 1097–1105 (2012)
11. Liu, Z., Luo, P., Qiu, S., et al.: DeepFashion: powering robust clothes recognition and retrieval with rich annotations. In: CVPR, pp. 1096–1104 (2016)
12. Manfredi, M., Grana, C., Calderara, S., Cucchiara, R.: A complete system for garment segmentation and color classification. Mach. Vis. Appl. **25**(4), 955–969 (2014)
13. Omohundro, S.M.: Bumptrees for efficient function, constraint and classification learning. In: NIPS, pp. 693–699 (1991)

14. Prince, S.: Computer Vision: Models, Learning, and Inference. Cambridge University Press, Cambridge (2012)
15. Schafer, J.B., Konstan, J.A., Riedl, J.: E-commerce recommendation applications. Data Min. Knowl. Discov. **5**(1–2), 115–153 (2001)
16. Shankar, D., Narumanchi, S., Ananya, H.A., et al.: Deep learning based large scale visual recommendation and search for e-commerce. arXiv:1703.02344 (2017)
17. Srinivasan, S.S., Anderson, R., Ponnavolu, K.: Customer loyalty in e-commerce: an exploration of its antecedents and consequences. J. Retail. **78**(1), 41–50 (2002)
18. Szegedy, C., Liu, W., Jia, Y., et al.: Going deeper with convolutions. In: CVPR, pp. 1–9 (2015)
19. Tajbakhsh, N., Shin, J.Y., Gurudu, S.R., et al.: Convolutional neural networks for medical image analysis: full training or fine tuning? IEEE Trans. Med. Imag. **35**(5), 1299–1312 (2016)
20. Zhang, S., Yao, L., Sun, A.: Deep learning based recommender system: a survey and new perspectives. arXiv:1707.07435 (2017)

A Machine Learning Approach for Line Outage Identification in Power Systems

Jia He[1], Maggie X. Cheng[1](✉), Yixin Fang[1], and Mariesa L. Crow[2]

[1] New Jersey Institute of Technology, Newark, NJ 07102-1982, USA
{jh495,maggie.cheng,yixin.fang}@njit.edu
[2] Missouri University of Science and Technology, Rolla, MO 65401, USA
crow@mst.edu

Abstract. This paper addresses power line topology change detection by using only measurement data. As Phasor Measurement Units (PMUs) become widely deployed, power system monitoring and real-time analysis can take advantage of the large amount of data provided by PMUs and leverage the advances in big data analytics. In this paper, we develop practical analytics that are not tightly coupled with the power flow analysis and state estimation, as these tasks require detailed and accurate information about the power system. We focus on power line outage identification, and use a machine learning framework to locate the outage(s). The same framework is used for both single line outage identification and multiple line outage identification. We first compute the features that are essential to capture the dynamic characteristics of the power system when the topology change happens, transform the time-domain data to frequency-domain, and then train the algorithms for the prediction of line outage based on frequency domain features. The proposed method uses only voltage phasor angles obtained by continuous monitoring of buses. The proposed method is tested by simulated PMU data from PSAT [1], and the prediction accuracy is comparable to the previous work that involves solving power flow equations or state estimation equations.

Keywords: Power systems · Machine learning · Logistic regression · Random forest

1 Introduction

According to the reports following the major blackout in August, 2003 in the United States and Canada, the primary cause for very costly large-scale power outage is inadequate system understanding and inadequate situational awareness [2,3]. How to improve system understanding and situational awareness becomes crucial and pertinent to protecting the grid from natural disasters as well as cyber and physical attacks. To improve situational awareness, continuous monitoring

M. Cheng is supported in part by National Science Foundation of USA.

© Springer Nature Switzerland AG 2019
G. Nicosia et al. (Eds.): LOD 2018, LNCS 11331, pp. 482–493, 2019.
https://doi.org/10.1007/978-3-030-13709-0_41

of the power system is important, and the status of the power lines, generators and transformers need to be updated in real-time.

The power grid generates tremendous amount of data every second from its monitoring devices. In recent years we have witnessed an increasing application of Phasor Management Units (PMUs) in power systems. A PMU measures the electrical waves on an electricity grid, called phasors. A phasor is a complex number that represents both the magnitude and phase angle of the waves (voltages and currents). When abnormal waveforms are observed, it is likely that anomaly has occurred in the power system configuration.

Mining of power grid measurement data is hopeful to give us an insight of what is hidden in the data. System understanding and real-time situational awareness comes naturally as a result of data analytics. Major power outage can be prevented if the control centers are equipped with the ability to interpret changes in the state of the network without cognitive overload of the operator and thus take appropriate control actions.

Among all disastrous events, power line outage hits most frequently. According to [4], in the five-year period from 2008–2012, weather-related outages accounted for 66% of power disruptions and affected up to 178 million customers. Power line outage identification becomes the first and most powerful tactic to improve situational awareness. Being aware of the changes on the power lines is paramount for several critical tasks, such as state estimation, power load flow analysis, real-time contingency analysis [5]. For instance, the state matrix used in the state estimation is based on an assumed topology and parameters. A topological change of the grid can completely overthrow the state estimation [6].

In this paper, we propose a data-driven approach for the task of line outage identification. Previous work on line outage detection and identification heavily rely on other critical tasks of the power system, such as state estimation, or load-flow analysis [5]. Some either use the residual of state-estimation [6], or carry out a joint outage identification and state estimation [7]. However, these tasks require a lot of global information that may not be available all the time. Measurement data from PMUs are more accessible than the system model information. We propose to use measurement data only to infer the status of power lines. The advantage of this approach is that there is no interlocking with other tasks, and therefore the line outage identification can be used independently and prior to the implement of state estimation or power flow analysis.

The proposed approach mainly uses the voltage phasor angles measured at buses. Three machine learning algorithms are used to estimate the prediction models. If PMU data are partially available at some buses but not available on all buses, the proposed approach can still be applied. For instance in the IEEE 9-bus test system, using measurements from 6 buses instead of all 9 buses can still do a reasonably good job on prediction with slightly reduced accuracy. This indicates that even the accurate topology information of the power grid is not essential. The data-driven inference approach compared to the previous approach that relies on accurate state information of the system is a fundamental breakthrough.

The rest of the paper is organized as follows: in Sect. 2, we survey the most related work in power line outage detection and identification; in Sect. 3, we cover the preliminaries underlying the proposed method; in Sect. 4, we provide detailed description of the line identification method and the machine learning algorithms used in this project; in Sect. 5, we provide results for identification of single and double line outages using IEEE standard test system; in Sect. 6, we conclude the paper with outlook for future work.

2   Related Work

As PMUs become increasingly deployed, power line outage detection and identification based on PMU measurements have received much attention recently. In early work [8], system topology information together with PMU phasor angle measurements are used to detect a single line outage. To further determine the broken line, the pre-outage flow is estimated, in which power flow equation is used and the system admittance matrix is needed.

Tate and Overby further extended the single line outage detection method in [8] to double line outage detection [9]. It is proved in [9] that there exists indistinguishable outages due to incomplete PMU deployment, and a method of recognizing these indistinguishable outages is presented. This conclusion is also consistent with the conclusion of our own study. The indistinguishable outages due to the limited PMU deployment is a major contributing factor of prediction errors.

Other recent work on outage identification based on PMU data include [5, 10]. In [10] trains, a linear multinomial regression model is estimated by solving maximum likelihood problem utilizing the sampled PMU data, however, the classifier is only effective in the presence of perfect dynamic simulation. [5] focuses on detecting multiple line outage at low complexity. [5] solved the sparse line outage identification problem by using the DC linear power flow model. Like [5], the proposed logistic regression and random forest models also avoid the combinatorial complexity issue; but different from [5], the proposed method only use PMU data and does not need the reactance information on the power lines. Similar work that also uses sparse overcomplete representation also includes [11] to address the fault estimation problem.

It is noted that most previous work are constrained to at most double line outage detection [8,9,12]. [5] may be the first that is not limited by the number of simultaneous line outages. Our work is a complete data driven approach, which does not use the system model, and is also not limited to the number of simultaneous line outages.

There are also notable work that solves outage detection, state estimation and optimization of sensor locations as coupled problems. In [7], a joint detection and estimation problem was studied for outage identification in power systems. The authors employed a Bayesian framework, in which the prior distributions on the outage events and the network states are assumed to be Gaussian, and developed closed form joint posterior distribution of the outage and the network states, which can be used to determine the optimal joint outage detector and state estimator.

3   Preliminaries

In order to find the features that can capture the signature changes in the power system when a line outage happens, it is necessary to look at what will be disturbed when a line is out.

The real power transfer from bus i to bus $j, j \neq i$ is given as follows:

$$P_{ij} = |V_i||V_j| \left(G_{ij}cos(\theta_i - \theta_j) + B_{ij}sin(\theta_i - \theta_j) \right)$$

If the transmission line has impedance of $z = r + jx$, then the admittance is $y = \frac{1}{z} = \frac{r}{r^2+x^2} - \frac{jx}{r^2+x^2} = g + jb$. Usually $r \ll x$, so the real part in y is close to zero, therefore the real part of the admittance matrix elements will be close to zero, i.e., $G_{ij} \rightarrow 0$, thus the power transfer equation can be simplified as:

$$P_{ij} \doteq |V_i||V_j| \left(B_{ij}sin(\theta_i - \theta_j) \right)$$

It is also observed that the difference in angles of the voltage phasor at two buses i and j connected by a circuit is usually very small, so $(\theta_i - \theta_j)$ is usually a very small number (usually less than $15°$, or 0.262 in radians, and $sin(0.262) = 0.259$.). Therefore we use $(\theta_i - \theta_j)$ to approximate $sin(\theta_i - \theta_j)$. The power transfer equation can be further simplified as:

$$P_{ij} \doteq |V_i||V_j| \left(B_{ij}(\theta_i - \theta_j) \right)$$

In the per-unit system, the numerical values of voltage magnitudes $|V_i|$ and $|V_j|$ are between 0.95 to 1.05, so it incurs very little error if we assume them to be 1.0. Therefore in the normal operation when two buses are connected by a transmission line, the power transfer amount is loosely proportional to the angle difference between the two buses. It is a dependent relationship. When the line between two buses is broken, the angles at two buses lose this dependent relationship and become conditionally independent given the voltage phasor angles at other buses. Upon line outage, we expect to see a transition from a relatively stable signal to a more dramatically changing signal. Based on this observation, we can develop a machine learning methodology to identify which line is out. The identified features will be used to train the machine learning algorithms, as well as predicting line outage.

4   Proposed Methods

This paper focuses on identification of power line outage, i.e., to locate which line is out. We use the measurement data from buses for this task. Such data are available from Phasor Measurement Units, which are widely available nowadays.

Throughout the implementation of the method, we assume no prior knowledge about the admittance matrix or state matrix of the system.

Early work has proposed a method of change point detection from time series [13], which can be used to detect changes from time series of measurements, but does not tell exactly which line is out. To further determine whether it is a single line outage or multiple line outages, and to identify which lines are out, we formulate the problem as a classification problem and use different statistical learning algorithms to solve the problem.

Based on the analysis above, the only information used in the classification problem is the phasor angles at buses. For a fair comparison among different methods, all methods are provided with the same data. Assume there are angle measurements from m buses, the data contains m-dimensional time series $\{\boldsymbol{\theta}_t\}$, where $\boldsymbol{\theta}_t \in \mathcal{R}^m$, and $t = 1 \dots n$.

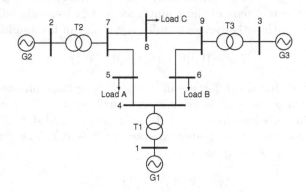

Fig. 1. Standard IEEE 9-bus test system.

Consider the standard IEEE 9-bus test system (see Fig. 1) as an example. Figure 2 shows the phasor angles at bus 8 and bus 9 as well as their differences. When a line outage occurs at the branch between the two buses, we observe a change in the difference of angles (see Fig. 2(a)). Although the change in the magnitude is small, the change in the dynamic feature is significant and visually detectable (see Fig. 2(b)).

It is noted that angle measurements in Fig. 2(a) are smoothed instead of mapped to $[0, 2\pi)$. This is to avoid angle oscillation, since as the angle increases with time slowly, the drastic change from $2\pi - \delta$ for a small δ to 0 is more significant than the changes caused by the line outage, and therefore can totally subvert the classifier.

4.1 Features Extraction

The difference of the voltage phasor angles at two buses is a low level feature for classification. However, the raw measurements won't make a good predictor,

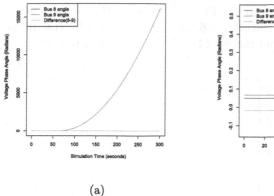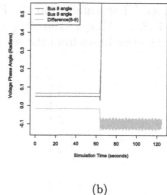

(a)   (b)

Fig. 2. The voltage phasor angles of bus 8 and bus 9 versus time. Line outage occurred at time 50 s. (a) shows the full range of angles, (b) shows details on angle differences.

since the angles are unbounded, and the length of the time series, n, can be very long or very short. The dimension of the predictors should not depend on the length of the time series.

To make the features independent of the length of the time series, we use Fourier Transform. Fast Fourier Transform (FFT) takes a discrete signal in the time domain and transforms that signal into its discrete frequency domain representation,

$$X_k = \sum_{j=0}^{n-1} x_j e^{-i2\pi kj/n}$$

where X_k is a complex number that encodes both amplitude and phase of a complex sinusoidal component $e^{i2\pi kj}$ of function x_j, and the frequency of the component is k cycles per n samples.

FFT results in a sequence of $\{X_k\}$, with $k = 1, \ldots, n$. FFT can also be evaluated for a specified number of points. We sort the amplitudes and consider the top K dominant frequencies so that the dynamic features of the time series are fully captured in the frequency domain. In this paper, we take the union of the dominant frequencies and use the amplitudes $|X_k|$ as predictors.

4.2   Classification Algorithms

We use supervised machine learning for the task of line outage identification. The data used by all classification algorithms have the following format (Fig. 3):

$$Y \sim \left(|X_k^{(l)}|, \ldots \right), \quad \text{for } k = 1 \ldots K, \text{ and } l = 1 \ldots W,$$

where k is among the first K dominant frequencies, $|X_k^{(l)}|$ is the amplitude of the corresponding frequency for line l. This is repeated for W times if W lines are observed.[1] For instance, in the Fig. 1, if we have PMUs installed on buses $\{4, 5, 6, 7, 8, 9\}$, then we have $W = 6$ time series formed by the differences of angles at the two buses from the set $\{4\text{–}5,\ 5\text{–}7,\ 7\text{–}8,\ 8\text{–}9,\ 9\text{–}6,\ 6\text{–}4\}$.

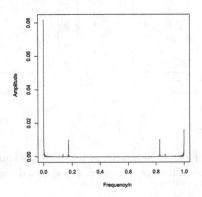

Fig. 3. The spectrum of the angle difference in Fig. 2 after line outage.

We consider three classification algorithms for identifying a single line outage.

– Logistic Regression
– Random Forest with Binary Outcome
– Multi-Categorical Random Forest.

Logistic Regression. In logistic regression, Y is the binary outcome variable, $X = (x_1, \ldots, x_p)$ is the set of predictor variables. Let q be the probability of Y being 1. The logit model is given by:

$$\log_e\left(\frac{q}{1-q}\right) = \beta_0 + \beta_1 x_1 + \beta_2 x_2 + \ldots \beta_p x_p = \boldsymbol{\beta} \cdot \mathbf{X}$$

where $\beta_1, \ldots, \beta_p$ are the regression coefficients indicating the relative influence of each particular predictor variable on the outcome, and β_0 is the intercept. Training data are used to estimate the coefficients $\boldsymbol{\beta}$.

A separate logit model is estimated for each line outage. If there are L lines, there are L models, i.e. L sets of $\boldsymbol{\beta}$. We use $\boldsymbol{\beta}^{(l)}$ to denote the coefficients for the line l. For line outage detection, we have a vector of $q^{(l)} = \mathrm{Prob}(y^{(l)} = 1)$ indicating the probability of line l being broken. Given the coefficient $\boldsymbol{\beta}^{(l)}$ and $\mathbf{X}$ from the new data point, the probability of outage on line l is computed as follows:

$$q^{(l)} = \frac{1}{1 + e^{-\boldsymbol{\beta}^{(l)} \cdot \mathbf{X}}}$$

The logistic regression method is used for both single line outage and multiple line outage. The process involves first estimating $\boldsymbol{\beta}^{(l)}$ and then predicting $q^{(l)}$ for

---

[1] Observed means PMUs are installed on the buses at the two end points of the line.

the new data, for $l = 1, \ldots, L$. The computational complexity does not increase with the number of outages.

Random Forest. Random Forest is a tree-based method. Different from regression models, Random Forest uses decision trees as building blocks to construct prediction models [14].

Random forest involves producing multiple classification trees which are then combined to yield a single consensus prediction by averaging all the predictions. When building individual decision trees, each time a split of the predictor space is considered, a random sample of m predictors are randomly chosen as split candidates out of the full set of p predictors. The split then uses only one of the m predictors. m is typically an integer close to $\sqrt{p}$.

Using random forest for the classification of line outage can be carried out in two ways:

1. Binary outcome. For each possible line outage, we train a separate model to decide if the line is broken, and then apply each of the models to the new observation. If multiple models yield "Yes", then there are multiple line outage.
2. Multi-categorical outcome. We build one training model for all possible classes. If there are L possible line outages, there are $L + 1$ classes, or $L + 1$ possible outcomes, with outcome $= 0$ indicating no line is broken, outcome $= l$ indicating line l is broken. This method can only be used for single line outage prediction or small systems. This is because in the case of multiple line outages, the number of classes grows exponentially with the number of lines. If all L lines are prone to break, there are a total of 2^L outcomes.

Experiments on single line outage detection indicate that multi-categorical Random Forest underperforms binary Random Forest by at least 20% in detection rate. For multiple line outage detection, the multi-categorical version cannot even detect 50% of the line outages due to having too many categories. Having too many categories with limited training data is detrimental to the method. Therefore in the following we will refer to the binary version by default whenever Random Forest is mentioned.

5 Results

The tests are implemented on IEEE standard 39-bus test system. Time domain simulation of the system is implemented in Power System Analysis Toolbox (PSAT) [1]. PSAT is a free open source package equipped with modules for solving power flow (PF), optimal power flow(OPF), continuation power flow (CPF) and time domain simulation (TDS). In this paper, we do not use power flow analysis, since we assume neither the admittance matrix nor the Jacobian matrix is available. We use the time domain analysis part of PSAT, and the only information needed for the proposed methods is the incidence matrix and the PMU measurements from a few buses.

The incidence matrix provides information for the link topology of the power system, i.e., which branches are connected to which bus, but does not have line impedance information. We use this information to estimate the models for line outage identification. In the simulation, we consider single line outages and double line outages. The simultaneous multiple outages for three or more lines are not simulated for this small network as some buses will be disconnected from the rest of the network, and bus voltages will have no variation after the outage, thus the problem of identifying broken lines becomes too trivial. We use the same assumption as in [9], i.e., we assume that the disconnected lines will not cause the underlying graph to become disconnected.

We load the 39-bus system into PSAT, and run the time domain analysis for a total of 300 s. Data are collected at the interval of 30 data points per second. For line outage simulation, a random number t between 1 to 300 is chosen, and line outage at time t is inserted in the simulation. The data set size is $16,215 \times 22$, with 50% for the training set and 50% for the validation set.

The PSAT simulation gives measurements at PMUs. For this study, we have used PMU measurements from all buses. Future study will address the situation when not all buses have PMUs installed. Missing data is an issue for machine learning algorithms, and will be addressed with modified algorithms.

Two prediction models are tested, i.e., Logistic Regression and Random Forest, for both single line and double line outage detection. For Random Forest, we use the algorithm with binary outcomes. The number of trees used in Random Forest is set to 100. The results are summarized in Table 1.

In this problem, precision is the fraction of true outages that are detected among the total detected outages, while recall (e.g., detection rate) is the fraction of true outages that are detected among all true outages.

Table 1. Line outage detection results

Methods	Precision		Recall (detection rate)	
	Single	Double	Single	Double
Logistic regression	0.972	0.935	0.839	0.794
Random forest	0.989	0.9996	1	0.983

Remarks:

1. The reported detection rate results from having time series aligned at the change point when the line is disconnected. This improves the detection rate for all models. For logistic regression, it shows 3% improvement; and for Random Forest, it shows up to 15% improvement in detection rate, as shown in Fig. 4(a).
2. Experiments are done with (1) using all lines in the network, and (2) using only the power lines from one-hop distance, which significantly reduces the

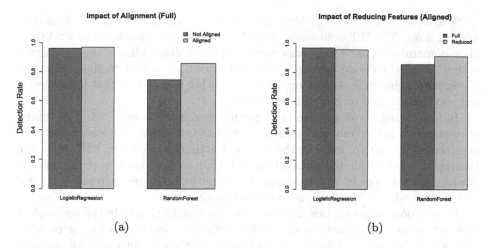

Fig. 4. (a) Having time series aligned around the change point improves detection rates for all models, (b) Using a smaller feature set improves Random Forest, but degrades Logistic Regression.

dimension of feature space. Using reduced feature space improves the detection rate of Random Forest, but degrades the performance of Logistic Regression, as shown in Fig. 4(b).

3. Comparing between two prediction models, Random Forest is superior to Logistic Regression overall regardless of feature selection. This can be explained by the fact that Logistic Regression faces the problem of collinearity, which leads to unstable estimates of coefficients. Random Forest, however, by dividing the predictors into regions consist of p-dimensional hyperrectangle, is free from collinearity.

4. Comparing the proposed methods with the method in [5], Random Forest with binary outcome developed in this paper has paramount advantage with prediction accuracy up to 100%, while accuracy by the method proposed in [5] is no greater than 96.6%. However, the method in [5] requires the knowledge of the Laplacian matrix and reactances of lines, while the proposed methods in this paper work like a blackbox and do not require the detailed Laplacian matrix and also skip the state estimation process.

6   Conclusion and Outlook

We have studied the line outage identification problem in power systems by using a machine learning framework. Under this framework, three learning algorithms are considered: Logistic Regression, Random Forest, and multi-categorical Random Forest. Logistic Regression and Random Forest with binary outcome can be applied to both single line outage and multiple line outage identification with acceptable detection rates, and multi-categorical Random Forest can only be used for single line outage detection.

The developed line outage identification algorithms have been tested via simulation data. The IEEE 39-bus standard test system is simulated in PSAT for time-domain analysis. Time series data from bus voltage phasor angle measurements are used to train the algorithms. Prediction on new data shows that the prediction accuracy is satisfactory and comparable to methods that involve solving state estimation or power flow analysis.

In the current work the tuning of parameters is not considered. The tuning of parameters in the frequency domain as well as in the machine learning models is expected to further improve the results. Further more, when PMU data are incomplete due to limited installation of PMUs in the power system, there is additional challenge for the algorithms to detect line outages with missing data. We plan to address this issue in the future work.

In addition, segmentation of time series by considering only the series after line outage can significantly improve the prediction accuracy, which is possible only when the timing of line outage is known. In the current work, we assumed that a separate procedure for change point detection [13] is applied to detect line outage without identifying the disconnected line. Future work will consider joint detection and identification.

References

1. Milano, F., Vanfretti, L., Morataya, J.C.: An open source power system virtual laboratory: the PSAT case and experience. IEEE Trans. Educ. **51**(1), 17–23 (2008)
2. Andersson, G., et al.: Causes of the 2003 major grid blackouts in north America and Europe, and recommended means to improve system dynamic performance. IEEE Trans. Power Syst. **20**(4), 1922–1928 (2005)
3. US Canada Power System Outage Task - Force: Final report on the august 14, 2003 blackout in the united states and canada: causes and recommendations, 5 April 2004
4. Amin, M.: Three questions about physical grid security. In: IEEE Smart Grid, March 2014
5. Zhu, H., Giannakis, G.B.: Sparse overcomplete representations for efficient identification of power line outages. IEEE Trans. Power Syst. **27**(4), 2215–2224 (2012)
6. Wu, W.B., Cheng, M.X., Gou, B.: A hypothesis testing approach for topology error detection in power grids. IEEE Internet Things J. **3**(6), 979–985 (2016)
7. Zhao, Y., Chen, J., Goldsmith, A., Poor, H.V.: Identification of outages in power systems with uncertain states and optimal sensor locations. IEEE J. Sel. Top. Sign. Proces. **8**(6), 1140–1153 (2014)
8. Tate, J.E., Overbye, T.J.: Line outage detection using phasor angle measurements. IEEE Trans. Power Syst. **23**(4), 1644–1652 (2008)
9. Tate, J.E., Overbye, T.J.: Double line outage detection using phasor angle measurements. In: 2009 IEEE Power Energy Society General Meeting, pp. 1–5, July 2009
10. Garcia, M., Catanach, T., Wiel, S.V., Bent, R., Lawrence, E.: Line outage localization using phasor measurement data in transient state. IEEE Trans. Power Syst. **31**(4), 3019–3027 (2016)
11. Gorinevsky, D., Boyd, S., Poll, S.: Estimation of faults in DC electrical power system. In: 2009 American Control Conference, pp. 4334–4339, June 2009

12. Emami, R., Abur, A.: Tracking changes in the external network model. In: North American Power Symposium 2010, pp. 1–6, September 2010
13. Cheng, M.X., Ling, Y., Wu, W.B.: In-band wormhole detection in wireless ad hoc networks using change point detection method. In: 2016 IEEE International Conference on Communications (ICC), pp. 1–6, May 2016
14. Breiman, L.: Random forests. Mach. Learn. 45(1), 5–32 (2001)

Sparse Feature Extraction Model with Independent Subspace Analysis

Radhika Nath and M. Manjunathaiah[✉]

Computer Science, SMPCS, University of Reading, Reading, UK
radhika.nath@gmail.com, m.manjunathaiah@reading.ac.uk

Abstract. Recent advances in deep learning models have demonstrated remarkable accuracy in object classification. However, the limitations of Convolutional Neural Networks such as the requirement for a large collection of labeled data for training and supervised learning process has called for enhanced feature representation and for unsupervised models.

In this paper we propose a novel unsupervised sparsity-based model using Independent Subspace Analysis (ISA) to implement a hierarchical network for feature extraction. The results of our empirical evaluation demonstrates an improved classification accuracy when max pooling is paired with square pooling within each layer. In addition to accuracy, we further show that it also reduces the data dimensions within the layers outperforming known sparsity-based models.

Keywords: Deep learning · Sparse models ·
Convolutional neural networks · Biologically inspired vision models

1 Introduction

Amongst the biologically inspired models of vision, the supervised Convolutional Neural Network models (ConvNets) have surpassed other models in terms of recognition accuracy [9]. However for scalable machine learning, ConvNets have some inherent limitations. They are supervised learning by design and do not scale as they require a large number of training data. Some generality has been achieved in recent years and issues of over-fitting and unstable gradient have also been addressed with *dropout* and *batch-normalization* techniques, respectively [8,23]. Such advancements have also lead to shorter training durations. However, these models are not specifically designed for robust feature representation. For example, Convolutional models (and others) are sensitive to rotations and thus require multiple rotated instances of the same image. Therefore, to achieve invariant response, the number of images required increases and puts a limit on computational complexity of the models for big data sets. Previous biologically inspired models such as HMAX have also used this technique of providing multiple translated versions of the same image [15]. The only type of invariance

---

R. Nath—Research performed whilst the author was at the University of Reading.

G. Nicosia et al. (Eds.): LOD 2018, LNCS 11331, pp. 494–505, 2019.
https://doi.org/10.1007/978-3-030-13709-0_42

that is encoded in these models is shift invariance which is accomplished with max-pooling.

The feature extraction model implemented in this paper is a hierarchical model where only the final layer is trained with a supervised method. This is similar to self-taught models where all the layers learn features of increasing complexity in an unsupervised manner. In place of the black-box type learning of deep learning models, we adopt a biologically inspired approach where some aspects of the visual cortex such as sparsity are considered to improve recognition accuracy. We demonstrate an improvement over previous unsupervised feature extraction methods and provide a solution for reducing data dimensions along the layers. The key to achieving these novelties is to apply sparsity based algorithms for learning features from the images. Instead of max-pooling as the only non-linear operation, there is an additional square pooling as a non-linearity step which further reduces the dimensions of the layer while improving the recognition accuracy. This step is inspired by the *retinotopic* arrangement of the cells in the visual cortex where dependent units are grouped together [6] which resemble a non-linear pooling operation.

2 Background: Independent Subspace Analysis

Evidence in various studies in neuroscience suggest that sparsity of response occurs in all layers of the visual cortex [2,3,16,17]. The non-Gaussianity in natural data was first represented in terms of sparse coding by Olhausen and Field, where an image is represented by linear combination of very small number of non-zero features [14]. The independent component analysis (ICA) is one of the more popular sparsity based algorithms. It generates features similar to sparse coding but they are statistically independent [6]. In an extension to the ICA, the independent subspace analysis (ISA) and topographical independent component analysis (TICA) were developed in which the components are grouped according to their energy dependencies [6]. One example of an ISA based hierarchical model is the deep learning framework for action recognition described in [10], where a convolution and stacking method is adopted. Another example (with TICA) is the multi-layer model with pooling and local contrast normalization described in [11]. This model simulates a large scale feature detection by training with un-labelled data.

The performance of these models greatly depends on its invariant feature representation which is generally achieved with a non-linearity function. In the convolutional neural networks, HMAX, and its sparsity regularized extension [3], translation invariance is achieved by a max-pooling function over neighbor locations on a feature map. Biological plausibility of max-pooling is also supported by studies that discovered similar functions in the V4 area of primate visual cortex and complex cells in cat visual cortex [21]. In the self-taught learning models described in [11] and [10], L2-pooling function over the feature maps is applied. Additionally, the original models also encode scale invariance by max-pooling over features of same orientations and positions, but slightly different spatial frequency [21]. Some convolutional neural networks have also extended scale invariance in their model [26]. There is always an aim to learn more than one type of invariance.

Phase and position invariance are known to be closely related [6]. Changes in phase for a spatially localized stimulus translates into small shifts in position (in the direction of its oscillations) such that it is termed as a special case of position invariance. Complex cell properties of the ISA and TICA therefore display phase invariance and limited shift invariance [4]. To obtain high level features with improved classification accuracy, both L2-pooling and max-pooling are applied in the proposed models in this paper.

2.1 Independent Subspace Analysis

In ISA, the dependent components are grouped into subspaces of pre-defined size. The neighborhood function in this case is defined as,

$$h(i,j) = \begin{cases} 1 & \text{if } \exists q : i, j \in S_q \\ 0 & \text{otherwise.} \end{cases} \tag{1}$$

where component $S \in \{s_1, ..., s_n\}$ from Eq. (1) is divided into n-tuples such that the s_i inside a tuple are dependent on each other, $i, j \in \{1, ..., q\}$ is the index of the n-tuple. S_q represents the set of indices of the component s_i that exists within that tuple [4]. With $W = (w_1, ..., w_n)^T = A^{-1}$ and input x, the cost function for maximum log likelihood estimation in this case is given by,

$$logL(W) = \sum_{t=1}^{T} \sum_{q=1}^{Q} G(\sum_{i \in S_q} (w_i^T x(t))^2) + Tlog|detW| \tag{2}$$

where $(w_i^T x)^2$ is the energy term, G is a function that gives the log probability density of s_i, Q denotes the number of subspaces and T is the number of realizations of input x [4].

The total response of each subspace is the squared sum of each component, also termed as the L2-pooling.

$$e_q = \sqrt{\sum_{i \in S_q} s_i^2} \tag{3}$$

Similarly, in the Topographic ICA, the arrangement of the learned units is in a way such that it reduces the distance between correlated components and thereby reducing the *wiring length* between two statistically related neurons. In neuroanatomy, it is explained as the length of the axons that connect the neurons [5]. This minimization of wiring length has been described in [6] as a model for the compactness of the brain volume and speed of signal processing. Here, the grouping of dependent components is determined by neighborhood function that defines the topography. Proximity within the topography indicates strength of its second order correlation. In this paper only ISA model is investigated in depth. While TICA can also be applied in similar manner, and achieve similar results, some factors such as neighborhood size and overlap area affect the model differently.

3   Feature Extraction Model with ISA

In this section, implementation of the new hierarchical feature extraction model is presented. The first simple and complex cell layers are denoted by S_1 and C_1. Here, the combination of S_1 and C_1 layer functions is referred to as V_1 layer. The V_i layer of this model comprises three sub layers: S_i is the response of orientation, spatial frequency and position selective linear filters, C_{i_a} represents the non linear L2-pooling of the S_i outputs within a subspace or topographic neighborhood by Eq. (3), and C_{i_b} denotes max-pooling output over neighboring locations for each C_{i_b} feature. Since these non-linearities correspond to phase and position invariance respectively, they are referred to as such in the model description.

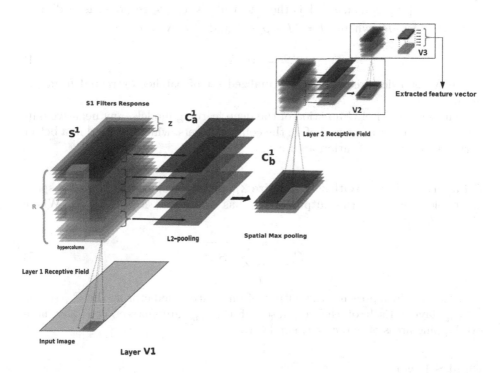

Fig. 1. Multiple V_i layers of the ISA feature extraction model

ISA Feature Extraction Model

In this model, the S layer filters are learned by applying ISA algorithm.

The structure of the model is in the form of *hypercolumns* (also illustrated in [19]) or feature maps which comprises all the filter outputs for a spatial location. The subspace size is denoted by Z_1 and the total number of filters or bases at S_1 is R_1. Figure 1, shows the full model with multiple V_1 layers. The receptive field size p_i of the V_i layer of the model indicates the width of the square area that is sampled from the $C_{(i-1)_b}$ layer (which is the input image for S_1).

S Layer. The filters generated from sampling random patches of images are grouped into subspaces based on higher order energy correlations. The impact of subspace size along the different layers has an effect on object classification results. For a fixed set of S_i filters, increasing subspace size Z strengthens phase invariance but decreases the number of features.

S_i **layer:** For the first layer V_1, each S_1 layer filter is of size $p_1 \times p_1$, where p_1 is the width of the square receptive field of the first layer. The S_1 filter is applied on a patch of $p_1 \times p_1$ of the input image X which is of size $M \times N$.

If $W_i = \{w_{i_1}, w_{i_2}, ..., w_{i_{R_i}}\}$ is the set of filters, the S_i response is of dimensions $\tilde{M} \times \tilde{N} \times R_i$, where $\tilde{M} = M - p_i + 1$ and $\tilde{N} = N - p_i + 1$.

$$S_i = \langle W, X_p \rangle \tag{4}$$

where X_p is a decorrelated and normalized set of patches extracted from the input image.

Similar to the ReLU function of the deep learning models, any negative output of S_i layer is set to zero before the complex layers, which resulted in a better performance in classification accuracy.

C Layer. C_{i_a} **layer:** With subspace size Z_i, all the S_i values within the subspace are pooled such that the output of C_{i_a} has dimensions $\tilde{M} \times \tilde{N} \times \tilde{R}_i$. Where $\tilde{R}_i = R_i / Z_i$.

$$C_{i_a} = \sqrt{\sum_{j \in Z_i} S_{ij}^2} \tag{5}$$

Equation (5) represents the output of one feature detector at the C_{i_a} stage.

C_{i_b} **layer:** Each of the responses of the C_{i_a} are max-pooled over non-overlapping areas of size $r_i \times r_i$ similar to [3].

Final S Layer

In the final S_n layer of the model, the square root of the sum of energies (or L2-pooling) of the values across all the location on each feature map is obtained as the feature vector.

$$C_n = \sqrt{\sum S_n^2} \tag{6}$$

The C_{i_a} step is generally not applied when obtaining the final feature vector (as illustrated in the V_3 layer of Fig. 1). The final feature vector of size $1 \times R_n$ forms the input for the classifier.

The model functions in two phases: The first phase is for learning the filters in which a limited set of images is used to learn all the filters in all the layers. During the second phase, all the images are passed through the above mentioned layer operations to extract features. These features are then tested for classification accuracy.

4 Empirical Evaluations

Part 1: Model Parameters

In the hierarchical model illustrated in Fig. 1, the factors that affect its performance are the subspace size and receptive field sizes. The experiments presented in this section are to evaluate the parameters that optimize the performance of the models. For convenience, a reduced dataset is used with even number of images[1]. All the images were resized such that the smallest side had dimension 140. The extracted features were classified using an SVM classifier with the average of 30 different splits of test and train data.

Subspace Size of the ISA Model. Generally, large number of feature detectors are optimal for recognition models as they capture image complexity more accurately. Reducing subspace size increases the final feature size of the V_i layer, whereas increasing subspace size improves processing speed by reducing the size of C_{i_a} output.

Subspace Size of Final Layer. The parameters for V_1 and V_2 are fixed, while changing subspace size for the V_3 layer in the ISA model from Fig. 1. The model specifications for this experiment are described in Table 1.

Table 1. Model specifications: The subspace size Z_3 of V_3 is varied

Models	$V_1, p_1 = 11$			$V_2, p_2 = 12$			$V_3, p_3 = 13$
	$S_1 (R_1)$	C_{1_a}		$S_2 (R_2)$	C_{2_a}		$S_3 (R_3)$
		Z_1	$\tilde{R}_1$		Z_2	$\tilde{R}_2$	
ISA	100	4	25	150	5	30	400

---

[1] Tested on a database of 10 different categories included: airplane, bonsai, butterfly, car-side, chandelier, faces, ketch, leopards, motorbikes, watch of objects from the *CalTech*101 dataset [1].

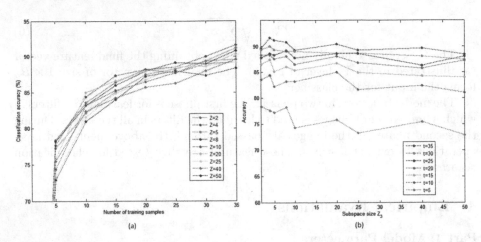

Fig. 2. Classification accuracy for 10 classes when subspace size Z_3 is changed with fixed number of R_3. L2-pooling at V_3 is not applied so the feature vector is of the same length for all the cases: (a) Accuracy with respect to number of training samples (b) Accuracy with respect to subspace size, where t represents the number of training samples

In Fig. 2, the S_3 layer filters are formed with different subspace sizes, but are not pooled with Eq. 5. So, after applying spatial pooling over all the locations of the features, the final feature vector is of size 1×400.

Figure 2a demonstrates when the overall performance of the features with smaller subspace size and with a fixed feature vector size perform better than the subspaces of largest sizes (45, 50). However, Fig. 2b indicates that $Z_3 = 4$ and $Z_3 = 20$ classifies with better accuracy for most of the training sample sizes.

Number of S2 Layer Filters. Here, the parameters for V_1 and V_3 are fixed, while changing subspace size for the V_2 layer in the ISA model from Fig. 1. The model specifications for this experiment are described in Table 2.

Table 2. Model specifications

Models	$V_1, p_1 = 11$			$V_2, p_2 = 12$			$V_3, p_3 = 13$
	$S_1\ (R_1)$	C_{1_a}		$S_2\ (R_2)$	C_{2_a}		$S_3\ (R_3)$
		Z_1	$\tilde{R}_1$		Z_2	$\tilde{R}_2$	
ISA	100	4	25	300	-	-	200

In Fig. 3a, the classification accuracy for the different subspace sizes at the V_2 layer is depicted. The model with largest $\tilde{R}$ shows highest accuracy, but from Fig. 3b, it is seen that the second largest $\tilde{R}$ which is 100 for $Z_2 = 3$ does not perform better than $Z_2 = \{4, 5\}$.

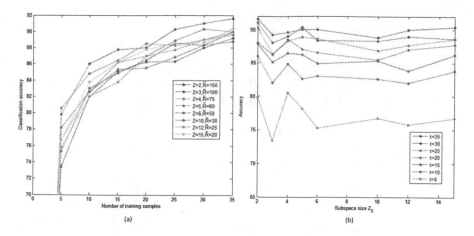

Fig. 3. Classification accuracy for 10 classes when subspace size Z_2 is changed with fixed value of $R_2 = 300$. Pooling at C_{2_a} is applied such the feature vector size $\tilde{R}_2$ changes for all the cases: (a) Accuracy with respect to number of training samples (b) Accuracy with respect to subspace size, where t represents the number of training samples

The above experiments indicate that although larger subspace sizes are preferred, simply increasing the number of filters or subspace sizes does not necessarily translate to a better model. For example, the results in Fig. 3b showing better accuracy for $Z_2 = 4, \tilde{R}_2 = 75$ than $Z_2 = 3, \tilde{R}_2 = 100$ indicate that larger subspace sizes represent the statistical properties of the data more accurately.

This highlights the drawback of applying ISA with prior assumption of pooling sizes since the probability of best data representation is not guaranteed. In [7] it is shown that a relatively large subspace size is optimal for representation of natural image statistics, depending on the size of the input patch. For higher complexity data, such as the input sample to the V_2 and V_3 layers, the most optimal subspace sizes are 2 and 5 for the V_2 layer and 4 for the V_3 layer. It is thus more beneficial for the subspace sizes to be estimated adaptively rather than being fixed.

Receptive Field Size. Studies have shown that the receptive field size increases as we go from lower to higher levels of the VC [20], where the cells of the first layer process local stimulus within a small localized area. Figure 4 shows the performance of the architecture in Fig. 1 with different receptive field sizes. The results in Fig. 4 indicate that increase in receptive field size also improves performance. The number of filters are given by Table 3 and p refers to width of the square patch.

Increasing receptive field size improved the performance only when the ratio of increase is not too large as seen from Fig. 4. The model with decreasing RF size ($p1 = 11, p2 = 10, p3 = 9$) is also more accurate than the ones with $p1 = 11, p2 = 13, p3 = 14$ and $p1 = 11, p2 = 13, p3 = 15$.

Table 3. Model specifications

Models	V_1			V_2			V_3
	$S_1(R_1)$	C_{1_a}		$S_2(R_2)$	C_{2_a}		$S_3(R_3)$
		h_1, Z_1	$\tilde{R}_1$		h_2, Z_2	$\tilde{R}_2$	
ISA	144	9	16	100	4	25	225

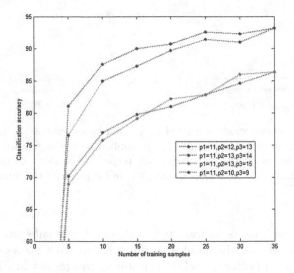

Fig. 4. Performance for different receptive field sizes p for ISA

Part 2: Multi-class Object Categorization on CalTech101 Dataset

To compare ISA model performance with other feature extraction models the *CalTech*101 database is used for multi-class object categorization [1]. The standard method of splitting the training set of images into 15 and 30 images per class is applied. The number of S_1 and S_2 layer filters are 144 and 300 respectively. The corresponding subspace sizes are $Z_1 = 9$ and $Z_2 = 5$. Although for better results a larger number of dictionaries in each layer is more beneficial these parameters allowed for a faster computation time. The sample size for learning dictionaries at each layer is $50,000$. The dictionaries (or filters) are learned from just 10 images from each category.

Current state of the art models have achieved good results for the *CalTech*101 database. The list of models here are mostly biologically motivated hierarchical models based on the unsupervised feature extraction models. Most of the reported accuracy of these models are a result of varying length of features. For example, in the HMAX model in [18, 22], classification using a dictionary of 4075 features had an accuracy of 54% [24]. In [24] by increasing the scale depth of the S_1 units, an accuracy of 61% with 4080 features is reported. In [3], an accuracy of 73.67% is achieved for training size of 30 for feature length of $21,504$. A further increase of 76.13% is reported for a feature length of $43,008$. In [27],

Table 4. Classification accuracy of the ISA model in comparison to similar feature extraction models for unsupervised learning.

Classification accuracy for number of 15 and 30 training images per category		
Model	15 images	30 images
Serre [22]	35	42
Mutch and Lowe [13]	48	54
HMAX-S [24]	54	61
HMAX-S (extended) [25]	68.49 ± 0.75	76.32 ± 0.97
Lee et al. [12]	57.7 ± 1.5	64.5 ± 0.5
Zeiler et al. [28]	-	71 ± 0.10
Yu et al. [27]	-	74.0
Sparsity regularised HMAX [3]	68.98 ± 0.64	76.13 ± 0.85
ISA model (dictionary size 4000)	**72.65% ± 1.08**	**79.70% ± 0.55**

an unsupervised two layer model with sparse coding and pooling is developed which also achieved a high classification accuracy of 74% with codebook of 4096 features. In most of the models listed here, they had dictionary sizes of at least 4000 or higher. Therefore, the last layer features of length 1000 are extracted 4 times and concatenated. The final accuracy value obtained is from averaging the results of 10 different splits of test and train data.

The resulting accuracy is higher than the unsupervised learning models in Table 4. Compared to the Adaptive Deconvolutional Networks [28], which uses 4 layers for feature extraction the accuracy of the ISA model with 4000 features is much higher.

5 Conclusion

The pathway to achieving biomimetic visual capabilities is through superior feature representation rather than having deeper hierarchies in models and from increasing the number of parameters. Inspiration from neuroscience research have led to breakthroughs in the computer vision field before and can deliver further advances in self-taught models that we investigate in this paper.

The main advantage of the ISA model is its accuracy and dimension reduction in comparison with other similar unsupervised or self-taught feature extraction models. Due to the drawbacks of current supervised models, the ISA based learning is a step towards sparsity oriented unsupervised learning models.

References

1. Learning generative visual models from few training examples: an incremental Bayesian approach tested on 101 object categories, vol. 106. https://doi.org/10. 1016/j.cviu.2005.09.012
2. Baddeley, R., et al.: Responses of neurons in primary and inferior temporal visual cortices to natural scenes. Proc. Biol. Sci. **264**(1389), 1775–1783 (1997). http://www.jstor.org/stable/51114
3. Hu, X., Zhang, J., Li, J., Zhang, B.: Sparsity-regularized HMAX for visual recognition. PLoS One **9**(1), e81813 (2014). http://journals.plos.org/plosone/ article?id=10.1371/journal.pone.0081813
4. Hyvärinen, A., Hoyer, P.: Emergence of phase- and shift-invariant features by decomposition of natural images into independent feature subspaces. Neural Comput. **12**(7), 1705–1720 (2000)
5. Hyvärinen, A., Hoyer, P.O., Inki, M.: Topographic independent component analysis. Neural Comput. **13**(7), 1527–1558 (2001). https://doi.org/10.1162/ 089976601750264992
6. Hyvärinen, A., Hurri, J., Hoyer, P.O.: Natural Image Statistics: A Probabilistic Approach to Early Computational Vision. Springer, Heidelberg (2009). https:// doi.org/10.1007/978-1-84882-491-1. Google-Books-ID: pq_Fr1eYr7cC
7. Hyvärinen, A., Köster, U.: Complex cell pooling and the statistics of natural images. Netw. Comput. Neural Syst. **18**(2), 81–100 (2007). https://doi.org/10. 1080/09548980701418942
8. Ioffe, S., Szegedy, C.: Batch normalization: accelerating deep network training by reducing internal covariate shift. CoRR abs/1502.03167 (2015). http://arxiv.org/ abs/1502.03167
9. Krizhevsky, A., Sutskever, I., Hinton, G.E.: ImageNet classification with deep convolutional neural networks. In: Pereira, F., Burges, C.J.C., Bottou, L., Weinberger, K.Q. (eds.) Advances in Neural Information Processing Systems 25, pp. 1097–1105. Curran Associates Inc. http://papers.nips.cc/paper/4824-imagenet-classification-with-deep-convolutional-neural-networks.pdf
10. Le, Q.V., Zou, W.Y., Yeung, S.Y., Ng, A.Y.: Learning hierarchical invariant spatio-temporal features for action recognition with independent subspace analysis. In: Proceedings of the 2011 IEEE Conference on Computer Vision and Pattern Recognition, CVPR 2011, pp. 3361–3368. IEEE Computer Society, Washington, DC (2011). https://doi.org/10.1109/CVPR.2011.5995496
11. Le, Q., et al.: Building high-level features using large scale unsupervised learning (2012). http://research.google.com/pubs/pub38115.html
12. Lee, H., Grosse, R., Ranganath, R., Ng, A.Y.: Convolutional deep belief networks for scalable unsupervised learning of hierarchical representations. In: Proceedings of the 26th Annual International Conference on Machine Learning, ICML 2009, pp. 609–616. ACM, New York (2009). https://doi.org/10.1145/1553374.1553453
13. Mutch, J., Lowe, D.G.: Object class recognition and localization using sparse features with limited receptive fields. Int. J. Comput. Vis. **80**(1), 45–57 (2008). http://link.springer.com/article/10.1007/s11263-007-0118-0
14. Olshausen, B.A., Field, D.J.: Sparse coding with an overcomplete basis set: a strategy employed by V1? Vis. Res. **37**(23), 3311–3325 (1997)
15. Riesenhuber, M., Poggio, T.: Hierarchical models of object recognition in cortex. Nat. Neurosci. **2**(11), 1019–1025 (1999). https://doi.org/10.1038/14819. PMID: 10526343

16. Rolls, E.T.: Invariant visual object and face recognition: neural and computational bases, and a model, VisNet. Front. Comput. Neurosci. **6**, 35 (2012). https://doi. org/10.3389/fncom.2012.00035, PMID: 22723777

17. Rolls, E.T., Treves, A.: The neuronal encoding of information in the brain. Prog. Neurobiol. **95**(3), 448–490 (2011). http://www.sciencedirect.com/ science/article/pii/S030100821100147X

18. Serre, T., Wolf, L., Poggio, T.: Object recognition with features inspired by visual cortex. In: IEEE Computer Society Conference on Computer Vision and Pattern Recognition, CVPR 2005, vol. 2, pp. 994–1000 (2005). https://doi.org/10.1109/ CVPR.2005.254

19. Serre, T.: Hierarchical models of the visual system. In: Jaeger, D., Jung, R. (eds.) Encyclopedia of Computational Neuroscience, pp. 1–12. Springer, New York (2014). https://doi.org/10.1007/978-1-4614-7320-6_345-1

20. Serre, T., Oliva, A., Poggio, T.: A feedforward architecture accounts for rapid categorization. Proc. Nat. Acad. Sci. **104**(15), 6424–6429 (2007). http://www.pnas. org/content/104/15/6424

21. Serre, T., Riesenhuber, M.: Realistic modeling of simple and complex cell tuning in the HMAX model, and implications for invariant object recognition in cortex (2004)

22. Serre, T., Wolf, L., Bileschi, S., Riesenhuber, M., Poggio, T.: Robust object recognition with cortex-like mechanisms. IEEE Trans. Pattern Anal. Mach. Intell. **29**(3), 411–426 (2007). https://doi.org/10.1109/TPAMI.2007.56

23. Srivastava, N., Hinton, G., Krizhevsky, A., Sutskever, I., Salakhutdinov, R.: Dropout: a simple way to prevent neural networks from overfitting. J. Mach. Learn. Res. **15**, 1929–1958 (2014). http://jmlr.org/papers/v15/srivastava14a.html

24. Theriault, C., Thome, N., Cord, M.: HMAX-S: deep scale representation for biologically inspired image categorization. In: 2011 18th IEEE International Conference on Image Processing, pp. 1261–1264 (2011). https://doi.org/10.1109/ICIP.2011. 6115663

25. Theriault, C., Thome, N., Cord, M.: Extended coding and pooling in the HMAX model. IEEE Trans. Image Process. **22**(2), 764–777 (2013). https://doi.org/10. 1109/TIP.2012.2222900

26. Xu, Y., Xiao, T., Zhang, J., Yang, K., Zhang, Z.: Scale-invariant convolutional neural networks. http://arxiv.org/abs/1411.6369

27. Yu, K., Lin, Y., Lafferty, J.: Learning image representations from the pixel level via hierarchical sparse coding. In: 2011 IEEE Conference on Computer Vision and Pattern Recognition (CVPR), pp. 1713–1720, June 2011. https://doi.org/10.1109/ CVPR.2011.5995732

28. Zeiler, M.D., Taylor, G.W., Fergus, R.: Adaptive deconvolutional networks for mid and high level feature learning. In: Proceedings of the 2011 International Conference on Computer Vision, ICCV 2011, pp. 2018–2025. IEEE Computer Society, Washington, DC (2011). https://doi.org/10.1109/ICCV.2011.6126474

Bayesian Clustering of Multivariate Immunological Data

Alberto Castellini[✉] and Giuditta Franco

Department of Computer Science, Verona University,
Strada Le Grazie 15, 37134 Verona, Italy
{alberto.castellini,giuditta.franco}@univr.it

Abstract. Given a dataset of B cell subpopulation quantities, for about six thousand patients, that is a cross-sectional immunological dataset, here we detect clusters representing models of immune system states in an unsupervised way (i.e., according only to their different statistical properties). Two time-evolving B cell networks are also generated from data-driven hidden Markov models, with four and five hidden states, respectively. Our interpretation from a biomedical viewpoint of the statistical parameters of the Bayesian models confirms an age related decline of some types of B cell functions and finds out a class of old patients with unexpected B cell values.

1 Introduction

The immune system may be assumed as a network of interacting cells, evolving during human life in terms of presence/absence and strength of type of interactions, which change through childhood, young, and mature adulthood, to the decline of old age [21]. Emergent research interests involving systems biology and knowledge discovery approaches, as well as biomedical data analysis, focus on the lifetime evolution of the immune system, in terms of changes of defence mechanisms of a human being during his/her infancy, growing/mature age and senescence [15,25]. A problem widely investigated in computational immunology is the role of B cell sub-populations (e.g., B cell memory, B cell activation) in the cellular and humoral response of the immune system [10,11,22].

Aging is a complex process which negatively impacts lymphocyte (in particular B-cell) biological variability [24], and an age-related decline, referred to as *immunosenescence*, seems to be characterized by a decrease in cell-mediated immune (T- and B-cell) functions [12,16]. Elucidate mechanisms underling immunosenescence is expected to have with a notable social and economical impact, for example, to design new therapies and vaccines for elderly people.

Fast and efficient computational techniques have been recently introduced in the literature of machine learning to infer new knowledge from data, in particular for multivariate clustering or time-series analysis, such as segmentation and change-point detection [2,17]. Segmentation of multiple time-series is a complex problem, since different data sequences may show different aspects of the

© Springer Nature Switzerland AG 2019
G. Nicosia et al. (Eds.): LOD 2018, LNCS 11331, pp. 506–519, 2019.
https://doi.org/10.1007/978-3-030-13709-0_43

underlying processes. Main methodologies in this field are based on motif clustering [13, 23].

In this paper, we provide and discuss some models of the immune system state based on B cell quantities and propose related time evolving networks of cell relationships, generated from Hidden Markov Models (HMMs) with four and five (hidden) states. Models are trained in an unsupervised way from an immunological dataset. Data were collected over about six thousands patients (whose age gave us a time line to sort the B cell quantities) that were given as a cross-sectional dataset. Eight different types of B-cell subpopulations are identified according to the expression configuration of three specific receptor clusters, and a multivariate dataset is analyzed as the system observation to define our probabilistic models. As an initial work, here we viewed the data of patients of different ages (given in days), as time-line data, by leaving undefined the temporal transition of distributions of eight B-cell subpopulations, and applied HMMs to model the given dataset. Finally, dynamical relationships between different types of B-cells has been measured by pairwise correlation, and resulted in an actual decline of B cells mean quantities by aging.

A first model proposed for this dataset describes a possible sequence of (ex-vivo observed) B cell maturation steps in human body [5]. It was based on Metabolic P Systems [4, 18], with linear regulation maps, generated by regression techniques based on genetic algorithms [7–9]. These models are discrete dynamical systems, introduced in the context of membrane computing and applied to immunological systems [14], to provide a deterministic multiset evolution by means of state functions (that give the quantities of transformed elements).

As in [26], our original goal has been to generate a segmentation of multivariate time series based on a changing correlation structure, as well as on changing mean and variance. The problem may be graphically explained as in Fig. 1: a univariate time series segmentation might be induced by a change in mean (a), variance (b), or model order in autoregressive processes (c). However, in the case of multivariate time series the correlation structure can also identify different segments (d). These changes are typically hard to recognize but they can contain key information about the generative process under investigation.

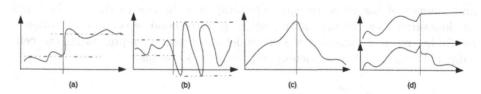

Fig. 1. Four measures of dissimilarity among time series segments: (a) mean, (b) standard deviation, (c) model order, (d) correlation structure.

In [6] we modeled each segment by piecewise multivariate linear regression, and generated networks of cells (for each segment) according to model coefficients. In [3] we computed the segments that maximize the differences in correlation matrices among segments themselves. The first method depends on parameters, such as the threshold of regression model errors in each segment, which may be difficult to tune, while the second method has limitations in the number of segments since it is a brute force algorithm. Along this research line of investigation to analyze and deeply understand the underlying information of our immunological dataset, here we use a more robust probabilistic framework, namely HMMs, having strong theoretical foundations and well-grounded algorithms for learning parameters from data, to infer the most likely sequence of hidden states that provide a sequence of observations.

A more refined approach could require the application of Inertial HMMs [19], where transition probabilities are defined to force clusters to form segments of the patients time-line, that is leaved as future work. In this initial work, we applied HMMs without any constraint on the transition matrix to find clusters of patients characterized by different statistical properties, and analyze the results also according to the patients age. Although the transition probabilities result in a random matrix, a different dynamical behavior for B cells is emerged in clusters corresponding to different age ranges.

In the next section we briefly describe the dataset and the probabilistic algorithms employed to define our models, while in Sect. 3 some results are discussed, also in biomedical terms, followed by a few conclusions.

2   Materials and Methods

2.1   Dataset

Data were collected at the University Hospital of Verona (Italy) from 2001 to 2012, as measures of amount of B cells exhibiting the combinations of receptor clusters *CD27*, *CD23* and *CD5* in 5,954 patients. There were 2,910 males and 3,045 females (male/female ratio: 0.95) and the median age of the patients was 37 years (range: 0–95 years). More details on the dataset and the clinical method used to collect it may be found in [5, 24].

In other terms, B cell phenotype of 8 subpopulations (indicated by presence and absence of the three receptor clusters), may be abstractly described by random variables accounting for quantities of corresponding cell in each patient. In Table 1 random variables of our model are reported, representing the cell subpopulation size of each phenotype present in our dataset.

Table 1. Model variables.

$X_1 = $ CD5+ CD23+ CD27−	$X_5 = $ CD5− CD23− CD27+
$X_2 = $ CD5− CD23+ CD27−	$X_6 = $ CD5+ CD23− CD27+
$X_3 = $ CD5− CD23− CD27−	$X_7 = $ CD5+ CD23+ CD27+
$X_4 = $ CD5+ CD23− CD27−	$X_8 = $ CD5− CD23+ CD27+

The cross-sectional dataset is a matrix of 5,954 rows and eight columns, in which rows (i.e., patients) can be sorted by age (given in days), so obtaining a multivariate time-line where patient age represents time. This artificial view of the data was advanced with our original goal to search for an optimal segmentation [3,6] and allow us here to cluster patients in an unsupervised way by HMMs with unconstrained transition probabilities.

This may be a particular case where it is reasonable to reduce cross-sectional data into multivariate time-series, even if having not auto-correlated values among patients of close age. From a different viewpoint, if we sort the data according to the age of patients, we have a screenshot of the human immune system (or, more specifically, of the B-cell network) along the lifetime of a meta-patient, who may be assumed to have a basic functioning system (the number of patients is high enough, to have a dense dataset on the time line and to be able to neglect possible known or unknown diseases in the system).

2.2 HMM-Based Clustering Model

Following the research line started in [3,5,6] here we present novel results about the application of dynamic Bayesian approaches, such as HMMs [20], for identifying age-related and data-driven properties of the immune system given the dataset introduced above.

HMMs enable to represent each segment by a multivariate Gaussian distribution whose parameters are interpretable as the main statistical properties (i.e., mean and variance of the number of cells for each cell type, and correlation among the number of cells of different cell types) of the data in the segment itself. Hence, we detect data segments in an unsupervised way, i.e., according only to their different statistical properties, and then we interpret these properties from a biomedical viewpoint.

An HMM is a probabilistic model able to describe Markovian stochastic processes, characterized by the following elements [20]:

- a set $S = \{S_1, \ldots, S_N\}$ of *hidden state values*, representing the hidden factors (i.e., the states of the immune system in our application) that generate observed data,
- a set $O \subseteq \mathbb{R}$ of *observed data values*, represented by observed B cell quantities in our case study,
- a *state transition probability distribution* $A = \{a_{ij}\}$, where $a_{ij} = P[q_{t+1} = S_j \mid q_t = S_i]$, $1 \leq i,j \leq N$, which represents the probability to switch from value S_i for an hidden state q_t, at time t, to the value S_j for the hidden state q_{t+1}, at time $t+1$;
- an *observation probability distribution* for each state S_j, namely the set $B = \{b_j(O)\}$ of probability distribution functions $b_j(O) = \mathcal{N}(O, \mu_j, \Sigma_j)$, where $j = 1, \ldots, N$, O is the set of observations, μ_j the mean and Σ_j the covariance matrix of the distribution b_j,
- an initial state distribution $\pi = \pi_i$, where $\pi_i = P[q_1 = S_i], 1 \leq i \leq N$.

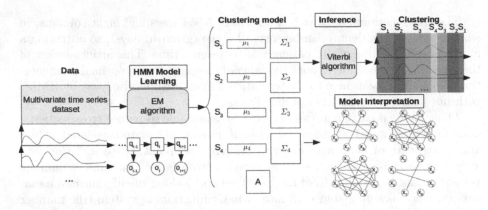

Fig. 2. Data analysis framework.

Figure 2 shows the key points of our data analysis framework based on HMMs. It includes *(i)* a *learning* stage in which HMM parameters are tuned to fit available data, *(ii)* an *inference* stage where the learned model is used to generate the segmentation of the dataset, *(iii)* the *interpretation* of segments given the parameters of related observation probability distribution. In the following the three steps are explained in more details.

HMM Model Learning. The Expectation-Maximization (EM) algorithm [1] was used to learn the HMM parameters, namely the state transition probability distribution A, the observation probability distributions B and the initial state distribution π. These parameters were tuned as to maximize the likelihood of the model to fit the dataset of patients' B-cell quantities. No data were associated to the hidden state, which represented the cluster learned in an unsupervised way. We generated two models having four and five hidden states respectively. We have chosen these small numbers of classes using a-priori knowledge because we aimed at generating coarse-grained models with interpretable clusters, however more complex models can be generated using a larger number of hidden states. Observation models were set to single component multivariate Gaussian distributions (with one dimension for each observed variable). The initial state distribution was set to uniform over the set of hidden states, the initial transition matrix was set to a uniform random stochastic matrix, initial means were computed by k-means and initial covariance matrices set to diagonal. The Matlab function *mhmm_em* (from the HMM Toolbox) was used to estimate model parameters. The maximum number of iterations for the EM algorithm was set to 20. The output of this stage is the HMM clustering model $\lambda = (A, B, \pi)$, shown in the middle of Fig. 2.

Inference. The Viterbi algorithm [1,20] (Matlab function *Viterbi_path*) was used to generate the most likely sequence of hidden states given the observed sequence of cell profiles. This sequence can be seen as an approximate segmentation of the dataset. We notice that classical HMMs, as used in this work, produce high rates of state transitioning, depending on the cluster size and patients age, since they do not consider any constraint on the transition probability (as some extended model do, e.g., Inertial HMMs). We used this method in order to investigate if age intervals (i.e., segments) can be naturally inferred from statistical properties of the data or if different structures (such as, clusters with heterogeneous ages) emerged. The output of this step is the clustering of the dataset presented in Sect. 3.

Model Interpretation. HMM parameters have the important feature of being interpretable in terms of statistical properties of the states (see Fig. 3). For each hidden state $S_i, i = 1, \ldots, K$, we have 8 parameters $\mu_{i1}, \ldots, \mu_{i8}$ related to variable means, 8 parameters $\sigma_{i1}, \ldots, \sigma_{i8}$ related to variable standard deviations, and 28 parameters $\rho_{i12}, \rho_{i13}, \ldots, \rho_{i78}$, where ρ_{ijk} is the correlation between variable X_j and variable X_k in state S_i (for instance, ρ_{i12} represents the correlation between variable X_1 and variable X_2 in class S_i). These values are computed by normalizing the covariance matrices Σ_i. In order to provide a first statistical and biological interpretation of these clusters, we compared their 44 parameters and generated correlation networks, where nodes represent variables, and edges are present if the correlation between two nodes (variables) is greater than a threshold. Moreover, we sorted the correlation parameters according to their variance among different states (see Fig. 3), since correlation parameters having larger variance have also a higher information content.

3   Results

Here we discuss the results emerged by the model with four states S_1, S_2, S_3, S_4, as in Fig. 3, and compare them with those obtained by the model with five states $S'_1, S'_2, S'_3, S'_4, S'_5$.

3.1   Cluster Characterization by Data Means

A first characterization may be discussed by analyzing the bar charts (and the means) reported in Fig. 4. They show that state S_2 includes 3085 patients, state S_3 involves 2222 patients, segment S_1 550 patients, and state S_4 97 patients. Box plots in Fig. 4c display the distribution of patient age over the states, and, for each variable, the distribution of its values around the means (represented by red lines). Even if a complete segmentation of patients was not obtained by our coarse-grain clustering, the four clusters may be ordered by increasing (mean) patients age, as S_1, S_3, S_2, and S_4, where infants mostly belong to segment S_1, S_3 includes patients falling into a second age range (approximately 20–40 years old), and S_2, S_4 cover more advanced age patients (respectively).

State S_1 *owns most of the maximal values, and is characterized by relatively high mean values*, maximal for variables X_1, X_2, X_3, X_4, whose expectation is

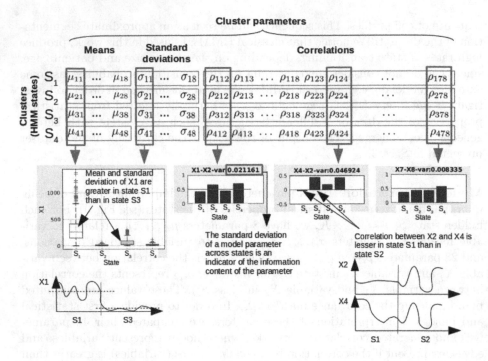

Fig. 3. Interpretation of cluster models: each cluster represents a recognized state of the immune system and it is identified by a group of patients having similar cell configurations. Cluster parameters are the means and standard deviation of each cell and the correlation between each couple of cells.

at least double of that in each other cluster. By looking at Table 1, we notice that these four variables are all those which do not express receptor CD27. Then, according to this model, we could hypothesize that all eight B cell subpopulations in an infant stage of human being life have relatively high mean values, which are even maximal for the subpopulations which do not express CD27.

State S_3 has average values of means for all variables. In Fig. 4, in fact we may notice that, in each box-plot, the mean value of this cluster is never maximal nor minimal. Relevantly, *state S_2 (which contain more than half patients) has minimal means*, with respect to the other clusters, for all the variables. Therefore, according to this probabilistic study, we may observe that expectations of the B cells subpopulations under consideration decreases with age, as also suggested by experimental observations in the medical literature [12,16].

In next section, we will analyze more in detail how correlations in B cell networks change during life time, beside the decrease of mean values for all variables, along the above three clusters which cover almost all patients. To conclude this first analysis, however, a comment should be devoted to the state S_4, including only the 97 generally old patients, that is characterized by maximally high mean values for variables X_5, X_6, X_7, X_8. This is a strange behavior, regarding all B

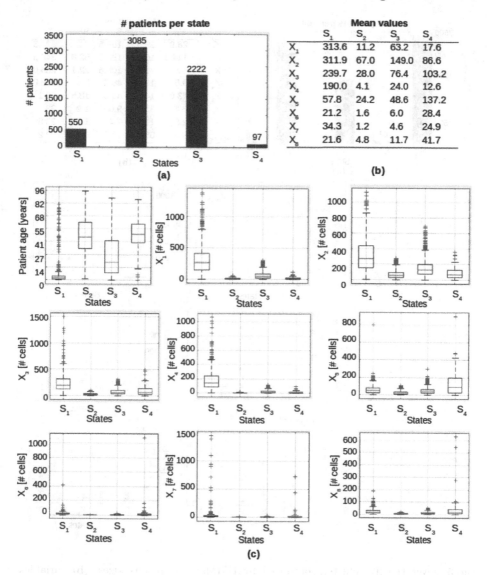

Fig. 4. Over the four clusters produced by HMMs: (a) clusters sizes; (b) variables median values; (c) box-plots with the age and the variable values distributions around the means. (Color figure online)

cell subpopulations having expressed the receptor clusters CD27, which identifies a class of outlier patients, possibly corresponding to an anomalous state of the immune system.

To conclude this section, we point out an aspect which confirms the analysis of our classification, also in the way it is interpreted. When a model with five clusters is considered, a correspondence may be found among states (although

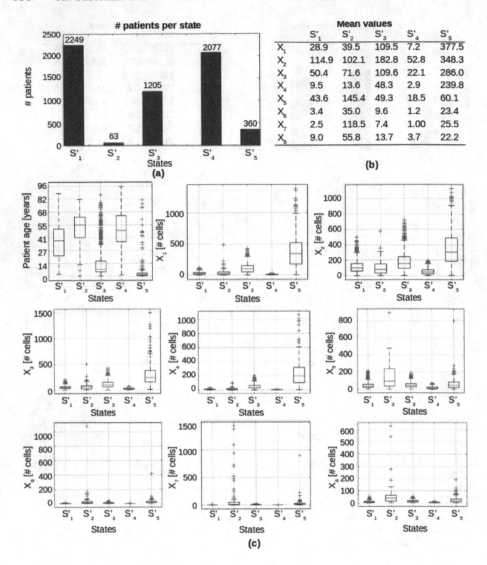

Fig. 5. Over the five clusters produced by HMMs: (a) clusters sizes; (b) variables median values; (c) box-plots with the age and the variable values distributions around the means.

different in number of patients), which keeps both age ranges and the expectation based characterizations described above. Indeed, if we call $S_1', S_2', \ldots, S_5'$ the states in the 5-model, then in Fig. 5 we may observe that S_1 corresponds to S_5' (in terms of young age range and characterization of having relatively high mean values, especially for variables X_1, X_2, X_3, X_4), S_3 is very similar to the union of S_3' and S_1' (young/mid age), S_2 to S_4' (mid/old age), and S_4 to S_2' (outliers) having scarce characterization and light similarities with the other states.

3.2   Correlation Based B Cell Networks, over Four and Five Clusters

Further interesting information may be deduced by observing correlation based connections between all eight different subpopulations of B cells, computed as an outcome of HMM based modeling. In Fig. 6 a bar chart is reported, for each couple of variables, to show correlation values between the two variables across different clusters. The diagrams of bar charts are sorted by decreasing variability of correlation among the states. Namely, the variables X_5 and X_6 (in both models) are those whose correlation is maximally different across the states, then the correlation between X_5 and X_6 is more informative in this state classification than that between another pairs of variables. On the other hand, couples as $\{X_1, X_4\}$ or $\{X_5, X_8\}$ (almost always correlated) and $\{X_2, X_3\}$ (almost never correlated) have a sort of constant interconnection across our clusters. In biomedical terms, these kind of observations on the model allow us to hypothesize that: (i) the B cell subpopulations having configuration with only one between CD5 and CD27 expressed (that is, X_1 and X_4, or X_5 and X_8) are correlated during the whole lifetime of a human being (no matter about the activation of CD23), while (ii) B cell subpopulations with no expression of both CD5 and CD27 are never correlated (in lifetime, and independently on the expression of CD23). Further experimental data on patients could validate our hypotheses from a medical viewpoint.

Let us here discuss more specifically the correlation based B-cell networks, obtained from our HMMs and reported in Fig. 7, where nodes represent the cell phenotypes and edges connect pair of nodes when corresponding variables have correlation greater or equal than 0.5). States S_1 (550 patients) and corresponding S_5' (360), including patients between 0 and about 6 years, are characterized by a clique of correlations among variables X_5, X_6, X_8, namely the couple $\{X_6, X_8\}$ has a correlation around 0.5 only for these clusters (then only for patients of infant age). Such a connection is the only case, in lifetime, when two B cell phenotypes having more than one (actually here it is a couple of) receptor(s) differently expressed are correlated.

The large cluster S_3 (with 2222 patients) and corresponding union of S_3' (1205) and S_1' (2249), including patients of approximate age range 10–45 years, have networks characterized by three correlations which are not all together present in other states: $\{X_1, X_7\}, \{X_6, X_7\}$, and unique $\{X_4, X_6\}$ (which correspond to activation of CD27, CD23, and CD27 respectively). Elder patients of states S_2 (3085 patients) and S_4' (2077) have a simple identical correlation network, with four edges, which includes a couple of pairs always present: $\{X_1, X_2\}, \{X_1, X_4\}$, and the couple of edges: $\{X_5, X_6\}, \{X_5, X_8\}$. State S_4 (with only 97 mainly elder patients), together with state S_2' (63 mainly elder patients) present a mix of high correlations which allow us to hypothesize that these clusters collect mostly "ill" patients (having some health disturb effecting B cell quantities and reciprocal dynamics). Some anomalies could correspond either to the presence of relatively high correlations for $\{X_3, X_4\}, \{X_7, X_8\}$ (which are never present in other states, and correspond to the activation of CD5, for the

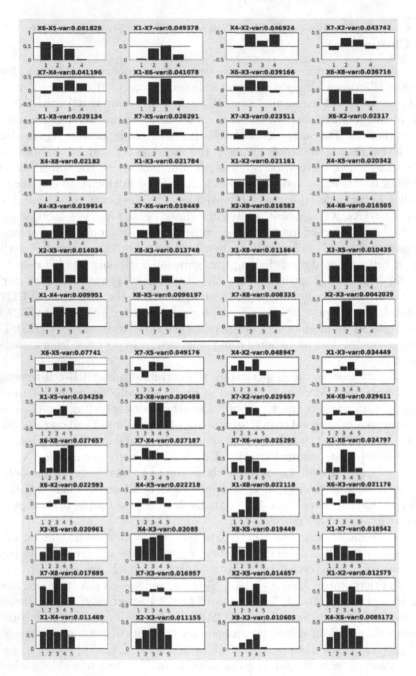

Fig. 6. B cell correlations by HMM based modeling, with four and five states respectively. Top: pair correlations across four clusters; bottom: pair correlations across five states. Horizontal red line denotes the 0.5 value. (Color figure online)

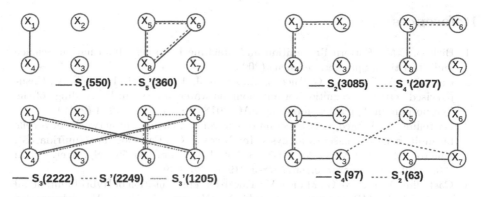

Fig. 7. Correlation based B cell networks, with four (straight edges) and five (dotted edges) states.

two cases of both either expressed or not expressed C23 and C27), or to the relatively high expectation of variables X_5, X_6, X_7, X_8 (the B cell subpopulations with CD27 expressed) yet noticed in previous section.

A final observation on the B cell networks in Fig. 7 is that the correlation between cells X_1 and X_2 (corresponding to the activation of CD5, when CD23 is expressed and CD27 does not) seems to hold only for adults. A biomedical validation of this model will be a next step for future work, in order to test, discuss, and eventually improve our hypotheses with new knowledge on the immune system functioning.

4   Conclusion and Ongoing Work

In this work we applied HMMs to infer a data-driven Bayesian model as a sequence of hidden states that provide given observations. The method is outlined and the computed model analyzed. This provides few clusters of patients with a significantly different structure in terms of B cell networks. Such states are described by multivariate Gaussian distributions, whose parameters are interpretable as main statistical properties of different dynamics and correlation based B cell populations interconnections. We finally advanced some hypotheses for a biological interpretation of these properties. Our observations seem to confirm that aging negatively impacts lymphocyte variability, in terms of an age-related decline, or decreasing mean values. A class of relatively few outliers has been also identified, where B cell subpopulations characterized by the expression of receptors CD27 increase their mean quantities in advanced age.

Acknowledgments. Authors would like to thank Antonio Vella (department of pathology and diagnostics, University Hospital of Verona) for providing the dataset used in this work and for interesting discussions on the role of B cells in the immune system.

References

1. Bishop, C.M.: Pattern Recognition and Machine Learning. (Information Science and Statistics). Springer, Secaucus (2006)
2. Castellini, A., Beltrame, G., Bicego, M., Blum, J., Denitto, M., Farinelli, A.: Unsupervised activity recognition for autonomous water drones. In: Proceedings of the Symposium on Applied Computing, SAC 2018, pp. 840–842. ACM (2018)
3. Castellini, A., Franco, G.: Detection of age-related changes in networks of b cells by multivariate time-series analysis. In: Nicosia, G., Pardalos, P., Giuffrida, G., Umeton, R. (eds.) MOD 2017. LNCS, vol. 10710, pp. 586–597. Springer, Cham (2018). https://doi.org/10.1007/978-3-319-72926-8_49
4. Castellini, A., Franco, G., Manca, V.: Toward a representation of Hybrid Functional Petri Nets by MP systems. In: Suzuki, Y., Hagiya, M., Umeo, H., Adamatzky, A. (eds.) Natural Computing, Proceedings in Information and Communications Technology, vol. 1, pp. 28–37. Springer, Tokyo (2009). https://doi.org/10.1007/978-4-431-88981-6_3
5. Castellini, A., Franco, G., Manca, V., Ortolani, R., Vella, A.: Towards an MP model for B lymphocytes maturation. In: Ibarra, O.H., Kari, L., Kopecki, S. (eds.) UCNC 2014. LNCS, vol. 8553, pp. 80–92. Springer, Cham (2014). https://doi.org/10.1007/978-3-319-08123-6_7
6. Castellini, A., Franco, G., Vella, A.: Age-related relationships among peripheral B lymphocyte subpopulations. In: 2017 IEEE Congress of Evolutionary Computation - CEC, pp. 1864–1871. Springer, Berlin (2017)
7. Castellini, A., Manca, V.: Learning regulation functions of metabolic systems by artificial neural networks. In: Proceedings of the 11th Annual Conference on Genetic and Evolutionary Computation, GECCO 2009, pp. 193–200. ACM Publisher, Montreal (2009)
8. Castellini, A., Paltrinieri, D., Manca, V.: MP-GeneticSynth: inferring biological network regulations from time series. Bioinformatics $31(5)$, 785–787 (2015)
9. Castellini, A., Zucchelli, M., Busato, M., Manca, V.: From time series to biological network regulations: an evolutionary approach. Mol. BioSyst. 9, 225–233 (2013)
10. Castiglione, F., Mannella, G., Motta, S., Nicosia, G.: A network of cellular automata for the simulation of the immune system. Int. J. Mod. Phys. C (Phys. Comput.) $10(4)$, 677–686 (1999)
11. Castiglione, F., Motta, S., Nicosia, G.: Pattern recognition by primary and secondary response of an artificial immune system. Theory Biosci. $120(2)$, 93–106 (2001)
12. Davey, F.R., Huntington, S.: Age-related variation in lymphocyte subpopulations. Gerontology 23, 381–389 (1977)
13. Duchêne, F., Garbay, C., Rialle, V.: Learning recurrent behaviors from heterogeneous multivariate time-series. Artif. Intell. Med. $39(1)$, 25–47 (2007)
14. Franco, G., Jonoska, N., Osborn, B., Plaas, A.: Knee joint injury and repair modeled by membrane systems. BioSystems $91(3)$, 473–488 (2008)
15. Gruver, A.L., Hudson, L.L., Sempowski, G.D.: Immunosenescence of ageing. J. Pathol. $211(2)$, 144–156 (2007)
16. Hicks, M.J., Jones, J.F., Minnich, L.L., Wigle, K.A., Thies, A.C., Layton, J.M.: Age-related changes in T- and B-lymphocyte subpopulations in the peripheral blood. Arch. Pathol. Lab. Med. $107(10)$, 518–523 (1983)
17. Lavielle, M., Teyssière, G.: Detection of multiple change-points in multivariate time series. Lith. Math. J. $46(3)$, 287–306 (2006)

18. Manca, V., Castellini, A., Franco, G., Marchetti, L., Pagliarini, R.: Metabolic P systems: a discrete model for biological dynamics. Chin. J. Electron. **22**(4), 717–723 (2013)
19. Montanez, G., Amizadeh, S., Laptev, N.: Inertial Hidden Markov models: modeling change in multivariate time series (2015)
20. Rabiner, L.R.: A tutorial on Hidden Markov models and selected applications in speech recognition. Proc. IEEE **77**(2), 257–286 (1989)
21. Simon, A.K., Hollander, G.A., McMichael, A.: Evolution of the immune system in humans from infancy to old age. Proc. Roy. Soc. B: Biol. Sci. **282**(1821), 20143085 (2015)
22. Stracquadanio, G., et al.: Large scale agent-based modeling of the humoral and cellular immune response. In: Liò, P., Nicosia, G., Stibor, T. (eds.) ICARIS 2011. LNCS, vol. 6825, pp. 15–29. Springer, Heidelberg (2011). https://doi.org/10.1007/978-3-642-22371-6_2
23. Vahdatpour, A., Amini, N., Sarrafzadeh, M.: Toward unsupervised activity discovery using multi-dimensional motif detection in time series. In: Proceedings of 21st International Joint Conference on Artificial Intelligence, IJCAI 2009, pp. 1261–1266 (2009)
24. Veneri, D., Ortolani, R., Franchini, M., Tridente, G., Pizzolo, G., Vella, A.: Expression of CD27 and CD23 on peripheral blood B lymphocytes in humans of different ages. Blood Transfus **7**, 29–34 (2009)
25. Weiskopf, D., Weinberger, B., Grubeck-Loebenstein, B.: The aging of the immune system. Transpl. Int. **22**, 1041–1050 (2009)
26. Xuan, X., Murphy, K.: Modeling changing dependency structure in multivariate time series. In: Proceedings of the 24th International Conference on Machine Learning, ICML 2007, pp. 1055–1062. ACM (2007)

Nonnegative Coupled Matrix Tensor Factorization for Smart City Spatiotemporal Pattern Mining

Thirunavukarasu Balasubramaniam[1]([⊠]), Richi Nayak[1], and Chau Yuen[2]

[1] Queensland University of Technology, Brisbane, QLD, Australia
thirunavukarasu.balas@qut.edu.au
[2] Singapore University of Technology and Design, Singapore, Singapore

Abstract. With the advancements in smartphones and inbuilt sensors, the day-to-day spatiotemporal activities of people can be recorded. With this available information, the automated extraction of spatiotemporal patterns is crucial to understand the people's mobility. These patterns can assist in improving the smart city environments like traffic control, urban planning, and transportation facilities. The smartphone generated spatiotemporal data is enriched with multiple contexts and efficiently utilizing them in a Machine Learning process is still a challenging task. In this paper, we propose a Nonnegative Coupled Matrix Tensor Factorization (CMTF) model to integrate and analyze additional contexts with spatiotemporal data to generate meaningful patterns. We also propose an efficient factorization algorithm based on variable selection to solve the Nonnegative CMTF model that yields accurate spatiotemporal patterns. Our empirical analysis highlights the efficiency of the proposed CMTF model in terms of accuracy and factor goodness.

Keywords: Nonnegative Coupled Matrix Tensor Factorization ·
Greedy coordinate descent · Variable selection · Smart city · Spatiotemporal ·
Pattern mining

1 Introduction

With more than half of the world population living in cities or urban areas, there is an increasing pressure to uplift the current status of infrastructure and resources available [1]. Urban cities should be accommodated with improved transportation and interconnectivity and improve social well-being by providing the environment and economic sustainability. The cities can be considered smart if it is instrumented, interconnected and intelligent according to Harrison et al. [2]. They also quote that the smart cities should use multiple devices like sensors, camera, smartphones, web, kiosks, personal devices to facilitate improved services. In some smart cities, these information and communication technologies (ICT) have been dominating to facilitate the city infrastructure, transportation, and mobility [3]. The role of ICT plays a critical role to define the smart cities. The economy and governance of a smart city are driven by innovation, communication, and smart people.

© Springer Nature Switzerland AG 2019
G. Nicosia et al. (Eds.): LOD 2018, LNCS 11331, pp. 520–532, 2019.
https://doi.org/10.1007/978-3-030-13709-0_44

The mounted sensing modules in smartphones are capable to collect a vast amount of data representing multiple contexts [4]. This sensor data generated from mobiles must be utilized to understand and make the smart cities intelligent and future ready. An example is the smartphones generated real-time Location-Based Social Networks (LBSNs) data like location, time and activity when the citizens navigate within the city [5]. Analyzing this real-world data is important to model the functionality of cities like traffic controlling, easy transportation and many others.

Understanding the spatiotemporal pattern is important and can help in urban planning. In the real world, any kind of incident can change the mobility pattern of people and the change in mobility pattern should be addressed properly [6, 7]. One such example is the Great East Japan Earthquake which triggered a change in the mobility pattern of the local commute [8].

Tensor modeling and tensor factorization (TF) is the generalization of matrix modeling and factorization for higher order. Tensor modeling has been successfully applied to model a spatiotemporal data and automatically derive the spatiotemporal patterns [6]. As shown in Fig. 1, a 3-mode *(user x location x time)* tensor model is factorized into 3 latent matrices with their respective latent factors. Each latent factor learns a pattern(s) hidden in the tensor data. However, the traditional TF struggles to efficiently utilize all the available contexts in processing because not all the contexts are interdependent. Some contexts are related only to one mode of the tensor.

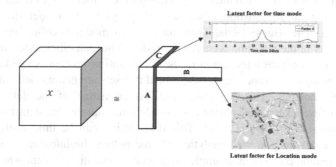

Fig. 1. Example of TF for analysis of spatiotemporal patterns using latent factors.

In this paper, we propose the Nonnegative Coupled Matrix Tensor Factorization (CMTF) model to efficiently utilize contexts in spatiotemporal pattern mining. We also introduce a novel fast and efficient variable selection based Coordinate Descent (CD) method to solve nonnegative CMTF. Firstly, we derive a single variable update rule for nonnegative CMTF and calculate the variable importance of all the variables based on the difference in the objective function. We then introduce a greedy CD (GCD) that traverses the factor matrix row-wise to select a single important variable to update.

Using synthetic and real-world datasets we demonstrate: (1) the speed of the proposed GCD algorithm to solve Nonnegative CMTF; and (2) the efficiency of the proposed nonnegative CMTF model in identifying the spatiotemporal patterns hidden in the smart city data.

To the best of our knowledge, this is the first work to model smart city data in the CMTF model to efficiently derive the spatiotemporal patterns.

2   Related Work

There exists only a handful of tensor based smart city applications that simply model the data available and attempt to understand the patterns. Recently, the spatial-temporal mobility data is modeled and factorized using probabilistic tensor factorization to understand the mobility of users to model the urban structure [6]. Tensor modeling has been applied to understanding the noise pollution across New York City to allow people to live in the better environment [9]. Here, the tensor is modeled as a third order with (*location x noise category x time slot*) information.

Researchers have also modified the traditional decomposition algorithms to learn the spatial and temporal context available in the data. The Orthogonal NTF method added the orthogonal constraints on the factor matrices [11]. Graph Laplacian Regularization was applied to Nonnegative Tensor Factorization (NTF) to enrich the tensor input with neighborhood information [10]. Though these methods were able to utilize spatial and temporal information efficiently, they ignore to model the additional information available with the data. Thus, an efficient model to collectively factorize multiple contexts in the real-time scenario is needed. Thus, an efficient tensor model to utilize additional available information in the factorization process is needed.

With technological advancements, real-world datasets are getting richer and exhibit multi-type relationships. This changing multifaceted nature of the dataset has led researchers to fuse matrix and tensor data models to capture this data and discover useful knowledge effectively [11–14]. This data fusion can be immensely useful in several applications such as web analytics, chemometrics, bioinformatics, signal processing, and metabolomics. For example, in a recommendation system where the task is to recommend an activity to a user at a location, the primary data source can be represented as a third order tensor (*user x location x activity*) with a supplementary matrix (*location x feature*) [15]. Inspired by the success of coupled matrix factorization (CMF), the Coupled Matrix Tensor Factorization (CMTF) method has been proposed to jointly analyze matrix and tensors [16]. The joint factorization will utilize the additional available information and hence can reveal more latent factors whereas a simple matrix or tensor factorization failed to reveal all the factors.

CMTF is traditionally solved using Alternating Least Square (ALS) where we update one factor matrix at a time by fixing the other factor matrices. As factorization is a nonconvex optimization problem, this alternating update is essential. The update rule of ALS involves time-consuming matrix products and is prone to poor convergence. To minimize this matrix product. Acar et al. introduced all at once optimization approach where all the factor matrices are solved together at once until convergence. This strategy improves the convergence speed of the factorization process [16]. However, it fails to identify the true factors. Hsieh et al. introduced the greedy coordinate descent (GCD) algorithm for *Matrix Factorization* where the matrix elements, to be updated, are greedily selected based on the importance measured as the gradient [17]. This results in fast convergence. In this paper, we derive a single element update rule in GCD and make it possible for nonnegative CMTF problem.

3   Understanding the Spatiotemporal Patterns Using CMTF

3.1   Coupled Matrix Factorization: Background

The coupled analysis initially started with the factorization of multiple matrices simultaneously. If $X \in \mathbb{R}^{I \times J}$ and $Y \in \mathbb{R}^{I \times K}$ are two matrices that share the first mode, the coupled Matrix factorization [18] problem can be represented as

$$f(V, U, T) = \left\| X - VU^T \right\|^2 + \left\| Y - VT^T \right\|^2 \tag{1}$$

Where $V \in \mathbb{R}^{I \times R}$ is the factor matrix that is shared by both the matrices with R being the Rank, and $U \in \mathbb{R}^{J \times R}$ and $T \in \mathbb{R}^{K \times R}$ are the second mode factor matrices of X and Y respectively. $\|\cdot\|$ denotes the Frobenius norm. Applications such as recommendation systems that require missing value predictions, matrix factorization using the Euclidean distance loss function is used to approximate the matrices with high accuracy.

3.2   The Proposed Coupled Matrix Tensor Factorization

Tensor Model Representation: Several senor-generated applications such as LBSNs uses smartphone GPS information of people and record their check-in behavior [5]. In addition to the location information, it will also track the timestamp associated with the check-in. Using the location as spatial context and time as the temporal context, a three-mode (*user x venue x time*) tensor model $\mathcal{X}$ can be generated with a number of check-ins as the value for each tuple.

Venue Categories as an Auxiliary Information: The tensor model can now be combined with a matrix model that adds another contextual information for analysis. There are multiple ways this additional information can be represented. For example, the social relationships between users can be added by using the user as a common mode between tensor and matrix. As another example, the additional location (or venue) information can be added by treating location as a common mode in the matrix model (*venue x feature*). In this paper, we propose to use the venue category as an auxiliary information.

The foursquare dataset which is an example of LBSN applications records the check-in behavior of restaurant customers. It also consists of restaurant/venue categories that are categorized based on multiple nature. As venue categories are dependent on the venue and independent of other factors, LBSNs researchers avoid this information in their analysis as the traditional NTF is not suitable to model it.

Hence, we represent the venue category (*venue x venue category*) matrix V_C and jointly analyse it with *venue x user x time* tensor.

The value for the matrix is interpreted as follows:

$$V_C(i,j) = \begin{cases} 1, & \text{if venue } i \text{ belongs to } j^{th} \text{ category} \\ 0, & \text{else} \end{cases}$$

As the only venue is shared between tensor and matrix, the factor learning process needs to be modified accordingly.

CMT Model Representation

We now have the tensor model $\mathcal{X}$ representing users' activities and a matrix model V_C representing venue category information. The generated V_C matrix can be jointly processed with the primary tensor by a fusion model as shown in Fig. 2 by sharing the common mode of venue. By adding the matrix V_C in the tensor model $\mathcal{X}$, we introduce this new information of venue categories in learning the factor matrices of tensor model afthe ter decomposition. The coupled model will not just be able to learn the relationship between venues in accordance with user and time slots, but also in accordance with each venue's category.

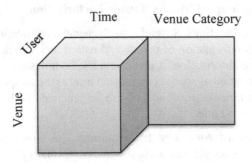

Fig. 2. A third order tensor and matrix coupled in venue mode.

Coupled Matrix Tensor Factorization (CMTF): The CMTF optimization is formulated as an extension of (1) with a given third-order tensor $\mathcal{X} \in \mathbb{R}^{I \times J \times K}$ and matrix $Y \in \mathbb{R}^{I \times L}$ coupled in the first mode as:

$$f(V, U, T, C) = \|\mathcal{X} - [\![V, U, T]\!]\|^2 + \|Y - VC^T\|^2 \tag{2}$$

Where tensor $\mathcal{X}$ is factorized as a CP model [19]. The factor matrix $V \in \mathbb{R}^{I \times R}$ is shared with matrix Y and other factor matrices $U \in \mathbb{R}^{J \times R}$, $T \in \mathbb{R}^{K \times R}$ and $C \in \mathbb{R}^{L \times R}$ are unshared factors.

4 Optimization Solution for Nonnegative CMTF

In this section, we derive the single element update rule to efficiently solve the non-negative CMTF optimization function as formulated in Eq. (2). The minimization function of CMTF as formulated in Eq. (2) can be represented as follows:

$$f = f_1(V, U, T) + f_2(V, C) \tag{3}$$

Where $f_1(V, U, T) = \|\mathcal{X} - [\![V, U, T]\!]\|^2$ and $f_2(V, C) = \|Y - VC^T\|^2$ are the minimization function for tensor and matrix respectively.

The goal is to solve the optimization problem Eq. (3) to find the accurate factor matrices V, U, T and $\mathbf{C}$.

We now explain the proposed algorithm for learning factor matrix V which is applicable to other factor matrices in principle. Let $\mathbf{G} = \frac{\partial f}{\partial V}$ be the gradient matrix that needs to be determined for finding the best factor matrix V.

To solve, we need to find the gradients of f_1 and f_2 as follows

$$\frac{\partial f}{\partial V} = \frac{\partial f_1}{\partial V} + \frac{\partial f_2}{\partial V} \tag{4}$$

$$\frac{\partial f_1}{\partial V} = V^{min} \|X_1 - V(T \odot U)^T\|_F^2 \tag{5}$$

$$\frac{\partial f_2}{\partial V} = V^{min} \|Y - VC^T\|_F^2 \tag{6}$$

where X_1 is the metricized tensor in the first mode and $\odot$ indicates the Khatri-Rao product [20].

Equations (5) and (6) can be rewritten after computing partial derivatives in Eq. (4) as:

$$\frac{\partial f}{\partial V} = \mathbf{G} = -2X_1(T \odot U) + 2V(U^T U * T^T T) + VC^T C - YC \tag{7}$$

Similarly, after computing second order partial derivatives as we obtain $\mathbf{H}$, the Hadamard product matrix:

$$\frac{\partial^2 f}{\partial V} = \mathbf{H} = \left(U^T U * T^T T\right) + C^T C \tag{8}$$

For each iteration, the values for Eqs. (7) and (8) can be calculated for entire matrix and the variables can be updated using the one variable update rule of CD [21] as,

$$\hat{v}_{ij} = \max\left(0, v_{ij} - \frac{-X_1(T \odot U)_{ij} + V\left(U^T U * T^T T\right)_{ij} + \left(VC^T C - YC\right)_{ij}}{\left(U^T U * T^T T\right)_{jj} + \left(C^T C\right)_{jj}}\right) - u_{ij} \tag{9}$$

where v_{ij} indicates i, j^{th} element of V anthe d $\hat{v}_{ij}$ indicates the computed variable.

$$v_{ij} = v_{ij} + \hat{v}_{ij} \tag{10}$$

The calculation of $X_1(T \odot U)$, $U^T U * T^T T$ and $VC^T C$ for every element is expensive and hence it should be calculated before updating the element.

For simplicity, let us represent

$$g_{ij} = -X_1(T \odot V)_{ij} + V\left(U^T U * T^T T\right)_{ij} + \left(VC^T C - YC\right)_{ij} \tag{11}$$

$$h_{ij} = \left(U^T U * T^T T\right)_{jj} + \left(C^T C\right)_{jj} \tag{12}$$

The factor matrices, U and T of the tensor and the factor matrix C are learned similarly using Eqs. 4 to 10 with the following changes. The matrix component exists only for shared factor matrix V, hence, f_2 is set to 0 when updating factor matrices, U and T of the tensor. Likewise, when updating the factor matrix C, the partial derivative of f_2 with respect to C alone can be calculated and f_1 is set to 0. To solve nonnegative CMTF based on variable selection, we next derive the greedy CD. GCD uses the greedy strategy in finding and updating single variable multiple times within a single inner iteration. Each time, the important variable is selected based on its importance calculated using gradients.

4.1   Variable Selection Using GCD – Row-Wise Update

In matrix factorization, variable selection has been proven to converge faster by updating the important variables repeatedly, instead of considering all variables [17]. In this paper, we propose to calculate the importance of a variable during coupled matrix tensor factorization using gradient principles [17] as,

$$O_{ij} = -\left(v_{ij} * g_{ij}\right) - 0.5 * \left(h_{ij} * v_{ij} * v_{ij}\right) \tag{13}$$

The rationale behind using Eq. (13) is that it calculates the difference in the objective function for each variable's update. Higher the difference value, higher is the variable's importance. We propose to employ the GCD algorithm for updating the elements based on the importance calculated as in Eq. (13). In GCD, the matrix O is computed initially that stores the variable importance of each element in factor matrix U. Once we have importance scores for all variables, the variable with the highest score is selected row-wise and updated repeatedly. In matrix factorization, this facilitates fast convergence with more accurate predictions. We expect the same in coupled matrix tensor factorization. With the calculated variable importance, the gradient of the factor matrix and update rule as described above, the update is performed according to algorithm 1. We suggest readers to refer [17] for theoretical analysis.

5 Empirical Analysis

In this section, we report the performance of GCD[1] in coupled matrix tensor factorization for generating factors for understanding spatio-temporal patterns. The nonnegative CMTF problem can also be considered as a solution to factor identification. Traditionally, the nonnegative CMTF algorithms are evaluated for the missing data estimation problem, it is also important to evaluate them in terms of how well they can capture the underlying factors. All the experiments were conducted on Intel (R) Core (TM) i7-6600U CPU @ 2.60 GHz model with 16 GB RAM.

5.1 Datasets

We used randomly generated synthetic datasets to evaluate the runtime performance of the proposed GCD for CMTF model. We also used the New York city foursquare data (NYC) [22] that consists of 928 users' check-in behavior at 14831 venues during April 2013. As discussed before, the dataset was added with an auxiliary information of 368 venue categories for different venues. We represent *(venue, user, time)* as a tensor and *(venue, venue category)* as an additional matrix. We set the time slots to 24 h to understand the temporal pattern over hours. Table 1. shows the statistics of various datasets used in this analysis. We set the rank of 4 for New York city foursquare dataset as we are interested to identify 4 patterns and 10 for all the other synthetic datasets after conducting a series of trials.

---

[1] https://github.com/thirubs/CMTF-GCD.

Algorithm 1: Greedy Coordinate Descent (GCD) Algorithm for CMTF

Input: *Tensor* $\mathcal{X}$; Matrix **Y**; *Randomly   initiated   factor ces* $\mathbf{V} \in \mathbb{R}^{I \times R}$, $\mathbf{U} \in \mathbb{R}^{J \times R}$, $\mathbf{T} \in \mathbb{R}^{K \times R}$, $\mathbf{C} \in \mathbb{R}^{L \times R}$; $\mathbf{O} = \emptyset$; *Rank* R; *init; inner_iter; tol: a value;*

Output: *Learned Factor matrices* $\mathbf{V}, \mathbf{U}, \mathbf{T}, \mathbf{C}$

Compute: $\mathbf{G}, \mathbf{H}$ using equations 7 & 8

$\mathbf{for}\ i = 1{:}I$

  $\mathbf{for}\ r = 1{:}R$

    $\hat{v}_{ir} \leftarrow g_{ir}/h_{ir}$;

    $\hat{v}_{ir} \leftarrow v_{ir} - \hat{v}_{ir}$;

    $\mathbf{if}\ (\ \hat{v}_{ir} < 0)$

      $\hat{v}_{ir} \leftarrow 0$;

    $\mathbf{end\ if}$

    $\hat{v}_{ir} \leftarrow \hat{v}_{ir} - v_{ir}$;

    $o_{ir} \leftarrow (-1) * \hat{v}_{ir} * g_{ir} - 0.5 * h_{rr} * \hat{v}_{ir} * \hat{v}_{ir}$;

    $\mathbf{if}(o_{ij} > init)$

      $init = o_{ir}$;

    $\mathbf{end\ if}$

  $\mathbf{end\ for}$

$\mathbf{end\ for}$

  $\mathbf{for}\ p = 1{:}I$

    $\mathbf{for}\ w_{inner} = 1{:}inner\_iter$

      $q = -1$; $bestvalue = 0$;

      $\mathbf{for}\ r = 1{:}R$

      $\hat{v}_{pr} \leftarrow g_{pr}/h_{pr}$; $\hat{v}_{pr} \leftarrow v_{pr} - \hat{v}_{pr}$;

      $\mathbf{if}\ (\ \hat{v}_{pr} < 0)\ \hat{v}_{pr} \leftarrow 0$;

      $\mathbf{end\ if}$

      $\hat{v}_{pr} \leftarrow \hat{v}_{pr} - v_{pr}$;

      $o_{pr} \leftarrow (-1) * v_{pr} * g_{pr} - 0.5 * h_{rr} * \hat{v}_{pr} * \hat{v}_{pr}$;

      $\mathbf{if}(o_{pr} > bestvalue)\ bestvalue = o_{pr}$; $q = r$; $\mathbf{end\ if}$

      $\mathbf{end\ for}$

      $\mathbf{if}\ (q == -1)\ break;\ \mathbf{end\ if}$

      $v_{pq} = v_{pq} + \hat{v}_{pq}$;

      $\mathbf{for}\ r = 1{:}R$

        $g_{pr} = g_{pr} + (\hat{v}_{pr} * h_{qr})$;

      $\mathbf{end\ for}$

      $\mathbf{if}\ (bestvalue < init * tol)$

        $break$;

      $\mathbf{end\ if}$

    $\mathbf{end\ for}$

  $\mathbf{end\ for}$

Repeat the above process for each of the matrices $\mathbf{U}, \mathbf{T}, \mathbf{C}$

Table 1. Dataset statistics

Dataset	Tensor size	Matrix size	Density of tensor
Syn1	$500 \times 500 \times 500$	500×500	0.0001
Syn2	$1000 \times 1000 \times 1000$	1000×1000	0.0001
Syn3	$1500 \times 1500 \times 1500$	1500×1500	0.0001
NYC	$14831 \times 928 \times 24$	14831×368	0.0001

5.2 Runtime Performance

As shown in Table 2, we use synthetic data and change the size of the input tensor to evaluate the runtime performance of GCD for CMTF with other benchmarks. ALS updates an entire factor matrix that involves complex matrix products for each iteration whereas TurboSMT-OPT converges slowly because of the information loss associated with the random projection of the input tensor. It can be evident that GCD is 33 times faster than ALS [16] and 1.5 times faster than TurboSMT-OPT [23]. This is because of the simplified single variable update in GCD.

Table 2. Runtime (in secs) vs scalability

Methods	Runtime in secs		
	Syn1	Syn2	Syn3
ALS	437.64	4949.4	19589
TurboSMT-OPT	62.39	334.67	904.44
GCD	**37.17**	**297.55**	**581.34**

5.3 Approximation Performance

We evaluate the quality of factorization by the approximation error and report the accuracy as Normalized Residual Value (NRV) [24] on real-world NYC dataset.

$$Approximation\ Error(NRV) = \frac{\left\| X_1 - V(T \odot U)^T \right\|_F^2}{\|\mathcal{X}\|_F^2} \tag{14}$$

We compared the proposed CMTF GCD model with the traditional tensor model with regularization (R_NTF) [25] and Orthogonal constraints (O_NTF) [26]. R_NTF imposes the neighborhood information as Graph Laplacian Regularization in the tensor model. It is clear from Table 3, that the CMTF GCD model outperforms R_NTF and O_NTF with a small margin. Also, CMTF based on OPT yields poorer performance because of the information loss during the random projection of input tensor. Both GCD and ALS yield 2 to 3 percentage improvement in the accuracy. While ALS and GCD perform equally better, ALS is not scalable to large datasets as shown in Table 2. Hence, CMTF with GCD is the best choice in mining the spatiotemporal patterns as it is able to include additional context in the dataset.

Table 3. Approximation performance

Methods	Approximation error
R_NTF	0.99
O_NTF	1.00
CMTF_ALS	0.97
CMTF_OPT	0.99
CMTF GCD	**0.97**

5.4 Spatiotemporal Pattern Mining

As we set the factorization rank to 4, each factor matrices have 4 latent/hidden factors. In Fig. 3, we plot the factors of time factor matrix to understand the temporal patterns. The red pattern indicates a peak at 12 pm and a small peak at 9 to 10 am. This clearly refers to the lunch time and dinner time. As the dataset is about the users' check-in at restaurants, this pattern is meaningful. The blue pattern indicates a peak at 1 pm showing that some set of people likely to have late lunch. On the other hand, the green pattern shows that some people are active during the afternoon and midnight. In Fig. 4, we plot the factors of venue factor matrix to understand the spatial patterns. The red and pink patterns in Fig. 3 do not show much distinction. However, comparing Fig. 3 with Fig. 4, red pattern and pink pattern refer to different locations. Similarly, by comparing Figs. 3 and 4, one can derive multiple spatiotemporal patterns. These spatiotemporal patterns are difficult to be extracted manually and it needs expert domain knowledge. CMTF facilitates the automated pattern elicitation.

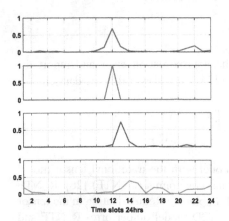

Fig. 3. Time factors. (Color figure online)

Fig. 4. Venue factors. (Color figure online)

6   Conclusion and Future Work

With the growing data generation and collection using smartphones and web applications, it is vital to utilize all the contexts in the machine learning process. Especially when smart city applications involve, it is important to perform the joint analysis task to facilitate the integration of contexts with the spatiotemporal dataset. In this paper, we introduce a Nonnegative CMTF model to facilitate this joint analysis. Nonnegative CMTF facilitates this through the factorization algorithm, however, existing algorithms are prone to slow convergence and false factor assessing. To overcome these challenges and to enhance the efficiency, we introduced a Greedy Coordinate Descent (GCD) for Nonnegative CMTF. Extensive empirical analysis with several state-of-the-art algorithms shows that the nonnegative CMTF algorithm is able to provide an effective solution for factor revealing problems. In future, we will extend the nonnegative CMTF for partially shared matrix tensor factorization. Also, we will apply coupled tensor factorization in relation to recommendation systems.

Acknowledgements. This work is supported by SUTD-MIT International Design Center and NSFC 61750110529.

References

1. Chourabi, H., et al.: Understanding smart cities: an integrative framework. In: 2012 45th Hawaii International Conference on System Science (HICSS). IEEE (2012)
2. Harrison, C., et al.: Foundations for smarter cities. IBM J. Res. Dev. **54**(4), 1–16 (2010)
3. Hancke, G.P., Hancke Jr., G.P.: The role of advanced sensing in smart cities. Sensors **13**(1), 393–425 (2012)
4. Lau, B.P.L., et al.: Extracting point of interest and classifying environment for low sampling crowd sensing smartphone sensor data. In: PerCom Workshops. IEEE (2017)
5. Kefalas, P., Symeonidis, P., Manolopoulos, Y.: A graph-based taxonomy of recommendation algorithms and systems in LBSNs. IEEE Trans. Knowl. Data Eng. **28**(3), 604–622 (2016)
6. Sun, L., Axhausen, K.W.: Understanding urban mobility patterns with a probabilistic tensor factorization framework. Transp. Res. Part B: Methodol. **91**, 511–524 (2016)
7. Kimura, T., et al.: Spatio-temporal factorization of log data for understanding network events. In: 2014 Proceedings IEEE on INFOCOM. IEEE (2014)
8. Fan, Z., Song, X., Shibasaki, R.: CitySpectrum: a non-negative tensor factorization approach. In: Proceedings of the 2014 ACM International Joint Conference on Pervasive and Ubiquitous Computing. ACM (2014)
9. Zheng, Y., et al.: Diagnosing New York city's noises with ubiquitous data. In: Proceedings of the 2014 ACM International Joint Conference on Pervasive and Ubiquitous Computing. ACM (2014)
10. Cai, D., et al.: Graph regularized nonnegative matrix factorization for data representation. IEEE Trans. Pattern Anal. Mach. Intell. **33**, 1548–1560 (2011)
11. Nion, D., Sidiropoulos, N.D.: Tensor algebra and multidimensional harmonic retrieval in signal processing for MIMO radar. IEEE Trans. Sig. Process. **58**(11), 5693–5705 (2010)
12. Symeonidis, P.: Matrix and tensor factorization with recommender system applications. Graph-Based Soc. Media Anal. **39**, 187 (2016)

13. Acar, E., Bro, R., Smilde, A.K.: Data fusion in metabolomics using coupled matrix and tensor factorizations. Proc. IEEE **103**(9), 1602–1620 (2015)
14. Bhargava, P., et al.: Who, what, when, and where: multi-dimensional collaborative recommendations using tensor factorization on sparse user-generated data. In: 24th International Conference on World Wide Web. ACM (2015)
15. Frolov, E., Oseledets, I.: Tensor methods and recommender systems. Wiley Interdiscip. Rev.: Data Min. Knowl. Discov. **7**(3), e1201 (2017)
16. Acar, E., Kolda, T.G., Dunlavy, D.M.: All-at-once optimization for coupled matrix and tensor factorizations. arXiv preprint arXiv:1105.3422 (2011)
17. Hsieh, C.-J., Dhillon, I.S.: Fast coordinate descent methods with variable selection for non-negative matrix factorization. In: KDD. ACM (2011)
18. Acar, E., et al.: Coupled matrix factorization with sparse factors to identify potential biomarkers in metabolomics. In: 2012 IEEE 12th International Conference on Data Mining Workshops (ICDMW). IEEE (2012)
19. Kiers, H.A.: Towards a standardized notation and terminology in multiway analysis. J. Chemom. **14**(3), 105–122 (2000)
20. Kolda, T.G., Bader, B.W.: Tensor decompositions and applications. SIAM Rev. **51**(3), 455–500 (2009)
21. Cichocki, A., Anh-Huy, P.: Fast local algorithms for large scale nonnegative matrix and tensor factorizations. IEICE Trans. Fundam. Electron. Commun. Comput. Sci. **92**(3), 708–721 (2009)
22. Yang, D., et al.: Fine-grained preference-aware location search leveraging crowdsourced digital footprints from LBSNs. In: Proceedings of the 2013 ACM International Joint Conference on Pervasive and Ubiquitous Computing. ACM (2013)
23. Papalexakis, E.E., et al.: Turbo-smt: accelerating coupled sparse matrix-tensor factorizations by 200x. In: Proceedings of the 2014 SIAM ICDM. SIAM (2014)
24. Kimura, K., Kudo, M.: Variable selection for efficient nonnegative tensor factorization. In: ICDM. IEEE (2015)
25. Han, Y., Moutarde, F.: Analysis of large-scale traffic dynamics in an urban transportation network using non-negative tensor factorization. Int. J. Intell. Transp. Syst. Res. **14**(1), 36–49 (2016)
26. Afshar, A., et al.: CP-ORTHO: an orthogonal tensor factorization framework for spatio-temporal data. In: Proceedings of the 25th ACM SIGSPATIAL International Conference on Advances in Geographic Information Systems. ACM, Redondo Beach (2017)

User Preferences in Bayesian Multi-objective Optimization: The Expected Weighted Hypervolume Improvement Criterion

Paul Feliot[1(✉)], Julien Bect[2], and Emmanuel Vazquez[2]

[1] Safran Aircraft Engines, Moissy-Cramayel, France
paul.feliot@safrangroup.com
[2] Laboratoire des Signaux et Systèmes (L2S), Centrale-Supélec,
Gif-sur-Yvette, France
{julien.bect,emmanuel.vazquez}@centralesupelec.fr

Abstract. In this article, we present a framework for taking into account user preferences in multi-objective Bayesian optimization in the case where the objectives are expensive-to-evaluate black-box functions. A novel *expected improvement* criterion to be used within Bayesian optimization algorithms is introduced. This criterion, which we call the *expected weighted hypervolume improvement* (EWHI) criterion, is a generalization of the popular *expected hypervolume improvement* to the case where the hypervolume of the dominated region is defined using a user-defined absolutely continuous measure instead of the Lebesgue measure. The EWHI criterion takes the form of an integral for which no closed form expression exists in the general case. To deal with its computation, we propose an importance sampling approximation method. A sampling density that is optimal for the computation of the EWHI for a predefined set of points is crafted and a sequential Monte-Carlo (SMC) approach is used to obtain a sample approximately distributed from this density. The ability of the criterion to produce optimization strategies oriented by user preferences is demonstrated on a simple bi-objective test problem in the cases of a preference for one objective and of a preference for certain regions of the Pareto front.

Keywords: Bayesian optimization · Multi-objective optimization ·
User preferences · Importance sampling · Sequential Monte-Carlo

1 Introduction

In this article, we present a Bayesian framework for taking into account user preferences in multi-objective optimization when evaluation results for the functions of the problem are obtained using a computationally intensive computer program. Such a setting is representative of engineering problems where structural analysis or fluid dynamics are used. The number of runs of the computer

© Springer Nature Switzerland AG 2019
G. Nicosia et al. (Eds.): LOD 2018, LNCS 11331, pp. 533–544, 2019.
https://doi.org/10.1007/978-3-030-13709-0_45

program that can be afforded is limited and the objective is to build a sequence of observation points that rapidly provides a "good" approximation of the set of *Pareto optimal solutions*, where "good" is measured using some user-defined loss function.

To this end, we formulate an *expected improvement* (EI) criterion (see, e.g., [20]) that uses the *weighted hypervolume indicator* (WHI) introduced by [29] as a loss function. This new criterion, which we call the *expected weighted hypervolume improvement* (EWHI) criterion, can be viewed as a generalization of the *expected hypervolume improvement* (EHVI) criterion of [14] that enables practitioners to tailor optimization strategies according to user preferences.

The article is structured as follows. First, we recall in Sect. 2 the framework of Bayesian optimization. Then, we detail in Sect. 3 the construction of the EWHI criterion and discuss computational aspects. The ability of the criterion to produce optimization strategies according to user preferences is then demonstrated on a simple bi-objective test problem in the cases of a preference for one objective and of a preference for certain regions of the Pareto front in Sect. 4. Finally, conclusions and perspectives are drawn in Sect. 5.

2   Bayesian Optimization

2.1   The Bayesian Approach to Optimization

Consider a continuous optimization problem $\mathcal{P}$ defined over a search space $\mathbb{X} \subset \mathbb{R}^d$ and let $\underline{X} = (X_1, X_2, X_3 \ldots)$ be a sequence of observation points in $\mathbb{X}$. The problem $\mathcal{P}$ can be, for example, an unconstrained single-objective optimization problem or a constrained multi-objective problem. The quality at time $n > 0$ of the sequence $\underline{X}$ viewed as an approximate solution to the optimization problem $\mathcal{P}$ can be measured using a positive loss function

$$\varepsilon_n : \underline{X} \mapsto \mathbb{R}^+, \tag{1}$$

such that $\varepsilon_n(\underline{X}) = 0$ if and only if the set $\{X_1, \ldots, X_n\}$ solves $\mathcal{P}$ and, given two optimization strategies $\underline{X}_1$ and $\underline{X}_2$, $\varepsilon_n(\underline{X}_1) < \varepsilon_n(\underline{X}_2)$ if and only if $\underline{X}_1$ offers a better solution to $\mathcal{P}$ than $\underline{X}_2$ at time n. Under this framework, one can formulate the notion of improvement as a measure of the loss reduction yielded by the observation of a new point X_{n+1}:

$$I_{n+1} = \varepsilon_n(\underline{X}) - \varepsilon_{n+1}(\underline{X}), \, n \geq 0. \tag{2}$$

The improvement is positive if X_{n+1} improves the quality of the solution at time $n + 1$ and zero otherwise.

Assume a statistical model with a vector-valued stochastic process model ξ with probability measure $\mathbb{P}_0$ representing prior knowledge over the functions involved in the optimization problem $\mathcal{P}$. Under the Bayesian paradigm, optimization algorithms are crafted to achieve, on average, a small value of $\varepsilon_n(\underline{X})$ when n increases; where the average is taken with respect to ξ. In this framework, the choice of the observation points X_i is a sequential decision problem.

The associated Bayesian-optimal strategy for a finite budget of N observations is, however, not tractable in the general case for N larger than a few units. To circumvent this difficulty, a common approach is to consider one-step look-ahead strategies (also referred to as myopic strategies, see, e.g., [22,24] and [8,18] for discussions about two-step look-ahead strategies) where observation points are chosen one at a time to minimize the conditional expectation of the future loss given past observations:

$$
\begin{aligned}
X_{n+1} &= \operatorname{argmin}_{x \in \mathbb{X}} \mathbb{E}_n\big(\varepsilon_{n+1}(\underline{X}) \mid X_{n+1} = x\big) \\
&= \operatorname{argmax}_{x \in \mathbb{X}} \mathbb{E}_n\big(\varepsilon_n(\underline{X}) - \varepsilon_{n+1}(\underline{X}) \mid X_{n+1} = x\big) \\
&= \operatorname{argmax}_{x \in \mathbb{X}} \mathbb{E}_n\big(I_{n+1}(\underline{X}) \mid X_{n+1} = x\big), \, n \geq 0,
\end{aligned} \tag{3}
$$

where $\mathbb{E}_n$ stands for the conditional expectation with respect to X_1, $\xi(X_1)$, ..., X_n, $\xi(X_n)$. The function

$$
\rho_n : x \mapsto \mathbb{E}_n\big(I_{n+1}(\underline{X}) \mid X_{n+1} = x\big), \, n \geq 0, \tag{4}
$$

is called the *expected improvement* (EI). It is a popular sampling criterion in the Bayesian optimization literature for designing optimization algorithms (see, e.g., [20,26] for applications to constrained and unconstrained global optimization problems).

2.2   Multi-objective Bayesian Optimization

We focus in this work on unconstrained multi-objective optimization problems. Given a set of objective functions $f_j : \mathbb{X} \to \mathbb{R}$, $j = 1, \ldots, p$, to be minimized, the objective is to build an approximation of the Pareto front and of the set of corresponding solutions

$$
\Gamma = \{x \in \mathbb{X} : \nexists x' \in \mathbb{X} \text{ such that } f(x') \prec f(x)\}, \tag{5}
$$

where $\prec$ stands for the Pareto domination rule defined on $\mathbb{R}^p$ by

$$
y = (y_1, \ldots, y_p) \prec z = (z_1, \ldots, z_p) \iff \begin{cases} \forall i \leq p, \ y_i \leq z_i, \\ \exists j \leq p, \ y_j < z_j. \end{cases} \tag{6}
$$

In this setting, it is common practice to measure the quality of optimization strategies using the hypervolume loss function (see, e.g., [21,23,30]) defined by

$$
\varepsilon_n(\underline{X}) = |H \setminus H_n|, \tag{7}
$$

where $|\cdot|$ denotes the usual (Lebesgue) volume measure in $\mathbb{R}^p$ and where, given an upper-bounded set $\mathbb{B}$ of the form $\mathbb{B} = \{y \in \mathbb{R}^p; \, y \leq y^{\mathrm{upp}}\}$ for some $y^{\mathrm{upp}} \in \mathbb{R}^p$, the subsets

$$
H = \{y \in \mathbb{B}; \, \exists x \in \mathbb{X}, \, f(x) \prec y\}, \tag{8}
$$

and

$$
H_n = \{y \in \mathbb{B}; \, \exists i \leq n, \, f(X_i) \prec y\}, \tag{9}
$$

denote respectively the subset of points of $\mathbb{B}$ dominated by the points of the Pareto front and the subset of points of $\mathbb{B}$ dominated by $(f(X_1), \ldots, f(X_n))$. The set $\mathbb{B}$ is introduced to ensure that the volumes of H and H_n are finite.

Using the loss function (7), the improvement function (2) takes the form

$$I_{n+1}(\underline{X}) = |H \setminus H_n| - |H \setminus H_{n+1}| = |H_{n+1} \setminus H_n|, \tag{10}$$

and an expected improvement criterion can be formulated as

$$\rho_n(x) = \mathbb{E}_n\left(I_{n+1}(\underline{X}) \mid X_{n+1} = x\right)$$

$$= \mathbb{E}_n\left(\int_{\mathbb{B} \setminus H_n} \mathbb{1}_{\xi(x) \prec y} \, \mathrm{d}y\right)$$

$$= \int_{\mathbb{B} \setminus H_n} \mathbb{P}_n\left(\xi(x) \prec y\right) \mathrm{d}y, \tag{11}$$

where $\mathbb{P}_n$ stands for the probability $\mathbb{P}_0$ conditioned on $X_1, \xi(X_1), \ldots, X_n, \xi(X_n)$. The multi-objective sampling criterion (11) is called the *expected hypervolume improvement* (EHVI) criterion. It has been proposed and studied by Emmerich and coworkers [12,14,15].

3   Expected Weighted Hypervolume Improvement (EWHI)

3.1   Formulation of the Criterion

To measure the quality of Pareto approximation sets according to user preferences, Zitzler et al. (2007) proposed to use a user-defined absolutely continuous measure in the definition of the hypervolume indicator[1] instead of the Lebesgue measure (see [29]):

$$\varepsilon_n(\underline{X}) = \mu(H \setminus H_n), \tag{12}$$

where the measure μ is defined by $\mu(\mathrm{d}y) = \omega(y) \, \mathrm{d}y$ using a positive weight function $\omega : \mathbb{R}^p \to \mathbb{R}^+$. The value $\omega(y)$ for some $y \in \mathbb{R}^p$ can be seen as a reward for dominating y that the user may specify. Optimization strategies crafted using the loss function (12) have been studied by [3,4,13,29].

Observe that, as discussed by [13], assuming that μ possesses the bounded improper integral property, (12) is well defined and upper-bounding values are no longer required in the definition of the sets H and H_n, which can be redefined as:

$$\begin{cases} H = \{y \in \mathbb{R}^p \,;\, \exists x \in \mathbb{X}, \, f(x) \prec y\}, \\ H_n = \{y \in \mathbb{R}^p \,;\, \exists i \leq n, \, f(X_i) \prec y\}. \end{cases} \tag{13}$$

---

[1] In the original definition, the authors introduce additional terms to weight the axis. In this work, one of our objective is to get rid of the bounding set $\mathbb{B}$, as proposed by [13]. Therefore we do not consider these terms.

Similarly to (7), the improvement function associated to the loss function (12) takes the form

$$I_{n+1}(\underline{X}) = \mu(H \setminus H_n) - \mu(H \setminus H_{n+1}) = \mu(H_{n+1} \setminus H_n), \qquad (14)$$

and an expected improvement criterion can be formulated as:

$$\begin{aligned}
\rho_n(x) &= \mathbb{E}_n\left(I_{n+1}(\underline{X}) \mid X_{n+1} = x\right) \\
&= \mathbb{E}_n\left(\int_{H_n^c} \mathbb{1}_{\xi(x) \prec y}\, \mu(dy)\right) \\
&= \int_{H_n^c} \mathbb{P}_n\left(\xi(x) \prec y\right) \omega(y)\, dy, \qquad (15)
\end{aligned}$$

where H_n^c denotes the complementary of H_n in $\mathbb{R}^p$. By analogy with the EHVI criterion, we call the expected improvement criterion (15) the *expected weighted hypervolume improvement* (EWHI) criterion.

3.2   Computation of the Criterion

Under the assumption that the components ξ_i of ξ are mutually independent Gaussian processes, which is a common modeling assumption in the Bayesian optimization literature (see, e.g., [25]), the term $\mathbb{P}_n\left(\xi(x) \prec y\right)$ in the expression (15) of the EWHI can be expressed in closed form: for all $x \in X$ and $y \in H_n^c$,

$$\mathbb{P}_n\left(\xi(x) \prec y\right) = \prod_{i=1}^{p} \Phi\left(\frac{y_i - \widehat{\xi}_{i,n}(x)}{\sigma_{i,n}(x)}\right), \qquad (16)$$

where Φ denotes the Gaussian cumulative distribution function and $\widehat{\xi}_{i,n}(x)$ and $\sigma_{i,n}^2(x)$ denote respectively the kriging mean and variance at x for the ith component of ξ (see, e.g., [25,28]).

The integration of (16) over H_n^c on the other hand, is a non-trivial problem. Besides, it has to be done several times to solve the optimization problem (3) and choose X_{n+1}. To address this issue, we propose to choose X_{n+1} among a set of predefined candidate points obtained using sequential Monte-Carlo techniques as in [17], and derive a method to compute approximations of (15) with arbitrary weight functions ω for this set.

Let then $\mathcal{X}_n = (x_{n,k})_{1 \le k \le m_x} \in X^{m_x}$ be a set of m_x points where ρ_n is to be evaluated and denote

$$\rho_{n,k} = \rho_n(x_{n,k}) = \int_{H_n^c} \omega(y)\, \mathbb{P}_n\left(\xi(x_{n,k}) \prec y\right) dy, \quad 1 \le k \le m_x. \qquad (17)$$

Using a sample $\mathcal{Y}_n = (y_{n,i})_{1 \le i \le m_y}$ of m_y points obtained from a density π_n on H_n^c with un-normalized density γ_n and with normalizing constant

$$Z_n = \int_{H_n^c} \gamma_n(y)\, dy, \qquad (18)$$

an importance sampling approximation of the $(\rho_{n,k})_{1 \le k \le m_x}$ can be written as

$$\widehat{\rho}_{n,k} = \frac{Z_n}{m_y} \sum_{i=1}^{m_y} \frac{\omega(y_{n,i}) \, \mathbb{P}_n \left(\xi(x_{n,k}) \prec y_{n,i} \right)}{\gamma_n(y_{n,i})}, \quad 1 \le k \le m_x. \tag{19}$$

To obtain a good approximation for all $\widehat{\rho}_{n,k}$ using a single sample $\mathcal{Y}_n$, the un-normalized density γ_n can be chosen to minimize the average sum of squared approximation errors:

$$\mathbb{E} \left(\sum_{k=1}^{m_x} \left(\widehat{\rho}_{n,k} - \rho_{n,k} \right)^2 \right)$$

$$= \frac{1}{m_y} \sum_{k=1}^{m_x} \left(Z_n \int_{H_n^c} \frac{\omega(y)^2 \, \mathbb{P}_n \left(\xi(x_{n,k}) \prec y \right)^2}{\gamma_n(y)^2} \gamma_n(y) \, \mathrm{d}y - \rho_{n,k}^2 \right) \tag{20}$$

$$= \frac{1}{m_y} \left(Z_n \int_{H_n^c} \frac{\sum_{k=1}^{m_x} \omega(y)^2 \, \mathbb{P}_n \left(\xi(x_{n,k}) \prec y \right)^2}{\gamma_n(y)^2} \gamma_n(y) \, \mathrm{d}y - \sum_{k=1}^{m_x} \rho_{n,k}^2 \right).$$

This leads, using the Cauchy-Schwarz inequality (see, e.g., [7]), to the definition of the following density on H_n^c:

$$L_2^{\mathrm{opt}}(y) \propto \gamma_n(y) = \sqrt{\sum_{k=1}^{m_x} \omega(y)^2 \, \mathbb{P}_n \left(\xi(x_{n,k}) \prec y \right)^2}. \tag{21}$$

To obtain a sample distributed from the L_2^{opt} density and carry out the approximate computation of the EWHI using (19), we resort to sequential Monte-Carlo techniques as well (see, e.g., [2,11,17]). The algorithm that we use is not detailed here for the sake of brevity. The reader is referred to Sect. 4 of [17] for a discussion about this aspect. Details about the computation of the normalizing constant Z_n and about the variance of the proposed estimator are given in Appendix A.

4   Numerical Experiments

In our experiments, we illustrate the operation of the EWHI criterion on the bi-objective BNH problem as defined in [10] for the following two weight functions adapted from [29]:

$$\begin{cases} \omega_1(y_1, y_2) = \dfrac{1}{15} e^{-\frac{y_1}{15}} \cdot \dfrac{\mathbb{1}_{[0,150]}(y_1)}{150} \cdot \dfrac{\mathbb{1}_{[0,60]}(y_2)}{60}, \\[2mm] \omega_2(y_1, y_2) = \dfrac{1}{2} \left(\varphi \left(y, \mu_1, C \right) + \varphi \left(y, \mu_2, C \right) \right), \end{cases} \tag{22}$$

where $\varphi(y, \mu, C)$ denotes the Gaussian probability density function with mean μ and covariance matrix C, evaluated at y. The ω_1 weight function is based on an exponential distribution and encodes preference for the minimization of

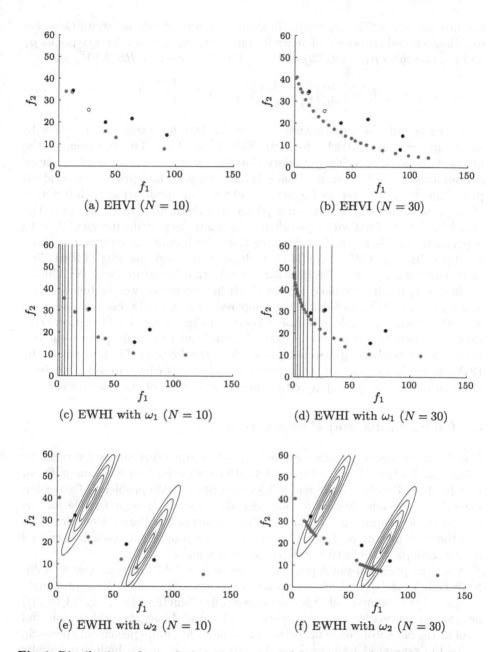

(a) EHVI ($N = 10$)

(b) EHVI ($N = 30$)

(c) EWHI with ω_1 ($N = 10$)

(d) EWHI with ω_1 ($N = 30$)

(e) EWHI with ω_2 ($N = 10$)

(f) EWHI with ω_2 ($N = 30$)

Fig. 1. Distributions obtained after 20 iterations of the optimization algorithm on the BNH problem when the weight functions ω_1 and ω_2 are used. The results obtained using the EHVI criterion are shown for reference. The contours of the weight functions are represented as black lines and the non-dominated solutions as red disks. Black disks indicate feasible dominated solutions and black circles indicate non-feasible solutions. (Color figure online)

the first objective. The ω_2 weight function is a sum of two bivariate Gaussian distributions and encodes preference for improving upon two reference points μ_1 and μ_2, chosen as $\mu_1 = (80, 20)$ and $\mu_2 = (30, 40)$ with $C = RS(RS)^T$, where

$$R = \begin{bmatrix} \cos\left(\frac{\pi}{4}\right) & -\sin\left(\frac{\pi}{4}\right) \\ \sin\left(\frac{\pi}{4}\right) & \cos\left(\frac{\pi}{4}\right) \end{bmatrix} \text{ and } S = \begin{bmatrix} 20 & 0 \\ 0 & 3 \end{bmatrix}. \tag{23}$$

To carry out the experiments, we use the BMOO algorithm of [16] with $m_x = m_y = 1000$ particles for both SMC algorithms. The functions of the problem are modeled using stationary Gaussian processes with a constant mean and an anisotropic Matérn covariance kernel. A log-normal prior distribution is placed on the parameters of the kernel and these are updated at each iteration of the algorithm using maximum a posteriori substitution (see, e.g., [5]). The algorithm is initialized with a pseudo-maximin latin hypercube design of $N = 10$ experiments and is iterated over 20 iterations. To handle the constraints of the BNH problem, the EWHI criterion is multiplied by the probability of feasibility, as is common practice in the Bayesian optimization literature (see, e.g., [26]).

In Fig. 1, results obtained by the algorithm using the weight functions ω_1 and ω_2 in the EWHI definition are compared to results obtained by the same algorithm using the EHVI criterion. Observe in Fig. 1(d) and 1(f) that observations are concentrated in regions of the Pareto front that correspond to high ω values, whereas observations are spread along the front in Fig. 1(b) where the EHVI is used. In practice, this means that less iterations would have been required to satisfyingly populate the interesting regions of the Pareto front.

5   Conclusions and Perspectives

It is shown in this paper how user-defined weight functions can be leveraged by a Bayesian framework to produce optimization strategies that focus on preferred regions of the Pareto front of multi-objective optimization problems. Two example weight functions from [29] which encode respectively a preference for one objective and a preference toward specific regions of the Pareto front are used, and the demonstration of the effectiveness of the proposed approach is carried out on a simple bi-objective optimization problem.

On more practical problems, crafting sensible weight functions can be a difficult task, especially when one has no prior knowledge about the approximate location of the Pareto front. The use of desirability functions (see, e.g. [13, 19, 27]) or utility functions (see, e.g., [1]) might provide useful insights on that issue and shall be the object of future investigations to provide a more principled approach.

In the presented framework, optimization strategies are built sequentially using an expected improvement sampling criterion called the expected weighted hypervolume improvement (EWHI) criterion. The exact computation of the criterion being intractable in general, an approximate computation procedure using importance sampling is proposed. A sampling density that is optimal for the simultaneous computation of the criterion for a set of candidate points is crafted and a sequential Monte-Carlo algorithm is used to produce samples from this density. This choice triggers an immediate question: What is the sample size m_y required by the algorithm? In fact, the problem is not so much to obtain a precise approximation of ρ_n for all $x \in X_n$, which would require a large sample size to distinguish very close points, but to deal with the optimization problem (3) and to identify with good confidence the points of X_n that correspond to high values of ρ_n. A first step toward a solution to this problem is to compute an approximation of the variance of $\widehat{\rho}_n$, as carried out in Appendix A. Further investigations on this issue are left for future work.

A   Approximate Variance of the EI Estimator

We derive in this appendix the variance of the SMC estimator for ρ_n. In the SMC procedure that we consider, the particles $(y_{n,i})_{1 \le i \le m}$ are obtained from a sequence of densities $(\pi_{n,t})_{0 \le t \le T}$, where $\pi_{n,0}$ is an easy-to-sample initial density and $\pi_{n,T} = \pi_n$ is the target density. Let $(\gamma_{n,t})_{0 \le t \le T}$ and $(Z_{n,t})_{0 \le t \le T}$ denote the corresponding sequences of un-normalized densities and normalizing constants.

First, observe that, for $1 \le t \le T$,

$$
\begin{aligned}
Z_{n,t} &= \int_{H_n^c} \gamma_{n,t}(y) \, dy \\
&= Z_{n,t-1} \int_{G_n} \frac{\gamma_{n,t}(y)}{\gamma_{n,t-1}(y)} \pi_{n,t-1}(y) \, dy.
\end{aligned}
\tag{24}
$$

Thus, we can derive a sequence of approximations $\widehat{Z}_{n,t}$ of $Z_{n,t}$, $t \ge 1$, using the following recursion formula:

$$
\begin{cases}
\widehat{Z}_{n,0} = Z_{n,0} = \int_{G_n} \gamma_{n,0}(y) \, dy, \\
\widehat{Z}_{n,t} = \widehat{Z}_{n,t-1} \left(\frac{1}{m} \sum_{i=1}^m \frac{\gamma_{n,t}(y_{n,t-1,i})}{\gamma_{n,t-1}(y_{n,t-1,i})} \right),
\end{cases}
\tag{25}
$$

where the particles $(y_{n,t-1,i})_{1 < i \le m} \sim \pi_{n,t-1}$ are obtained using an SMC procedure (see, e.g., [6]). The estimator of $\rho_n(x)$ that we actually consider is then

$$
\widehat{\rho}_n(x) = \frac{\widehat{Z}_n}{m} \sum_{i=1}^m \frac{\omega(y) \, \mathbb{P}_n \left(\xi(x) \prec y_{n,i} \right)}{\gamma_n(y_{n,i})} = \widehat{Z}_n \widehat{\alpha}_n(x)
\tag{26}
$$

where

$$
\widehat{\alpha}_n(x) = \frac{1}{m} \sum_{i=1}^m \frac{\omega(y) \, \mathbb{P}_n \left(\xi(x) \prec y_{n,i} \right)}{\gamma_n(y_{n,i})},
\tag{27}
$$

and

$$
\widehat{Z}_n = \widehat{Z}_{n,T} = Z_{n,0} \prod_{u=1}^T \widehat{\theta}_{n,u},
\tag{28}
$$

with

$$
\widehat{\theta}_{n,t} = \frac{1}{m} \sum_{i=1}^m \frac{\gamma_{n,t}(y_{n,t-1,i})}{\gamma_{n,t-1}(y_{n,t-1,i})}.
\tag{29}
$$

Now, assume the idealized setting, as usual in the SMC literature (see, e.g., [9]), where

(i) $y_{n,t,i} \overset{\text{i.i.d}}{\sim} \pi_{n,t}, 1 \le i \le m,$

(ii) the samples $\mathcal{Y}_{n,t} = (y_{n,t,i})_{1 \le i \le m}$ are independent, $0 \le t \le T.$

Observe from (19) and (24) that under (i), $\widehat{a}_n(x)$ is an unbiased estimator of $\alpha_n(x) = \frac{\rho_n(x)}{Z_n}$, and $\widehat{\theta}_{n,t}$ is an unbiased estimator of $\theta_{n,t} = \frac{Z_{n,t}}{Z_{n,t-1}}, 1 \le t \le T.$ Moreover, under (ii), $\widehat{a}_n(x)$ and the $(\widehat{\theta}_{n,t})_{1 \le t \le T}$ are independent. Thus,

$$\text{Var } \widehat{\rho}_n(x) = \mathbb{E}(\widehat{a}_n^2)\mathbb{E}(\widehat{Z}_n^2) - \mathbb{E}(\widehat{a}_n(x))^2 \mathbb{E}(\widehat{Z}_n)^2$$
$$= (\text{Var } \widehat{a}_n(x) + \alpha_n(x)^2)(\text{Var } \widehat{Z}_n + Z_n^2) - \alpha_n(x)^2 Z_n^2$$
$$= \text{Var } \widehat{a}_n(x)\text{Var } \widehat{Z}_n + \alpha_n(x)^2\text{Var } \widehat{Z}_n + Z_n^2\text{Var } \widehat{a}_n(x).$$

We obtain the coefficient of variation of $\widehat{\rho}_n(x)$

$$\frac{\text{Var } \widehat{\rho}_n(x)}{\rho_n(x)^2} = \Lambda_n(x)^2 + (1 + \Lambda_n(x)^2) \Delta_{n,T}^2, \tag{30}$$

where $\Lambda_n(x)^2 = \frac{\text{Var } \widehat{a}_n(x)}{\alpha_n(x)^2}$ and $\Delta_{n,t}^2 = \frac{\text{Var} \widehat{Z}_{n,t}}{Z_{n,t}^2}$ are the coefficients of variation of $\widehat{a}_n(x)$ and $\widehat{Z}_{n,t}$ respectively.

Using the same ideas as above, we have

$$\Delta_{n,t}^2 = \delta_{n,t}^2 + (1 + \delta_{n,t}^2) \Delta_{n,t-1}^2, \tag{31}$$

where $\delta_{n,t}^2 = \frac{\text{Var } \widehat{\theta}_{n,t}}{\theta_{n,t}^2}$ is the coefficient of variation of $\widehat{\theta}_{n,t}$.

Estimators of $\Lambda_n(x)^2$, $\Delta_{n,t}^2$ and $\delta_{n,t}^2$ can be derived under (ii). For instance, observe that

$$\delta_{n,t}^2 = \frac{1}{m} \frac{\text{Var}\left(\frac{\gamma_{n,t}(y_{n,t-1,1})}{\gamma_{n,t-1}(y_{n,t-1,1})}\right)}{\mathbb{E}\left(\frac{\gamma_{n,t}(y_{n,t-1,1})}{\gamma_{n,t-1}(y_{n,t-1,1})}\right)^2}. \tag{32}$$

Thus, an estimator of $\delta_{n,t}^2$ is

$$\widehat{\delta}_{n,t}^2 = \frac{\sum_{i=1}^m \frac{\gamma_{n,t}(y_{n,t-1,i})^2}{\gamma_{n,t-1}(y_{n,t-1,i})^2}}{\left(\sum_{i=1}^m \frac{\gamma_{n,t}(y_{n,t-1,i})}{\gamma_{n,t-1}(y_{n,t-1,i})}\right)^2} - \frac{1}{m}. \tag{33}$$

Plugging (33) in (31), we obtain an estimator of $\Delta_{n,t}^2$:

$$\widehat{\Delta}_{n,t}^2 = \widehat{\delta}_{n,t}^2 + (1 + \widehat{\delta}_{n,t}^2) \cdot \widehat{\Delta}_{n,t-1}^2. \tag{34}$$

Similarly, an estimator of $\Lambda_n(x)^2$ is

$$\widehat{\Lambda}_n(x)^2 = \frac{\sum_{i=1}^m \frac{w(y)^2 \, \mathbb{P}_n(\xi(x) \prec y_{n,i})^2}{\gamma_n(y_{n,i})^2}}{\left(\sum_{i=1}^m \frac{w(y) \, \mathbb{P}_n(\xi(x) \prec y_{n,i})}{\gamma_n(y_{n,i})}\right)^2} - \frac{1}{m}. \tag{35}$$

As a result, we obtain the following numerically tractable approximation of the variance of $\widehat{\rho}_n(x)$:

$$\text{Var}\left(\widehat{\rho}_n(x)\right) \approx \widehat{\rho}_n(x)^2 \cdot \left(\widehat{\Lambda}_n(x)^2 + \left(1 + \widehat{\Lambda}_n(x)^2\right) \cdot \widehat{\Delta}_{n,t}^2\right), \tag{36}$$

where $\widehat{Z}_{n,t}$ and $\widehat{\Delta}_{n,t}^2$ are obtained recursively using (25) and (34), $\widehat{\Lambda}_n(x)^2$ is computed using (35) and $\widehat{\rho}_n(x)$ is computed using (26).

References

1. Astudillo, R., Frazier, P.: Multi-attribute Bayesian optimization under utility uncertainty. In: Proceedings of the NIPS Workshop on Bayesian Optimization, December 2017, Long Beach, USA (To appear)
2. Au, S.K., Beck, J.L.: Estimation of small failure probabilities in high dimensions by subset simulation. Probab. Eng. Mech. **16**(4), 263–277 (2001)
3. Auger, A., Bader, J., Brockhoff, D., Zitzler, E.: Articulating user preferences in many-objective problems by sampling the weighted hypervolume. In: Proceedings of the 11th Annual Conference on Genetic and Evolutionary Computation, pp. 555–562. ACM (2009)
4. Auger, A., Bader, J., Brockhoff, D., Zitzler, E.: Investigating and exploiting the bias of the weighted hypervolume to articulate user preferences. In: Proceedings of the 11th Annual Conference on Genetic and Evolutionary Computation, pp. 563–570. ACM (2009)
5. Bect, J., Ginsbourger, D., Li, L., Picheny, V., Vazquez, E.: Sequential design of computer experiments for the estimation of a probability of failure. Stat. Comput. **22**(3), 773–793 (2012)
6. Bect, J., Li, L., Vazquez, E.: Bayesian subset simulation. SIAM/ASA J. Uncertain. Quantif. **5**(1), 762–786 (2017)
7. Bect, J., Sueur, R., Gérossier, A., Mongellaz, L., Petit, S., Vazquez, E.: Échantillonnage préférentiel et méta-modèles: méthodes bayésiennes optimale et défensive. In: 47èmes Journées de Statistique de la SFdS-JdS 2015 (2015)
8. Benassi, R.: Nouvel algorithme d'optimisation bayésien utilisant une approche Monte-Carlo séquentielle. Ph.D. thesis, Supélec (2013)
9. Cérou, F., Del Moral, P., Furon, T., Guyader, A.: Sequential Monte Carlo for rare event estimation. Stat. Comput. **22**(3), 795–808 (2012)
10. Chafekar, D., Xuan, J., Rasheed, K.: Constrained multi-objective optimization using steady state genetic algorithms. In: Cantú-Paz, E., et al. (eds.) GECCO 2003. LNCS, vol. 2723, pp. 813–824. Springer, Heidelberg (2003). https://doi.org/10.1007/3-540-45105-6_95
11. Del Moral, P., Doucet, A., Jasra, A.: Sequential monte carlo samplers. J. R. Stat. Soc.: Ser. B (Stat. Methodol.) **68**(3), 411–436 (2006)
12. Emmerich, M.: Single - and multiobjective evolutionary design optimization assisted by Gaussian random field metamodels. Ph.D. thesis, Technical University Dortmund (2005)
13. Emmerich, M., Deutz, A.H., Yevseyeva, I.: On reference point free weighted hypervolume indicators based on desirability functions and their probabilistic interpretation. Proc. Technol. **16**, 532–541 (2014)
14. Emmerich, M., Giannakoglou, K.C., Naujoks, B.: Single - and multi-objective evolutionary optimization assisted by Gaussian random field metamodels. IEEE Trans. Evol. Comput. **10**(4), 421–439 (2006)

15. Emmerich, M., Klinkenberg, J.W.: The computation of the expected improvement in dominated hypervolume of Pareto front approximations. Technical report, Leiden University (2008)

16. Feliot, P.: A Bayesian approach to constrained multi-objective optimization. Ph.D. thesis, IRT SystemX and Centrale Supélec (2017)

17. Feliot, P., Bect, J., Vazquez, E.: A Bayesian approach to constrained single-and multi-objective optimization. J. Glob. Optim. **67**(1–2), 97–133 (2017)

18. Ginsbourger, D., Le Riche, R.: Towards Gaussian process-based optimization with finite time horizon. In: Giovagnoli, A., Atkinson, A., Torsney, B., May, C. (eds.) mODa 9 – Advances in Model-Oriented Design and Analysis. Contributions to Statistics, pp. 89–96. Springer, Heidelberg (2010). https://doi.org/10.1007/978-3-7908-2410-0_12

19. Harrington, E.C.: The desirability function. Ind. Qual. Control. **21**(10), 494–498 (1965)

20. Jones, D.R., Schonlau, M., Welch, W.J.: Efficient global optimization of expensive black-box functions. J. Glob. Optim. **13**(4), 455–492 (1998)

21. Knowles, J., Corne, D.: On metrics for comparing nondominated sets. In: Proceedings of the 2002 Congress on Evolutionary Computation, CEC 2002, vol. 1, pp. 711–716. IEEE (2002)

22. Kushner, H.J.: A new method of locating the maximum point of an arbitrary multipeak curve in the presence of noise. J. Fluids Eng. **86**(1), 97–106 (1964)

23. Laumanns, M., Rudolph, G., Schwefel, H.P.: Approximating the pareto set: concepts, diversity issues, and performance assessment. Secretary of the SFB 531 (1999)

24. Mockus, J., Tiesis, V., Žilinskas, A.: The application of Bayesian methods for seeking the extremum. In: Dixon, L.C.W., Szegö, G.P. (eds.) Towards Global Optimization, vol. 2, pp. 117–129. North Holland, New York (1978)

25. Santner, T.J., Williams, B.J., Notz, W.: The Design and Analysis of Computer Experiments. Springer Series in Statistics. Springer, New York (2003). https://doi.org/10.1007/978-1-4757-3799-8

26. Schonlau, M., Welch, W.J., Jones, D.R.: Global versus local search in constrained optimization of computer models. In: New Developments and Applications in Experimental Design: Selected Proceedings of a 1997 Joint AMS-IMS-SIAM Summer Conference. IMS Lecture Notes-Monographs Series, vol. 34, pp. 11–25. Institute of Mathematical Statistics (1998)

27. Wagner, T., Trautmann, H.: Integration of preferences in hypervolume-based multiobjective evolutionary algorithms by means of desirability functions. IEEE Trans. Evol. Comput. **14**(5), 688–701 (2010)

28. Williams, C.K.I., Rasmussen, C.: Gaussian Processes for Machine Learning, vol. 2(3), p. 4. The MIT Press, Cambridge (2006)

29. Zitzler, E., Brockhoff, D., Thiele, L.: The hypervolume indicator revisited: on the design of pareto-compliant indicators via weighted integration. In: Obayashi, S., Deb, K., Poloni, C., Hiroyasu, T., Murata, T. (eds.) EMO 2007. LNCS, vol. 4403, pp. 862–876. Springer, Heidelberg (2007). https://doi.org/10.1007/978-3-540-70928-2_64

30. Zitzler, E., Thiele, L.: Multiobjective optimization using evolutionary algorithms — a comparative case study. In: Eiben, A.E., Bäck, T., Schoenauer, M., Schwefel, H.-P. (eds.) PPSN 1998. LNCS, vol. 1498, pp. 292–301. Springer, Heidelberg (1998). https://doi.org/10.1007/BFb0056872

Reinforcement Learning Methods for Operations Research Applications: The Order Release Problem

Manuel Schneckenreither[✉][iD] and Stefan Haeussler[iD]

Department of Information Systems, Production and Logistics Management,
University of Innsbruck, Innsbruck, Austria
{manuel.schneckenreither,stefan.haeussler}@uibk.ac.at

Abstract. An important goal in Manufacturing Planning and Control systems is to achieve short and predictable flow times, especially where high flexibility in meeting customer demand is required. Besides achieving short flow times, one should also maintain high output and due-date performance. One approach to address this problem is the use of an order release mechanism which collects all incoming orders in an order-pool and thereafter determines when to release the orders to the shop-floor. A major disadvantage of traditional order release mechanisms is their inability to consider the nonlinear relationship between resource utilization and flow times which is well known from practice and queuing theory. Therefore, we propose a novel adaptive order release mechanism which utilizes deep reinforcement learning to set release times of the orders and provide several techniques for challenging operations research problems with reinforcement learning. We use a simulation model of a two-stage flow-shop and show that our approach outperforms well-known order release mechanism.

Keywords: Operations research · Production planning ·
Order release · Machine learning · Reinforcement learning

1 Introduction

Manufacturing planning and control (MPC) systems play an important role in managing flow of material through manufacturing organizations. An important goal in MPC systems are short and predictable flow times, especially in environments where high flexibility in meeting customer demand is required. Due to complexity reduction purposes MPC systems are often hierarchically structured into two levels [8]. The top level (goods flow control) coordinates the production units that constitute the logistic chain by coordinated releases of production orders and thus sets the targets for the base level (production unit control; for these terms, see [8]). The base level performs detailed scheduling within the production units. The interface between the top level and the base level is *order*

This research is partly supported by the Aktion D. Swarovski KG (2016) grant.

G. Nicosia et al. (Eds.): LOD 2018, LNCS 11331, pp. 545–559, 2019.
https://doi.org/10.1007/978-3-030-13709-0_46

release which is defined as the transfer of the control over the respective work orders from the top to the base level, which are the decision making units within the production departments.

Within the MPC task the setting of planning parameters plays a crucial role. One of the key parameters is the planned lead time which is defined as the *planned* time that elapses between the release of an order and its completion (hereinafter denoted as lead time). In contrast, the *actual* time an order takes to make it through the production system is called flow time and is used as a performance measure. Flow times consist of processing, setup, control, transport, and waiting times, whereas the latter is the governing factor. Waiting times are a result from queuing (e.g. jobs queue before and after processing). Queuing in production systems depends heavily on the amount of jobs in the system (WIP). Thus, waiting times are relatively difficult to estimate which makes the setting of favorable lead times so difficult (e.g., [41,47]).

A common approach to set lead times is either to use fixed (also referred to static) lead times which is mostly used in a Material Requirements Planning (MRP) context (e.g., [29,42,49]) or a workload limit for the resources or the whole system under study (e.g., workload control or CONWIP; see [43] or [39] respectively) where companies and researchers assume that the set lead times or limits are constant and fixed over the planning horizon. Hoyt was the first to criticize the use of fixed lead times by arguing that lead times should be set dynamically in order to react to the dynamic operational characteristics of the production process [14]. However, a solely *reactive* approach often generates an erratic order release pattern which is generally denoted as the lead time syndrome (LTS; see e.g., [24]). The LTS can be described as a positive feedback loop in which the flow times increase the lead times via lead time updating [17]. As the lead times increase, more orders are released to the system which results in higher inventory levels and thus in higher flow times again which closes the vicious cycle.

Therefore, in order to overcome the LTS one needs an anticipation function that predicts the flow times of the work orders as a function of the order release decisions which leads to a *predictive* lead time management approach [36]. Within this stream of research the problem of setting lead times is seen as a forecasting problem. This means that one tries to find lead times that best fit the dynamically changing production environment. Enns et al. develop methods for setting dynamically planned lead times which are based on exponentially smoothed feedback on the flow times of each order at each stage [10]. Similarly, Selcuk et al. [37] apply exponentially smoothed lead times to a capacitated multi-stage make-to-stock system. However, the latter study shows that exponentially smoothed lead times may induce the LTS in the case of demand variations and high utilization levels. Therefore, a suitable predictive order release model has to forecast the future performance, identify pathological behavior and then generate necessary corrective actions to prevent the incurrence of the LTS.

Among a wide variety of prediction methods, machine learning algorithms (e.g., artificial neural networks) are considered the most effective because of their flexible non-linear and interaction effects modeling capability, and consequently have been recommended as a decision support tool for other production planning problems (e.g., due date assignment; see e.g., [15] or [3,48] for a literature review on this topic). To the best of our knowledge, there exist only two papers that use machine learning for making order release decisions [19,31]. However, [19] only decides on the sequence of the releases and [31] limit themselves to a very simple production system. Furthermore, both use a continuous order release method (CONWIP) although in practice order release decisions often need to be made on a daily basis (see [11]).

Therefore, we contribute to this stream of research by developing a periodic adaptive order release mechanism using reinforcement learning which dynamically decides on whether to release orders based on a deep reinforcement learning algorithm. The viability of the developed approach is tested on a multi product, two-stage hypothetical flow shop and is compared to the performance of conventional order release mechanisms. The tested production system setup is characteristic for problems in the operations research domain in the sense of periodic and discrete decisions with the aim to maximize profits and a high degree of complexity. Therefore, the presented reinforcement learning methods can be easily adapted and applied to various problems in operations research.

Clearly, several properties of the production system under investigation (e.g., the work load, utilization, etc.) are to a large extend determined by the used order release mechanism. This means that one experiences 'sampling issues' since the state of the system depends on the order release quantities and thus a training set which is based on a certain order release mechanism is distorted. This makes supervised learning techniques unsuited, since they use the generated data to infer its knowledge on how to release orders and will always follow the characteristics of the chosen release technique for data generation. Therefore, reinforcement learning is the appropriate AI technique to use, since it bases each learning step on its current knowledge and thus is able to explore the whole solution space. Additionally, reinforcement learning takes into account queuing effects which influence system states over multiple periods, since it is designed to connect the output of consecutive periods whereas standard supervised learning techniques would not take care of these interrelations over multiple periods.

The rest of the paper is structured as follows. The next section reviews the relevant literature and thereafter in Sect. 3 we describe the used methodology. In Sect. 4 we present the results and conclude our findings in Sect. 5.

2   Literature Review

This section is structured into two related fields of research which are the basis for our paper. First, there exist a large body of literature on order release models which constitutes the foundation of this paper. Secondly, studies that use techniques based on AI in the field of operations research with a special focus on production planning.

The literature on order release models can be divided into two main streams, namely the conventional rule based order release mechanisms (e.g., [6,13,47]) and multi-period optimization models. Note that a review of optimization based order release models is out of the scope of this paper (see [30] for a review on this topic). The literature on rule based order release mechanisms can be divided into two groups: continuous methods, which may trigger a release at any moment in time; and, periodic release methods, for which release decisions take place at the start of each period. This paper focuses on the latter approach. Several rule based order release mechanisms have been proposed over the last decades ranging from very simple (e.g., immediate release) to more sophisticated ones (e.g., backward infinite loading). The backward infinite loading (BIL) technique was introduced by Ackerman in [1] and calculates the release date (RD) of an order j by subtracting its lead time (LT) of its due date (DD):

$$RD_j = DD_j - LT_j \tag{1}$$

Besides the two above mentioned uncapacitated approaches also capacitated mechanism were developed. The two most prominent methods are the probabilistic and atemporal approach. The probabilistic approach was introduced by Bechte [5], and is known as Load Oriented Manufacturing Control (LOMC) concept (see also [47]). This order release mechanism estimates the input from jobs upstream to the direct load of a work centre using a depreciation factor based on historical data. The atemporal approach (also called LUMS method) was introduced in [7] and [12] and simply adds direct and indirect load (so called aggregate load).

We divide the second literature stream into three categories:

1. Application of AI techniques to scheduling problems,
2. Application of AI techniques to support order release decisions,
3. Application of AI techniques to make decisions on the order entry level (order inquiry, due date assignment).

Within the first category, several different AI techniques are used to address scheduling problems (for a review see [3]) ranging from expert systems (e.g., [35]) to decision trees (e.g., [26]) or methods using ANNs or ANNs in conjunction with other methods (e.g., [20,34]; see [2] for a recent literature review on scheduling with ANNs).

To the best of the authors knowledge there are only two studies that belong to the second category (e.g., [19,31]). [19] combines a genetic algorithm and an induced decision tree. They test their approach on a small and a large job shop (with three and seven machines respectively) and use a CONWIP release rule to determine *when* to release an order and use the decision tree to find the sequence for order release and a genetic algorithm to find a sequence for dispatching at each machine. [31] uses a reinforcement learning method for order release in a single product, serial flow line and compares its performance (WIP costs) with conventional order release policies (e.g., Kanban and CONWIP).Both use a continuous order release method (CONWIP) although in practice order release decisions often need to be made on a daily basis (see [11]).

Table 1. Operation times (upper half) and costs setup (in monetary unit).

Work center	WC1	WC2	WC3
Operation time	$\mathcal{U}(70, 130)$	$\mathcal{U}(130, 170)$	$\mathcal{U}(180, 200)$
Cost	Wc p. Order/Period	Fc p. Order/Period	Bc p. Order/Period
Value	3	10	20

The third category of literature in this review uses AI techniques to make decisions on the order entry level. Here, ANN or ANN in combination with genetic algorithms are used to predict flow times in order to set due dates (e.g., [15, 16, 20, 32, 33]).

As described in the literature review above, there is a lack of studies that use AI techniques for making order release decisions. This is quite interesting since early studies (e.g., [25]) state that dispatching becomes less important when combined with an appropriate order release mechanism. Therefore, we develop a periodic order release model based on deep reinforcement learning and show its viability on a multi-product, two-stage hypothetical flow shop.

3 Methodology

This section describes the simulation model, the used conventional order release methods, and the reinforcement learning algorithm including parameter setup and tested variants.

3.1 Simulation Model

To ensure generalizability, we use a hypothetical flow shop make-to-order manufacturing system similar to the system analyzed in [19]. The simulation model consists of three work centres; each consists of a single machine and can process only one order at a time. The number of orders arriving is uniformly distributed between 3 and 15 order per period and thus with a mean of 9 order per period. In other words, every 106.67 min one order arrives in the order pool. Incoming orders are queued at the order pool until released. The due date slack (dds), that is the periods until incoming orders are due, is set to 7. Once released they are queued at each work center and wait until being processed by the machine. The queuing priority is first-come-first-serve. No preemptions are allowed. Order routings are deterministic and embrace two production stages with diverging shape and no return visits. This results in 2 different products. The operation times of the work centres are uniformly distributed, cf. Table 1. These characteristics lead to a utilization rate of 90% for the bottleneck work center (WC3) in steady state. Planning periods were set to 960 min (16 h).

To evaluate the performance of the different order release models we define following performance measures similar to literature (e.g., [4, 44]):

- **Cost related measures:** average total holding costs for WIP and finished goods inventory, costs for backorders. There are no earnings, thus the algorithms minimizes the costs. The setup for the actual values of the costs is given in Table 1 and shows that late deliveries are especially expensive. All costs are per order and period, and are measured and reported at the end of each period.
- **Delivery related measures:** mean tardiness of late orders (TA), standard deviation of lateness (σTA).
- **Flow time related measures:** mean shop floor throughput time (time duration from release until entry of finished goods inventory; SFTT).

The length of each simulation run to evaluate the performance was 1750 periods including a warm-up period of 750 periods. Welch's procedure was applied to approximate the length of the warm-up period (see [18]). Each order release method was evaluated on 25 predefined demand streams.

3.2   Conventional Order Release Rules

As external benchmark for comparison we use different parameterized backward infinite loading (BIL) techniques, cf. Eq. 1. Note that we set lead times for each product type and not for every order as done by Ackerman [1].

3.3   Reinforcement Learning Order Release Algorithm

Q-Learning in a Nutshell. We use the ideas of the Q-Learning agent as introduced in [40] to set lead times LT_p for each product type p. Furthermore, inspired by [27,28,38] the algorithm is extended and adapted, as well as lifted to an actor-critic model. The action-value function $Q(s, a)$ (also known as Q-function) for a state $s \in S$ and an action $a \in A$, as well as the policy π, are represented by feed-forward artificial neural networks, where S is the set of all states and A the set of available actions. Furthermore, we use a target network parameters θ_Q^T and θ_π^T which are softly updated using the worker networks θ_Q and θ_π respectively as in [21].

In a nutshell the Q-Learning algorithms work as follows. An agent explores the environment by consecutively taking an action $a \in A$ and observing rewards r as it traverses from one state $s \in S$ to another environmental state $s' \in S$. According to the observed reward r and future state s' the state-action tuple (s, a) is assessed. This value is then stored in the Q-function before repeating the process. Nonetheless, Q-Learning assumes an underlying Markov Decision Processes (MDP). Informally, that is a memory-less task with states, actions, rewards and transition probabilities. For a formal definition see for instance [40, p. 61ff]. Under a certain policy $\pi = P(a|s)$ the optimal action-value function for an observed state s and action a is given by $Q^\star(s, a) = \mathbb{E}_\pi[\sum_{i=0}^{\infty} \gamma^i \cdot r_{t+i}|s_t = s, a_t = a]$ which is the expected sum of rewards r_t discounted by γ for each time-step t [40]. Q-learning approximates the optimal action-value function by iteratively updating the state-action values while exploring the solution space.

Convergence and Actor-Critic Model. Even though shown to converge to an optimal action-value function $Q^*(s,a)$ if the Q-function is represented discretely [46] for most applications a tabular representation is infeasible due to the exponential growth of the memory requirement. Thus, often (deep) ANNs are used in lieu of actual tables to approximate the Q-function, which however, results in the fact that the algorithm is unstable or may even diverge [45]. Inspired by [22,28] we partly overcome this issue by using experience replay, which randomly re-trains a batch out of already visited state-action pairs. Therefore, our algorithm stores a set of experiences $e_t = (s_t, a_t, r_t, s_{t+1})$ and re-trains the Q-function of a randomly picked subset of predetermined size of experiences on each time step t. Adding to this we additionally extend the algorithm as in A3C [27] to simultaneously learn a policy. Thus we lift the algorithm to be actor-critic and therefore a second neural network θ_π is added.

Markov Decision Process. The underlying MDP is unichain and looks as follows. Both, state space S and action space A are discrete. Any state $s \in S$ of the *state space* is composed of the following information for each product p:

- The currently set lead time $LT_p \in \{1, 2, \ldots, \text{dds}\}$ (Recall: Due Date Period − Lead Time = Release Period). Note that we bound the maximum lead time with due date slack dds, which is 7 in our setup.
- Counters $OP_{p,d} \in \mathbb{N}$ for the number of orders in the order pool grouped by time buckets $d \in \{1, 2, \ldots, \text{dds}\}$, which stands for the number of periods until the due date.
- Counters $Q_i \in \mathbb{N}$ standing for the number of orders for each queue i.
- Counters $FGI_{p,d} \in \mathbb{N}$ of orders in the finished goods inventory grouped by time buckets $d \in \{-5, -4, -3, \ldots, \text{dds}\}$, which stands for the number of periods until the due date. Orders with a due date with more than 5 periods ago are listed in the counter FGI_{-5}.
- Counters $S_{p,d} \in \mathbb{N}$ of shipped orders from the last period grouped by time buckets $d \in \{-5, -4, -3, \ldots, \text{dds}\}$, which stands for the number of periods until the due date. Orders with a due date with more than 5 periods ago are listed in the counter S_{-5}.

The algorithm implicitly learns a function which maps the current state of the production system to a release decision. In an optimal situation it does so by multiply exploring every action for each state and assessing its economic viability.

Recall that orders are released once the due date is within the interval $[t, t +$ lead time], whereas t is the current period. Clearly, this yields bulk releases (either all or no orders with the same due date and product type are released).

The *actions space* is composed of two independent decisions. These are the relative changes of the lead times to the currently set lead times $LT_p \in \{1, 2, \ldots, \text{dds}\}$ for each product p. Recall that $p = 2$. Furthermore, we restrict the action space for each state s_{t+1} according to the last set lead time LT_p from state s_t by restricting the change of the lead time for consecutive periods to a maximum of 1. Thus, if LT_p is the current lead time for product p the action space for this product is given by $\{1, 2, \ldots, \mathsf{dds}\} \cap \{LT_p - 1, LT_p, LT_p + 1\}$. Put differently the algorithm can increase or decrease the lead time by 1 or leave it as it is, as long as it acts within the discrete action space given by the set $\{1, 2, \ldots, \mathsf{dds}\}$ for each type of product. Thus the action space over all products is given by the full enumeration of available actions of the individual products.

Reward. In each period the agent chooses an action which generates a reward while traversing to the next period by simulating the production system. The rewards are the accumulated costs at the end of the period. These costs are consist of the number of backorders, the current WIP level and the number of orders in the inventory. The costs are divided by the normalizing parameter η (see Table 2) and then clipped (cut off) to the interval $[-0.5, 0.5]$.

Neural Network Setup. After manually testing various depths and layer widths we found two three layer fully-connected networks to be the most satisfying setup of which both have the same shape. The number of nodes and activation functions are 41-ReLU[1]-89-ReLU-20-ReLU-9, with the output activations being Softmax (normalized exponential function) for the policy networks θ_π, θ_π^T and Tanh (hyperbolic tangent function) for the Q-networks θ_Q, θ_Q^T. The output for both networks consist of $3^2 = 9$ nodes. This is due to the fact that all combinations of increase, decrease and no change of the lead time for each product type have to be represented. As expected the policy network outputs 9% for any given state which specify action probabilities. The Q-network output is a value representing the expected discounted (and average) reward.

Adapted Algorithm. Algorithm 1 shows the adapted Q-Learner algorithm. The experience replay memory is filled by executing random actions before the learning process starts. Note that we do not learn state values as done by A3C but rather state-action values like in Q-Learning, as we use the average of these values for the policy loss (Line 12). Nonetheless we also use a measure of entropy, in particular the Gini-impurity, with parameter $\beta = 0.03$ for preventing early convergence (see [27] for implementation details). The network θ_π^T represents the target policy network, whereas the network θ_Q^T represents the target Q-function. All workers operate on the worker networks θ_π and θ_Q. The binary variable alg is used to determine the used algorithm. In case of alg $= 1$ the average reward value is subtracted for the policy loss, while in case of alg $= 0$ average reinforcement learning techniques as presented in the algorithm R-learning [23] are used. The idea in average reinforcement learning is that only the biases of the average reward are used as Q-values.

Although the task of setting the lead times for the orders is done periodically there are no episodes. Further, we observed that standard reinforcement learning algorithms have problems when presented with rewards in every step, as it is the case in our application as every period the occurring costs are shown to the agent. For standard algorithms this results in ongoing increasing Q-values, until

---

[1] ReLU stands for rectified linear unit.

Algorithm 1. Adapted Q-Learning Algorithm.

1: Initialize $\theta_Q, \theta_\pi, \theta_Q^T, \theta_\pi^T$ arbitrarily, Initialize state s_t, Set time $t = 0$, Initialize experience replay memory with size N, Set $fork \leftarrow False$, set $\rho = 0$

2: **repeat** (for each step and worker $w_1, \ldots, w_W$)

3:     Take action a_t according to $\pi(a_t \mid s_t; \theta_\pi^T)$, observe reward r_t and state s_{t+1}

4:     With probability p_f fork the worker which copies the current environment and take action $a_{t,f} \in (\mathcal{A}_{s_t}/\{a_t\})$, observe reward $r_{t,f}$ and state $s_{t+1,f}$, set $T \leftarrow t + T_f$ and repeat steps $3, 17, 18$ until done, report experience-tuples

5:     Store all arising experiences (s_t, a_t, r_t, s_{t+1}) to the experience replay memory $\mathcal{E}$

6:     Update average reward: $\rho \leftarrow (1 - \alpha)\rho + \alpha[r_t + \max_a Q(s_{t+1}, a) - \max_a Q(s_t, a)]$

7:     Randomly choose a subset $\mathcal{E}_t \subset \mathcal{E}$ of size n

8:     Reset gradients $d\theta_Q \leftarrow 0, \, d\theta_\pi \leftarrow 0$

9:     **repeat** (for each experience instance $(s_e, a_e, r_e, s_{e+1}) \in \mathcal{E}_t$)

10:         Save target value $Q_{\text{target}} \leftarrow r_e + \gamma \cdot \max_a Q(s_{e+1}, a; \theta_Q^T) - (1 - \text{alg})\rho$

11:         Accumulate gradients: $d\theta_Q \leftarrow d\theta_Q + \partial(Q_{\text{target}} - Q(s_e, a_e; \theta_Q))^2/\partial\theta_Q$

12:         Accumulate gradients: $d\theta_\pi \leftarrow d\theta_\pi + \frac{\partial \log \pi(a_e \mid s_e; \theta_\pi)(Q_{\text{target}} - \text{alg} \, \text{avg}_a \, Q(s_e, a; \theta_Q))}{\partial\theta_\pi}$

13:     **until** all experiences in $\mathcal{E}_t$ processed

14:     [Synchronized] Perform update of worker networks θ_Q, θ_π using $d\theta_Q$ and $d\theta_\pi$

15:     [Only if w_1] Softly update network: $\theta_Q^T(s, a) \leftarrow \tau \cdot \theta_Q^T(s, a) + (1 - \tau)\theta_Q(s, a)$

16:     [Only if w_1] Softly update network: $\theta_\pi^T(s, a) \leftarrow \tau \cdot \theta_\pi^T(s, a) + (1 - \tau)\theta_\pi(s, a)$

17:     $t \leftarrow t + 1$

18: **until** $t \geqslant T$

---

all state-actions values are close to the maximum value. We tackle this by using the average state-value (over all actions) as base for the policy error.

Forking. We have added, so called, forks which are used to search the solution space by copying the current environment of the agent. Thus, at each state the agent explores another randomly chosen action with probability p_f (Line 4). That means, it copies the current state and takes actions until the T_f periods have been observed. All experiences are then returned to the origin worker, which saves them to the experience replay memory. This allows additionally exploring the solution space. We observed a decrease in learning time and an increase in stability of the algorithm with this technique.

Workers. The main worker w_1, which is the one that is used to evaluate the policy, operates on the learned policy π. The other 20 workers choose a random action with 20% probability and otherwise select the actions according to the learned policy π. Using additionally workers with different or even randomly changing BIL-policies did not yield better results.

Variants. We experiment with the way we represent actual rewards to the algorithm. First we use the standard way which presents the actual costs occurring in one period to the algorithm (MLShipped). However, this results in the fact that the algorithm has to learn two delays. On one hand the standard delay on what actions to choose to traverse to a good state in which high rewards are

Table 2. Variants tested.

Variant	Description	η
MLShipped	Rewards are actual costs of the shipped orders observed at the end of the period	2500.00
MLShippedAvg	Same as MLShipped but the rewards are averaged over the orders	$2500/9 = 277.78$
MLShippedAvg$_{AvgRL}$	Same as MLShippedAvg but uses average reinforcement learning, thus alg $= 0$ and $\gamma = 0.5$	277.78
MLRelOrds	The released orders are used to calculate the costs	2500.00
MLOrdPool	The orders in the order pool at period $t + 1$ are used to sum up the costs. Averaged over all orders	277.78

Table 3. Parameter setup.

Parameter	p_f	γ	α	τ	N	B	Train iterations	Learn. rate	L2
Value	0.2	0.995[a]	0.95	0.001	30000	128	4	0.01	0.0001

[a]For MLShippedAvg$_{AvgRL}$ we use $\gamma = 0.5$ to allow balancing of average and discounted reinforcement learning.

expected, but on the other hand it must also learn the delay that these rewards are not observed directly, but rather after several periods when the production system outputs the results for the action taken in that state. Therefore, we added the variations of keeping track of either the released orders (MLRelOrds) or the orders in the order pool (MLOrdPool) for each action and present the resulting tuple of action taken, orders affected and costs to the algorithm once the costs are known. Note that as this only affects the policy this variation does not have an effect on the Markov property.

The variants are summarized in Table 2. The mean number of orders arriving at one period is 9. The values for the min-max normalization η is shown on the right and were picked manually after testing multiple values. If not specified then alg $= 1$.

4   Results

This section gives an overview of the performance of the algorithm compared to the conventional static lead time setting algorithms.

Parameters. Table 3 gives the parameters used for the Q-Learning algorithm and the ANN. The fork probability p_f was set to 0.2 and the discount factor to 0.995, except for the MLShippedAvg$_{AvgRL}$ variant where $\gamma = 0.5$ to let the two methods balance out. In each period 128 experiences are retrained out of a set of experiences with size 30000. The gradients are trained 4 times, with an ANN learning rate of 0.01 and no momentum. The soft update procedure allows this high value for the learning rate, but the momentum was disabled as the networks otherwise diverge. These values were manually optimized beforehand by experimenting with multiple setups for all variants. We decided to stick to the values given above as it seems to be the most appropriate setting for the problem. At the beginning the production system is empty, thus a cold start is performed by the agent. The algorithm learned for 150k periods. The runtime for the evaluations was 1000 periods, with 750 periods startup phase and all evaluations were repeated 25 times.

Table 4. Evaluation results.

Algorithm/KPI	SUM	BOC	FGIC	WIPC	TARD	σTARD	SFTT
MLShippedAvg$_{AvgRL}$	464.84	201.47	153.49	109.89	1.80	2.57	4.35
MLShippedAvg	473.46	326.41	63.24	83.81	1.87	2.79	3.13
MLOrdPool	497.83*	331.57	52.18	114.08	1.92	2.84	4.23
BIL3	535.45	400.97	28.86	105.62	2.06	2.99	4.41
BIL2	644.85	532.97	6.78	105.11	2.28	3.27	4.39
MLShipped	657.85	516.27	24.13	117.45	2.33	3.33	4.84
BIL1	803.74	698.66	0.00	105.08	2.49	3.65	4.39
MLRelOrds	896.51	677.89	113.38	105.24	2.46	3.64	4.21

*The p-value of the comparison between MLShippedAvg and MLOrdPool is 0.07364

Results. Table 4 shows the results of the evaluations. The first column denotes the tested order release approaches, namely (i) the static lead time approaches – that calculate the release time of the orders according to (Eq. 1) – with different static lead times of 1, 2 and 3 (denoted as BIL1, BIL2 and BIL3 respectively) and (ii) the different tested machine learning algorithms (summarized in Table 2) denoted as MLShipped, MLShippedAvg, MLRelOrds, MLOrdPool, and MLShippedAvg$_{AvgRL}$.

Column two to five show the cost-based performance measures ($\cdot 10^3$): the sum over all costs (SUM), backorder costs (BOC), finished goods inventory costs (FGIC) and the costs for held WIP (WIPC). Finally, column six to eight depict the average tardiness of all late orders (denotes as TARD), the standard deviation of tardy orders (σ TARD) and the mean shop floor throughput time (SFTT) which is the time an order takes from order release to completion. The measures TARD, σTARD and SFTT are all measured in number of periods.

Table 4 compares the mean of SUM values of all algorithms at a significance level of $p = 0.05$ using the Friedman Test [9]. The models shown in grey cells are not statistically distinguishable from each other.

The results in Table 4 show that the standard application of reinforcement learning (MLShipped) is unable to learn the two delays of future rewards. The first one being the standard future reward for taking actions to increase the expected return, whereas the second delay is introduced by the production system which delays the reward according to the current load of the system. Thus the agent would have to learn both delays of rewards, whereas the later one dynamically adapts to the system state. Clearly additional delay of reward opposed by the production system increases the complexity of linking chosen actions to observed rewards. This coincides with the suggestion by Zhang to use different neural networks for different lead times [50].

We overcome this problem twofold. On one hand we reduce the complexity the agent has to learn by using future rewards, while on the other hand we provide the agent with the average reward. In case of MLShippedAvg, which only rewards on the average of the shipped orders of the specific period, a continuous output with rather low variance is required for the agent to be able to correctly link the action to the result. To clarify, consider one period with many orders being finished and low average costs to a period with only one order being finished but high costs. In case of MLShippedAvg the agent observes only the average and thus cannot infer the actual average costs over the periods. In our setting however, MLShippedAvg performs well as the output of the production system has low variance.

As can be seen in Table 4, the machine learning algorithms using the average reward (MLShippedAvg$_{AvgRL}$, MLShippedAvg, MLOrdPool) yield the lowest total costs outperforming the conventional static order release models. Furthermore, it is noteworthy that the while the MLShippedAvg$_{AvgRL}$ yields the lowest lateness measures (TARD and σTARD) the MLShippedAvg algorithm yields the lowest SFTT. The MLShipped algorithm takes the third last position together with BIL2 and is (next to the MLRelOrds) outperformed by the static lead time approach BIL3.

This shows that the way of representing actual rewards to the machine learning algorithm has a major influence on the performance of the approach. Furthermore, the results show that the average reinforcement learning methods work well in periodic but non-episodic operations research problems, and should be considered a viable research direction.

5   Conclusions

This paper adds to the growing body of evidence that machine learning algorithms can contribute positively to a company's performance. The paper describes a successful application of an order release model based on reinforcement learning. The performance is tested on a multi-product, two-stage hypothetical flow shop and is measured by cost/profit, delivery and lead time related measures. We show that our developed machine learning approach outperforms all other tested order release approaches by yielding lower total costs, less mean and standard deviation of tardiness and a shorter shop floor throughput time (SFTT).

The study provides important insights, but we are aware of its limitations. Firstly, the results are limited to the simulated case and the validity of the results for other MTO production systems (e.g., job shop production systems) must be assessed in future studies. Secondly, adding further experimental factors, like machine failures, might contribute in improving the system. It would also be interesting to test the performance of the algorithm to multi-stage production systems and to include scenarios with seasonal demand where one might expect even greater benefits from setting the lead times dynamically. Furthermore, a comparison with other order release mechanisms or models (e.g., LUMS/LOMC or optimization based order release models) are an interesting direction for future research.

References

1. Ackerman, S.: Even-flow a scheduling method for reducing lateness in job shops. Manag. Technol. **3**, 20–32 (1963)
2. Akyol, D.E., Bayhan, G.M.: A review on evolution of production scheduling with neural networks. Comput. Ind. Eng. **53**(1), 95–122 (2007). http://www.sciencedirect.com/science/article/pii/S0360835207000666
3. Aytug, H., Bhattacharyya, S., Koehler, G.J., Snowdon, J.L.: A review of machine learning in scheduling. IEEE Trans. Eng. Manag. **41**, 165–171 (1994)
4. Baykasoglu, A., Gocken, M.: A simulation based approach to analyse the effects of job release on the performance of a multi-stage job-shop with processing flexibility. Int. J. Prod. Res. **49**(2), 585–610 (2011). <GotoISI>://WOS:000284413100015
5. Bechte, W.: Theory and practice of load-oriented manufacturing control. Int. J. Prod. Res. **26**(3), 375–395 (1988)
6. Bechte, W.: Load-oriented manufacturing control just-in-time production for job shops. Prod. Plan. Control **5**(3), 292–307 (1994)
7. Bertrand, J.W.M., Wortmann, J.C.: Production Control and Information Systems for Component Manufacturing Shops. Elsevier Science Inc., New York (1981)
8. Bertrand, J., Wortmann, J., Wijngaard, J.: Production Control: A Structural and Design Oriented Approach. Elsevier, Amsterdam (1990)
9. Conover, W.: Practical Nonparametric Statistics. Wiley Series in Probability and Statistics, 3rd edn. Wiley, New York (1999)
10. Enns, S., Suwanruji, P.: Work load responsive adjustment of planned lead times. J. Manuf. Technol. Manag. **15**(1), 90–100 (2004)
11. Gelders, L., Van Wassenhove, L.N.: Hierarchical integration in production planning: theory and practice. J. Oper. Manag. **3**(1), 27–35 (1982)
12. Hendry, L., Kingsman, B.: Production planning systems and their applicability to make-to-order companies. Eur. J. Oper. Res. **40**(1), 1–15 (1989). http://www.sciencedirect.com/science/article/pii/037722178990266X
13. Hendry, L., Kingsman, B.: A decision support system for job release in make-to-order companies. Int. J. Ope. Prod. Manag. **11**(6), 6–16 (1991)
14. Hoyt, J.: Dynamic lead times that fit today's dynamic planning (quoat lead times). Prod. Inventory Manag. **19**(1), 63–71 (1978)
15. Hsu, S.Y., Sha, D.Y.: Due date assignment using artificial neural networks under different shop floor control strategies. Int. J. Prod. Res. **42**(9), 1727–1745 (2004). https://doi.org/10.1080/00207540310001624375
16. Karaoglan, A.D., Karademir, O.: Flow time and product cost estimation by using an artificial neural network (ANN): a case study for transformer orders. Eng. Econ. **62**(3), 272–292 (2017). https://doi.org/10.1080/0013791X.2016.1185808

17. Knollmann, M., Windt, K.: Control-theoretic analysis of the lead time syndrome and its impact on the logistic target achievement. Procedia CIRP **7**, 97–102 (2013)
18. Law, A.M., Kelton, W.D.: Simulation Modeling & Analysis, 3rd edn. McGraw-Hill Inc., New York (2000)
19. Lee, C.Y., Piramuthu, S., Tsai, Y.K.: Job shop scheduling with a genetic algorithm and machine learning. Int. J. Prod. Res. **35**(4), 1171–1191 (1997). https://doi.org/10.1080/002075497195605
20. Li, S., Li, Y., Liu, Y., Xu, Y.: A GA-based NN approach for makespan estimation. Appl. Math. Comput. **185**(2), 1003–1014 (2007). Special Issue on Intelligent Computing Theory and Methodology. http://www.sciencedirect.com/science/article/pii/S0096300306008253
21. Lillicrap, T.P., et al.: Continuous control with deep reinforcement learning. arXiv preprint arXiv:1509.02971 (2015)
22. Lin, L.J.: Reinforcement learning for robots using neural networks. Technical report, School of Computer Science, Carnegie-Mellon University, Pittsburgh, PA (1993)
23. Mahadevan, S.: Average reward reinforcement learning: foundations, algorithms, and empirical results. Mach. Learn. **22**(1), 159–195 (1996). https://doi.org/10.1007/BF00114727
24. Mather, H., Plossl, G.W.: Priority fixation versus throughput planning. Prod. Inventory Manag. **19**, 27–51 (1978)
25. Melnyk, S.A., Ragatz, G.L.: Order review release - research issues and perspectives. Int. J. Prod. Res. **27**(7), 1081–1096 (1989). <GotoISI>://WOS:A1989AC60400003
26. Metan, G., Sabuncuoglu, I., Pierreval, H.: Real time selection of scheduling rules and knowledge extraction via dynamically controlled data mining. Int. J. Prod. Res. **48**(23), 6909–6938 (2010). https://doi.org/10.1080/00207540903307581
27. Mnih, V., et al.: Asynchronous methods for deep reinforcement learning. In: International Conference on Machine Learning, pp. 1928–1937 (2016)
28. Mnih, V., et al.: Human-level control through deep reinforcement learning. Nature **518**(7540), 529–533 (2015)
29. Molinder, A.: Joint optimization of lot-sizes, safety stocks and safety lead times in a MRP system. Int. J. Prod. Res. **35**(4), 983–994 (1997)
30. Pahl, J., Voß, S., Woodruff, D.L.: Production planning with load dependent lead times: an update of research. Ann. Oper. Res. **153**(1), 297–345 (2007). https://doi.org/10.1007/s10479-007-0173-5
31. Paternina-Arboleda, C.D., Das, T.K.: Intelligent dynamic control policies for serial production lines. IIE Trans. **33**(1), 65–77 (2001). https://doi.org/10.1080/07408170108936807
32. Patil, R.: Using ensemble and metaheuristics learning principles with artificial neural networks to improve due date prediction performance. Int. J. Prod. Res. **46**(21), 6009–6027 (2008)
33. Philipoom, P.R., Rees, L.P., Wiegmann, L.: Using neural networks to determine internally-set due-date assignments for shop scheduling. Decis. Sci. **25**(5–6), 825–851 (1994). http://dx.doi.org/10.1111/j.1540-5915.1994.tb01871.x
34. Raaymakers, W., Weijters, A.: Makespan estimation in batch process industries: a comparison between regression analysis and neural networks. Eur. J. Oper. Res. **145**(1), 14–30 (2003). http://www.sciencedirect.com/science/article/pii/S037722170200173X
35. Savell, D.V., Perez, R.A., Koh, S.W.: Scheduling semiconductor wafer production: an expert system implementation. IEEE Expert **4**(3), 9–15 (1989). (Fall)

36. Schneeweiss, C.: Distributed decision making–a unified approach. Eur. J. Oper. Res. **150**(2), 237–252 (2003)
37. Selcuk, B., Fransoo, J.C., De Kok, A.: The effect of updating lead times on the performance of hierarchical planning systems. Int. J. Prod. Econ. **104**(2), 427–440 (2006)
38. Silver, D., et al.: Mastering chess and shogi by self-play with a general reinforcement learning algorithm. arXiv preprint arXiv:1712.01815 (2017)
39. Spearman, M.L., Woodruff, D.L., Hopp, W.J.: CONWIP: a pull alternative to Kanban. Int. J. Prod. Res. **28**(5), 879–894 (1990). http://www.tandfonline.com/doi/abs/10.1080/00207549008942761
40. Sutton, R.S., Barto, A.G.: Reinforcement Learning: An Introduction, vol. 1. MIT Press, Cambridge (1998)
41. Tatsiopoulos, I., Kingsman, B.: Lead time management. Eur. J. Oper. Res. **14**(4), 351–358 (1983)
42. Teo, C.C., Bhatnagar, R., Graves, S.C.: An application of master schedule smoothing and planned lead time control. Prod. Oper. Manag. **21**(2), 211–223 (2012)
43. Thuerer, M., Stevenson, M., Silva, C.: Three decades of workload control research: a systematic review of the literature. Int. J. Prod. Res. **49**(23), 6905–6935 (2011)
44. Thuerer, M., Stevenson, M., Silva, C., Land, M.J., Fredendall, L.D.: Workload control and order release: a lean solution for make-to-order companies. Prod. Oper. Manag. **21**(5), 939–953 (2012)
45. Tsitsiklis, J.N., Van Roy, B.: Analysis of temporal-difference learning with function approximation. In: Advances in Neural Information Processing Systems, pp. 1075–1081 (1997)
46. Watkins, C.J.C.H., Dayan, P.: Q-learning. Mach. Learn. **8**(3), 279–292 (1992). https://doi.org/10.1007/BF00992698
47. Wiendahl, H.: Load-Oriented Manufacturing Control, 1st edn. Springer, Berlin (1995). https://doi.org/10.1007/978-3-642-57743-7. http://books.google.at/books-id=e66fmQEACAAJ
48. Wuest, T., Weimer, D., Irgens, C., Thoben, K.D.: Machine learning in manufacturing: advantages, challenges, and applications. Prod. Manuf. Res. **4**(1), 23–45 (2016). https://doi.org/10.1080/21693277.2016.1192517
49. Yano, C.: Setting planning lead times in serial production systems with earliness costs. Manag. Sci. **33**(1), 95–106 (1987)
50. Zhang, G.P.: Avoiding pitfalls in neural network research. IEEE Trans. Syst. Man Cybern. Part C (Appl. Rev.) **37**(1), 3–16 (2007)

Author Index

Printed in the United States
By Bookmasters